Book of Abstracts of the 56th Annual Meeting of the European Association for Animal Production

The EAAP Book of Abstracts is published under the direction of Ynze van der Honing

EAAP - European Association for Animal Production

The European Association for Animal Production wishes to express its appreciation to the
Ministero delle Politiche Agricole e Forestali (Italy) and the
Associazione Italiana Allevatori (Italy)
for their valuable support of its activities.

Book of Abstracts of the 56th Annual Meeting of the European Association for Animal Production

Uppsala, Sweden, 5 - 8 June 2005

Ynze van der Honing, Editor-in-chief

E. Strandberg, E. Cenkvári, E. von Borell, B. Kemp, C. Lazzaroni, M. Gauly, C. Wenk, W. Martin-Rosset, A. Bernués Jal and C. Thomas

Subject headings:
Animal production
Book of abstracts

ISBN 9076998663
ISSN 1382-6077

First published, 2005

Wageningen Academic Publishers
The Netherlands, 2005

Preface

The 56th annual meeting of the European Association for Animal Production (EAAP) is held in Uppsala, Sweden, 5 to 8 June 2005. The annual EAAP meeting gives the opportunity to present new scientific results and discuss their potential applicability in animal production practices. This year's meeting is of particular interest for participants from a wide range of animal production organisations and institutions. Discussions stimulate developments in animal production and encourage research on relevant topics. In Uppsala the main theme is " **Impact and Challenges of a widening Europe for Animal Production and Research**". Of the 35 sessions in total, 17 joint sessions of two or more Study Commissions are planned on:

- Implications of EU restructuring and free trade on feed quality and safety, disease and food quality and safety
- Specialised ruminant products to sustain systems and genetic resources
- Alternative low input / organic production methods
- Nutrition and management strategies to improve resource use in livestock systems
- Functional genomics of reproduction and disease resistance
- Functional traits in cattle
- Feed evaluation systems for ruminants, horses and pigs
- Physiology of pregnancy
- High health pig systems
- Coping with new regulation: Alternatives to antibiotic growth promoters and castration
- Performance and health in young horses

The book of abstracts is the main publication of the scientific contributions to this meeting; it covers a wide range of disciplines and livestock species. It contains the full programme and abstracts of the invited as well as the contributing speakers, including posters, of all 35 sessions. The number of abstracts submitted for presentation at this meeting (773) is a true challenge for the different study commissions and chairpersons to put together a scientific programme. In addition to the theatre presentations, there will be a large number of poster presentations during the conference.

Several persons have been involved in the development of the book of abstracts. Wageningen Academic Publishers has been responsible for organising the administrative and editing work and the production of the book. The contact persons of the study commissions have been responsible for organising the scientific programme and communicating with the chairpersons and invited speakers. Their help and that of Cled Thomas in the programme of all free communications is highly appreciated.

The programme is very interesting and I trust we will have a good meeting in Uppsala. I hope that you will find this book a useful reference source as well as a reminder of a good meeting during which a large number of people actively involved in livestock science and production will meet and exchange ideas.

Ynze van der Honing
Editor-in Chief

EAAP Program Foundation

Aim

EAAP aims to bring to our annual meetings, speakers who can present the latest findings and views on developments in the various fields of science relevant to animal production and its allied industries. In order to sustain the quality of the scientific program that will continue to entice the broad interest in EAAP meetings we have created the "EAAP Program Foundation". This Foundation aims to support:

- Invited speakers by funding part or all of registration and travel costs.
- Delegates from less favoured areas by offering scholarships to attend EAAP meetings.
- Young scientists by providing prizes for best presentations.

The "EAAP Program Foundation" is an initiative of the Scientific Committee (SC) of EAAP. The Foundation is aimed at stimulating the quality of the scientific program of the EAAP meetings and to ensure that the science meets societal needs. In its first year (2003), the "EAAP Program Foundation" concentrated on the program of the Genetics commission. For the coming years the activities will be broadened to the entire meeting. The Foundation Board of Trustees oversees theses aims and seeks to recruit sponsors to support its activities.

Sponsorships

We distinguish three categories of sponsorship: Student award sponsor, Gold sponsor, and Sponsor. The sponsors will be acknowledged during the scientific sessions. The names of the Student Award sponsors will be linked to the awards given to young scientists with the best presentation. Gold Sponsors and Student Award sponsors will have the opportunity to advertise their activities during the meeting and their support for EAAP.

Contact and further information

If you are interested in becoming a sponsor of the "EAAP Program Foundation" or want to have further information, please contact the secretary of the Foundation, Dr Andrea Rosati (e-mail: rosati@eaap.org).

Sponsors of 2005 meeting

The board of Trustees wants to thank the following organisations for their support:

Sponsors:
Alltech
www.alltech.com/

Gold Sponsor:
Janssen Animal Health
www.janssenanimalhealth.com
janah@janbe.jnj.com

European Association for Animal Production (EAAP)

President	Jim Flanagan
Secretary General	Andrea Rosati
Address	Via G.Tomassetti 3 A/1 I-00161 Rome, Italy
Phone	+39 06 44202639
Telefax	+39 06 86329263
E-Mail	eaap@eaap.org

In cooperation with

Swedish University of Agricultural Sciences, Uppsala, Sweden; www.slu.se
Swedish Dairy Association, Stockholm, Sweden , www.svenskmjolk.se

Organising Committee 2005

- Bengt Everitt, President, Swedish Dairy Association, Stockholm
- Hans Gustafsson, Vice President, Swedish Dairy Association, Uppsala
- Britt Grandin, Secretary, Swedish Dairy Association, Eskilstuna
- Tia Eriksson, Academic Conferences, Uppsala

Members

- Jan Philipsson, Swedish University of Agricultural Sciences (SLU), Uppsala
- Hans-Erik Pettersson, Swedish Dairy Association, Lund
- Göran Dalin, SLU, Uppsala
- Ulf Emanuelson, SLU, Uppsala
- Lotta Rydhmer, SLU, Uppsala
- Hans Wiktorsson, SLU, Uppsala
- Peter Kallings, Swedish Horse Council Foundation, Stockholm

Scientific Committee

- President Jan Philipsson, SLU, Uppsala
- Secretary Karl-Ivar Kumm, SLU, Uppsala

Members

- Erling Strandberg, SLU, Uppsala
- Jan Erik Lindberg, SLU, Uppsala
- Lotta Berg, Swedish Animal Welfare Agency, Skara
- Kjell Holtenius, SLU, Uppsala
- Jan Bertilsson, SLU, Uppsala
- Anna Näsholm, SLU, Uppsala
- Maria Neil, SLU, Uppsala
- Karin Ericson, The Swedish Educational Centre of Harness Racing, Wången
- Krister Sällvik, SLU, Alnarp

Congress Secretariat

Academic Conferences
P.O Box 7059
SE-750 07 Uppsala, Sweden

Phone: +46 18 67 20 84
Fax: + 46 18 67 35 30
E-mail: EAAP2005@slu.se

57th Annual Meeting of the European Association for Animal Production

Antalya ,TURKEY, September 17-20, 2006

Organizing Committee

President: S.Metin YENER
Ankara University, Faculty of Agriculture,
Department of Animal Science, Ankara
yener@agri.ankara.edu.tr

Secretary: Fatin CEDDEN
Ankara University, Faculty of Agriculture,
Department of Animal Science, Ankara
cedden@agri.ankara.edu.tr

Congress Secretariat

SALTUR Tourism
Atatürk Blv. No:175/8 Kavaklıdere
Ankara - TURKEY
Phone: +90 312 419 84 80
Fax: +90 312 419 84 79
E-mail: saltur@eaap2006.gen.tr
Web: www.eaap2006.gen.tr

Abstract submission
Please find the abstract form on the internet:
www.WageningenAcademic.com/eaap
Submission of the completed abstract form before March 31, 2006 to:
EAAP2006@WageningenAcademic.com

Mediterranean Symposium

"Comparative advantages for typical animal products from the Mediterranean area"

Boa Fonte, Portugal, September 25-27, 2005

Contact person
J.M.C Ramalho Ribeiro
Estação Zootécnica Nacional
Fonte Boa
2005- 048 Vale de Santarém
Portugal

Tel: + 351-243-760202/3/4
Fax: + 351-243-760540
E-mail: dir.ezn@mail.telepac.pt

35th ICAR Session and INTERBULL Meeting

June 5-10, 2006, Kuopio, Finland

Contact: ICAR Secretariat, Rome, Italy
Email: icar@eaap.org

Scientific programme EAAP-2005

Plenary – Sunday 5 June – 08.30 – **European research programme and animal health** – John Caxton (EU)

Sunday 5 June 09.30 - 13.00

Session 1 (M*, N, L, P, S, C, OIE)
Implications of EU restructuring and free trade on feed quality and safety, disease and food quality and safety *followed by* Quality assurance systems to ensure compliance
Chair: Noordhuizen (NL)

Session 2 (G*, Ph)
Functional genomics of reproduction and disease resistance
Chair: Wimmers (D)

Session 3
Systems of identification in horses (H) Chair: Guibert (F)
followed by Free communications - Horses (H)
Chair: Curik (CRO)

Sunday 5 June 14.00 - 18.00

Session 4 (L)
Adaptation of livestock farming systems to EU reform and restructuring
Chair: Gibon (F)

Session 5 (C*, G and Interbull)
Functional traits in cattle
Chair: Crettenand (CH) and Emanuelson (Swe)

Session 6 (N*, C, S, H)
Feed evaluation systems for ruminants and horses
Chair: Sauvant (F)
followed by
Free Communications in Nutrition
Chair: Lindberg (SWE)

Session 7 (M)
Animal health and welfare: costs and benefits
Chair: Stott (UK)
followed by
Free Communications in Management and Health (M)
Chair: Metz (NL)

Session 8 (P)
Robust Pigs
Chairs: Lundeheim and Knap (D)

Session 9 (G)
Genetics of variability / Free Communications in Genetics
Chair: San Cristobal (F) and Eriksson (SWE)

Session 10 (Ph)
Physiology of stress and reproduction
Chair: Kemp (NL)

Monday 6 June 08.30 - 12.30

Session 11 (S*, C, N, L,)
Specialised ruminant products to sustain systems and genetic resources
Chair: Schneeberger (CH)

Session 12 (G)
Developments in quantitative genetics
Chair: Zengting Liu (D)

Session 13 (Ph* and H)
Physiology of pregnancy
Chair: Rátky, (Hun)

Session 14 (P*, M)
High health pig systems
Chair: Chadd (UK)

Key - G Genetics; N, Nutrition; Ph, Physiology; P, Pig production; C, Cattle Production; S, Sheep and Goat Production; M, Management and Health; H, Horse production; L, Livestock Farming Systems. (*) denotes organising commission. **Bold - Sessions contributing to the theme of the meeting** 'Impact and challenges of a widening Europe for Animal Production and Research

Monday 6 June 14.00 - 18.00	Tuesday 7 June 08.30 - 12.30	Wednesday 8 June 08.30 - 12.30
Session 15 (N) Free Communications (Ruminant nutrition and physiology) Chair: Crovetto (I)	Session 25 (C*, S, L, M, P,) **Alternative low input / organic production methods** **Chair : Keane (IRL)**	Session 30 (N*, P, L) **Nutrition and management strategies to improve resource use in livestock systems** **Chair: Milne (UK)**
Session 16 (M) Free Communications (Management and health of ruminants) Chair: Metz, (NL)	Session 26 (G) **Breeding programmes for a wide range of systems** **Chair: Wickham (IRL)**	Session 31 (H) Horse production in Sweden Chair: Karlander (SWE)
Session 17 (G) Free Communications (Animal genetics) Chair: Grandinson (SWE)	Session 27 (P*, N, Ph) Coping with new regulation: Alternatives to antibiotic growth promoters and castration Chair: Tollarrardona (ES)	Session 32 (G) Increased understanding of the genetics of quantitative traits - theory and applications Chair: Maki-Tanila (Fin)
Session 18 (P) Free Communications (Pig production, management and health) Chair: Kovac (Slo)	Session 28 (H*+M) Performance and health in young horses Chair: Dalin (SWE)	Session 33 (S) Progress towards reduction of disease in sheep and goats Chair: Gavier-Widen (SWE)
Session 19 (P) Free communications (Pig and poultry nutrition) Chair:	Session 29 (G) Free communications Animal Genetics Chair: Andersson-Eklund (SWE)	Session 34 (M and ECPLF) Utilisation of records to manage health Chair: Geers (B)
Session 20 (C) Free Communications (Cattle production) Chair: Hocquette (F)		Session 35 (C* and Ph) Free Communication (Reproduction in cattle) Chair: Lazzaroni (IT)
Session 21 (S) Free Communications (Sheep and goat production) Chair: Gabiña (ES)		
Session 22 (Ph) Free Communications (Reproduction) Chair: Sejrsen (DK)		
Session 23 (L) Free Communications (Livestock farming systems) Chair: Hermansen (DK)		
Session 24 (H) Equine science education Chair: Martin-Rosset (F) *followed by* Free Communications - Horses (H) Chair: Habe (SLO)		

Key - G Genetics; N, Nutrition; Ph, Physiology; P, Pig production; C, Cattle Production; S, Sheep and Goat Production; M, Management and Health; H, Horse production; L, Livestock Farming Systems. (*) denotes organising commission. **Bold - Sessions contributing to the theme of the meeting** 'Impact and challenges of a widening Europe for Animal Production and Research

Commission on Animal Genetics

Dr Ducrocq	President	INRA
	France	Vincent.Ducrocq@dga.jouy.inra.fr
Prof. Dr Simianer	Vice-President	University of Goettingen
	Germany	simianer@genetics-network.de
Prof. Mäki-Tanila	Vice-President	MTT Agrifood
	Finland	asko.maki-tanila@mtt.fi
Dr Strandberg	Secretary	SLU
	Sweden	Erling.Strandberg@hgen.slu.se
Dr Martyniuk	Secretary	National Research Institute
	Poland	martyniuk@alpha.sggw.waw.pl

Commission on Animal Nutrition

Prof. Crovetto	President	University of Milano
	Italy	matteo.crovetto@unimi.it
Dr Lindberg	Vice-President	SLU
	Sweden	Jan-Eric.Lindberg@huv.slu.se
Dr Ortigues-Marty	Vice-President	INRA
	France	ortigues@sancy.clermont.inra.fr
Dr Moreira	Secretary	University of the Azores
	Portugal	jsilva@notes.angra.uac.pt
Dr Cenkvàri	Secretary	Szent Istvan University
	Hungary	Czenkvari.Eva@aotk.szie.hu

Commission on Animal Management & Health

Prof. von Borell	President	University Halle-Wittenberg
	Germany	borell@landw.uni-halle.de
Dr Sorensen	Vice-President	DIAS
	Denmark	jantind.sorensen@agrsci.dk
Prof. Metz	Vice-President	Wageningen University
	Netherlands	jos.metz@user.aenf.wag-ur.nl
Prof. Formigoni	Secretary	University Bologna
	Italy	aformigoni@alma.unibo.it
Prof. Fourichon	Secretary	INRA
	France	fourichon@vet-nantes.fr

Commission on Animal Physiology

Dr Sejrsen	President	DIAS
	Denmark	Kr.Sejrsen@agrsci.dk
Dr Knight	Vice-President	Hannah Institute
	UK	Knightc@hri.sari.ac.uk
Dr Royal	Vice-President	University Liverpool
	UK	mdroyal@liverpool.ac.uk
Dr Ratky	Vice-President	Research Institute
	Hungary	jozsef.ratky@atk.hu
Dr Chilliard	Secretary	INRA
	France	chilliar@clermont.inra.fr
Dr. Kemp	Secretary	Wageningen University
	Netherlands	bas.kemp@GENR.VH.WAU.NL

Commission on Cattle Production

Dr Gigli	President	ISZ
	Italy	sergio.gigli@isz.it
Dr Keane	Vice-President	TEAGASC
	Ireland	gkeane@grange.teagasc.ie
Dr Hocquette	Vice-President	INRA
	France	hocquet@clermont.inra.fr
Prof. Kalm	Vice-President	University of Kiel
	Germany	ekalm@tierzucht.uni-kiel.de
Dr Lazzaroni	Secretary	University of Torino
	Italy	carla.lazzaroni@unito.it
Dr Kuipers	Secretary	Wageningen University
	Netherlands	Abele.Kuipers@wur.nl

Commission on Sheep and Goat Production

Dr Gabiña	President	CIHEAM
	Spain	gabina@iamz.ciheam.org
Dr Rihani	Vice-President	DPA
	Morocco	n.rihani@iav.ac.ma
Dr Dýrmundsson	Vice-President	Farmers Assoc.
	Iceland	ord@bondi.is
Prof. Gauly	Vice-President	University Göttingen
	Germany	mgauly@gwdg.de
Prof. Niznikowski	Vice-President	Warsaw Agricultural University
	Poland	niznikowski@alpha.sggw.waw.pl
Dr Bodin	Secretary	INRA-SAGA
	France	bodin@toulouse.inra.fr

Commission on Pig Production

Prof. Dr Wenk	President Switzerland	ETH Zentrum caspar.wenk@inw.agrl.ethz.ch
Dr Chadd	Vice-President UK	Royal Agric. College steve.chadd@royagcol.ac.uk
Dr Knap	Vice-President Germany	PIC Deutschland KnaP@de.pic.co.uk
Dr Kovac	Secretary Slovenia	University Ljubljana milena.kovac@uni.lj.si
Dr Torrallardona	Secretary Spain	IRTA David.Torrallardona@irta.es

Commission on Horse Production

Dr Martin-Rosset	President France	INRA wrosset@clermont.inra.fr
Dr Kennedy	Vice-President UK	Writtle College mjk@writtle.ac.uk
Dr Verini	Vice-President Italy	University Perugia vete2@unipg.it
Dr Koenen	Vice-President Netherlands	NRS BV Koenen.E@cr-delta.nl
Dr Saastamoinen	Secretary Finland	MTT Equines markku.saastamoinen@mtt.fi
Dr Sondergaard	Secretary Denmark	DIAS eva.sondergaard@agrsci.dk

Commission on Livestock Farming Systems

Dr Gibon	President France	UMR INRA gibon@toulouse.inra.fr
Prof. Peters	Vice-President Germany	University of Berlin k.peters@rz.hu-berlin.de
Prof. Zervas	Vice-President Greece	Agricultural University Athens gzervas@aua.gr
Prof. Mihina	Vice-President Slovakia	Research Institute Nitra mihina@vuzv.sk
Dr Hermansen	Secretary Denmark	DIAS john.hermansen@agrsci.dk
Dr Bernues Jal	Secretary Spain	C.I.T.A. abernues@aragon.es

Genetics	Session 2 (page 9 - 14)	Session 5 (page 42 - 55)
	Session 9 (page 79 - 102)	Session 12 (page 117 - 126)
	Session 17 (page 190 - 217)	Session 26 (page 318 - 323)
	Session 29 (page 336 - 345)	Session 32 (page 358 - 374)
Nutrition	Session 1 (page 1 - 9)	Session 6 (page 56 - 64)
	Session 11 (page 114 - 117)	Session 15 (page 133 - 172)
	Session 27 (page 323 - 330)	Session 30 (page 345 - 356)
Management and Health	Session 1 (page 1- 9)	Session 7 (page 65 - 73)
	Session 14 (page 128 - 133)	Session 16 (page 173 - 189)
	Session 25 (page 309 - 317)	Session 28 (page 330 - 336)
	Session 34 (page 378 - 383)	
Physiology	Session 2 (page 9 - 14)	Session 10 (page 103 - 114)
	Session 13 (page 126 - 128)	Session 22 (page 281 - 292)
	Session 27 (page 323 - 330)	
Cattle	Session 1 (page 1- 9)	Session 5 (page 42 - 55)
	Session 6 (page 56 - 64)	Session 11 (page 114 - 117)
	Session 20 (page 245 - 262)	Session 25 (page 309 - 317)
	Session 35 (page 383 - 387)	
Sheep and Goats	Session 1 (page 1- 9)	Session 6 (page 56 - 64)
	Session 11 (page 114 - 117)	Session 21 (page 263 - 281)
	Session 25 (page 309 - 317)	Session 33 (page 374 - 378)
Pigs	Session 1 (page 1- 9)	Session 8 (page 74 - 79)
	Session 14 (page 128 - 133)	Session 18 (page 218 - 239)
	Session 19 (page 240 - 245)	Session 25 (page 309 - 317)
	Session 27 (page 323 - 330)	Session 30 (page 345 - 356)
Horses	Session 3 (page 15 - 33)	Session 6 (page 56 - 64)
	Session 13 (page 126 - 128)	Session 24 (page 302 - 309)
	Session 28 (page 330 - 336)	Session 31 (page 357 - 357)
Livestock farming systems	Session 1 (page 1- 9)	Session 4 (page 34 - 41)
	Session 11 (page 114 - 117)	Session 23 (page 292 - 301)
	Session 25 (page 309 - 317)	Session 30 (page 345 - 356)

Session 1: Implications of EU restructuring and free trade on feed quality and safety, disease and food quality and safety + Quality assurance systems to ensure compliance

Date: 05 June '05; 9:30 - 13:00 hours
Chairperson: Noordhuizen (NL)

Theatre	Session 1 no.	Page
The EU policy on food and feed safety: an overview *Vanthemsche, P.*	1	1
The Codex Alimentarius and consumer protection related to foods of animal origin *Slorach, S.A.*	2	1
Feed and feed safety in livestock production *Heeschen, W. and A. Blüthgen*	3	2
Suisse Garantie: more than a label *Beglinger, C., P. Gresch and R. Kennel-Hess*	4	2
Quality assurance programs for the USA pig industry: focus on food safety and animal welfare *McGlone, J.*	5	3
Animal Health, welfare and public health issues *Pastoret, P.P.*	6	3
Balancing public and private interests in EU food production *Stott, A.W., C. Milne, S. Peddie, F. Williams and G.J. Gunn*	7	4
Control of salmonella in food of animal origin: the role of animal feed *Häggblom, P.E.*	8	4
Utilising sustainable data in an effective way *Kuipers, A. and F.J.H.M. Verhees*	9	5

Poster	Session 1 no.	Page
Development of quality indicators for e-Learning in the domain of farm animal welfare *Sossidou, E., D. Stamatis and E. Szücs*	10	5
Comparison of conventional and electronic identification in beef cows under rangeland and long term conditions *Ghirardi, J.J., G. Caja, C. Conill and M. Hernández-Jover*	11	6

Improving skills on food quality and safety issues in the Leonardo da Vinci Vocational Training Project WELFOOD in virtual environment 12 6
Szücs, E., R. Geers, J. Praks, V. Poikalainen, E. Sossidou and T. Jezierski

Subcutaneous nasal transponder implant, an alternative method for electronic identification of pigs - preliminary results 13 7
Lungu, S., G. Fiore and C. Korn

Tracing pigs by using conventional and electronic identification devices 14 7
Babot, D., M. Hernández-Jover, G. Caja, C. Santamarina and J.J. Ghirardi

Electronic identification of swine: comparison of results obtained using intra-peritoneal transponders and ear tags 15 8
Marchi, E., N. Ferri and F. Comellini

Effect of the transportation on the readability and retention of the endo-reticulum transponder in goats Sarda breed 16 8
Pinna, W., P. Sedda, G. Delogu, G. Moniello, M.P.L. Bitti and I.L. Solinas

Effect of intraperitoneal transponder on the productive performances of suckling piglets 17 9
Pinna, W., P. Sedda, G. Delogu, G. Moniello, M.G. Cappai and I.L. Solinas

Session 2: Functional genomics of reproduction and disease resistance

Date: 05 June '05; 9:30 - 13:00 hours
Chairperson: Wimmers (D)

Theatre **Session 2 no. Page**

Expression of the swine MHC genes 1 9
Rogel Gaillard, C., L. Flori, C. Renard and P. Chardon

Expression patters of developmentally relevant genes in cattle preimplantation embryos 2 10
Schellander, K. and D. Tesfaye

Abnormal gene expression results in the immotile short tail sperm defect in Finnish Large White 3 10
Sironen, A., B. Thomsen, M. Andersson, V. Ahola and J. Vilkki

Quantitative trait loci affecting fertility and calving traits in Swedish dairy cattle 4 11
Holmberg, M. and L. Andersson

Fine mapping of QTL for primary antibody responses to KLH in laying hens 5 11
Siwek, M., S. Cornelissen, L.L. Janss, E.F. Knol, M. Nieuwland, H. Bovenhuis, M.A.M. Groenen, H.K. Parmentier, J.A.M. van Arendonk and J. van der Poel

Candidate SNPs for chicken immune response genes 6 12
Sironi, L., B. Lazzari, P. Ramelli, S. Cerolini and P. Mariani

Exclusion of the SLA as candidate region and reduction of the position interval for the porcine chromosome 7 QTL affecting growth and fatness 7 12
Demeure, O., M.P. Sanchez, J. Riquet, N. Iannuccelli, K. Fève, J. Gogué, Y. Billon, J.C. Caritez, D. Milan and J.P. Bidanel

SNP's within exons of candidate genes for fertility traits 8 13
Hastings, N., S. Donn, K. Derecka, A.P.F. Flint and J.A. Woolliams

Association of IGF1 and IGF1 receptor genes polymorphism to fertility traits in Thai indigenous cattle 9 13
Boonyanuwat, K., S. Suwanmajo and A. Suklim

Poster **Session 2 no. Page**

SNP detection on LHB gene and association analysis with litter size in pigs 10 14
Muñoz, G, A. Fernández, C. Barragán, L. Silió, C. Óvilo and C. Rodríguez

Investigation of Genetics involved in Boar Taint 11 14
Moe, M., S. Lien, H. Tajet, T.H.E. Meuwissen and E. Grindflek

Session 3: Systems of identification in horses / Free communications

Date: 05 June '05; 9:30 - 13:00 hours
Chairperson: Guibert (F) / Curik (CRO)

Theatre **Session 3 no. Page**

Identification methods of horses in Germany 1 15
Dohms, T. and K. Miesner

Identification of horses in Great Britain, including the role of horse passports and the establishement of the national equine database 2 15
Newman, P., D. Holmes and D. Yates

Identification and registration of equines in the Netherlands 3 16
Klaver, J., A. de Vries and M. van Lent

Identification of horses in France 4 16
Guibert, X. and D. De Cadolle

Identification of horses in Ireland 5 17
Finnerty, N.

Identification system of horses in Hungary 6 17
Bodó, I., S. Mihok, B. Pataki, J. Posta and Zs. Tóth

Identification of equidae in the European Community 7 18
Füssel, A.-E. and K.U. Sprenger

International horse identification tools 8 18
Guibert, X. and D. De Cadolle

The Universal Equine Life Number (UELN) 9 19
Guibert, X. and S. Gautier

Data exchanges with XML format 10 19
Gautier, S.

FEI new database, link with the original stud books 11 20
Sluyter, F. and C. de Coulon

Practical and methodological views on identification and parentage control of horses 12 20
Mikko, S. and K. Sandberg

Connectedness among five European sport horse populations 13 21
Thorén, E., H. Jorjani and J. Philipsson

Genetic correlations between movement and free-jumping traits and performance in show-jumping and dressage competition of Dutch warmblood horses 14 21
Ducro, B.J., E.P.C. Koenen and J.M.F.M. van Tartwijk

Efficiency of past selection of French Sport Horse : the Selle Français and suggestions for the future 15 22
Dubois, C. and A. Ricard

Is the freestyle dressage competition a reliable test of the horse's performance? 16 22
Stachurska, A., M. Pieta, W. Markowski and K. Czyrska

Analysis of body shape variation among different horse breeds via Generalised Procrustes Analysis (GPA) 17 23
Druml, T., I. Curik and J. Sölkner

Genetic evaluation of station performance test results of Dutch Friesian horses 18 23
Koenen, E.P.C. and I. Hellinga

Selection strategies in an endangered Norwegian horse breed 19 24
Olsen, H.F., G. Klemetsdal, M. Holtsmark and J. Ruane

Dry matter production and energy content of a natural pasture grazed by wild horses over a two years period 20 24
Costantini, M., N. Miraglia, M. Polidori and G. Meineri

Analysis of the daily biorhythm from horses, measured on movement activity and rest periods, with pedometer in different horse keeping systems 21 25
Rose, S., U. Brehme, U. Stollberg, Y. Buchor and R. v. Niederhäusern

Poster **Session 3 no. Page**

DNS microsatellite test of Hutsul horses in Hungary 22 25
Józsa, Cs., B. Bán, S. Mihok, I. Bodó and F. Husvéth

Assesment of Old Kladruby horse´s constitution 23 26
Sobotkova, E., I. Jiskrová, D. Misar and M. Holecova

Preliminary analysis of the morphofunctional evaluation in horse-show of Andalusian Horse 24 26
Valera, M., J.A. Gessa, M.D. Gómez, A. Horcada, C. Medina, I. Cervantes, F. Goyache and A. Molina

The estimation of the breeding value of English thoroughbreds in the Czech Republic 25 27
Svobodova, S., C. Blouin and B. Langlois

Modelling breeding schemes for two Bavarian horse populations 26 27
Edel, C. and L. Dempfle

Genetic analysis of caliber index in Lipizzan horse using random regression 27 28
Kaps, M., I. Curik and M. Baban

Characterisation of triacylglycerols in donkey milk using HPLC/MS 28 28
Chiofalo, B., P. Dugo, E. Salimei, L. Mondello and V. Chiofalo

The raising system for Marismeña equine breed in the Natural Park of Doñana (Spain) 29 29
Luque, M., E. Rodero, F. Peña, A. Molina, F. Goyache and M. Herrera

Feeding, riding properties and behaviour 30 29
Herlin, A.H., M. Rundgren and M. Lundberg

Relationship between maintenance conditions and elimination behaviour in horses 31 30
Nedelcu, C.I., M.M. Nedelcu, Gh. Georgescu and A. Marmandiu

Human-mare relationships and behaviour of foals toward humans 32 30
Henry, S., D. Hemery, M.-A. Richard and M. Hausberger

Evaluation of horse keeping in Schleswig-Holstein 33 31
Petersen, S., K.H. Tolle, K. Blobel, A. Grabner and J. Krieter

An estimation of reliability of judging the horse dressage competitions 34 31
Stachurska, A., J. Niewczas and W. Markowski

Which movements of freestyle dressage programs are difficult for horses to execute and for judges to assess? 35 32
Stachurska, A., W. Markowski, M. Pieta and M. Wesolowska-Janczarek

Evaluation for riding and driving purposes of Bardigiano Horse stallions and mares 36 32
Catalano, A.L., F. Martuzzi, S. Filippini and F. Vaccari Simonini

On breeding value of the Tori Horses 37 33
Peterson, H.

Evaluation of linear conformation traits of the autochthonous horse breeds in Croatia 38 33
Ivankovic, A., P. Caput, P. Mijic and M. Konjacic

Session 4: Adaptation of livestock farming systems to EU reform and restructuring

Date: 05 June '05; 14:00 - 18:00 hours
Chairperson: Gibon (F)

Theatre **Session 4 no. Page**

Diversity of livestock farming systems in Europe and prospective impacts of the CAP reform 1 34
Pflimlin, A. and C. Perrot

Rapid structural change in Danish dairy production 2 34
Clausen, S.

Decoupling: Irish farmers attitudes and planning intentions 3 35
Dunne, W. and M. Cushion

Consequences of the CAP reform for the Bavarian agriculture 4 35
Heissenhuber, A. and H. Hoffmann

Prospective economic incidence of the CAP reform on Spanish sheep farming systems 5 36
Manrique, E. and A. Olaizola

Sheep breed and system dynamics over the last thirty years: responding to policy and economic changes in the Britain 6 36
Pollot, G.E. and D.G. Stone

Short term impact of the EU membership on the Hungarian livestock sector 7 37
Wagenhoffer, Zs.

Economic position of beef sector in Hungary 8 37
Szabó, F., Gy. Buzás, Zs. Vincze, M. Török and Zs. Wagenhoffer

Strategies and management practices of part-time livestock farmers: an example of sheep farming in a French grassland region 9 38
Fiorelli, C., J.-Y. Pailleux and B. Dedieu

Poster **Session 4 no. Page**

The impact of the EU direct payments on beef production systems and feed costs 10 38
Dunne, W.

Beef cattle sector in Hungary as affected by the enlargement 11 39
Wagenhoffer, Zs. and D. Mezoszentgyorgyi

New grassland management and quality adjustment strategies for dairy farmers in less-favoured areas 12 39
Thénard, V., J.P. Theau and M. Duru

Cost evaluation of the use of conventional and electronic identification for the national sheep and goat populations in Spain 13 40
Saa, C., M.J. Milán, G. Caja and J.J. Ghirardi

Comparison of cattle identification costs using conventional or electronic systems in Spain 14 40
Saa, C., M.J. Milán, G. Caja and J.J. Ghirardi

Effect of frequency of manure removal and drying on ammonia, methane, nitrous oxide and carbon dioxide emissions from laying hen houses 15 41
Piñeiro, C., G. Montalvo and M. Bigeriego

Best available techniques to reduce ammonia, methane and nitrous oxide emissions control from piglets facilities 16 41
Piñeiro, C., G. Montalvo and M. Bigeriego

Session 5: Functional traits in cattle

Date: 05 June '05; 14:00 - 18:00 hours
Chairperson: Crettenand (CH) and Emanuelson (SE)

Theatre **Session 5 no. Page**

International trends in genetic evaluation of functional traits in dairy cattle 1 42
Mark, T., J.H. Jakobsen, H. Jorjani, W.F. Fiske and J. Philipsson

Functional traits in cattle breeding programs: implementation issues 2 42
Sölkner, J., A. Willam, C. Egger-Danner and H. Schwarzenbacher

US perspective: the importance of functional traits and crossbreeding in dairy cattle 3 43
Rogers, G.W.

Threshold versus Linear model estimates of genetic and environmental cor relations between udder health and production traits 4 43
Negussie, E., I. Strandén and E.A. Mäntsysaari

Survival analysis for genetic evaluation of mastitis in dairy cattle: A simulation study 5 44
Carlén, E., U. Emanuelson and E. Strandberg

Genetic analysis of cases of subclinical mastitis 6 44
Schafberg, R., F. Rosner, G. Anacker and H.H. Swalve

Genetic correlations between clinical mastitis, milk fever, ketosis, and retained placenta within and between the first three lactations 7 45
Heringstad, B., Y.M. Chang, D. Gianola and G. Klemetsdal

Genetic parameters for predictors of body weight, production traits and somatic cell count in Swiss dairy cows 8 45
Haas, Y. de and H.N. Kadarmideen

Does persistency of lactation influence the disease liability in German Holstein dairy cattle? 9 46
Harder, B., J. Bennewitz, D. Hinrichs and E. Kalm

Genetic effects on stillbirth and calving difficulty in Swedish Red and White dairy cattle at first and second calving 10 46
Steinbock, L., K. Johansson, A. Näsholm, B. Berglund and J. Philipsson

International genetic evaluation of female fertility traits in 11 Holstein populations 11 47
Jorjani, H.

Poster **Session 5 no.** **Page**

Genotype by environment interaction for udder health traits in Swedish Holstein cows 12 47
Carlén, E., K. Jansson and E. Strandberg

Relationship of somatic cell count and udder conformation traits in Iranian Holstein cattle 13 48
Sanjabi, M.R., A. Ghoibaighi and R.V. Torshizi

Genetic analysis of somatic cell score in Danish dairy cattle using random regression test-day model 14 48
Elsaid, R., A. Sabry, M. Lund and P. Madsen

Comparison of different strategies of quantitative genetic analysis of health traits in dairy cattle 15 49
Bergfeld, U., R. Fischer, C. Kehr and M. Klunker

Longitudinal genetic analyses of functional traits in dairy cows using daily random regression methodology 16 49
Karacaören, B., F. Jaffrézic and H.N. Kadarmideen

Legendre polynomials for genetic evaluations of persistency of milk production of Holstein cows 17 50
Cobuci, J.A., C.N. Costa, N.M. Teixeira and A.F. Freitas

Relationships between milkability traits in Brown Swiss 18 50
Dodenhoff, J., R. Emmerling and D. Sprengel

Factors affecting days open in Polish Holsteins 19 51
Jagusiak, W. and A. Zarnecki

Phenotypic and genetic variability of reproduction traits of black and white cows 20 51
Djedovic, R., V. Bogdanovic and P. Perisic

Gestation length for genetic evaluation of calving traits in dairy cattle 21 52
Stamer, E., W. Junge, W. Brade and E. Kalm

Analysis of genetic and environmental effects on claw disorders diagnosed at hoof trimming 22 52
Swalve, H.H., R. Pijl, M. Bethge, F. Rosner and M. Wensch-Dorendorf

Phenotypic relationships between type traits and survival in New Zealand dairy cattle 23 53
Berry, D.P., B.L. Harris, A.M. Winkelman and W. Montgomerie

Relationships between bodyweight, milk yield, and longevity of Estonian test cows 24 53
Saveli, O. and M. Voore

Estimation of economic values of longevity and other functional traits in Finnish dairy cattle 25 54
Toivakka, M., J.I. Nousiainen and E.A. Mäntsysaari

Joint genetic evaluation for functional longevity for Pinzgau cattle 26 54
Egger-Danner, C., O. Kadlecik and C. Fuerst and R. Kasarda

Study on the longevity of beef cows of different breeds 27 55
Dákay, I., D. Márton, Z. Lengyel, M. Török and Zs. Vincze

Multiple breed evaluation for cow survival and fertility in Irish beef cattle 28 55
Pool, M.H., V.E. Olori, A.R. Cromie and R.F. Veerkamp

Session 6: Feed evaluation systems for ruminants and horses / Free communications

Date: 05 June '05; 14:00 - 18:00 hours
Chairperson: Sauvant (F) / Lindberg (SWE)

Theatre	**Session 6 no.**	**Page**
NorFor: the new Nordic feed evaluation system! *Gustafsson, A.H., M. Mehlqvist, H. Volden, M. Larsen and G. Gudmundsson*	1	56
Karoline: The Nordic dairy cow model *Danfær, A., P. Huhtanen and H. Volden*	2	56
Moving from a range of systems, currently assessing practical workloads in equines to a common system *Ellis, A.D.*	3	57
A model to predict nitrogen excretion from dairy cattle using the PDI protein system *Olsson, V., J.J. Murphy, F. O'Mara, M.A. O'Donovan and F.J. Mulligan*	4	57

Poster	**Session 6 no.**	**Page**
The effects of urea treatment on in vitro gas production of pomegranate peel *Feizi, R., A. Ghodratnama, M. Zahedifar, M. Danesh Mesgaran and M. Raisianzadeh*	5	58
Performance prediction using ARC, ME System for different breeds of beef cattle fed with two different feeding periods grown under feedlot conditions *Bozkurt, Y. and S. Ozkaya*	6	58
Prediction of nutritional content of silages by analyses of green chop *Bertilsson, J.*	7	59
The effect of grain type and processing on chewing activity in horses *Brøkner, C., P. Nørgaard, L. Eriksen and T.M. Søland*	8	59
Investigations on protein requirements of fattening bulls of the German Holstein breed *Meyer, U., P. Lebzien, G. Flachowsky and H. Boehme*	9	60
Evaluation of buffering capacity for ruminant ration formulation and its effects on rumen fluid *Moharrery, A.*	10	60
Effect on milk production when maize silage from two hybrids harvested at two stages of maturity is partly substituted with grass silage *Hymøller, L. and M.R. Weisbjerg*	11	61

Relative energy requirements of beef bull breeds 12 61
Drennan, M.J., M. McGee and A. Grogan

In vitro gas production profile of non-washable, insoluble washable and soluble washable fractions in some concentrate ingredients 13 62
Azarfar, A., S. Tamminga and B.A. Williams

Degradability characteristics of dry matter of some feedstuffs using In Vitro Technique 14 62
Taghizadeh, A., H. Abdoli and A. Tahmasbi

Monitoring the fate of untreated and micronized rapeseed meal proteins in the rumen using SDS-PAGE 15 63
Sadeghi, A.A. and P. Shawrang

Ruminal starch and protein degradation kinetics of untreated and microwave treated barley grain 16 63
Sadeghi, A.A., P. Shawrang and A. Nikkhah

Chemical composition, in vitro DM and OM digestibility of ten pasture species 17 64
Shawrang, P. and A. Nikkhah

Manure evaluation in dairy cows 18 64
Steen, K., T. Eriksson and M. Emanuelson

Session 7: Animal health and welfare: costs and benefits

Date: 05 June '05; 14:00 - 18:00 hours
Chairperson: Stott (UK)

Theatre **Session 7 no. Page**

Economic changes under the CAP: the implications for animal science and animal welfare 1 65
Seabrook, M.F.

The challenge of competing goals: animal welfare, the environment, human health and the profitability of livestock production 2 65
Milne, C.E., A.W. Stott and G.E. Dalton

Integration of animal welfare in the food quality chain: from public concern to improved welfare and transparent quality 3 66
Blokhuis, H.J., R.B. Jones, R. Geers, M. Miele and I. Veissier

Assessing the costs of infection by Bovine Viral Diarrheoa Virus (BVDV) in dairy herds 4 66
Fourichon, C., F. Beaudeau, N. Bareille and H. Seegers

Economic value of mastitis incidence in dairy herds in the Czech Republic *Wolfová, M., M. Stípková and J. Wolf*	5	67
Variation in milk yield associated with the cow-status to *Mycobacterium avium* subspecies paratuberculosis (Map) infection in French dairy herds *Beaudeau, F., M. Belliard, A. Joly and H. Seegers*	6	67
Economic analysis of foot-and-mouth disease in Turkey-I: acquisition of required data via Delphi expert opinion survey *Senturk, B. and C. Yalcin*	7	68
Economic analysis of foot-and-mouth disease in Turkey- II: an assessment of financial losses in infected animals and cost of disease at national level *Yalcin, C. and B. Senturk*	8	68
Economics of sub-clinical helminthosis control through anthelmintics and supplementation in Menz and Awassi-Menz crossbred sheep in Ethiopia *Tibbo, M., K. Aragaw, J. Philipsson, B. Malmfors, A. Näsholm, W. Ayalew and J.E.O. Rege*	9	69

Poster	**Session 7 no.**	**Page**
Solid dosage medicine for poultry *Ustyanich, A.E., M.A. Ustyanich and E.P. Ustyanich*	10	69
Clinicopathological studies on Theileria Annulata infection in Siwa Oasis in Egypt *Abdou, T.A., T.R. Abou-El-naga and M.A. Mahmoud*	11	70
Economic aspects of cystic ovarian disease treatment *Farhoodi, M. and K. Valipoor*	12	70
Duration effect of propylene glycol oral drenching on animal health and production in transition dairy cows *Malek Mohammadi, H.A., H.R. Rahmani and G.R. Ghorbani*	13	71
Effect of two type of stress on the plasmatic levels of cortisol and some haematic parameters *Mura, M.C., V. Carcangiu, G.M. Vacca, A. Parmeggiani and P.P. Bini*	14	71
Dairy cattle show preferences between different types of cow-brushes *Tuyttens, F., K. Van den Bossche, L. Lens, J. Mertens and B. Sonck*	15	72
Yield losses associated with clinical mastitis in Swedish dairy cows *Hagnestam, C., U. Emanuelson, H. Andersson, J. Philipsson and B. Berglund*	16	72
Teat closure and condition after dry-off in high producing dairy cows *Odensten, M.O., K. Holtenius and K. Persson Waller*	17	73
The map of hip dysplasia of the Hungarian police dog population *Hegedüs, E. and M. Horvai Szabó*	18	73

Session 8: Robust pigs

Date: 05 June '05; 14:00 - 18:00 hours
Chairperson: Lundeheim and Knap (D)

Theatre	Session 8 no.	Page
Biological robustness of pigs *Napel, J. ten*	1	74
Traits associated with sow stayability *Knauer, M., T. Serenius, K.J. Stalder, T.J. Baas, J.W. Mabry and R.N. Goodwin*	2	74
Analysis of true sow longevity *Stalder, K.J., T. Serenius, T.J. Baas, J.W. Mabry and R.N. Goodwin*	3	75
Estimates of additive and dominance genetic effects for sow longevity *Serenius, T. and K.J. Stalder*	4	75
Breeding against osteochondrosis in swine: the Swedish experience *Lundeheim, N.*	5	76
Removal of Swedish sows *Engblom, L., N. Lundeheim, A.-M. Dalin and K. Andersson*	6	76
Environmental sensitivity and robustness *Strandberg, E.*	7	77
Developing the breeding goal for Norwegian Landrace: aiming at a robust and superior sow *Holm, B., E. Gjerlaug-Enger and D. Olsen*	8	77
Investigations on the impact of genetic resistance to oedema disease on performance traits and its relation to stress susceptibility in pigs of different breeds *Binder, S., K.U. Götz, G. Thaller and R. Fries*	9	78
Enzymes, proteins, *Escherichia coli* and Ryanodin receptor gene polymorphisms and their association with osteochondrosis in pigs *Kadarmideen, H.N. and P. Voegeli*	10	78

Poster	Session 8 no.	Page
Parity and production in sows *Neil, M.*	11	79

Session 9: Genetics of variability / Free communications animal genetics

Date: 05 June '05; 14:00 - 18:00 hours
Chairperson: San Cristobal (F) / Eriksson (SWE)

Theatre	**Session 9 no.**	**Page**
Genetic control of environmental variability: evidence from snails, pigs and rabbits *Sorensen, D.*	1	79
Genetic homogenization of birth weight in rabbits: evolution of the characteristics of the genital tract after two generations of selection *Bolet, G., H. Garreau, T. Joly, M. Theau-Clement, J. Hurtaud and L. Bodin*	2	80
Possibilities for selection for uniformity in pig carcasses *Knol, E.F., P.R.T. Bonekamp and P.E. Zetteler*	3	80
A weighted regression approach for the detection of QTL effects on within-subject variability *Wittenburg, D., V. Guiard and N. Reinsch*	4	81
Using the reaction norm approach to investigate genotype-by-environment interactions in the UK Suffolk Sire Referencing Scheme *Pollot, G.E.*	5	81
Evidence of genotype by time interaction for protein production in dairy cattle *Madsen, P., J. Pedersen, U.S. Nielsen and J. Jensen*	6	82
A test of quantitative genetic theory using *Drosophila*: effects of inbreeding and rate of inbreeding on heritabilities and variance components *Kristensen, T.N., A.C. Sørensen, D. Sorensen, K.S. Pedersen, J.G. Sørensen and V. Loeschcke*	7	82
Estimation of ancestral inbreeding coefficients *Suwanlee, S., R. Baumung, J. Sölkner and I. Curik*	8	83
Modelling of the inbreeding depression as a function of the age of the inbreeding *Hinrichs, D., M. Holt, J. Ødegård, O. Vangen, J.A. Woolliams and T.H.E. Meuwissen*	9	83
Evaluation of the epistatic kinship based on genotyping fullsib pairs of three Göttingen Minipig populations *Flury, C. and H. Simianer*	10	84
The role of gene banks as a safe guard in scrapie genotype eradication schemes *Roughsedge, T., B. Villanueva and J.A. Woolliams*	11	84
Individual-based assessment of population structure and admixture levels among Austrian, Croatian and German draught horses *Curik, I., T. Druml, R. Baumung, K. Aberle, O. Distl and J. Sölkner*	12	85

The estimation of extinction probabilities of five German cattle breeds 13 85
Bennewitz, J. and T.H.E. Meuwissen

Poster **Session 9 no. Page**

Estimation of genetic parameters for fertility and hatchability in the two laying cycle (four periods) of three White Leghorn strains (repeated measurement) 14 86
Tazari, M., I. McMillan, J.W. Wilton and V.M. Quinton

The evaluation of genetic variability of wild ancestors, founders and autochthonous cattle breeds in Serbia 15 86
Savic, M., J. Jovanovic and R. Trailovic

Estimation of inbreeding in Iberian pigs using microsatellites 16 87
Alves, E., C. Barragán and M.A. Toro

Sequence variation of the PRKAG3 gene in two cattle breeds: a phylogenetic analysis 17 87
Ciani, E., M. Roux, R. Ciampolini, E. Mazzanti, F. Cecchi, H. Leveziel, S. Presciuttini and V. Amarger

Genetic relationships between populations of Andalusian bovine local breeds in danger of extinction from the polymorphism of microsatellites of ADN 18 88
Rodero, E., P.J. Azor, M. Luque, A. Molina, M. Herrera, M. Valera and A. Rodero

Preliminary study on microsatellite variation of Baluchi sheep breeding flock 19 88
Qanbari, S., M.P. Eskandari Nasab, S.E. Khanian and R. Osfoori

The pool of erythrocyte antigenes in Yakutsky cattle 20 89
Kamaldinov, E.V., V.L. Petukhov, O.S. Korotkevich and A.I. Zheltikov

Analysis of bovine PIT1 gene polymorphism in Iranian Sarabi cattle (*Bos taurus*) using PCR based RFLP 21 89
Tavakolian, J., S. Zenali, B. Azimifar, N. Asadzadeh and A. Javanmard

Molecular analysis of the bovine leptin gene in Iranian Sarabi cattle (Iranian Bos taurus) 22 90
Javanmard, A., G. Elyasi-Zarringabayi, A.A. Gharadaghi, M.R. Nassiry and A. Javadmanash

Genetic polymorphism of ovine calpastatain locus in Iranian kordi sheep by PCR-RFLP 23 90
Nassiry, M.R., A. Javadmanash, M. Nosrati, A. Mohamadi and A. Javanmard

Polymorphism of growth hormone and growth hormone receptor genes in two Iranian Sarabi (*Bos taurus*) and sistani (*Bos indicus*) cattle breeds 24 91
Javanmard, A., R. Miraei-Ashtiani, A. Torkamanzehi, M. Moradi Sharbabak, M. Esmailzadeh, M.R. Nassiry and G. Elyasi-Zarringabayi

Polymorphism of bovine lymphocyte antigen DRB3.2 alleles in Iranian Holstein cattle 25 91
Pashmi, M., A.R. Salehi, A. Ghorasi, M.R. Mollasallalehi, A. Javanmard, S. Qanbari and R. Salehi-Taba

Genetics parameters of variability for litter size and litter weight at birth in *Mus musculus* 26 92
Gutiérrez, J.P., B. Nieto, P. Piqueras and C. Salgado

Study of canalization in an experiment of divergent selection for uterine capacity in rabbit 27 92
Ibáñez, N., D. Sorensen, R. Waagepetersen and A. Blasco

Distinct genotypes from different origins for black coat color in pigs 28 93
Han, S.H., Y.L. Choi, M.S. Ko, M.Y. Oh and I.C. Cho

Genetic diversity and maternal origins of Jeju (Korea) native pigs 29 93
Cho, I.C., Y.L. Choi, M.S. Ko, J.T. Jeon, H.S. Kang and S.H. Han

Allele frequencies of Stearoyl CoA desaturase genetic variants in various cattle breeds 30 94
Moioli, B., F. Napolitano, G. Congiu, L. Orrù and G. Catillo

Ascertainment of evolutionary processes from the genetic variation associated to each geographic point in a map 31 94
Quevedo, J.R., E. Fernández-Combarro, L.J. Royo, I. Álvarez, A. Beja-Pereira, I. Fernández, A. Bahamonde and F. Goyache

Combined use of genealogy and microsatellites in the endangered Xalda sheep 32 95
Álvarez, I., J.P. Gutiérrez, L.J. Royo, I. Fernández, E. Gómez and F. Goyache

Molecular analysis of MC1R gene in the Italian pig breed Mora Romagnola 33 95
Marilli, M., F. Fornarelli, E. Delmonte and P. Crepaldi

Genetic characterization of indigenous Southern African sheep breeds using DNA markers 34 96
Buduram, P., J.B. van Wyk and A. Kotze

Genetic variation among six Iranian goat breeds using RAPD markers 35 96
Javanrouh Aliabad, A., S. Esmaeelkhanian, N. Dinparast and R. Vaez Torshizi

Effects of different mating systems on genetic variance and the average of breeding value in dairy cattle 36 97
Aminafshar, M., A.A. Sadeghi and P. Shawrang

Pairwise comparison of mtDNA sequences in two Croatian sheep populations 37 97
Bradic, M., B. Mioc, V. Pavic and Z. Barac

Characterisation of the founder matrilines in Asturcón pony via mitochondrial DNA 38 98
Royo, L.J., I. Álvarez, I. Fernández, E. Gómez and F. Goyache

Genetic variability in pigs assessed by pedigree analysis: the case of Belgian Landrace NN and Pietrain in Flanders 39 98
Janssens, S., J. Depuydt, S. Serlet and W. Vandepitte

Survey of milk protein polymorphism in the "Rossa Siciliana" dairy cattle 40 99
Zumbo, A., R. Finocchiaro, M.T. Sardina, J.B.C.H.M. van Kaam, A. Rundo Sotera, E. Budelli and B. Portolano

Development of variation in a random bred mouse population 41 99
Schlote, W., A. Wolc, T.A. Schmidt and T. Szwaczkowski

Investigation of the Russian sheep breeds using DNA microsatellites 42 100
Gladyr, E., N. Zinovieva, M. Müller and G. Brem

A genetic linkage map of the blue fox (*Alopex lagopus*) 43 100
Keski-Nisula, S., K. Elo, J. Tähtinen and M. Ojala

Canalising selection on ultimate pH in pig muscle: consequences on meat quality 44 101
Larzul, C., P. Le Roy, T. Tribout, J. Gogué and M. SanCristobal

Phylogenetic relationship and divergence time between two Iranian Kordi sheep populations using microsatellite markers 45 101
Banabazi, M.H., S.R. Miraei Ashtiani and S. Esmaeelkhanian

Breed and typological composition of cattle in townships Prozor-Rama and Konjic 46 102
Ivankovic, S., A. Zelenika and J. Pavlovic

Different types of sheep breeds in Algeria: further molecular characterizations 47 102
Gaouar, S., M. Aouissat, L. Dhimi, A. Routel, B. Kouar and N. Saidi-Mehtar

Session 10: Physiology of stress and reproduction

Date: 05 June '05; 14:00 - 18:00 hours
Chairperson: Kemp (NL)

Theatre **Session 10 no. Page**

Unravelling stress-induced subfertility in ruminants 1 103
Dobson, H., S.P. Ghuman, S. Prabhaker, R. Morris, S. Walker, V. Gandotra and R.F. Smith

Studies on stress and reproduction in the mare: effect of ACTH on adrenal steroid hormone levels in mares 2 103
Hedberg, Y., A.-M. Dalin and H. Kindahl

Effects of stress during pregnancy on endocrine and immune responses in pigs 3 104
Kanitz, E., W. Otten, K.P. Bruessow, M. Tuchscherer and F. Schneider

Effect of social stress on embryonic survival in indoor and outdoor housed gilts and sows 4 104
Hazeleger, W. and N.M. Soede

Effect of summer stress on quantity and quality of semen of Holstein-Friesian and Jersey bulls under subtropical environments 5 105
Fiaz, M. and R.H. Usmani

Factors describing stress of dairy cows around insemination and their effect on non-return rate 6 105
Schrooten, C. and W.A.J. Veldman

Plasma cortisol level in relationship to welfare conditions in dairy farms 7 106
Trevisi, E., R. Lombardelli, M. Bionaz and G. Bertoni

Glutathione peroxidase activity and its relathionship with cortisol and oppiod responses to training in trotters 8 106
Diverio, S., A. Barone, G. Tami, D. Beghelli and C. Pelliccia

Poster **Session 10 no. Page**

Physiological and hematological responses of Baladi goats to tree- sheltering in summer 9 107
Shaker, Y.M., S.A. Kandil and A.A. Azamel

Calving season affects reproductive performance of high yielding but not low yielding Jersey cows 10 107
Soydan, E., E. Sirin, Z. Ulutas and M. Kuran

Application of eustress and distress to pigs and the effect on meat quality 11 108
Küchenmeister, U., I. Fiedler, B. Puppe, K. Ernst, G. Manteuffel and K. Ender

Impact of stress during oestrus on the sperm reservoir and on progesterone concentrations in the sow 12 108
Brandt, Y., A. Lang, H. Rodriguez-Martinez and S. Einarsson

Search for regulatory DNA variation in genes related to stress response in pigs 13 109
Murani, E., S. Ponsuksili, K. Schellander and K. Wimmers

Assessment of Fas and FasL immunoreactivity in ejaculated bull spermatozoa 14 109
Meggiolaro, D., F. Porcelli, A. Carnevali, P. Crepaldi, M. Marilli and B. Ferrandi

Fertility and productive postpartum traits in crossbreds from Pelibuey sheep 15 110
Avendaño, L., F.D. Alvarez, L. Molina, R. Santos, A. Correa and N. Pérez

Effect of a spray and fans cooling system on productive and physiological response of Holstein steers under heat stress 16 110
Correa, A., M. Morales, L. Avendaño, C. Leyva, A. Pérez, R. Díaz and F. Rivera

The effect of hormone treatment on reproductive performance of pigs 17 111
Siukscius, A.

The effect of early wearing on the reproductive performances on Romanian sows 18 111
Parvu, M., C. Dinu and A. Marmandiu

The influence of some transport associated factors on heart rate response in sheep 19 112
Barone, A., S. Diverio, R. Cavallina and N. Falocci

Reproductive performance evaluation of different prostaglandins for repeated synchronization program in postpartum dairy cows: preliminary results 20 112
Pérez, C.C., J.M. Sánchez, L. Molina, M. Luque and J. Perea

The effect of training and competition on the endocrine-metabolic response to stress in trotters 21 113
Diverio, S., A. Barone, G. Tami, D. Beghelli and N. Falocci

Effects of two different exercises on physiological stress responses of training trotters 22 113
Diverio, S., A. Barone, C. Pelliccia, L. Moscati and N. Falocci

Evaluation of ovarian follicular growth patterns between the left and the right ovary in control line gilts and selected for high ovulation rate gilts 23 114
Vatzias, G., G. Maglaras, E. Asmini, R.V. Knox, C.H. Naber and D.R. Zimmerman

Session 11: Specialised ruminant products to sustain systems and genetic resources

Date: 06 June '05; 8:30 - 12:30 hours
Chairperson: Schneeberger (CH)

Theatre **Session 11 no. Page**

Local products for genetic resources sustainability in Southern Europe: a solution or a problem? 1 114
Casabianca, F.

Examples of successful commercialisation of sheep and goat products in Alpine regions 2 115
Ringdorfer, F.

Specialised small ruminant products and genetic resources in the Middle and Eastern European countries 3 115
Kukovics, S., K. Kume, D. Dimov, E. Gyarmathy, J. Dubravska, E. Martyniuk, S. Stojanovic, D. Kompan, V. Matlova and I. Padeanu

Genetic relationships between growth and pelt quality traits in the Gotland sheep breed 4 116
Näsholm, A.

Fatty acid composition of lamb's meat from two different genotypes 5 116
Vacca, G.M., M.L. Dettori, L. Cengarle, G. Tillocca and V. Carcangiu

Relationship between tissue thicknesses measured on live Pinzgau bulls by ultrasound and weight of hot carcass 6 117
Polák, P., E.N. Blanco Roa, E. Krupa, J. Huba, D. Peskovicová and M. Oravcová

Session 12: Developments in quantitative genetics

Date: 06 June '05; 8:30 - 12:30 hours
Chairperson: Zengting Liu (D)

Theatre **Session 12 no. Page**

Use of genomic information for genetic improvement of livestock 1 117
Meuwissen, T.H.E.

An approximate interval mapping procedure for selective DNA pooling 2 118
Dolezal, M., H. Schwarzenbacher, J. Sölkner and P.M. Visscher

Background bias on cDNA micro-arrays 3 118
Pool, M.H., B. Hulsegge and L.L.G. Janss

Modelling liveweight change over a lactation in Irish dairy cows 4 119
Quinn, N., L. Killen and F. Buckley

Genetic selection against cannibalism related mortality in layer chicken 5 119
Ellen, E.D., W.M. Muir and P. Bijma

Selection for intramuscular fat in Duroc pigs using real-time ultrasound 6 120
Baas, T.J., C.R. Schwab and K.J. Stalder

From single loci to chromosome segments: a different quantitative genetic perspective 7 120
Simianer, H., C. Flury, M. Tietze and H. Täubert

Genetic analyses of traits affected by interaction among individuals 8 121
Bijma, P. and W.M. Muir

Validation of an approximate approach to compute genetic correlations between longevity and linear traits 9 121
Tarres, J., J. Piedrafita and V. Ducrocq

Different selection strategies for improving lactation milk yield and persistency *Lin, C.Y. and K.Togashi*	10	122
Estimation of realised genetic trends in French Large White pigs from 1977 to 1998 using frozen semen: farrowing and early lactation periods *Canario, L., T.Tribout, J. Gogué and J.P. Bidanel*	11	122
Graphic explanation of response prediction in long-term selection program *Nishida, A., K. Suzuki and Y. Ohtomo*	12	123

Poster	**Session 12 no.**	**Page**
The construction of indexes with constant restrictions by iterative procedure *Lin, C.Y.*	13	123
Multiple traits Bayesian analysis of birth, weaning and yearling body weights of Egyptian Zaraibi goat *Shaat, I., L.Varona and W. Mekkawy*	14	124
Development of epistatic variance components in the two locus bi-allelic model *Curik, I., J. Sölkner and M. Kaps*	15	124
LDLA, a package to compute IBD matrices for QTL fine mapping by variance component methods *Janss, L.L.G. and H.C.M. Heuven*	16	125
Genetic evaluation using markers completely linked to QTLs *Liu, Y. and P.K. Mathur*	17	125
Estimate of genetic parameters for competition effect in selected line of Duroc pigs *Oikawa, T., A. Nagata, M.Tomiyama, K. Suzuki, H. Kadowaki and T. Shibata*	18	126

Session 13: Physiology of pregnancy

Date: 06 June '05; 8:30 - 12:30 hours
Chairperson: Rátky (H)

Theatre	**Session 13 no.**	**Page**
A protective role of seminal plasma in sperm-induced endometritis *Troedsson, M.H.T., A.S. Alghamdi and A. Desvousges*	1	126
The cow in endocrine focus before and after calving *Kindahl, H., B. Kornmatitsuk and H. Gustafsson*	2	127

Current physiological aspects of pregnancy between implantation and partus in the pig | 3 | 127
Wähner, M.

Poster **Session 13 no. Page**

Arachidonic acid activates the matrix metalloproteinase of placental fibroblast cells | 4 | 128
Kamada, H., Y. Ueda and M. Murai

Session 14: High health pig systems

Date: 06 June '05; 8:30 - 12:30 hours
Chairperson: Chadd (UK)

Theatre **Session 14 no. Page**

High health pig systems: the Danish approach | 1 | 128
Bækbo, P.

Porcine embryo vitrification and transfer: a way to maintain high health | 2 | 129
Cuello, C., J.M. Vázquez, J. Roca and E.A. Martínez

Mortality pattern and causes in crossbred pigs | 3 | 129
Srinivasa Rao, D.

The relationship between birth weight, the condition of the umbilical cord and the time interval at birth | 4 | 130
Fischer, K. and M. Wähner

Relation between (the breeding value for) weaning survival and periparturient sow behaviour | 5 | 130
Uitdehaag, K.A., E.D. Ekkel, E. Kanis, E.F. Knol and T. van der Lende

Development of a protocol to record functional traits and inherited disorders affecting welfare in pigs | 6 | 131
Fernàndez, X., J. Tibau, J. Piedrafita and E. Fàbrega

Effects of rearing system on performance, animal welfare and meat quality in two pig genotypes | 7 | 131
Lebret, B., M.C. Meunier-Salaün, A. Foury, E. Dransfield and J.-Y. Dourmad

The influence of rearing conditions on meat and back fat quality of large white breed | 8 | 132
Holló, G., J. Seregi, J. Csapó, E. Varga-Visi and I. Holló

Effects of reduced phosphorus levels during growth and pregnancy on leg weakness, osteochondrosis and longevity in sows 9 132
Jørgensen, B. and H.D. Poulsen

Poster **Session 14 no. Page**

Effect of zeolite clinoptilolite on biochemical and hematological parameters in weaned piglets fed with increased zearalenone level 10 133
Speranda, M., B. Liker, T. Speranda, V. Seric, Z. Antunovic, D. Sencic, Z. Grabarevic and Z. Steiner

Session 15: Free communications animal nutrition (Ruminant nutrition and physiology)

Date: 06 June '05; 14:00 - 18:00 hours
Chairperson: Crovetto (I)

Theatre **Session 15 no. Page**

Milk urea content as affected by roughage type 1 133
Campeneere, S. de and D.L. de Brabander and J.M. Vanacker

Effect of rumen escape starch in maize silage based diets for dairy cattle 2 134
Brabander, D.L. de, S. de Campeneere, J.M. Vanacker and N.E. Geerts

Nutritional value and effect of Vicia ervilia seed on Holstein dairy cow performance 3 134
Moeini, M., H. Amanlo, M. Azari and M. Souri

Effect of Ca-soap of linseed oil on rumen fermentation pattern and on the characteristics of goat milk 4 135
Cenkvári, E., S. Fekete, H. Fébel, T. Veresegyházi and E. Andrásofszky

Increasing amounts of sunflower seeds increase CLA and vaccenic acid content in milk fat from dairy cows 5 135
Nielsen, T.S., E.M. Straarup, M.T. Sørensen and K. Sejrsen

Kinetics of trans and conjugated fatty acids (FA) concentrations in cow milk after addition of plant oils to different basal diets 6 136
Roy, A., A. Ferlay, A. Ollier and Y. Chilliard

Lamb vigour is affected by DHA supplementation of ewe diets during late pregnancy 7 136
Pickard, R.M., A.P. Beard, C.J. Seal and S.A. Edwards

Influence of feed withdrawal on plasma leptin concentrations in lambs of different carcass composition 8 137
Borell, E. von, H. Sauerwein and M. Altmann

Central effects of histamine on food intake, and kind of histamine receptors in sheep brain 9 137
Rahmani, H.R., M. Mohammadalipour and C.D. Ingram

Selenium status around peripartum in beef cows and calves offered grass silage and barley produced with selenium enriched fertilizers 10 138
Cabaraux, J.F., J.-L. Hornick, N. Schoonheere, L. Istasse and I. Dufrasne

Analysis of n-alkanes in kidney fat for tracing feeding systems in meat producing animals 11 138
Smet, S. de, K. Raes, E. Claeys, M.J. Petron and K. Vervaele

Poster **Session 15 no. Page**

Effect of starting time of feeding milk replacer on the performance of Holstein calves 12 139
Ghassemi, Sh.J., Y. Rouzbehan and A. Nikkhah

The influence of feeding level and milk replacer protein content on growth and blood protein levels of Holstein-Friesian calves 13 139
Wicks, H.C.F., R.J. Fallon, J. Twigge, L.E.R. Dawson and M.A. McCoy

The effect of by-pass methionine supplement served before and after calving on milk yield and physiological parameters in dairy cows 14 140
Kudrna, V., J. Illik, P. Lang and P. Mlázovská

Effect of L-glutamine-containing oral rehydratation solution on the absorptive function of small intestine in diarrhoeal calves infected with *Cryptosporidium parvum* 15 140
Klein, P., H. Lelkova, J. Lastovkova and M. Soch

The effect of addition of selenium to a milk diet of calves on the meat quality 16 141
Skrivanova, V., Y. Tyrolova, M. Marounek and M. Houska

Effect of feed blocks on the growth of grazing heifers 17 141
Chaudhry, A.S., C.J. Lister and W. Taylor

Effect of offering two levels of crude protein and two feeding levels of milk replacer on calf performance 18 142
Fallon, R.J., H.C.F. Wicks and J. Twigge

Bioelectrical impedance analysis for the prediction of saleable products in buffalo 19 142
De Lorenzo, A., F. Sarubbi, F. Polimeno, R. Baculo, M. Servidio, P. Abrescia and L. Ferrara

Intestinal digestibility of rumen undegraded protein determined by mobile bag method in rapeseed, rapeseed meal and extracted rapeseed meal 20 143
Homolka, P. and V. Koukolová

Ruminal degradability and mobile bag intestinal digestibility of individual amino acids of pasture forage 21 143
Homolka, P., J. Trinácy and A. Skeríková

Characteristic size dimensions of washed faeces particles from dairy cows fed different concentrate/forage ratios 22 144
Nørgaard, P., M.R. Weisbjerg, K.F. Jørgensen and D. Bossen

Effect of plant phenolic compounds on growth of some rumen bacteria 23 144
Rullo, R., A. Tava, L. Ferrara and G. Maglione

Comparing metabolic traits glucose and insulin in their relationship to milk production 24 145
Panicke, L., G. Freyer, R. Staufenbiel and E. Fischer

Effect of feeding pistachio hulls on performance of lactating dairy cows 25 145
Vahmani, P., A.A. Naserian, J. Arshami and H. Nasirimoghadam

Nutritive evaluation of processed cottonseed fed Holstein dairy cows 26 146
Foroughi, A.R., R. Valizadeh, A.A. Naserian and M. Danesh Mesgaran

Evaluation of processed cottonseed and ruminally protected lysine and methionine for lactating dairy cows 27 146
Foroughi, A.R., A.A. Naserian, R. Valizadeh and M. Danesh Mesgaran

The effect of application a dry feed additive on the fermentation characteristics of lupin silage 28 147
Dolezal, P., L. Zeman, J. Dolezal and V. Pyrochta

Development of a quick method of evaluating flavour preferences in concentrates for lactating cows 29 147
Roura, E., C. Ossensi, R. Mantovani and L. Bailoni

Milk production and composition as affected by feeding supplemental fat in Sahiwal cattle 30 148
Iqbal, A., J. Akbar, M. Abdullah and M. Sarwar

Degradation of dry matter and fiber of five feeds by rumen anaerobic fungi of sheep 31 148
Ghoorchi, T., S. Rahimi, M. Rezaeian and G.R. Ghorbani

The effect of Polyethylen Glycol (PEG) addition on in vitro organic matter digestibility (IVOMD) of grape pomace 32 149
Alipour, D. and Y. Rouzbehan

In vitro enzymatic proteolysis of different protein sources 33 149
Chaudry, A.S.

Nutritional assessment of genetically modified rape seed and potatoes, differing in their output traits 34 150
Boehme, H., B. Hommel, E. Rudloff and L. Huether

Investigation of some preparation procedures of fatty acid methyl esters for capillary gas-liquid chromatographic analysis of conjugated linoleic acid in feed 35 150
Wágner, L., K. Dublecz, F. Husvéth, L. Pál, Á. Bartos, G. Kovács, Z. Garádi, Sz. Stiller and J. Karnóth

Effect of vitamin B6 on lysine synthesis from 2,6-diaminopimelic acid by mixed rumen protozoa and bacteria 36 151
El-Waziry, A.M.

Pollen composition of six plant species in some nutrients, microminerals and aromatic substances and potential use in animal nutrition 37 151
Liamadis, D., T. Zisis, Ch. Milis and A. Thrasivoulou

Apparent ileal amino acid digestibility in diets supplemented with phytase and pancreatine for pigs 38 152
Copado, F., M. Cervantes, J. Yánez, J.L. Figueroa and W. Sauer

Mustard seed (*Sinapis Alba*): Nutrient content and digestibility in swine 39 152
Ács, T., A. Hermán, M. Szelényi, A. Regius and J. Gundel

Influence of dietary fibre on the gut morphology and pancreatic and intestinal enzyme activities in the weaned piglet 40 153
Trevisi, P., J.P. Lallès, I. Luron and B. Sève

Mannan oligosaccharide enhances absorption of colostral Ig G in newborn calves and piglets 41 153
Lazarevic, M.

Growth promoting effects of Rare Earth Elements in piglets 42 154
Wehr, U., C. Knebel and W.A. Rambeck

An hydrolyzed protein concentrate (Palbio 62(r)) increases feed intake and villus height in early weaning pigs 43 154
Borda, E., D. Martinez-Puig and F. Perez

Effects of supplemented dietary L-tryptophan on growth performance of 15- to 30-kilogram pigs 44 155
Buraczewska, L., E. Swiech, L. Le Bellego and D. Melchior

Growth performance, nutrient digestibility and intestinal morphology in weanling pigs fed Insoluble Fibre Concentration(IFC) and direct-fed microbials(DFM) 45 155
Han, Y.K., J.H. Lee and K.M. Park

Influence of herbal feed additives on intestinal microflora of weanling piglets 46 156
Galletti, S., S. Stella and D. Tedesco

Effect of protein level variation of feed ration on slaughtering results of broiler chicken 47 156
Tudorache, M., I. Custura, E. Popescu-Miclosanu and G. Dinita

The effect of different dietary unsaturated to saturated fatty acids ratios on the performance and serum lipids in broiler chickens under feed restriction 48 157
Navidshad, B., M. Shivazad, A. Zareh Shahne and G. Rahimi

Use of long chain calcium salt of fatty acid plant in Holstein calves rations on performance and plasma concentrations of thyroid hormones 49 157
Ahangari, Y.J. and Y. Roozbahan

The effect of different levels of calcium and phosphorous on broiler performance 50 158
Kheiri, F. and H.R. Rahmani

Effect of *Vicia villosa* Roth inclusion on the performance of rabbits 51 158
Camacho-Morfin, D., L. Morfín-Loyden, J.I. Pérez Dosta and D.G. López Rodriguez

The health status and growth performance of growing rabbits receiving either one or two diets during fattening period 52 159
Volek, Z., V. Skrivanova and M. Marounek

Statistical prediction of rumen volume in Friesian bulls using four different markers and live body weight 53 159
Salem, S.M.

Estimation of the energy value of high lactating ewes' milk at early lactation 54 160
Milis, Ch. and D. Liamadis

Effect of Physalis alkekengi fruit feeding on the fertility and reproduction of ewes 55 160
Yousef Elahi, M. and E. Baghaei

Effects of concentrate level and starch degradability on milk yield and fatty acid (FA) composition in goats receiving a diet supplemented in sunflower oil 56 161
Bernard, L., J. Rouel, A. Ferlay and Y. Chilliard

Effects of concentrate level and starch degradability on expression of mammary lipogenic genes in goats receiving a diet supplemented in sunflower oil 57 161
Bernard, L., C. Leroux, Y. Faulconnier, D. Durand and Y. Chilliard

Chemical characteristics of Kuruma prawn (*Marsupenaeus japonicus*) from two different farming systems 58 162
Ragni, M., L. Turi, L. Melodia, A. Caputi Jambrenghi, F. Giannico, A. Vicenti and G. Vonghia

Investigating the compensatory growth in finishing lambs 59 162
Ghoorchi, T. and H. Safarzadeh Torghabeh

Study of fattening potential of lambs fed by grazing barley forage 60 163
Ghoorchi, T., Z. Karimi and A, Zeinali

Milk composition in llamas (*Lama glama*) and the effect of lactational stage 61 163
Riek, A. and M. Gerken

Effect of subterranean clover feeding on sheep milk yield and quality 62 164
Orrù, L., A. Scossa, B. Moioli, F. Spirito, G. Catillo, C. Tripaldi and V. Pace

Nutritional value and effect of rapeseed meal on Sanjabi lamb performance 63 164
Souri, M., M. Moeini and M. Kheirabadi

The effects of dietary inclusion of organic selenium (Sel-Plex) on ewe performance and milk selenium level in sheep 64 165
Crosby, F., M. Fooley and S. Andrieu

Effects of Silymarin administration to middle-lactating dairy goats 65 165
Tedesco, D., J. Turini and S. Galletti

Cactus pear (Opuntia ficus-indica) as a complement to urea-treated straw in dry season feeding systems of ruminants 66 166
Tegegne, F., K.J. Peters and C. Kijora

The effect of lasalocid and monensin on rumen metabolites of sheep 67 166
Abdoli, H., A. Taghizadeh and A. Tahmasbi

Effect of biotin supplementation on milk performance in dairy cows 68 167
Brabander, D.L. de and V. Wouters

Fibrolytic enzymes in dairy calf starter 69 167
Ghorbani, G.R., A. Jafari and A. Nikkhah

Influence of dry period shortening on colostrum quality and Holstein calves average daily gain 70 168
Amini, J., H.R. Rahmani and G.R. Ghorbani

Influence of Baker's yeast (*Saccharomyces cerevisiae*) on digestion and fermentation patterns in the rumen of sheep and milk response in dairy cattle 71 168
Allam, A.M.

Essential amino acid requirements of Dorper lambs estimated by the whole empty body essential amino acid profile 72 169
Ferreira, A.V. and A.H. Jurgens

Effect of graded levels of threonine on gimmizah layer hens performance 73 169
Khalifah, M. and A. Abdella

Effects of pre-partum body condition score on milk yield of Holstein dairy cows 74 170
Pezeshki, A., G.R. Ghorbani and H.R. Rahmani

Effect of physical form of the starter on performance of Holstein calves 75 170
Bagheri, M. and G.R. Ghorbani

The effect of direct-fed fibrolytic enzyme on mid-lactating dairy cow's performance and digestibility 76 171
Alikhani, M., M. Shahzeidi, G.R. Ghorbani and H. Rahmany

The effects of reducing fish meal of diets supplemented with DL-methionine and L-lysine hydrochloride on female broiler performance 77 171
Eila, N., B. Hemati and R. Jalilian

Degradability of corn silage in ruminal ambient with different additives 78 172
Katsuki, P.A., E.S. Pereira, B.M.O. Ramos, E.L.A. Ribeiro, M.A. Rocha, A.P. Pinto, R. Salmazo, T.R. Casimiro, T.C. Alves, M.N. Bonin and I.Y. Mizubuti

Effect of different feed additives in sugar-cane diets on the productive performance of Limousin x Nelore crossbred heifers 79 172
Katsuki, P.A., E.S. Pereira, B.M.O. Ramos, E.L.A. Ribeiro, F.B. Moreira, M.A. Rocha, A.P. Pinto, R. Salmazo, T.R. Casimiro, T.C. Alves and I.Y. Mizubuti

Session 16: Free communications animal management and health (Management and health of ruminants)

Date: 06 June '05; 14:00 - 18:00 hours
Chairperson: Metz (NL)

Theatre **Session 16 no. Page**

Performance and feeding behaviour of dairy cows in an automatic milking systems with controlled cow traffic 1 173
Melin, M., K. Svennersten-Sjaunja and H. Wiktorsson

Effect of milking frequency on lactation persistency in an automated milking system 2 173
Svennersten-Sjaunja, K. and G. Pettersson

Effect of mastitis on culling in Swedish dairy cattle 3 174
Schneider, M. del P., E. Strandberg and A. Roth

Dairy cow health and the effects of genetic merit for milk production, management and interactions between these: udder health parameters 4 174
Ouweltjes, W., B. Beerda, J.J. Windig, M.P.L. Calus and R.F. Veerkamp

Osteochondrosis in beef sires in Sweden 5 175
Persson, Y. and S. Ekman

On the development of asymmetry between lateral and medial rear claws in dairy cows 6 175
Telezhenko, E., C. Bergsten, M. Magnusson, M. Ventorp, J. Hultgren and C. Nilsson

Evaluation of AFB1/AFM1 carry-over in lactating goats exposed to different levels of AFB1 contamination 7 176
Ronchi, B., P. Danieli, A. Vitali, A. Sabatini, U. Bernabucci and A. Nardone

The ACTH challenge test to evaluate the individual welfare condition 8 176
Bertoni, G., E. Trevisi, R. Lombardelli and L. Calamari

Effect of transport for up to 24 hours followed by twenty-four hours recovery on liveweight, physiological and haematological responses of bulls 9 177
Early, B., D.J. Prendiville and E.G. O'Riordan

Poster **Session 16 no. Page**

The effect of herbal medicine (RHAM) (r) on animal dermatic and claw infection treatment 10 177
Ahadi, A.H., M.R. Sanjabi, M. Moeini and A. Ghahramani

Free cortisol in milk as an indicator of stress in dairy cows held in barns with automated milking and control gates 11 178
Melin, M., K. Svennersten-Sjaunja and H. Wiktorsson

Effect of milking frequency on milk quality 12 178
Svennersten-Sjaunja, K., L. Wiking, A. Edvardsson, A.K. Båvius and I. Andersson

The welfare of weanling bulls transported from Ireland to Italy 13 179
Early, B., D.J. Prendiville and E.G. O'Riordan

Cattle metabolic diseases and changes in central nervous system 14 179
Pilmane, M., I. Zitare, A. Jemeljanovs and I.H. Konosonoka

Investigations on microbiological spectrum in subclinical and clinical mastitis of dairy cows in Latvia 15 180
Konosonoka, I.H., A. Jemeljanovs and I. Ciprovica

Comparative research on the impact of the maintenance system upon milk cows' performance and health 16 180
Vidu, L., A. Udroiu, I. Raducuta, I. Calin and V. Bacila

Cubicles height over the floors in passages: implications for hygiene 17 181
Herlin, A.H., M. Magnusson, M. Ventorp and S. Lorentzon

The effect of short dry periods on health disorders in dairy cattle 18 181
Pezeshki, A., H.R. Rahmani, G.R. Ghorbani, M. Alikhani and M. Mohammadalipour

Longitudinal slope of the cubicle for the dairy cow 19 182
Herlin, A.H., M. Magnusson and C. Hagberg

Analysis of the factors affecting somatic cell count in milk 20 182
Kiiman, H., T. Kaart, M. Henno and O. Saveli

Characteristics for mastitis incidence in dairy herds in the Czech Republic 21 183
Stipková, M., M. Wolfová and J. Wolf

Milk let-down parameters' association with udder health problem incidence: a case study 22 183
Livshin, N., E. Maltz, M. Tinsky and E. Aizinbud

Teat and teat-end type in three dairy cattle breeds in the Tropics 23 184
Riera, M., A. Cefis, O. Pedron and J.C. Alvarez

Comparing two concentrate allowances in an automatic milking system 24 184
Halachmi, I.

Analysis of the synchronisation of passages out to the pasture in an automatically milked herd with day and night access to the grazing area 25 185
Spörndly, E. and M. Bergman

Relationships between temperature, precipitation, and carcass weights of the red deer (Cervus elaphus L.) in North-Eastern Poland 26 185
Janiszewski, P. and T. Daszkiewicz

Identification of strains of mink Aleutian disease virus in Nova Scotia 27 186
Farid, A., B.F. Benkel, F.S.B. Kibenge and G.G. Finley

The influence of outdoor raising of pigs on their growth rate and behaviour 28 186
Juskiene, V. and R. Juska

Short and long slaughter transports increase mortality rates in pigs 29 187
Werner, C., K. Reiners and M. Wicke

Dust spatial distribution and seasonal concentration of windowless broiler building 30 187
Choi, H.C., K.Y. Yeon, J.I. Song, H.S. Kang, D.J. Kwon, Y.H. Yoo, C.B. Yang, S.S. Cheon and Y.K. Kim

Age-related differences of Ascaridia galli egg output and worm burden in chickens following a single dose infection 31 188
Gauly, M., T. Homann and G. Erhardt

Comparison of two force molting methods on performance of laying hens in second phase of egg production 32 188
Hassanabadi, A. and H. Kermanshahi

The functional condition of the stomach in goats infected with the alimentary tract nematodes 33 189
Birgele, E., D. Keidane, A. Mugurevics and J. Jegorova

The functional condition of stomach and some indices of meat quality in bulls in their ontogenesis 34 189
Ilgaza, A., E. Birgele, A. Mugurevics, D. Keidane and J. Jegorova

Session 17: Free communications animal genetics

Date: 06 June '05; 14:00 - 18:00 hours
Chairperson: Grandinson (SE)

Theatre	Session 17 no.	Page
Are time-budgets of dairy cows affected by genetic improvement of milk yield? *Lovendahl, P. and L. Munksgaard*	1	190
Estimates of genetic parameters for milkability from automatic milking *Gäde, S., E. Stamer, W. Junge and E. Kalm*	2	190
Genetic parameters for conception rate and days open in Holsteins *Tsuruta, S., I. Misztal and T.J. Lawlor*	3	191
When to farrow? Genetic correlation between gestation length and piglet survival *Rydhmer, L. and N. Lundeheim*	4	191
A method to define sustainable breeding goals for livestock breeding programmes *Nielsen, H.M. and L.G. Christensen*	5	192
Validation of an approximate multitrait model for prediction of breeding values in dairy cattle - a stochastic simulation study *Lassen, J., M.K. Sørensen and P. Madsen*	6	192
Another useful reparameterisation to obtain samples from conditional inverse Wishart distributions *Korsgaard., I.R., A.H. Andersen, P. Madsen and J. Ødegård*	7	193
Properties of random regression models using linear splines *Misztal, I.*	8	193
Genetic improvement in broilers using indirect carcass measurements *Zerehdaran, S., A.L.J. Vereijken, H. Bovenhuis, J.A.M. van Arendonk and E.H. van der Waaij*	9	194
Base populations in fish breeding programs: a simulation study *Holtsmark, M., J.A. Woolliams, A.K. Sonesson and G. Klemetsdal*	10	194

Poster	Session 17 no.	Page
Evaluation of heterosis, general and special combining ability for some biological characters in six silkworm lines *Qotbi, A., A.R. Seidavi, M.R. Gholami and A.R. Bizhannia*	11	195
Non-additive breed effects on milk production in cattle *Wolf, J., L. Zavadilová and E. Nemcová*	12	195

Estimates of genetic parameters of final weight at slaughter, yield grade and marbling scores in beef cattle 13 196
Márquez, A.P., A. Correa, J.F. Ponce, L. Avendaño and S.C. Ochoa

Genetic parameters for type traits of Brazilian Holstein cattle 14 196
Costa, C.N., J.A. Cobuci, A.F. Freitas, N.M. Teixeira, R.B. Barra and A.A. Valloto

Effect of genetic potential for immunocompetence on vaccination efficiency in broilers 15 197
Ask, B.

A PC program to analyse the genetic and environmental trends in the Italian Holstein dairy herds 16 197
Biffani, S., F. Canavesi and M. Marusi

Gene flow in animal genetic resources: a study on status, impacts, trends from exchange of breeding animals 17 198
Musavaya, K. and A. Valle Zárate

Optimization of the Bavarian PIG Testing and breeding scheme 18 198
Habier, D. and L. Dempfle

Genetic variability of growth traits in performance testing of bulls 19 199
Bogdanovic, V., R. Djedovic and P. Perisic

Genetic analysis of the fertility in Hanoverian Warmblood horses 20 199
Hamann, H., H. Sieme and O. Distl

Genetic relationship between different measures of feed efficiency and its component traits in Wagyu (Japanese Black) bulls 21 200
Hoque, A., T. Kunieda and R. Oikawa

Genetic and environmental relationships between milk yields at different parts of lactation in Iranian Holsteins 22 200
Farhangfar, H. and H. Naeemipour

Non-genetic influence on test day milk and fat yields data of Moroccan Holstein-friesian cows 23 201
Tijani, A.

Genetic parameters of fur coat and reproduction traits for Polish arctic foxes 24 201
Wierzbicki, H. and W. Jagusiak

Effect of mating ratio on response for a selection index 25 202
Campo, J.L., S.G. Dávila and I. Pena

An investigation on the erythrocyte potassium polymorphism and relation between several Mohair characteristics in Angora Goat *(Capra hircus)* 26 202
Soysal, M.I., E.K. Gürcan, E. Özkan, M. Aytac and S. Özkan

Genetic parameters for carcass traits of field progeny and their relations with feed efficiency traits of their sire population for Japanese Black cattle 27 203
Hoque, A., T. Oikawa and K. Hiramoto

Genetic parameters of body weights and carcass traits in two quail strains 28 203
Vali, N., M.A. Edriss, H.R. Rahmani and A. Samie

Weaning performance of Hungarian Grey calves 29 204
Nagy, B., Z. Lengyel, I. Bodó, I. Gera, M. Török and F. Szabó

Polymorphism detection in bovine Stearoyl-CoA desaturase (SCD) locus by means of microarray analysis 30 204
Conte, G., B. Castiglioni, M. Mele, A. Serra, S. Chessa, G. Pagnacco and P. Secchiari

Value of traits in beef cattle breeding 31 205
Pribyl, J., P. Safus, J. Pribylova, L. Stadník, M. Stipkova, Z. Vesela and M. Wofova

New microsatellites assignment using a hamster-sheep cell hybrid panel 32 205
Tabet-Aoul, N., R. Ait-Yahia, A. Derrar, N. Boushaba and N. Saidi-Mehtar

General combining ability estimates in silkworm inbreed lines (*Bombyx mori l.*) 33 206
Dinita, G., C. Antonescu, A. Marmandiu and M. Tudorache

Algebraical and geometrical interpretation of restricted best linear unbiased prediction of breeding values 34 206
Satoh, M. and M. Takeya

Direct genetic, maternal genetic and common environmental effects on Landrace and Duroc piglet growth 35 207
Suzuki, K., H. Kadowaki, T. Shibata, H. Uchida and A. Nishida

Simulation of multiple trait data for testing breeding value estimation programs 36 207
Wensch-Dorendorf, M., H.H. Swalve and J. Wensch

Estimation of the milk urea course during lactation 37 208
Nemcová, E., M. Stipkova, F. Jílek and M. Krejcová

Estimates of genetic parameters for reproduction and production traits of purebred Berkshire in Japan 38 208
Tomiyama, M., T. Oikawa, T. Sano, T. Arakane and H. Mori

Non-genetic factors effect on body weights and grease fleece weight in Loribakhtiary sheep flock 39 209
Asadi-khoshoei, E. and R. Miraei-Ashtiani

Genetic parameters of calving interval in Japanese Holstein 40 209
Sasaki, O., M. Aihara, K. Hagiya, K. Ishii and Y. Nagamine

Estimation of genetic parameters for early growth traits in the Fat-tailed Lori-Bakhtiary lambs 41 210
Asadi-khoshoei, E.

Prediction of the genetic progress applying embryo transfer into the breeding programs in cattle 42 210
Marmandiu, A., M. Nedelcu, M. Pîrvu, G. Dinita and I. Raducuta

Genetics of size traits, fur quality traits and litter size in blue fox 43 211
Peura, J., I. Strandén and E.A. Mäntsysaari

Effect of small-herd clustering on the genetic connectedness of the Portuguese Holstein cattle population 44 211
Vasconcelos, J. and J. Carvalheira

The interdependence of some morphological traits in Frasinet carp breed 45 212
Nicolae, C., L.D. Urdes and I. Cringanu

Impact of pedigree errors on the genetic gain in a dairy cattle population 46 212
Sanders, K., J. Bennewitz and E. Kalm

Modelling post-weaning growth of the Avileña Negra-Ibérica beef cattle breed under commercial using random regression 47 213
Díaz, C., A. Moreno, A.M. Serrano and M.J. Carabaño

Responses to seven generations of selection for ovulation rate or prenatal survival in Large White pigs 48 213
Rosendo, A., J. Gogué, T. Druet and J.P. Bidanel

New nucleotide sequence polymorphism within 5'noncoding region of the bovine receptor estrogen α (ERα) gene 49 214
Szreder, T. and L. Zwierzchowski

Effect of breeding value of bulls and performance traits of their daughters 50 214
Sitkowska, B. and S. Mroczkowski

Influence of paternal country origin on chosen performance traits of their daughters 51 215
Sitkowska, B. and S. Mroczkowski

Carcass analysis in Japanese quail lines divergently selected for shape of growth curve 52 215
Hyánková, L., J. Lastovková and Z. Szebestová

Evaluation of imported gene, direct and maternal heterosis, and estimation of genetic parameter and in the Iranian crossbred dairy cattle population 53 216
Ghorbani, A., S. Miraii Ashtiani and M. Moradi Shahrebabak

Estimation of genetic parameters of growth traits in Golpayegan's calves 54 216
Harighi, M.F.

Selection Indices and sub indices for improvement milk composition traits in Friesian cattle in Egypt 55 217
Hussein, A.M., M.N. El-Arian, A.S. Khattab, E.A. Omer and F.H. Farrag

The genetic and environmental effects on milk yield and fat percentage in Isfahan dairy farms 56 217
Babaei, M.

Session 18: Free communications pig production (Pig production, management and health)

Date: 06 June '05; 14:00 - 18:00 hours
Chairperson: Kovac (SLO)

Theatre **Session 18 no. Page**

Effect of sex and slaughter weight on pig performance and carcass quality 1 218
Mullane, J., P.G. Lawlor, P.B. Lynch, J.P. Kerry and P. Allen

Estimation of whole body lipid mass in finishing pigs 2 218
Kloareg, M., J. Noblet, J. Mourot and J. van Milgen

Preliminary study of the effect of the IGF-II genotype on meat quality in pigs 3 219
Maagdenberg, K. van den, A. Stinckens, E. Claeys, N. Buys and S. de Smet

Genetic parameters for meat quality and production traits in Finnish Landrace and Large White pigs 4 219
Sevón-Aimonen, M-L. and A. Mäki-Tanila

Influence of genetic markers on the drip loss development of case-ready pig meat 5 220
Otto, G., R. Roehe, G.S. Plastow, P.W. Knap, H. Looft and E. Kalm

Genetic parameters of direct and ratio traits of Hungarian pig populations 6 220
Nagy, I., L. Csató, J. Farkas, K. Tisza and L. Radnóczi

Does the Swedish animal welfare legalisation influence the working-hours spent in pig production? 7 221
Mattsson, B., Z. Susic and N. Lundeheim

Sow behaviour and litter size in first parity sows kept outdoors 8 221
Wallenbeck, A., L. Rydhmer and K. Thodberg

Behaviour, health and performance of piglets exposed to atmospheric ammonia 9 222
Borell, E. von, A. Özpinar, K.M. Eslinger, A.L. Schnitz, Y. Zhao and F.M. Mitloehner

Ranking of discrete choice | 10 | 222
Halekoh, U., E. Jørgensen, M.B. Jensen, L.J. Pedersen and M. Studnitz

Influence of emergency vaccination on the course of classical swine fever epidemics | 11 | 223
Witte, I., J. Teuffert, G. Rave and J. Krieter

Poster **Session 18 no. Page**

Performance traits of improved Lithuanian White pigs | 12 | 223
Klimas, R. and A. Klimiene

The sequence of myostatin in double-muscled pigs | 13 | 224
Stinckens, A., J. Bijttebier, T. Luyten, K. van den Maagdenberg, N. Harmegnies, S. de Smet, M. Georges and N. Buys

Implementation of a selection and mating strategy to optimize genetic gain and rate of inbreeding in the Swiss pig breeding program | 14 | 224
Luther, H. and A. Hofer

Results concerning the use of the compactness index in boars selection | 15 | 225
Bacila, V., M. Vladu, I. Calin, L. Vidu, P. Tapaloaga and R.A. Popa

The growth rate of piglets from primiparous and multiparous litters with equalized pre- and postnatal effect of maternal environment conditions related to litter size | 16 | 225
Czarnecki, R., M. Rózycki, M. Kamyczek, A. Pietruszka and B. Delikator

Pig osteochondrosis in Lithuania and its relationship with fattening traits and meatiness | 17 | 226
Klimiene, A. and R. Klimas

The comparison of prediction abilities of pig carcass dissection methods | 18 | 226
Pulkrabek, J., J. Pavlík, L. Valis and M. Vítek

MRI as a reference technique to assess carcass composition in pig performance testing | 19 | 227
Baulain, U., E. Tholen, R. Hoereth and M. Wiese

Effect of vitamin E administration on the qualitative characteristics of the meat of Nero Siciliano pig | 20 | 227
Chiofalo, B., D. Piccolo, L. Liotta, A. Zumbo and V. Chiofalo

Ability of fresh thigh evaluation to predict cured ham quality | 21 | 228
Restelli, G.L., A. Stella and G. Pagnacco

Influence of lairage time and genotype on levels of stress in pigs | 22 | 228
Salajpal, K., M. Dikic, D. Karolyi, Z. Sinjeri, B. Liker and I. Juric

The effect of outdoor raising on meat quality of pigs 23 229
Juska, R. and V. Juskiene

Fatty acid composition in different tissues of Mangalitsa crossbreds 24 229
Seenger, J., Cs. Ábrahám, M. Mézes, H. Fébel and E. Szücs

The effect of driving pigs to stunning prior to slaughter on their stress status and meat quality 25 230
Ábrahám, Cs., M. Weber, K. Balogh, J. Seenger, M. Mézes, H. Fébel and E. Szücs

Effect of sex and slaughter weight on carcass measurements of pigs 26 230
Mullane, J., P.G. Lawlor, P.B. Lynch, J.P. Kerry and P. Allen

Genetic correlation of intramuscular fat content with performance traits and litter size in Duroc pigs 27 231
Solanes, F.X., J. Reixach, M. Tor, J. Tibau and J. Estany

Correlation coefficients between productive traits of Polish Large White, Duroc and Pietrain gilts 28 231
Orzechowska, B. and M. Tyra

Relationships between ultrasound measurements of the *m. longissimus dorsi* in pigs using Piglog 105 and Aloka SDD 500 devices 29 232
Tyra, M. and B. Orzechowska

Cd level in organs and tissues of pigs of early ripe meat breed (SM-1) 30 232
Patrashkov, S.P.A., V.L. Petukhov and O.S. Korotkevich

The influence of heavy metals on some biochemical indices in pigs 31 233
Korotkevich, O.S., S.P.A. Patrashkov, V.L. Petukhov and V.V. Gart

Digestibility and net energy value of wheat bran and sunflower meal: pregnant sows versus fattening pigs 32 233
Oeckel, M.J. van, J.M. Vanacker, N. Warnants, M. De Paepe and D.L. de Brabander

Energy intake limiting effects of wheat bran and sunflower meal in gestation diets of sows 33 234
Oeckel, M.J. van, J.M. Vanacker, N. Warnants, M. De Paepe and D.L. de Brabander

Comparison between wet/dry feeders and liquid feeding in a trough for growing-finishing pigs 34 234
Botermans, J.A.M., L. Meijer, M.A. Andersson and D. Rantzer

Effect of diet supplementation with grass-meal on performance and carcass composition of Duroc and Landrace cross bred pigs 35 235
Lawlor, P.G., P.B. Lynch, J.P. Kerry and S. Hogan

An in vivo model development to know feed ingredient preferences in weanling pigs 36 235
Solá-Oriol, D., E. Roura and D. Torrallardona

Reduction of the crude protein content in diets for growing-finishing pigs 37 236
Torrallardona, D., M. Cirera, D. Melchior and L. Le Bellego

Effects of amino acid type on odour from pig manure 38 236
Le, P.D., A.J.A. Aarnink, R. Jongbloed, C.M.C. van der Peet-Schwering, M.W.A. Verstegen and N.W.M. Ogink

Water consumption of pigs at weaning: effect of mixing water with pelleted feed and of using a water nipple or water bowl 39 237
Rantzer, D., M. Andersson, L. Stålhandske and J. Svendsen

Relation between activity of enzymes in boar sperm plasma and breeding value of the boar 40 237
Owsianny, J., B. Matysiak, A. Konik and A. Sosnowska

The reproductive value of 990 line young boars after primiparae or multiparae sows considering equalization of "maternal effect" related to litter size both before and after the birth 41 238
Czarnecki, R., M. Rózycki, M. Kawecka, A. Pietruszka, B. Delikator and M. Kamyczek

Workers influence on reproductive performance of gestating sows 42 238
Morales, J., L.M. Ramírez, S. Ayllón, M. Aparicio and C. Piñeiro

Optimal standardised ileal digestible (SID) lysine level in hybrid meat pigs (70-110 kg) 43 239
Warnants, N., M.J. van Oeckel, M. De Paepe and D.L. de Brabander

Feeding activity and other activities during the grazing journey of the iberian pig in the montanera fattening period 44 239
Rodríguez-Estévez, V., A. Martínez, A.G. Gomez and C. Mata

Session 19: Free communications pig production (Pig and poultry nutrition)

Date: 06 June '05; 14:00 - 18:00 hours
Chairperson:

Theatre **Session 19 no. Page**

Ideal protein to improve protein utilization of sows during gestation and lactation 1 240
Kim, S.W., F. Ji, G. Wu and R.D. Mateo

Effect of organic selenium on tissue Se content in growing-finishing pigs 2 240
Mateo, R.D., J.E. Spallholz, F. Ji, R. Elder, I.K. Yoon and S.W. Kim

Preference of piglets for mash or granulated cereal diets 3 241
Solà-Oriol, D., E. Roura and D. Torrallardona

Time-length of postweaning starvation period affects adaptation of pancreatic secretion in piglets 4 241
Huguet, A., G. Savary, E. Bobillier, Y. Lebreton and I. Le Huerou-Luron

Increased litter size as an effect of fat and polyunsaturated fat in sow diets 5 242
Wigren, I.J., M. Neil and J.E. Lindberg

Effect of dietary fat of different sources on growth and slaughter performance of growing pigs and fatty acid pattern of back fat and intramuscular fat 6 242
Boehme, H., R. Kratz, E. Schulz, G. Flachowsky and P. Glodek

Effect of feed processing on size of (washed) faeces particles from pigs measured by image analysis 7 243
Nørgaard, P., L.F. Kornfelt, C.F. Hansen and T. Thymann

Investigations on the tryptophan requirements of weaned piglets 8 243
Roth, F.X., T. Ettle, C. Relandeau, L. Le Bellego and J. Bartelt

Effects of breeder age and egg storage on hatchability, time of hatch and supply organ weights in quails 9 244
Yildirim, I. and A. Aygun

Growth performance, meat yield and abdominal fat of broilers subjected to early feed restriction 10 244
Mohamed, M.A., A.E. El-Sherbiny, A.S. Hamza and T.M. El-Afifi

Effect of nonphytate-P level and phytase supplementation on performance and bone measurements of broiler chicks 11 245
El-Sherbiny, A.E., H.M.A. Hassan, Y.A.F. Hammuda and M.A. Mohamed

Session 20: Free communications cattle production

Date: 06 June '05; 14:00 - 18:00 hours
Chairperson: Hocquette (F)

Theatre **Session 20 no. Page**

Dairy cow health and the effects of genetic merit for milk production, management and interactions between these: blood metabolites and enzymes 1 245
Beerda, B., W. Ouweltjes, J.J. Windig, M.P.L. Calus and R.F. Veerkamp

Methodology of breeding value estimation for functional longevity in Czech Republic 2 246
Páchová, E., L. Zavadilová and J. Sölkner

Relationship of metabolic traits and breeding values for milk production traits 3 246
Freyer, G., R. Staufenbiel, E. Fischer and L. Panicke

Genetic effects of embryo transfer in dairy cows 4 247
Bosselman, F., S. König and H. Simianer

Duration of mobilization period in dairy cows 5 247
Bossen, D. and M.R. Weisbjerg

Fatty acid composition of intramuscular and subcutaneous fat from Limousin and Charolais heifers supplemented with extruded linseed 6 248
Barton, L., V. Teslík, V. Kudrna, M. Krejcová, R. Zahrádková and D. Bures

Bovine intramuscular adipose tissue has a higher potential for fatty acid synthesis from glucose than subcutaneous adipose tissue 7 248
Hocquette, J.F., C. Jurie, M. Bonnet and D. Pethick

Lean meat estimate on half carcass crossbreed bulls 8 249
Iacurto, M. and D. Settineri

Body condition scoring in double-muscled Belgian Blue beef cows 9 249
Fiems, L.O., W. van Caelenbergh, S. de Campeneere and D.L. de Brabander

Poster **Session 20 no. Page**

A long-period study for the improvement of quantitative traits in a Hungarian Holstein-Friesian herd 10 250
Györkös, I., E. Báder, K. Kovács, J. Völgyi Csík, E. Szücs, T. Pétro and A. Kovács

Growth and mature weight of female beef cattle of different breeds 11 250
Bene, Sz., M. Török and F. Szabó

Preliminary data on body measurements and temperament of Aubrac heifers in Hungary 12 251
Szentléleki, A., J. Tőszér, Z. Domokos, R. Zándoki, C. Bottura, A. Massimiliano and Cs. Ábrahám

Ethnic study of Minhota cattle breed: Biometric and liveweight analysis 13 251
Araújo, J.P., H. Machado, J. Pires, J. Cantalapiedra, A. Iglesias, F. Petim-Batista, J. Colaço and L. Sánchez

Genetic improvement of cattle population of Yayladag district of hatay province by crossing of south yellow x brown swiss breed and first results obtained 14 252
Sekerden, Ö. and Y.Z. Guzey

Progeny performance of bulls differing in genetic index 15 252
Keane, M.G. and M.G. Diskin

Comparison of productivity in mixed herds of local and commercial dairy cattle breeds 16 253
Rizzi, R., C. Maltecca, F. Pizzi, A. Bagnato and G. Gandini

Functional polymorphism within the bovine growth hormone receptor (GHR) gene 17 253
Zwierzchowski, L., A. Maj and A. Gajewska

The effect of different calf rearing systems on calf's growth, feed consumption and behaviour and the cow's milk production 18 254
Fröberg, S., L. Lidfors, I. Olsson, A. Herrloff and K. Svennersten-Sjaunja

Evaluation of fat depth of rump (P8) measured by real-time ultrasound machine in polled and horned Charolais young bulls 19 254
Tőszér, J., Z. Domokos, I. Holló, G. Holló, A. Szentléleki, R. Zándoki, M. Bujdosó and M.L. Wolcott

Weaning effect on the Galician veal quality acceptable for the protected geographical indication "Ternera Gallega" 20 255
Moreno, T., L. Monserrat, J.A. Carballo, L. Sánchez and N. Pérez

Physical, compositional and organoleptic properties of beef from Charolais and Limousin heifers fed different diets 21 255
Bures, D., L. Barton, V. Teslík, M. Krejcová and R. Zahrádková

Carcass and meat quality of "Vitellone Bianco dell'Appennino Centrale" (IGP) 22 256
Preziuso, G., C. Russo, M. D'Agata and P. Verità

Effects of storage on colour and water holding capacity of meat 23 256
Russo, C., G. Preziuso, M. D'Agata and P. Verità

Castration effect on fatty acid profile of intramuscular fat in Charolais cattle 24 257
Holló, G., R. Zándoki, G. Pohn, E. Varga-Visi, J. Tőszér, Z. Domokos and J. Csapó

Glucose-6-Phosphate Dehydrogenase and leptin are linked to marbling differences between Limousin and Angus or Japanese Black x Angus steers 25 257
Bonnet, M., Y. Faulconnier, C. Leroux, C. Jurie, I. Cassar-Malek, D. Bauchart, P. Boulesteix, D. Pethick, J.F. Hocquette and Y. Chilliard

The influence of linseed supplementation on carcass quality traits of Hungarian Simmental and Holstein-Friesian young bulls 26 258
Holló, G., J. Seregi, J. Tőszér, L. Nagy, Cs. Ábrahám, B. Húth and I. Holló

Comparison of concentrate feeding strategies for growing bulls 27 258
Manni, K., M. Rinne and P. Huhtanen

Production and economic comparison of milking F1 Holstein x Gir cows with and without the stimulus of the calf 28 259
Junqueira, F.S., F.E. Madalena and G.L. Reis

Influence of κ-casein and β-lactoglobulin genotypes on the milk coagulation properties 29 259
Kübarsepp, I., M. Henno, H. Viinalass, D. Sabre and O. Saveli

Differences in milk urea content in dependency on selected non-nutritive factors 30 260
Jílek, F., M. Stipková, M. Fiedlerova, D. Rehák, J. Volek and E. Nemcová

The causes and extent of variability in lactation curve parameters in British commercial dairy herds 31 260
Albarrán-Portillo, B. and G.E. Pollott

Comparative study of body condition scores and blood parameters in dairy cows 32 261
Báder, E., Z. Gergácz, E. Bryedl, I. Györkös and A. Kovács

Effect of sensory stimulation, feeding or brushing, during milking on milk yield and hormone release in dairy cows 33 261
Wredle, E., M.S. Herskin, L. Munksgaard, R.M. Bruckmaier and K. Svennersten-Sjaunja

Time of first insemination in Hungarian dairy herds 34 262
Báder, E., Z. Gergácz, I. Györkös, P. Báder, A. Kovács and N. Boros

Growth rate of scrotal circumference in males of the Galician Blond cattle breed 35 262
Ferreiro, J.M., J.M.G. Pires, J.P.P. Araújo, J. Cantalapiedra, L. Sánchez and A. Iglesias

Session 21: Free communications sheep and goat production

Date: 06 June '05; 14:00 - 18:00 hours
Chairperson: Gabiña (ES)

Theatre **Session 21 no. Page**

Selection for lean weight based on CT and ultrasound in a meat line of sheep 1 263
Kvame, T. and O. Vangen

Changes in, and relationships among, lamb growth and carcass traits measured by computer tomography (CT) 2 263
Lambe, N.R., E.A. Navajas, K.A. McLean, L. Bünger and G. Simm

Using X-ray computed tomography to predict intramuscular fat content in sheep 3 264
Macfarlane, J.M., M.J. Young, R.M. Lewis, G.C. Emmans and G. Simm

Genetic parameters for birth weight, growth and litter size for Danish Texel and Shropshire 4 264
Maxa, J., E. Norberg, P. Berg and J. Pedersen

Effect of birth type on milk production traits in East Friesian ewes 5 265
Fuerst-Waltl, B. and J. Sölkner and R. Baumung

Year-round lamb production systems on pasture: theory and practice 6 265
Morel, P.C.H., S.T. Morris, P.R. Kenyon, G. de Nicolo and D.M. West

Evaluation of in-farm versus weather station data as heat stress indicator in Mediterranean dairy sheep 7 266
Finocchiaro, R., A. Di Grigoli, J.B.C.H.M. van Kaam, A. Bonanno and B. Portolano

Flight test as detector of transport stress in sheep 8 266
Diverio, S., A. Barone, G. Tami and N. Falocci

Body posture and orientation in sheep during transportation 9 267
Diverio, S., A. Barone, G. Tami and N. Falocci

Low-cost feeding strategies for fattening Awassi lambs in Syria 10 267
Hartwell, B.W., L. Iniguez, M. Wurzinger and W. Knaus

The effect of supplemental feeding duration on performance of Balouchi ewes 11 268
Kashki, V., M.R. Kianzad, M. Raisianzadeh, M. Nowrozi, H. Tavakoli and A. Davtalabzarghi

Kinetics of responses of goat milk fatty acids to dietary forage:concentrate ratio and/or high doses of sunflower or linseed oil, or extruded mixture of seeds 12 268
Chilliard, Y., J. Rouel, P. Guillouet, K. Raynal-Ljutovac, L. Leloutre and A. Ferlay

Poster **Session 21 no. Page**

Research regarding the morpho-productive parameters of F1 crossbreds resulted from crossing Lacaune breed with Tigaia breed 13 269
Raducuta, I., L. Vidu and A. Marmandiu

Heterosis analysis of Haemonchus contortus resistance and productions traits in Rhönschaf, Merinoland sheep and crossbred lambs 14 269
Hielscher, A., M. Gauly, H. Brandt and G. Erhardt

Alpine network for sheep and goat promotion for a sustainable territory development 15 270
Bigaran, F., D. Kompan, C. Mendel, A. Feldmann, F. Ringdorfer, G. de Ros, S. Venerus and E. Piasentier

Alternative sheep breeding schemes in Norway 16 270
Eikje, L.S., L.R. Schaeffer, T. Ådnøy and G. Klemetsdal

Morphologic characterization and body measurement of Hungarian goats 17 271
Németh, T., A. Molnár, G. Baranyai and S. Kukovics

Evaluation of boer goat performances in two climatic environments 18 271
Láczó, E., P. Póti and F. Pajor

Effect of winter shearing during late pregancy in the Latxa dairy sheep 19 272
Ruiz, R., J. Arranz and I. Beltrán de Heredia

Influence of weaning age on lamb growth and animal health in Boer goats 20 272
Das, G., E. Moors and M. Gauly

Changes in somatic cell counts of sheep milk and their effect on rennetability and quality of rennet curdling during lactation 21 273
Zajicova, P. and J. Kuchtik

Breeding for scrapie resistance and controll strategie in Hungary 22 273
Nagy, B., L. Fésüs and L. Sáfár

Development of a generic database for sheep and goat in Germany 23 274
Müller, U., R. Fischer, F. Rikabi, R. Walther, E. Groeneveld and U. Bergfeld

Effect of parity and number of kids in yield and milk composition in Sarda 24 274
Vacca, G.M., V. Carcangiu, M.C. Mura, M.L. Dettori and P.P. Bini

Estimation of lamb carcass composition using real-time ultrasound 25 275
Pajor, F., P. Póti, E. Láczó and J. Tőszér

Evaluation of carcass quality of different sheep breeds in performance testing 26 275
Baulain, U., W. Brade, A. Schoen and S. von Korn

Quality of meat from Pomeranian lambs and crossbreeds by Berrichon du Cher and Charolaise rams, stored under modified atmosphere conditions 27 276
Brzostowski, H. and Z. Tanski

Ram odour - does it affect Meat quality in Norwegian lambs? 28 276
Eik, L.O., J.E. Haugen, O. Sørheim and T. Ådnøy

Growth performance of Awassi, Charollais-Awassi and Romanov-Awassi ram and ewe lambs 29 277
Abdullah, A.Y., R.T. Kridli, A.Q. Al-Momani and M. Momani-Shaker

Experiences on S/EUROP meat qualification system on sheep breeds in Hungary 30 277
Molnár, A., Gy. Toldi, S. Kukovics and T. Németh

The effect of mannanoligosaccharide supplementation on the performance of ewes in late pregnancy and on lamb performance 31 278
Crosby, E.J., M. Fooley and S. Andrieu

Effect of feeding pistachio skins on performance of lactating dairy goats 32 278
Naserian, A.A. and P. Vahmani

The effect of roasted cereals on growth and blood parameters of lambs fattening 33 279
Antunovic, Z., M. Domacinovic, M. Speranda, B. Liker, D. Sencic and Z. Steiner

Influence of the diet on the productive performances of Gentile di Puglia lambs 34 279
Vicenti, A., M. Ragni, C. Cocca, L. Turi, F. Toteda and G. Vonghia

Influence of zinc-methionine supplementation on milk composition, somatic cell counts and udder health in sheep 35 280
El-Barody, M. and S. Abd El-Razek

Goat dairy performances according to dietary forage:concentrate ratio and/or high doses of sunflower or linseed oil, or extruded mixture of seeds 36 280
Rouel, J., E. Bruneteau, P. Guillouet, A. Ferlay, P. Gaborit, L. Leloutre and Y. Chilliard

Milk yield and it's constituents of local barbary sheep in Libyan Arab Jamahirya 37 281
Azaga, I.A. and K.M. Marzouk

Session 22: Free communications animal physiology (Reproduction)

Date: 06 June '05; 14:00 - 18:00 hours
Chairperson: Sejresen (DK)

Theatre **Session 22 no. Page**

Estrus synchronization by $PGF_{2\alpha}$ and $PGF_{2\alpha}$ + PMSG applications on Awassi sheep and their effect on fertility 1 281
Yavuzer, Ü. and F. Aral

Cryopreservation and insemination of ejaculated and epididymal semen from Dutch rare sheep breeds 2 282
Woelders, H., C.A. Zuidberg, H. Sulkers, M. Pieterse, K. Peterson and S.J. Hiemstra

Divergent selection lines for spontaneous spring ovulatory activity in Mérinos d'Arles sheep: results of the first generations 3 282
Bodin, L., J. Teyssier, B. Malpaux, M. Migaud, S. Canepa and P. Chemineau

Honeybee royal jelly: an alternative source to serum for in vitro maturation of ovine oocyte 4 283
Onal, A.G., M. Kuran, I. Tapki, E. Sirin and M. Gorgulu

Puberty occurrence in Awassi (A), F1 Charollais-Awassi (CA) and F1 Romanov-Awassi (RA) ram and ewe lambs 5 283
Kridli, R.T., A.Y. Abdullah, A.Q. Al-Momani and M. Momani-Shaker

An investigation on the effects of gossypol in cottonseed meal on scrotal circumference and spermatozoa quality in Atabay rams 6 284
Ghanbari, F., Y.J. Ahangari, T. Ghoorchi and S. Hasani

Sugarcane bagasse silage treated with different levels of urea for improvement sheep production: II. Body weight changes and ewes` reproductive performance 7 284
Kobeisy, M., M. Zenhom, I.A. Salem and M. Hayder

Reproductive and endocrine characteristics of delayed pubertal ewe lambs after melatonin and L-Tyrosine administration | 8 | 285
El-Battawy, K.A.

Poster **Session 22 no. Page**

The ventricullar ependyma fine structure at *Gallus domesticus*, in Scanning Electron Microscopy | 9 | 285
Urdes, L.D. and C. Nicolae

Peripheral blood growth hormone level can influence temperament in mithun (Bos frontalis) | 10 | 286
Mondal, M., A. Dhali, B. Prakash, C. Rajkhowa and B.S. Prakash

A multi-threshold approach to analyze sensory panel data: An example of rabbit meat quality | 11 | 286
Hernández, P. and L. Varona

Effect of Silymarin on plasma and milk redox status in lactating goats | 12 | 287
Spagnuolo, M.S., P. Abrescia, S. Galletti, D. Tedesco, F. Sarubbi and L. Ferrara

Peripheral blood growth hormone level can influence temperament in mithun (*Bos frontalis*) | 13 | 287
Mondal, M., A. Dhali, B. Prakash, C. Rajkhowa and B.S. Prakash

Expression of CD147 and monocarboxylate transporters MCT1, MCT2 and MCT4 in porcine small intestine and colon | 14 | 288
Sepponen, K., M. Ruusunen and A.R. Pösö

Interrelation between milk urea concentration and reproductive performances in dairy cows under farm condition | 15 | 288
Dhali, A., D.P. Mishra, R.K. Mehla and S.K. Sirohi

The association of first 60-d cumulative milk yield and embryo survival in Iranian Holstein dairy cows | 16 | 289
Heravi Moussavi, A., M. Jamchi, R. Noorbakhsh, M. Danesh Mesgaran and M.E. Moussavi

The relationship between plasma Leptin and FSH concentrations in sheep with different lambing rate | 17 | 289
Moeini, M., A. Towhidi, H. Solgi and M.R. Sanjabi

Embryo survival and its association with first 60-d cumulative milk yield in Holstein dairy cows | 18 | 290
Heravi Moussavi, A., M. Jamchi, R. Noorbakhsh, M. Danesh Mesgaran and M.E. Moussavi

Effect of media and presence of corpus luteum on in vitro maturation of buffalo (*bubalus bubalis*) oocytes | 19 | 290
Faheem, M.S., A.H. Barkawi, G. Ashour and Y. Hafez

Male-female interactions on pubertal events in female goats | 20 | 291
Papachristoforou, C., A. Koumas, C. Photiou and C. Christofides

Survey of blood electrolytes changes two weeks before to two weeks after delivery luri Bakhtiari race sheep | 21 | 291
Faghani, M. and F. Kheiri

Effects of bombesin neurotransmitter injection on the thyroid hormones concentraction in Sarabi cows | 22 | 292
Yousef Elahi, M. and E. Baghaei

Session 23: Free communications livestock farming systems

Date: 06 June '05; 14:00 - 18:00 hours
Chairperson: Hermansen (DK)

Theatre **Session 23 no. Page**

Characterization of farms in winter cattle fattening, Argentine | 1 | 292
Castaldo, A., J. Martos, D. Valerio, R. Acero, A. García and J. Pamio

Sustainable technical index for bio-economic farm models in Pampean region Argentine | 2 | 293
Martos, J., A. García, R. Acero, D. Valerio, V. Rodríguez and J. Perea

Animal production systems in Algeria: transformation and tendencies in the area of Sétif | 3 | 293
Abbas, K.

Evaluation of smallholder pig production systems in North Vietnam considering input, management, output and comparing economic and biological efficiency | 4 | 294
Lemke, U., L.T. Thuy, B. Kaufmann and A. Valle Zárate

Analysis of the beef cattle growth curves in the main livestock systems of the Basque Country | 5 | 294
Ruiz, R., A. Igarzabal, N. Mandaluniz, M.E. Amenabar and L.M. Oregui

Indigenous selection criteria in Ankole cattle and different production systems in Uganda | 6 | 295
Wurzinger, M., D. Ndumu, R. Baumung, A. Drucker, O. Mwai and J. Sölkner

Concentration of conjugated linoleic acid in grazing sheep and goat milk | 7 | 295
Tsiplakou, E., K. Mountzouris and G. Zervas

Revitalizing livelihoods of tsunami victims in Aceh, Indonesia through flood risk sensitive livestock development | 8 | 296
Wollny, C.B.A. and G. Tesfahun

Poster	Session 23 no.	Page
The development of milk production in dairy farms in Saudi Arabia *Alshaikh, M.A.*	9	296
An investigation on the angora goat raising in Ankara of Turkey *Soysal, M.I., E.K. Gürcan, E. Özkan, M.Aytac and S. Özkan*	10	297
HERDYN: a dynamic model to simulate herd dynamics in beef cattle extensive systems *Bernués, A., R. Ruiz and D. Villalba*	11	297
Comparative study on the behaviour of four goose genotypes during the preconditioning for laying *Molnár, M., I. Nagy, T. Molnár, K. Tisza and F. Bogenfürst*	12	298
Herbage from wet semi-natural meadows: 2. Nutrient content of some dominating species during different parts of the season *Spörndly, E., Z. Lifvendahl, Å. Berg and T. Gustafson*	13	298
Herbage from wet semi-natural meadows: 1. Digestibility of hay harvested late to protect bird life *Spörndly, E., I. Olsson and K. Holtenius*	14	299
Reproductive, survival and growth traits of the crossbreeding Belgian Texel x Moroccan local breeds of sheep *El Fadili, M. and P.L. Leroy*	15	299
Comparison of small and large size dairy cows in a pasture-based production system *Steiger Burgos, M., R. Petermann, P. Hofstetter, P. Thomet, S. Kohler, A. Muenger, J.W. Blum and P. Kunz*	16	300
Daytime maintenance and social behaviours during suckling period in jennies reared under semi-extensive conditions *D'Alessandro, A.G., D. Casamassima, G. Martemucci, N. Simone and G.E. Colella*	17	300
Application of discriminant analysis to the morphostructural differentiation of 7 extensive goat breeds of extensive *Luque, M., E. Rodero, F. Peña, A. García and M. Herrera*	18	301
Size versus beauty: farmers' choices in a ranking experiment with African Ankole Long-Horned Cattle *Ndumu, D., M. Wurzinger, R. Baumung, A. Drucker, O. Mwai and J. Sölkner*	19	301

Session 24: Equine science education

Date: 06 June '05; 14:00 - 18:00 hours
Chairperson: Habe (Slo)

Theatre	Session 24 no.	Page
Equine science education in Sweden: ten years of experience of the Equine Studies Program *Holgersson, A.-L., L. Roepstorff, J. Philipsson, G. Dalin, S. Lundesjö-Öhrström, K. Morgan, K. Ericson, M. Gottlieb-Vedi and A. Forslid*	1	302
Equine science education in Finland *Saastamoinen, M.T.*	2	302
Equine science education in Norway *Austbø, D.*	3	303
Equine science education at universities and higher technical colleges in Germany *Bruns, E.W.*	4	303
Equine science education in the UK *Ellis, A.D., S.V. Tracey and H.C. Owen*	5	304
Equine science education in France *Drogoul, C., R. Beaufrère, A. Rousselière, D. Perrin and V. Julliand*	6	304
Equine science education in Italy *Miraglia, N.*	7	305
Equine science education in Switzerland *Burger, D., I. Imboden and P.-A. Poncet*	8	305
Equine science education in Hungary *Mihok, S., I. Bodó, J. Posta and W. Hecker*	9	306
HorseConnexion: translating scientific knowledge into easily accessible information for riders, riding teachers and horse owners *Zetterqvist, M. and A.D. Ellis*	10	306

Poster	Session 24 no.	Page
Equine higher education in the United Kingdom *Kennedy, M.J.*	11	307
Equine education on bachelor level in Finland *Thuneberg-Selonen, T. and K. Paakkolanvaara*	12	307

Equine science education in Slovenia | 13 | 308
Habe, F.

Equine science education in Croatia | 14 | 308
Ivankovic, A., Z. Petrovic and M. Baban

EuroRide: an international education for riding instructors on level 2 | 15 | 309
Zetterqvist, M.

Session 25: Alternative low input / organic production methods

Date: 07 June '05; 8:30 - 12:30 hours
Chairperson: Keane (IRL)

Theatre **Session 25 no.** **Page**

Organic Livestock Systems: characteristics and challenges for improvement | 1 | 309
Hermansen, J.E., T. Kristensen and B. Ronchi

Environmental ethics and organic farming | 2 | 310
Jensen, K.K.

Factors causing a higher level of liver abscesses in organic compared with conventional dairy herds | 3 | 310
Jorgensen, K.F., A.M. Kjeldsen, F. Strudsholm and M. Vestergaard

Comparing sheep for meat production of organic versus conventional farms: structures, functioning, technical and economic results | 4 | 311
Benoit, M. and G. Laignel

Alternative low input system in sheep milk production: competitiveness to intensive production system | 5 | 311
Kukovics, S., P. Kovács, S. Nagy, G. Csatári and J. Jávor

Low input dairy systems: balancing environment and animal performance | 6 | 312
Sebek, L.B.J., R.L.M. Schils, J. Verloop, H.F.M. Aarts and Z. van der Vegte

Genotype environment interaction for milk production traits between conventional and organic dairy farming in the Netherlands | 7 | 312
Nauta, W.J., T. Baars and H. Bovenhuis

Nitrogen self-sufficiency at the suckler cattle farm scale: adaptation of the farming systems, economic consequences | 8 | 313
Veysset, P., M. Lherm and D. Bébin

Effects of chicory roots on finishing performance and CLA and fatty acid composition in longissimus muscle of Friesian steers | 9 | 313
Vestergaard, M., H.R. Andersen, P. Lund, T. Kristensen, L.L. Hansen and K. Sejrsen

Influence of amino acid levels to indoor and outdoor growing/finishing pigs on performance, carcass quality and behaviour 10 314
Høøk Presto, M., K. Andersson and J.E. Lindberg

Poster **Session 25 no. Page**

Future scenarios for sustainable beef production in Sweden 11 314
Kumm, K.-I., S. Stern, S. Gunnarsson, U. Sonesson, I. Öborn and T. Nybrant

Living and slaughtering performance of Piemontese young bulls reared according to organic farming method 12 315
Lazzaroni, C. and D. Biagini

Feeding of layers of different genotypes in an organic feed environment 13 315
Lagerkvist, G., K. Elwinger and R. Tauson

Housing systems for organic slaughter pigs 14 316
Olsson, A.-C., J. Svendsen, J. Botermans and M. Andersson

Effect on milk production and vitamin status in cows fed without synthetic vitamins 15 316
Johansson, B. and E. Nadeau

A historical perspective on low input dairy production 16 317
Israelsson, C.

Control of fitness in low input socio-ecological systems 17 317
Kaufmann, B.A.

Session 26: Breeding programmes for a wide range of systems

Date: 07 June '05; 8:30 - 12:30 hours
Chairperson: Wickham (IRL)

Theatre **Session 26 no. Page**

A systematic approach to the design and enhancement of breeding programmes 1 318
Garrick, D

Practical aspects in setting up a national cattle breeding program for Ireland 2 318
Olori, V.E., A.R. Cromie, A. Grogan and B. Wickham

Developments in international pig breeding programmes 3 319
Merks, J.W.M., E.H.A.T. Hanenberg and E.F. Knol

Collaboration of breeding programs with genotype by environment interaction: possibilities and limitations 4 319
Mulder, H.A., R.F. Veerkamp and P. Bijma

Cattle breeding programs for environments with poor formal information: Integrating traditional breeding knowledge and local breeding strategies - the case of The Gambia 5 320
Steglich, M. and K.-J. Peters

Improving carcass quality of UK hill sheep using Computerised Tomography (CT) 6 320
Conington, J., N. Lambe, P. Amer, S. Bishop, L. Bünger and G. Simm

Genetic evaluation of multibreed dairy sires in five environmental clusters within New Zealand 7 321
Bryant, J.R., N. Lopez-Villalobos, J.E. Pryce and C.W. Holmes

A selection index for Ontario organic dairy farmers 8 321
Rozzi, P. and F. Miglior

Environmental influences on genetic and phenotypic relationships between production and health and fertility in Dutch dairy cows 9 322
Windig, J.J., M.P.L. Calus, B. Beerda, W. Ouweltjes and R.F. Veerkamp

Use of structured antedependence models to estimate genotype by environment interaction 10 322
Calus, M.P.L., F. Jaffrézic and R.F. Veerkamp

Possibility and profitability of crossbreeding in Holstein dairy cattle 11 323
König, S. and H. Simianer

Session 27: Coping with new regulation: Alternatives to antibiotic growth promoters and castration

Date: 07 June '05; 8:30 - 12:30 hours
Chairperson: Tollarrardona (ES)

Theatre **Session 27 no. Page**

Alternatives to piglet castration 1 323
Bonneau, M. and A. Prunier

Experiences with use of local anaesthesia for piglet castration 2 324
Fredriksen, B., O. Nafstad, B.M. Lium and C.H. Marka

Estimation of genetic parameters of boar taint; skatol and androstenon and their correlations with sexual maturation 3 324
Tajet, H. and T.H.E. Meuwissen

Inhibition of CYP2E1 expression by androstenone: relation to boar taint 4 325
Doran, O., J.D. McGivan, F.M. Whittington, W.S. Tambyrajah and J.D. Wood

Rearing of entire male pigs in a "farrow-to-finish-system": effects on boar taint substances and animal welfare 5 325
Fredriksen, B., O. Nafstad, B.M. Lium, C.H. Marka, B. Heier and E. Dahl

Seasonal production of small entire male pigs raised in one unit system 6 326
Jensen, H.F and B.H. Andersen

Boar taint in pigs fed raw potato starch 7 326
Chen, G., G. Zamaratskaia, H.K. Andersson, K. Andersson and K. Lundström

Influence of chicory roots (*Cichorium intybus L*) on boar taint in entire male and female pigs 8 327
Hansen, L.L., M.T. Jensen, J. Mejer, A. Roepstorff, S.M. Thamsborg, D.V. Byrne, A.H. Karlsson, J. Hansen-Møller and M. Tuomola

Alternatives to antimicrobial growth promoters (AGP) 9 327
Wenk, C.

Holo-analysis of the effects of genetic, managemental, chronological and dietary variables on the efficacy of a pronutrient mannanoligosaccharide in piglets 10 328
Rosen, G.D.

Compatibility of B. toyoi and colistin in post-weaned piglets medicated diets 11 328
Morales, J., C. Piñeiro, G. Jiménez and A. Blanch

Efficacy of benzoic acid in the feeding of weanling pigs 12 329
Torrallardona, D. and J. Broz

Poster **Session 27 no. Page**

Season for pigs: New quality of pig meat. Experiences with a demonstration sale 13 329
Andersen, B.H., A.C. Bech and H.F Jensen

The effects of different concentrations of hops on the performance, gut morphology, microflora and liver enzyme activity of newly weaned piglets 14 330
Williams, J., A.H. Stewart, J. Powles, S.P. Rose and A.M. Mackenzie

Session 28: Performance and health in young horses

Date: 07 June '05; 8:30 - 12:30 hours
Chairperson:

Theatre **Session 28 no. Page**

Testing young horses for sport and for genetic evaluations of Swedish riding horses 1 330
Viklund, Å., J. Philipsson, Å. Wikström, Th. Arnason, E. Thorén, A. Näsholm, E. Strandberg and I. Fredricson

Estimation of genetic parameters for two different performance tests of young stallions in Germany | 2 | 331
Harder, B., T. Dohms and E. Kalm

Genetic parameters for competition traits at different ages of Swedish riding horses | 3 | 331
Wikström, Å., Å. Viklund, A. Näsholm and J. Philipsson

Application of a 3D morphometric method to the follow-up of conformational changes with growth and to the study of the correlations between morphology and performance | 4 | 332
Crevier-Denoix, N., P. Pourcelot, D. Concordet, D. Erlinger, A. Ricard, L. Tavernier and J.-M. Denoix

Group housing exerts a positive effect on the behaviour of young horses | 5 | 332
Søndergaard, E. and J.W. Christensen

Learning performance in relation to fear in young horses | 6 | 333
Christensen, J.W., K. Olsson, M. Rundgren and L. Keeling

Evaluation of breeding strategies against osteochondrosis (OC) in warmblood horses | 7 | 333
Busche, M. and E. Bruns

Breed variations in the distribution of osteo-articular lesions in horses at weaning | 8 | 334
Robert, C., J.P. Valette, S. Jacquet and J.M. Denoix

Genetic analyses of radiographic appearance of navicular bones in the Warmblood horse | 9 | 334
Stock, K.F. and O. Distl

Effect of maximal vs moderate growth on osteoarticular development in the yearling | 10 | 335
Donabedian, M., G. Fleurance, G. Perona, C. Trillaud-Geyl, C. Robert, S. Jacquet, J.-M. Denoix, O. Lepage, D. Bergero and W. Martin-Rosset

Bone spavin in Icelandic horses | 11 | 335
Björnsdóttir, S.

Poster **Session 28 no. Page**

Evolution of haematological and biochemical reference values in the growing horse | 12 | 336
Valette, J.P., C. Robert, G. Fortier, M.P. Toquet and J.M. Denoix

Session 29: Free communications animal genetics

Date: 07 June '05; 8:30 - 12:30 hours
Chairperson: Andersson-Eklund (SWE)

Theatre	Session 29 no.	Page
Search for finding ovulation rate responsible mutation in the Booroola fecundity (FecB) gene in Iranian Afshari sheep *Qanbari, S., M.P. Eskandari Nasab, R. Osfoori and A. Javanmard*	1	336
Selection strategies for body weight and reduced ascites susceptibility in broilers *Pakdel, A., P. Bijma, B.J. Ducro and H. Bovenhuis*	2	337
Detection of quantitative trait loci for meat quality and carcass composition traits in blackface sheep *Karamichou, E., G.R. Nute, R.I. Richardson, K. McLean and S.C. Bishop*	3	337
Statistical aspects of QTL detection in small samples based on a data set from selective DNA pooling in blue fox *Szyda, J., M. Zaton-Dobrowolska, H. Wierzbicki and A. Rzasa*	4	338
The map expansion obtained with recombinant inbred strains and intermated recombinant inbred populations for finite generation designs *Teuscher, F. and V. Guiard*	5	338
Optimal haplotype size for combined linkage disequilibrium and co-segregation based fine mapping of quantitative trait loci *Firat, M.Z., H. Gilbert, R.L. Fernando and J.C.M. Dekkers*	6	339
A comparison of regression interval mapping and multiple interval mapping for linked QTL *Mayer, M.*	7	339
The BovMAS Consortium: analysis of BTA14 for QTL influencing milk yield, milk composition and health traits in the Italian Holstein-Friesian cattle breed *Fontanesi, L., E. Scotti, D. Pecorari, M. Dolezal, P. Zambonelli, S. Dall'Olio, D. Bigi, R. Davoli, E. Lipkin, M. Soller and V. Russo*	8	340
Bayesian analysis of selection with restrictions in beef cattle *Melucci, L.M., A.N. Birchmeier, E.P. Cappa and R.J.C. Cantet*	9	340
Selection of commercial boars and replacement sows for variable grid pricing systems *Quinton, V.M., J.W. Wilton and J.A.B. Robinson*	10	341
Genetic analysis of survival data from challenge testing of furunculosis in Atlantic salmon: model comparison using field survival data *Ødegård, J., I. Olesen, B. Gjerde and G. Klemetsdal*	11	341

Genetic analysis of racing performance in Irish Greyhounds 12 342
Täubert, H., D. Agena and H. Simianer

Poster **Session 29 no. Page**

Cumulative discounted expressions of dairy and beef traits in Ireland 13 342
Berry, D.P., F.E. Madalena, A.R. Cromie and P.R. Amer

Effectiveness of selection for lower somatic cell count (SCC) in herds with different levels of SCC 14 343
Calus, M.P.L., L.L.G. Janss, J.J. Windig, B. Beerda and R.F. Veerkamp

Impact of management and recording quality on the success of community breeding programmes for smallholder dairying in the tropics 15 343
Zumbach, B. and K.J. Peters

Genetic evaluation of mothering ability for multiple parities in Iberian pigs 16 344
Fernández, A., J. Rodrigáñez, M.C. Rodríguez and L. Silió

Characterization of four Ethiopian cattle breeds for typical phenotypic features, productivity and trypanotolerance 17 344
Stein, J., W. Ayalew, B. Malmfors, E. Rege and J. Philipsson

Systems of breeding in Algeria: transformation and tendencies of evolution. Case of the area of Sétif 18 345
Abbas, K.

Session 30: Nutrition and management strategies to improve resource use in livestock systems

Date: 08 June '05; 8:30 - 12:30 hours
Chairperson: Milne (UK)

Theatre **Session 30 no. Page**

Improving resource use in ruminant systems 1 345
Milne, J.A.

Nutrition and animal management as part of a global strategy for reducing the environmental impact of pig production 2 346
Bonneau, M., J.-Y. Dourmad, C. Jondreville and P. Robin

Impact of alternative dairy systems on greenhouse gas emission 3 346
Fredeen, A.H., M. Main, S. Juurlink, S. Cooper and R. Martin

Effect of group size on feed intake and growth rate in kids and lambs 4 347
Van, D.T.T. and I. Ledin

Effect of feeding different levels of foliage from *Moringa oleifera* to creole dairy cows on intake, digestibility, milk production and composition 5 347
Sánchez, N.R. and I. Ledin

Effect of milking frequency and nutritional level on milk production characteristics and reproductive performance of dairy cows 6 348
O'brien, B.

Effect of milking frequency and nutrition on cow welfare 7 348
Boyle, L., B. O'brien and D. Gleeson

Dealcoholized beer replacement for water in poultry as a novel alternative approach 8 349
Parlat, S.S. and I. Yildirim

Replacing barley with soy hulls in an automatic milking system 9 349
Halachmi, I., E. Shoshani, R. Salomon, E. Maltz and J. Miron

Development of a deterministic model to create a dairy herd 10 350
Nousiainen, J.I., L. Jauhiainen, M. Toivakka and P. Huhtanen

Poster **Session 30 no. Page**

Performance of imported dairy cows under Libyan hot weather conditions 11 350
Belgasem, B.M. and A.B. El-Magdub

Efficiency of using grape marc in rations of Rahmani lambs 12 351
Awadalla, I.M.

The effect of yeast culture (*Saccharomyces cervisiae*) on digestibility of sugar cane bagasse in sheep 13 351
Kafilzadeh, F. and A. Paryad

Effects of substitution of cottonseed meal by canola meal on milk yield and apparent digestibility of dry matter, organic matter and crud protein in diets of dairy cow 14 352
Ghorbani, G.R. and S.M. Masumi

Effects of short dry periods on milk yield of Holstein dairy cattle 15 352
Ghorbani, G.R. and H.R. Rahmani and A. Pezeshki

Influence of phosphorus level and soaking on phosphorus availability and performance in growing-finishing pigs 16 353
Lyberg, K., A. Simonsson and J.E. Lindberg

Effect of ruminant livestock systems on grassland condition in Patagonia, Argentina 17 353
Villagra, S., C.B.A. Wollny, C. Giraudo and G. Siffredi

Effect of two diets on the growth of the *Helix aspersa Müller* during the juvenile stage 18 354
García, A., J. Perea, R. Martín, R. Acero, A. Mayoral and M. Luque

Effect of milking frequency and nutritional level on milking characteristics and teat condition of dairy cows 19 354
Gleeson, D., B. O'brien and L. Boyle

Comparison of pelleted vs mash-feed form administration in nursery-finisher pigs 20 355
Morales, J., L.M. Ramírez and C. Piñeiro

Performance and carcass quality of broilers fed diets based on three fibre types 21 355
Iyayi, E.A., O. Ogunsola and R.A. Iyaya

Dynamics of pastures and fodder crops for Mirandesa cattle breed 22 356
Galvao, L., O.C. Moreira, R. Valentim, J. Ramalho Ribeiro and V. Alves

Reduction of nitrogen environmental impact by diet manipulation in swine 23 356
Moreira, O.C., O. Oliveira, J. Martins Santos, M.A. Castelo Branco, F. Calouro, S. Sousa, A.S. Monteiro and J. Ramalho Ribeiro

Session 31: Horse production in Sweden

Date: 08 June '05; 8:30 - 12:30 hours
Chairperson: Karlander (SE)

Theatre **Sessio 31 no. Page**

Swedish horse breeding and sport 1 357
Olsson, C., D.-A. Danielsson and J. Philipsson

The economic importance of the Swedish horse industry 2 357
Johansson, D., A. Hedberg and H. Andersson

Session 32: Increased understanding of the genetics of quantitative traits - theory and applications

Date: 08 June '05; 8:30 - 12:30 hours
Chairperson: Maki-Tanila (FIN)

Theatre **Session 32 no. Page**

How important are regulatory mutations for genetic variation in multifactorial traits? 1 358
Andersson, L.

Bioinformatic tools in analysing molecular genetic data for breeding and genetics 2 358
Janss, L.L.G.

Reassessing quantitative genetic theory in the light of modern molecular genetic ideas 3 359
Pollot, G.E.

Estimable genetic variance components under mixed additive mendelian and imprinted inheritance 4 359
Reinsch, N. and V. Guiard

Strategies for selective DNA pooling for multiple traits 5 360
Schwarzenbacher, H., P. Visscher, M. Dolezal, M. Soller and J. Sölkner

Testing candidate genes using a Bayes Factor 6 360
Varona, L.

Benefits from marker assisted selection under an infinitesimal model 7 361
Villanueva, B., R. Pong-Wong, J. Fernández and M.A. Toro

Gene expression profiling for meat quality in swine 8 361
Gorni, C., C. Garino, S. Iacuaniello, B. Castiglioni, G.L. Restelli, A. Stella, G. Pagnacco and P. Mariani

Haplotype structure of casein genes in Norwegian goats and effects on production traits 9 362
Hagesæther, N., B. Hayes, T. Ådnøy, G. Pellerud and S. Lien

Effect of IGF2 on growth characteristics of F2 Meishan X White crossbreds 10 362
Heuven, H.C.M. and H. Bovenhuis

Fishy taint in chicken eggs is associated with a substitution within a conserved motif of the FMO3 gene 11 363
Honkatukia, M.S., K. Reese, R. Preisinger, M. Tuiskula-Haavisto, S. Weigend, J. Roito, A. Mäki-Tanila and J. Vilkki

Poster **Session 32 no. Page**

Mapping of quantitative trait loci for leg conformation traits in Danish Holstein 12 363
Buitenhuis, A.J., M.S. Lund, B. Thomsen and B. Guldbrandtsen

QTL analysis for eight milk production traits in the German Angeln dairy cattle population 13 364
Sanders, K., J. Bennewitz, N. Reinsch, E.-M. Prinzenberg and E. Kalm

Selection of the habitat in the rest phase of the Helix aspersa under laboratorial conditions 14 364
Perea, J., M. Herrera, A. García, A. Mayoral, M. Luque, E. Felix and C. Pérez

The association between CSN3 genotypes and milk production parameters in Czech Pied cattle 15 365
Kucerová, J., E. Nemcová, M. Stipková, O. Jandurova, A. Matejícek and J. Bouska

Genetic variability of MYF3 and MYF4 genes in Large White and Landrace breeds of pigs in the Czech Republic 16 365
Verner, J. and T. Urban

SNPs analysis in selected candidate genes in pigs using resequencing 17 366
Knoll, A., J. Verner and Z. Vykoukalová

Relationship between the myogenin gene (myf4) and litter size of large white sows 18 366
Humpolicek, P., T. Urban and J. Verner

Isolation of differentially expressed genes related with ham salting loss in Italian heavy pig 19 367
Russo, V., D. Bigi, P. Zambonelli and M. Colombo

Allelic frequencies of MC1r and ASIP genes in Iberian horses 20 367
Royo, L.J., I. Álvarez, I. Fernández, M. Valera, J. Jordana, A. Beja-Pereira, E. Gómez and F. Goyache

In silico inference of multi-locus genotypes from SSCP markers 21 368
Panzitta, F., P. Mariani, G.C. Gandini, P.J. Boettcher and A. Stella

The use of phenotypic information for refinement of haplotype reconstruction using the Expectation-Maximization algorithm 22 368
Boettcher, P.J. and A. Stella

A pipeline for automatic detection of SNPs in goat polymorphic sequences 23 369
Lazzari, B., J. Nardelli-Costa, F. Panzitta, A. Stella, A.R. Caetano and P. Mariani

QTL mapping for teat number in an Iberian by Meishan pig intercross 24 369
Rodríguez, M.C., A. Tomás, E. Alves, O. Ramírez, M. Arque, C. Barragán, L. Varona, L. Silió, G Muñoz, M. Amills and J.L. Noguera

Analysis of effects of genes differentially expressed during myogenesis on pork quality 25 370
Murani, E., M.F.W. Te Pas, K.C. Chang, R. Davoli, J.W.M. Merks, H. Henne, R. Wörner, H. Eping, S. Ponsuksili, K. Schellander, N. da Costa, D. Prins, B. Harlizius, E. Knol, M. Cagnazzo, S. Braglia and K. Wimmers

Leptin polymorphism and its association with milk production and plasma glucose in early lactating cows 26 370
Heravi Moussavi, A., M. Ahouei, M.R. Nassiry, E. Jorjany, M. Salary and A. Javadmanesh

Effects of the bovine DGAT1 (K232A) polymorphism in Swedish dairy cattle 27 371
Näslund, J., F. Fikse, G. Pielberg and A. Lundén

Identification of mutations and mapping of candidate genes for lysosomal proteinase and esopeptidase activities of dry cured hams in pigs 28 371
Fontanesi, L., R. Davoli, S. Galli and V. Russo

Creating high density Radiation Hybrid maps to fine map QTLs in cattle 29 372
Williams, J.L., S.D. McKay, N. Hastings, O. Jahn, M. Janitz, S. Hennig, S. Floriot, P. Ajmone-Marsan, E. Milanesi, A. Valentini, C. Marchitelli, M. Savarese, S.S. Moore and A. Eggen

Genetic and genomic approaches to improve production in water buffalo 30 372
Strazzullo, M., C. Campanile, M. D'Esposito and L. Ferrara

Analysis of allelic variants in exon 5 of the growth hormone gene in Sistani and Dashtiari breeds of cattle 31 373
Emrani, H. and A. Torkamanzehi

Type I DNA markers (ATP1A2, CA3 and DECR1) on porcine chromosome 4 for carcass and meat quality traits in pigs 32 373
Davoli, R., S. Braglia, I. Nisi, L. Fontanesi, L. Buttazzoni and V. Russo

Molecular charactrization of CD18 gene and identification carriers of BLAD disorder in Iranian Holstein bulls 33 374
Asadzadeh, N., M. Esmaeilzadeh, F. Sarhaddi, A. Javanmard and Mollasalehi

Session 33: Progress towards reduction of disease in sheep and goats

Date: 08 June '05; 8:30 - 12:30 hours
Chairperson: Gavier-Widen (SE)

Theatre **Session 33 no. Page**

Scrapie: an overview; policy issues and potential eradication measures 1 374
Matthews, D.

Rare sheep breeds and breeding for scrapie resistance in the Netherlands 2 375
Kaal, L.M.T.E. and J.J. Windig

Simulation of different strategies for breeding towards scrapie resistance 3 375
Vries, F. de, J. Hamann, C. Drögemüller and O. Distl

PrP allele frequencies in non-infected Valle del Belice and infected cross-bred flocks 4 376
Kaam, J.B.C.H.M. van, R. Finocchiaro, M. Vitale, B. Portolano, F. Vitale and S. Caracappa

Blue tongue in sheep: brief overview of the disease, impact on production and current epidemiological situation in Europe, control and prevention 5 376
Casal, J.

The problem of anthelmintic resistance in nematode parasites of sheep and goats, and the prospects for non-chemotherapeutic methods of control 6 377
Waller, P.J.

Sustainable internal parasite control in Australian Merino sheep 7 377
Karlsson, L.J.E. and J.C. Greeff

Analyses of udder health in Valle del Belice dairy sheep using SCC 8 378
Portolano, B., V. Riggio, M.T. Sardina, R. Finocchiaro and J.B.C.H.M. van Kaam

Session 34: Utilisation of records to manage health

Date: 08 June '05; 8:30 - 12:30 hours
Chairperson: Geers (B)

Theatre **Session 34 no. Page**

The link between animal health planning and record keeping for UK farm assurance schemes 1 378
Butterworth, A.

New behaviour based developments to collect health information in dairy cattle farming 2 379
Winckler, C.

Health and welfare management of pigs based on slaughterline records 3 379
Velarde, A., E. Fàbrega and X. Manteca

Pathogen records as a tool to manage udder health 4 380
Koivula, M. and E.A. Mäntsysaari

Can changes in milk composition at udder quarter level be used to detect udder disturbances? 5 380
Svennersten-Sjaunja, K., I. Berglund, G. Pettersson and K. Östensson

The use of a database for genetic evaluation and to manage health in dairy cows 6 381
Karsten, S., E. Stamer, W. Junge, W. Lüpping and E. Kalm

Mastitis detection in dairy cows by application of Fuzzy Logic 7 381
Cavero, D., K.H. Tolle, C. Buxadé and J. Krieter

Poster **Session 34 no. Page**

The relative day-to-day variation in milk yield and composition for cows milked two or three times daily 8 382
Svennersten-Sjaunja, K., J. Bertilsson, U. Larsson and L-O. Sjaunja

Milk flow and udder health in dairy cows 9 382
Tolle, K.H., U. Tölle and J. Krieter

Oestrus detection in dairy cattle using ALT-Pedometer and Fuzzy-Logic 10 383
Pache, St., C. Ammon, U. Brehme, H.J. Rudovsky, R. Brunsch, J. Spilke and U. Bergfeld

Session 35: Free communications cattle production

Date: 08 June '05; 8:30 - 12:30 hours
Chairperson: Lazzaroni (IT)

Theatre	Session 35 no.	Page
Modelling repeated measures of ejaculate volume of Holstein bulls using a Bayesian random regression approach *Serrano, M., M.J. Carabaño and C. Díaz*	1	383
Genetic and environmental effects on semen quality of Austrian Simmental bulls *Gredler, B., C. Fuerst, B. Fuerst-Waltl, H. Schwarzenbacher and J. Sölkner*	2	384
Effect of body condition on dairy and reproductive performance in Holstein-Friesian cows *Szücs, E., J. Püski, Tran Anh Tuan, A. Gáspárdi and J. Völgyi-Csik*	3	384
The relationship between lactation persistency and reproductive performance in New Zealand dairy cattle *Lopez-Villalobos, N., L.R. McNaughton and R.J. Spelman*	4	385
Normal and atypical progesterone profiles in Swedish dairy cows *Petersson, K.-J., H. Gustafsson, E. Strandberg and B. Berglund*	5	385
Oestrus detection in dairy cows using control charts and neural networks *Krieter, J., E. Stamer and W. Junge*	6	386
Bioenergetic factors affecting conception rate in Holstein cows *Patton, J., D. Kenny, F. O'Mara, J.F. Mee and J.J. Murphy*	7	386
Effect of honeybee royal jelly on the nuclear maturation of bovine oocytes in vitro *Kuran, M., E. Sirin, E. Soydan and A.G. Onal*	8	387

Session 1 Theatre 1

The EU policy on food and feed safety: an overview

P. Vanthemsche, Chief Executive Officer of the Belgian Food Safety Agency, Simon Bolivarlaan 30, 1000 Brussels, Belgium

In January 2000 the EU-Commission published the White Paper on food safety. In this white paper the Commission announced the development of a legal framework covering the entire food chain from farm to fork. The general objectives of the European food safety policy are to ensure a high level of protection of human and animal health, place quality at the forefront of concerns, and restore consumer confidence. To accomplish this goal, five general principles have been accepted: an integrated approach of the food chain as a whole, food safety is based on risk analysis, responsibility lies with all operators of the food sector, products (feed and food) must be traceable in all stages of the food chain, authorities inform the public clearly and accurately. In order to assure a consistent scientific risk assessment, the European Food Safety Authority (EFSA) was established. The development of this new legal framework, based on regulations rather than directives, is on-going and on schedule. It encompasses various items such as food safety, animal health, plant health and animal welfare. It also provides a framework for the management of incidents in the agro-food chain. The new food safety policy of the EU is more in line with the rules of international trade than before. This paper gives, as far as food safety policy is concerned, an overview of the principles and the structure of the new EU legislation, the organisation of the EU and the Member States and of its impact on society and industry, both on the local and global level.

Session 1 Theatre 2

Codex Alimentarius and consumer protection related to foods of animal origin

S.A.Slorach, National Food Administration, Box 622, SE-751Uppsala, Sweden

The main purpose of the Joint FAO/WHO Food Standards Programme implemented by the Codex Alimentarius Commission is to protect consumer health and ensure fair food trade practices. To this end, the Commission produces the Codex Alimentarius or " food code" comprising standards, guidelines and recommendations, while the scientific basis is provided by international independent expert groups.

Both the Agreement on the Application of Sanitary and Phytosanitary Measures (SPS), and the Agreement on Technical Barriers to Trade (TBT) recognize the importance of harmonizing standards internationally to minimize the risks of standards becoming barriers to international trade. The Codex standards and guidelines are recognized by these agreements as international benchmarks. Developing countries participate in Codex meetings.

This paper covers the latest developments in Codex related to foods of animal origin, including a code of hygiene practice for meat, a code of practice to minimize and contain microbial resistance, and maximum residue limits for veterinary drugs in food. Moreover, the code of practice on good animal feeding is described, and current work on microbiological risk management, control of *Listeria monocytogenes* in foods, and a revised code of hygienic practice for egg products is addressed.

Codex recognizes the relevance of a whole food chain approach to food safety and quality, and the need to apply good agricultural and veterinary practices on farms. Closer cooperation with the OIE is strived for to avoid gaps and duplication.

Session I Theatre 3

Feed and feed safety in livestock production

W. Heeschen[1] and A. Blüthgen[2], [1]former Federal Dairy Research Centre, Kiel, Germany, [2]Federal Research Centre for Nutrition and Food, Kiel, Germany

A series of feed scandals associated with contaminants in the past with the possibility to jeopardize consumers' health through polluted food of animal origin have alerted the public as well as governmental actors. Albeit the standard of feed safety was high within the European Union before its expansion in 2004, a new approach to improve food and feed safety as an unconditional entity was launched with the Regulative (EC) No. 178/2002. With respect to uniform standards for food and feed safety in the community a framework of risk analysis, precautionary principles, transparency, traceability, responsibilities, and liability was elaborated and came into force since 2005 at the latest. Undesirable substances from the environment and production sites may contaminate feedstuffs for livestock and occasionally lead to residues of concern in animal derived food. Feed safety strategies must therefore cover the whole feed supply from the farm to industrially manufactured formula feed. Whereas feed safety on the farm can be maintained by a combination of codes of good practices in the agricultural feed production chain, the instrumentalized Total Quality Management Systems in the feed industry culminate in the application of the HACCP concept. This concept provides a maximum of feed safety through a systematic hazard analysis and risk minimization approach during processing instead of randomised end product testing.

Session I Theatre 4

Suisse Garantie: more than a label

C. Beglinger[1], P. Gresch[2], R. Kennel-Hess[3], [1]Tierverkehrsdatenbank AG, Bern, Switzerland, [2]Environmental sciences dept., ETH Zurich, Switzerland, [3]Proviande, Bern, Switzerland

Like many other Swiss products, Swiss meat competes globally. Surveys show that transparency, declarations and traceability are what Swiss consumers want and Swiss suppliers welcome. This led in 2004 to creation of "SUISSE GARANTIE" for Swiss agricultural products - a distinctive mark to show Swiss origin and processing. Products carrying the SUISSE GARANTIE mark undergo strict checks throughout the entire production chain. A comprehensive, independent control and certification system guarantees compliance with international ISO standards 14020 and 14024. Meat and meat products from cows, sheep, pigs, goats, horses and domestic birds are subject to a special regulation, in which requirements for SUISSE GARANTIE are stipulated. Evidence of a quality assurance programme for production, humane animal transport and full traceability of meat back to the earmark of the animal are all requirements of the first production stage. Animals are marked at birth. They are simultaneously registered in the Swiss national animal movement database, which then tracks the animal's life through to its slaughter. At the production stage, checks are carried out in accordance with the SUISSE GARANTIE requirements as part of animal production QA programmes. Over 90 % of cattle production already takes place in a controlled environment. SUISSE GARANTIE meat products must be produced from 100 % home-produced constituents. Certification to SN/EN/ISO 45011 ensures compliance with the requirements at the processing stage.

Quality assurance programs for the USA pig industry: focus on food safety and animal welfare

J.J. McGlone, Pork Industry Institute, Texas Tech University, Lubbock, Texas, USA 79409-2141

Quality assurance programs include educational programs, assessments, certification, and audits. The initial focus was and remains in the area of food safety. All major pork processors who purchase live pigs require participation in the Pork Quality Assurance (PQA) program. This program involves attendance in an educational session that covers 10 main topics including, antibiotic use, identifying treated animals, needle safety, drug inventories, veterinarian-client-patient relationship, animal welfare, and documentation. The PQA program is delivered to pork producers and they receive a certification that is valid for 3 years. The animal care section of the PQA program was expanded into a 9-part animal welfare program called the Swine Welfare Assurance Program (SWAP). The SWAP program includes herd health and nutrition, caretaker training, animal observation, euthanasia, handling, facilities and continuing education and assessment. The SWAP program is being used as an internal assessment and education program. On-farm audits by third parties are relatively new and are rapidly growing on commercial farms. In one audited system, 10% of the 4 million pigs produced per year are subjected to a thorough audit and the data used to verify compliance with company animal welfare policies. Food retailers are requiring third-party animal welfare audits and the process is being refined to be science-based, effective and efficient.

Animal Health, welfare and public health issues

P-P. Pastoret, Biotechnology and Biological Sciences Research Council, Institute for Animal Health , United Kingdom

Globalisation has increased the risk from infectious diseases of man and animals through the five Ts: Trade, Travel, Transport, Tourism and Terrorism.

Global warming is also responsible for the extension of some vectors of animal infectious diseases such as Blue tongue in Europe.

In recent years we have experienced the emergence or reemergence of many animal diseases, some of which are zoonotic, like West Nile in Europe and the United States.

The modifications of biological conditions are also conducive to the transmission of pathogens from animals to man, such as the molecular evolution of viruses and the opening of previously concealed ecosystems.

A peculiar problem is linked to the extension of the European Union to new Member States with unequivalent animal and public health status.

In the European Union , public opinion is more and more hostile to mass slaughtering of livestock for the control of animal infectious diseases, for instance foot and mouth disease or classical swine fever. There is nowadays a trend to implement instead a policy of vaccination for live using marker vaccines.

Vaccination is without doubt the most useful single measure to prevent animal infectious diseases. It is even the only available method, apart from sanitary prophylaxis, to prevent, or sometimes cure, viral infections in the absence of broad spectrum antivirals.

Session I Theatre 7

Balancing public and private interests in EU food production

A.W. Stott, C. Milne, S. Peddie, F. Williams and G.J. Gunn. SAC, West Mains Road, Edinburgh, United Kingdom*

Reform of the CAP aims to de-couple subsidy from food production. It will also release funds for direct support of public goods that flow from agriculture. This highlights the need to ensure value for money through appropriate allocation of subsidy between competing public goods. Such allocations will depend not only on the relative value placed on a public good and the extent to which it is rewarded in existing markets, but also the ease with which it can be dealt with through policy mechanisms. Public goods that are understood, easy to identify, quantify, verify and fit into existing support frameworks are likely to be well safeguarded. Others that lack these attributes such as some animal diseases and animal welfare may be less well provided for. This point is illustrated using the example of bovine paratuberculosis (BPTB), a possible zoonosis. Using a survey of farmers and of vets and an adaptive conjoint analysis of risk factors we showed that the financial incentive to control BPTB at the farm level (private benefits) was relatively small and yet farmers were identified as the most important group for reducing any potential food safety risks. We therefore outline further research linking animal health/welfare with environmental benefits, reducing the relative weakness of the former in competition for policy support.

Session I Theatre 8

Control of salmonella in food of animal origin: the role of animal feed

P. Häggblom, National Veterinary Institute, Uppsala, Sweden

Ensuring the safety of animal feed has become a key issue in the present EG-legislation in order to improve the safety of the human food. Sweden has achieved an efficient control of Salmonella in the food chain much attributed to the work carried out in the primary production. The comprehensive control programme recognizes important control points in the food chain. Surveillance, according to the Swedish salmonella control programme indicates an overall prevalence of salmonella below 0.1% in red and white meat and eggs.

The control of feed is an essential element of the control programme and commercial feed is monitored for salmonella. The most important risk factors in feed production are raw materials of plant origin, inappropriate heat treatment, condensation, recontamination and poor cleaning of the premises. The monitoring of imported plant raw materials especially from the crushing industry, have shown that new serotypes of Salmonella are appearing which may possess an increased risk for contamination of the food chain. Salmonella strains, isolated from feed raw materials and recent outbreaks in animal herds, carry new genetic properties which may limit the efficiency of the present technology for salmonella decontamination.

Session 1 Theatre 9

Utilising sustainable data in an effective way

A.Kuipers[1] and F.J.H.M. Verhees[2], [1]Expertise Centre for Farm Management and Knowledge Transfer, Box 35, 6700AA, Wageningen and [2]Marketing and Consumer Behaviour group, Wageningen University, Hollandseweg 1, 6706KN, Wageningen, The Netherlands*

Supported by 7 organisations in the dairy sector chain in Netherlands, efficiency of sustainable data streams and marketing opportunities for sustainable data based on needs and availability were examined. Interviews with 20 leading persons from all levels of dairy chain and quality assurance organisations were held. Selective and smart use of existing data bases was often mentioned as a major challenge. Lack of transparency in medicine use was mentioned as a weak point but also identified as a marketing opportunity. Data about energy utilisation are scarce. Also a questionnaire was distributed under consumers and farmers using the same research model. Some opportunities were indicated. A distinct consumer need exists for guaranties of no pollution and no medicines in food, while the information is lacking. If asked, farmers are prepared to provide it. Consumers appreciate a healthy and animal friendly kept herd and farmers are prepared to give info about this. However, a complication is that animal health data are difficult to present to the public in an understandable way. Consumers have not much interest in nutrient management issues. But farmers are requested by regulation to administer the use of minerals extensively. Therefore, administrative costs are not in agreement with market wishes. Results were compared with the elements included in a certification scheme (KKM) of which the practical application regularly results in disputes.

Session 1 Poster 10

Development of quality indicators for e-Learning in the domain of farm animal welfare

E. Sossidou[1], D. Stamatis[2] and E.Szücs[3], [1]National Agricultural Research Foundation, Animal Research Institute,58 100 Giannitsa, Greece, [2]Technological Educational Institute of Thessaloniki, P.O.Box 14561, GR-54101, [3]Szent István University, H-2103, Gödöllö, Hungary*

The aim of the present study is to analyze virtual learning environments and to provide a framework for developing a set of quality indicators for e-Learning in the domain of farm animal welfare. The framework is constructed according to the experimental learning for a case study: An e-learning course which is developed in the context of the Leonado da Vinci Community Vocational Training Action Pilot Project entitled 'WELFOOD-Promoting quality assurance in animal welfare-environment-food quality interaction studies through upgraded e-Learning'. First, general issues and known questions regarding Quality Assurance for Open Distance Learning are examined. Then, the domain specific needs for e-Learning based education are studied in relation to the target groups/sectors addressed (education, agriculture, food industry). The framework also takes into consideration the fact that the course is developed and should be offered and tested, for its educational value, collaboratively by the project partners.

Session 1 Poster 11

Comparison of conventional and electronic identification in beef cows under rangeland and long term conditions

J.J. Ghirardi, G. Caja, C. Conill and M. Hernández-Jover. Universitat Autònoma de Barcelona, 08193 Bellaterra, Spain.

A total of 161 'Bruna dels Pirineus' beef cows under Pyrenees Mountain conditions were used to compare conventional and electronic identification devices for a period of 8 years (1997-2004). All cows wore two official plastic ear tags (6 g; Azasa, Spain), according to CE 1760/2000, and were also identified with two types of ISO half-duplex radiofrequency devices: 1) electronic button ear tags (10 g and 30 mmØ; Allflex, France) on the left ear; and, 2) electronic ceramic boluses (75 g and 21x68 mm; specific gravity, 3.36; Rumitag, Spain) orally applied. Reading controls of all devices were carried out at the annual brucellosis and tuberculosis testing and drugging. Electronic devices were read using a Gesreader-2S handheld reader (Rumitag). After 8 years, the 50 cows remaining in the herd retained 89.0% of official plastic ear tags. Electronic ear tags showed a 88.4% readability (1.6% failed and 10.0% were lost). Annual ear tag losses was 1.4% on average, but dramatically increased after the 6th year. Two electronic boluses were lost at application and were reapplied on the same day. No new losses or failures of boluses occurred during the experiment. In conclusion, bolus readability in beef cows (100%) was not affected in the long term and was greater than official and electronic ear tags, recommending their use under rangeland conditions.

Session 1 Poster 12

Improving skills on food quality and safety issues in the Leonardo da Vinci Vocational Training Project WELFOOD in virtual environment

E. Szücs[1], R. Geers[2], J. Praks[3], V. Poikalainen[3], E. Sossidou[4], T. Jezierski[5], [1] Szent István University, Gödöllő Hungary [2]Catholic University, Leuven, Belgium, [3]Estonian Agricultural University, Tartu, Estonia, [4]National Agricultural Research Foundation, Animal Research Institute, Giannitsa, Greece, [5]Institute of Genetics and Animal Breeding, Polish Academy of Sciences, Jastrzebiec, Poland*

Interest of public in Europe has recently focussed on three main points in food production systems such as animal welfare; environmental impacts on and of animals; as well as quality and safety of foods. Thus, significance of knowledge on quality assurance has been increasing in all phases of food production of animal origin. The topic requires special skills in the sectors involved including animal husbandry and food industry. In line with the principle "from fork-to-farm" in foods of animal origin, i.e. traceability, transparency and labelling, research on production methods should aim to meet the consumer's requirements. Recent experiences and requests from commercial practice reveal expanding demand for experts who are skilled at state-of-the art and up-to-date level in the subject. Thus, in the Pilot Project WELFOOD "Promoting quality assurance in animal welfare-environment-food quality interaction studies through upgraded e-Learning" multidisciplinary approach is aimed at for the inclusion and transfer of recent scientific findings and knowledge into curricula in virtual environment. Driving force for application of ICT is due its high didactic and added value and rapid transfer of knowledge in an interactive and collaborative way.

Session 1 Poster 13

Subcutaneous nasal transponder implant, an alternative method for electronic identification of pigs: preliminary results

S. Lungu, G. Fiore, C. Korn, EC Commission, Joint Research Centre, Institute for Protection and Safety of the Citizen

For the effective control of the production system, collecting data, prevention and control of notifiable diseases, it is indispensable to have a reliable method for identifying individuals and possibly also groups of pigs within the herd. Traceability of meat to farm of origin is a consumer requirement also.

Both those bondages would be greatly facilitated by electronic identification, because the current identification and registration system is not fully reliable, due to the loss of ear tags and the lack of effective control possibilities.

A small size transponder injected subcutaneously 24 hours after the birth, readable throughout its life (from the birth to the slaughtering), can be a good solution, in comparison with other injection sites (intra-peritoneal, outer ear, neck, ear base, dewclaw, head top and lower jaw).

The objective of this work was to to determinate the practical working distance, shape and areas at which a common stationary or portable read-out unit can read passive implantable transponders used in animal identification. This injection site, never explored before, allow certain advantages: a very low risk for transponder migration, the possibility to tag piglets in the first 24 h after birth, the very precisely defined injection area irrelevant from the commercial point of view.

Session 1 Poster 14

Tracing pigs by using conventional and electronic identification devices

D. Babot[1,2], M. Hernández-Jover[3], G. Caja[3], C. Santamarina[2] and J.J. Ghirardi[3], [1]Àrea de Producció Animal, Centre UdL-IRTA, Lleida, Spain, [2]Departament de Producció Animal, UdL, Lleida, Spain, [3]Departament de Ciència Animal i dels Aliments, UAB, Bellaterra, Spain

A total of 1032 pigs were used to evaluate the traceability of different identification systems from birth to slaughter. Devices were: 1) plastic button ear tags (PET, n = 352); 2) half-duplex (EEH, n = 333) and full-duplex B (EEF, n = 347) electronic button ear tags; and, 3) half-duplex (IPH, 32 mm, n = 340) and full-duplex B (IPF, 34 mm, n = 335) intraperitoneal transponders. No negative effects were reported after the identification. No differences ($P > 0.05$) in losses were observed during fattening (PET, 2.7%; EEH, 3.8%; EEF, 3.8%; IPH, 1.3%; and, IPF, 2.1%). Losses during the slaughter line were greater ($P < 0.05$) for the EEF (5.7%) than the EEH (4.4%) and PET (1.7%), the two last not differing. Electronic failures of electronic ear tags during slaughtering averaged 1.5%. Intraperitoneal transponders were not affected by the slaughter line, 97.1% were recovered at evisceration and no transponders were found in carcasses. Intraperitoneal transponders showed the greatest traceability (IPH, 98.7%; and, IPF, 97.9%), the last not differing from PET (95.7%). Electronic ear tags showed the lowest traceability (EEH, 90.6%; and, EEF, 88.6%). In conclusion, the intraperitoneal injection seems to be an efficient system for the traceability of pigs in the food chain.

Session I — Poster 15

Electronic identification of swine: comparison of results obtained using intra-peritoneal transponders and ear tags

Enrico Marchi[1], Nicola Ferri[1] and Federico Comellini[2], [1]Istituto Zooprofilattico Sperimentale dell'Abruzzo e del Molise "G. Caporale", via Campo Boario, 64100 Teramo, Italy, [2]Coop Italia, via del Lavoro 8, Casalecchio diReno, Bologna, Italy

The Italian official swine identification system is founded on utilization of a tattoo on the left ear, or, only for those farms included in a PDO (protected designation of origin) consortium, on the external part of hams.
The target of this work is to evaluate both electronic identification systems (intra-peritoneal and ear tag) for swine species, in some farms located in Lombardia and Emilia-Romagna regions and to compare the results obtained.
The first identification system uses a transponder intraperitoneally injected by a five centimeter steel needle in lactating piglets; the other one uses an electronic ear tag. Preliminary results show that intra-peritoneal transponders have a high reading performance (min. 96.8% up to100%), with low recovery percentage within slaughterhouses (up to 70%) while ear tags show either a high reading efficiency (up to 98.5%) or recovery (up to 92.5%), in all identified swine.

Session I — Poster 16

Effect of the transportation on the readability and retention of the endo- reticulum transponder in goats Sarda breed

W. Pinna[1], P. Sedda[1], G. Delogu[1], G. Moniello[1], M.P.L. Bitti[2], I.L. Solinas[3], [1]University of Sassari, Via Vienna, 2 - 07100 Sassari, Italy, [2]Provincial Breeders Association of Nuoro (Sardinia), Via Alghero, 6 - 08100 Nuoro, Italy, [3]EU - Joint Research Centre - IPSC - Via Enrico Fermi 1, 21020 Ispra, Italy*

The authors report the results of an experiment developed with the aim of: a) verify the ceramic bolus retention in goats during the transportation and immediately after; b) verify the transponder readability after the transportation. During the VII goats provincial fair (Sardinia) has been evaluated the transponder retention and readability of 104 goats. 63 pluriparas and 41 primiparas came form 8 breeding where the whole flock has been electronically identified. The retention and readability of transponder has been verified, immediately before leaving the breed and after coming back in it, using a portable reader (static reading). The animals at the moment of arrival in the fair has been read using a stationary reader supplied by an antenna located at one side of a corridor where the animals were obligated to run through (dynamic reading). No cases of loss or malfunctioning of the identifiers were observed. The readability of transponders resulted to be 100% during the three different controls.

Session 1 Poster 17

Effect of intraperitoneal transponder on the productive performances of suckling piglets

W. Pinna[1], P. Sedda[1], G. Delogu[1], G. Moniello[1], M.G. Cappai[1], I.L. Solinas[2], [1]University of Sassari, Via Vienna, 2 - 07100 Sassari, Italy, [2]EU - Joint Research Centre - IPSC - Via Enrico Fermi 1, 21020 Ispra, Italy*

32 piglets, 1 - 4 days old and weighting between 1,400 kg and 3,950 kg, have been separated into 2 groups: 16 piglets (group T) have been electronically identified by intraperitoneal injection of a bio-glass encapsulated transponder (HDX technology, TIRIS™); 16 piglets (group C) were the control. The time requested to apply a single transponder and record the data was of 2' 30", on average. The health and the welfare status of the animals have been evaluated during all the experiment. The readability of transponders in group T was evaluated both in live animals at the moment of injection using a portable reader supplied by a stick antenna and at slaughterhouse, by static reading in *ante mortem* visit and by dynamic reading in *post mortem* visit. The weight of animals at the moment of injection was 2651±786 kg vs. 2705±581 kg and, at slaughterhouse, 9467±1762 kg vs. 9645±1804 kg respectively for T and C group. No statistic difference on productive performances of the two groups rose. The transponder showed 100% readability and recovery. The intraperitoneal electronic identification provided a foolproof method to check animal identity in productive chain, both in farm and at slaughterhouse.

Session 2 Theatre 1

Expression of the swine MHC genes

C. Rogel Gaillard, L. Flori, C. Renard and P. Chardon. Institut National de la Recherche Agronomique (INRA) and Commissariat à l'Energie Atomique (CEA) Jouy en Josas, France.*

During the last decade, numerous significant associations have been reported between the region harbouring the swine major histocompatibility complex (SLA) and physiological traits, including immune responsiveness to a variety of parasites, and male and female production performances, making the SLA region an obvious candidate for marker-assisted selection. In order to have a comprehensive understanding of the function of the SLA complex and its surrounding region, we developed a chromosomal region expression array approach that integrates comparative mapping, bioinformatics and microarray technology. From the complete sequence of the SLA complex we generated 140 probes corresponding to the complete set of genes located into the SLA complex. Besides, about 440 additional gene sequences have been identified in swine BAC contigs assigned to the two flanking chromosomal segments of the SLA complex, using both sequence alignments and comparative mapping. Corresponding cDNAs were identified from EST libraries and in absence of positive ESTs, exon sub-clones were generated from BACs with primers derived from swine sequences or from human orthologues. Thus, a total of approximately 600 PCR amplified products have been printed onto glass slides. Microrarrays will be used to determine differential gene expression patterns between lymphocytes after immune challenges and to dissect QTLs located in the SLA area.

Session 2 Theatre 2

Expression patters of developmentally relevant genes in cattle preimplantation embryos

K. Schellander and D. Tesfaye, Institute of Animal Breeding Science, University Bonn, Endenicher Allee 15, D53115 Bonn, Germany*

"Fertility" is a very important trait and comprises a complex interaction of environmental, paternal, maternal and embryonal/fetal factors. A main requirement is the generation of a fully developmental competent embryo. It is evident, that this normal development depends on proper genetic programming in the preimplantation period starting even before fertilization during gametogenesis. Embryogenomics approaches have shed light on the genetic interplay during this period and identified a number of genes crucially involved in determining the fate of a fertilized zygote. Although embryotranscriptomics is hampered by the scarcity of the material, molecular methods have been adopted to be used with minute amounts of material. ESTs have been identified from single stage cDNA libraries from oocyte up to blastocyst stage and paved the way for more advanced expression analyses systems. Further, differential display RT-PCR techniques and subtractions-hydrisation techniques allowed to analyse factor-specific (embryostage, embryo quality, external influences on embryos) gene expression patterns and thus to identify genes involved in the complex interaction of embryos with the environment. Recently, the development of preimplantation cattle embryos specific cDNA arrays allowed the simultaneous analyses of gene expression patterns from a large number of genes. In the present paper the recent developments in analyses of global expression patterns of genes in cattle preimplantation stage embryos will be reviewed.

Session 2 Theatre 3

Abnormal gene expression results in the immotile short tail sperm defect in Finnish Large White

A. Sironen[1], B. Thomsen[4], M. Andersson[2], V. Ahola[3], J. Vilkki[1], [1]MTT Agrifood Research Finland, Animal Production Research, FIN-31600 Jokioinen, Finland, [2]Department of Clinical Veterinary Sciences, University of Helsinki, Finland, [3]MTT Agrifood Research Finland, Food Research, FIN-31600 Jokioinen, Finland, [4]Danish Institute of Agricultural Sciences, PO Box 50, Tjele DK-8830, Denmark*

The immotile short tail sperm defect has been identified as a hereditary fertility disorder within the Finnish Yorkshire pig population. The syndrome is inherited as an autosomal recessive disease exclusively expressed in male individuals as shorter sperm tail length and immotile spermatozoa. The disease-causing mutation was recently mapped by homozygosity mapping on porcine chromosome 16 within a 3 cM region between markers SW2411 and SW419. For fine-mapping porcine BAC clones were picked up with markers within this area. Comparative mapping with porcine BAC-end sequences and human genome mapped the disease-associated area to a 2 cM region on human chromosome 5p13.2. Sequence analysis of porcine ESTs from orthologous genes within the region revealed polymorphisms (SNPs) in three genes. These SNPs defined the location of the disease-associated area and the most probable candidate gene within the area was sequenced. Sequencing revealed a mutation causing skipping of one exon of the candidate gene in the testis of affected individuals, leading to a premature termination of the translation. Expression analyses indicated that the skipped exon is only expressed in testis explaining the testis specific symptoms of affected individuals.

Session 2 Theatre 4

Quantitative trait loci affecting fertility and calving traits in Swedish dairy cattle

M. Holmberg and L. Andersson-Eklund, Department of Animal Breeding and Genetics, Swedish University of Agricultural Sciences (SLU), Box 7023, 75007 Uppsala, Sweden*

Impaired fertility is the main reason for involuntary culling of dairy cows in Sweden. The objective of this study was to map quantitative trait loci (QTL) influencing fertility and calving traits in the Swedish dairy cattle population. The traits analysed were number of inseminations, 56 days non-return rate, interval calving to first insemination, fertility treatments, heat intensity, stillbirths, and calving ease. A genome scan covering 20 bovine chromosomes was performed using 145 microsatellite markers. The mapping population consisted of 10 sires and their 427 sons in a granddaughter design. Nine of the sires belong to the Swedish Red and White breed, and one is of the Swedish Holstein breed. Least squares regression was used to map loci affecting the analysed traits, and permutation tests were used to set significance thresholds. Cofactors were used in the analyses of individual chromosomes to adjust for QTL found on other chromosomes. The use of cofactors increased both the number of QTL found and the significance level. In total we found 30 putative QTL when including cofactors in the analyses versus 13 without cofactors. The QTL found without cofactors were mapped to chromosomes 6, 7, 9, 11, 13, 15, 20, and 29. In conclusion, we were able to map several QTL affecting fertility and calving traits in Swedish dairy cattle.

Session 2 Theatre 5

Fine mapping of QTL for primary antibody responses to KLH in laying hens

M. Siwek[1#], S. Cornelissen[1], L.L. Janss[1], E.F. Knol[2], M. Nieuwland[1], H. Bovenhuis[1], M.A. M. Groenen[1], H.K. Parmentier[1], J.A.M. van Arendonk[1], J. van der Poel[1], [1]Animal Breeding and Genetics Group, Adaptation Physiology Group, Wageningen Insitute of Animal Sciences, Wageningen University, P.O. Box 338, 6700AH Wageningen, The Netherlands, [2]IPG, Institute for Pig Genetics, P.O. Box 43, 6640 AA Beuningen, The Netherlands, [#]Current address: Department of Biochemistry and Biotechnology, University of Technology and Agriculture, Mazowiecka St 28, 85-225 Bydgoszcz*

The aim of this study was to fine map QTL associated with primary antibody responses to KLH on GGA14. The QTL was detected in F_2 cross of lines selected for primary antibody responses to Sheep Red Blood Cells (H/L lines). The QTL region spanned over 50cM. To be able to narrow down the QTL region, additional SNP markers were detected. The set of 16 SNP markers was used to genotype F_2 generation of H/L population. The SNP information was combined with previously collected microsatellite data and analyzed using joined linkage disequilibrium/ linkage analysis (LDLA). Successful application of relevant SNP markers and combined LDLA analysis allowed us significantly narrow down QTL region to the marker bracket around 1cM. The haplotype block related with high KLH effect was defined and uncovered in the F_1 sires. The study suggests novel analysis approach to evaluate the potential relevance of particular SNP to the dissected trait.

Session 2 — Theatre 6

Candidate SNPs for chicken immune response genes

L. Sironi[1], B. Lazzari[1], P. Ramelli[1], S. Cerolini[2], P. Mariani[1], [1]PTP, Livestock Genomics 2, Lodi, Italy, [2]VSA Department, University of Milan, Milan, Italy*

Chicken MHC is located on microchromosome 16 (GGA16). On the same chromosome, class I-*like* and IIβ-*like* genes are clustered within the *Rfp-Y* region. From previous studies, both loci seem to be associated to some viral disease positive outcomes. In the present study the MHC genes *Tapasin*, *TAP1* and *TAP2* and the *Rfp-Y* genes *YFV* and *YFVI* were analised in several commercial lines and indigenous breeds to identify new SNPs within the two regions. DNA sequences were edited and processed using Phred. Database searches (EMBL, GenBank) and initial sequence comparisons were carried out with BLASTn. The polymorphism identification analyses were carried out by the CAP3 and ClustalW multiple alignment programs. The sequence data here produced were further analysed with the PolyBayes SNP discovery pipeline established at PTP, Italy. *Tapasin*, *TAP1* and *TAP2* genes showed a low level of within/between breed polymorphism. On the other hand, *YFV* and *YFVI* genes showed an higher level of polymorphism with within/between breed nucleotide substitution rates ranging between 6% in the exons and 20% in the intronic regions. Several of these base substitutions are non-synonimous and when located within exonic regions can be of great consequence as they could affect the final phenotype. Therefore, most of the here presented SNPs will likely be invaluable tools in future association studies.

Session 2 — Theatre 7

Exclusion of the SLA as candidate region and reduction of the position interval for the porcine chromosome 7 QTL affecting growth and fatness

O. Demeure[1], M.P. Sanchez[2], J. Riquet[1], N. Iannuccelli[1], K. Fève[1], J. Gogué[3], Y. Billon[4], J.C. Caritez[4], D. Milan[1] and J.P. Bidanel[2], [1,2,3,4]INRA, [1]Laboratoire de Génétique Cellulaire, BP27, 31326 Castanet-Tolosan, [2] Station de Génétique Quantitative et Appliquée, 78352 Jouy-en-Josas, [3]Domaine expérimental de Bourges, 18390 Osmoy, [4]Domaine expérimental du Magneraud, Saint-Pierre d'Amilly, 17700 Surgères, France

In different studies, pig chromosome 7 (SSC7) has been shown to be rich in quantitative trait loci (QTL) affecting numerous traits such as growth, backfat thickness, carcass composition and meat quality. Meishan alleles were associated with better performances and had dominant effects over European alleles. Most studies mapped the QTL close to the Swine Leukocyte Antigens (SLA), which has a large impact on adaptability and natural selection. Previous comparative mapping studies suggested that the 15 cM region limited by *LRA1* (mapped at 55 cM) and *S0102* (mapped at 70 cM) contains hundreds of genes. To reduce the number of candidate genes, we improved mapping resolution with a genetic chromosome dissection through a backcrossing program. During this study, 48 growth, fatness, carcass and meat quality traits were measured. We reduced the QTL interval from 15-20 cM to less than 6 cM and excluded the SLA as a candidate region. We also showed that the QTL we studied has limited effects on meat quality and maternal quality traits.

Session 2 Theatre 8

SNP's within exons of candidate genes for fertility traits

N. Hastings[1], S. Donn[1], K. Derecka[2], A.P.F. Flint[2] and J.A. Woolliams[1], [1]Roslin Institute, Midlothian, Scotland, EH25 9PS. [2]University of Nottingham, Sutton Bonington Campus, Leicestershire, LE12 5RD, United Kingdom*

The increase in milk yield of dairy cows over the last 25 years has been accompanied by a decline in conception rate to first service approaching 1% per year. Although poorly understood, this relationship has a genetic component thought to be mediated through the competitive demands for energy and nutrients. Our project aimed to discover single nucleotide polymorphisms (SNP's) within eight candidate genes associated with endocrine pathways regulating fertility. Sequencing primarily focused on the coding regions of the candidate genes, 28 exons covering 11,500 base pairs in total. Within the eight candidate genes 20 SNP's with frequencies greater than 2% were identified. This represents an average of 1.74 SNP's per Kb within exons. Four of these SNPs, located within 3 candidate genes, resulted in changes to the transcribed amino acid sequences.

Session 2 Theatre 9

Association of IGF1 and IGF1 receptor genes polymorphism to fertility traits in Thai indigenous cattle

K. Boonyanuwat[1], S. Suwanmajo[1] and A. Suklim[2], [1]Division of Livestock Breeding, Department of Livestock Development, Ministry of Agriculture and Cooperative, Rachathevee, Bangkok, 10400, Thailand, [2]Phrae Research and Testing station, Phrae Province, 54000, Thailand*

The insulin-like growth factor 1 (IGF1) and insulin-like growth factor 1 receptor genes were assessed as candidate genes for fertility trait (calving interval) in Thai indigenous cattle. Three hundred and fifty-nine Thai indigenous cows (184 Khou Lumphoon, 31 Kho-Lan, 97 Kho-Esan, and 47 Kho-Chon) from Research and Breeding Center, Research and Testing Station of Department of Livestock Development, Ministry of Agriculture and Cooperative were used in this study. Blood samples of these cows were individually withdrawn from the jugular vein and precipitated for the DNA. Diluted DNA was amplified by primer of IGF1 and IGF1 receptor genes, designed from cds of IGF1 and IGF1 receptor genes in Genbank, in PCR machine. The PCR products (350 bp and 300 bp) were run by SSCP method to investigate of SSCP pattern. Calving interval data of these cows were collected from data based system to analyze association of IGF1 and IGF1 Receptor Genes Polymorphism to fertility trait by generalized linear model analysis. The results were represented 3 groups of IGF1 SSCP patterns and 3 groups of IGF1 receptor SSCP patterns. There were associations between calving interval, IGF1, and IGF1 receptor SSCP patterns ($P<0.01$). This result is expected using IGF1 and IGF1 receptor genes to develop marker assisted selection, MAS for fertility traits in cattle.

Session 2 Poster 10

SNP detection on *LHB* gene and association analysis with litter size in pigs

G. Muñoz, A. Fernández, C. Barragán, L. Silió, C. Óvilo and C. Rodríguez, Dpto Mejora Genética Animal, INIA, Madrid, Spain.*

The *LHB* (luteinizing hormone beta polypeptide) gene is a member of the glycoprotein hormone beta chain family and encodes the beta subunit of luteinizing hormone (LH). LH is expressed in the pituitary gland and promotes spermatogenesis, ovulation and luteinization of the ovarian follicle by stimulating the testes and ovaries to synthesize steroids. In mammals this gene can be considered candidate for traits related to fertility and prolificacy. We performed the sequence analysis of 1383 bp of this gene (positions 1030 to 2427, Genbank D00579), including the three exons, by amplifying three overlapping fragments from genomic DNA samples of 9 pigs (5 Iberian and 4 from a Chinese-European synthetic line). The alignment of the sequences allowed the detection of 6 SNPs, four of them located on intron 1, one on exon 2 and the last one on intron 2. Three out of the six SNPs were polymorphic in the synthetic line and one of them, on position 1549, was genotyped on 292 sows from this population using a PCR-RFLP *Tsp45*I protocol. Allelic frequencies were 0.6 and 0.4 for G and C alleles, respectively. An statistical association analysis was performed for this SNP fitting an animal model with repeatability, using records of total number of piglets born and number of piglets born alive from 1,230 litters. No statistically significant association between the *LHB* polymorphism and litter size was found.

Session 2 Poster 11

Investigation of genetics involved in boar taint

M. Moe[1], S. Lien[2], H. Tajet[1], T. Meuwissen[2] and E. Grindflek[1], [1]Norsvin, 2304 Hamar, Norway, [2]Cigene, University of Life Science, Norway*

Male pigs are normally castrated early in life to prevent boar taint in the meat. In Norway, castration will be prohibited from 2009, which claims for a reduction of boar taint in the meat of uncastrated pigs. The goal of this project is to promote a mechanistic understanding of the complex genetic system controlling boar taint and to use this knowledge in practical breeding to reduce boar taint. A characterisation of candidate genes involved in the biochemical pathway of androstenone and the metabolism of skatole have been started to identify functional SNPs. Furthermore, a genome scan will be performed to identify QTLs affecting androstenone, skatole and some fertility traits, using 2400 Norwegian Landrace and Duroc boars from Norsvin's boar testing stations. Genome scan will be followed by finemapping using additional SNPs in interesting regions. A supplementary approach is to study gene expression patterns in animals with extreme androstenone values, and to combine the data from the genome scan. Two methods for gene expression will be applied. First the microarray method which provides the possibilities to analyse the expression patterns for thousands of genes in testis and liver samples, and second we will follow up the results with the sensitive real competitive PCR (rcPCR) method by using MassARRAY technology.

Session 3 Theatre 1

Identification methods of horses in Germany

T. Dohms and K. Miesner, German Equestrian Federation, Freiherr-von-Langen-Str. 13, 48231 Warendorf, Germany

In Germany the identification of registered horses by the breeding organisations takes place by means of the four methods: at first the indication of the sex and the description of colour and markings of horses including the diagram in the passport. The document (passport) for the identification of registered horses is issued in accordance with the regulations of the Commission Decision over the document for the identification (93/623/EEC). The next identification method is in addition to the brand mark of the breeding organisation the number brand. All registered horses are identified with hot branding which is burned exclusively outside on the left thigh. In the year of the birth the assignment of branding is done by the breeding organisation. Additionally at the birth registration each foal gets the unique equine life number with 15 digits. This number is important for the identification as well as for evaluations regarding performance tests and estimation of breeding values. And at the inscription into the studbook a name for the horse must be assigned. This name must be maintained during the complete life.
The breeding organisation can require a pedigree check to save the identity. For the inscription of stallion a DNA-profile has to be submitted for each stallion. Furthermore some breeding organisations provide as well a DNA-profile for all registered foals and mares. National database with information about identification and with all dates from Stud Book exists.

Session 3 Theatre 2

Identification of horses in Great Britain, including the role of horse passports and the establishement of the national equine database

Paul Newman and Duncan Holmes, Department for Environment, Food and Rural Afairs, Rabies and Equine Division, 1A Page Street, London, SW1P 4PQ, England

The government has laid down rules on the identification of horses and the issue of horse passports, and the recording of an Unique Equine Life Number for that horse in passports. Scotland, Wales and Northern Ireland are responsible for separate legislation in their own territories.
The law requires all horse owners to register with an approved organisation in order to get a passport for their horse. Government ensures that Passport Issuing Organisations rules do not contravene the general registration rules in the EU Zootechnics legislation.
The minimum identification is by completion of a silhouette of the horse. Although there is no legal requirement, many Breed Society horses as well as racing foals are micro chipped. Additional methods of identification such as branding or freeze branding are also utilised.
Estimates vary but it is likely to range between 600,000 and I million horses. The National Equine Database will assist in producing a more definitive view.
The National Equine Database. NED is a joint industry and government database which is populated by information supplied by Passport Issuing Organisations and competition disciplines. It will help Government to implement the horse passports legislation, plan disease control work and develop an overall strategy for the horse. It will be used by the industry to help improve the breeding and performance of British horses.

Session 3 Theatre 3

Identification and registration of equines in the Netherlands

J. Klaver, A. de Vries, M. van Lent, Product Boards for Livestock, Meat and Eggs (PVE), P.O. Box 460, 2700 AL Zoetermeer, The Netherlands*

Since January 1st 2004, Dutch legislation concerning the identification and registration of equines came into force. On behalf of this legislation, all equines must have a uniformed horse-passport and a unique life number (15 digits UELN-number).The owner of the horse has to indicate in the passport whether he wants the horse finally to be slaughtered for human consumption or not. Veterinarians have to indicate the medical treatments of the horse in the passport. The PVE have authorised the approved studbooks and the KNHS (the national hippic organisation) to issue horse-passports. To be 100% sure that a certain passport belongs to a certain horse, all horses must be identified with a micro-chip. The micro-chip number is registered in the passport and in the database of the organisation issuing the passports. With a central search-facility on the web, it is possible to find with the help of the micro-chip number the passport-issuing organisation were a horse has been registered. Only micro-chips which have been produced in accordance with ISO-standards and which number contains a country-code are allowed. Only veterinarians and well trained horse-identification specialists are allowed to implant micro-chips.

Session 3 Theatre 4

Identification of Horses in France

X. Guibert[1], D. De Cadolle[2], Les Haras Nationaux, 19230 Pompadour (France), [1]European and International department, [2]Direction des connaissances, France

Identification became compulsory in France in 2002; "Les Haras Nationaux" (HN) have been given the management of identification and registration of equidae into central database SIRE, established in 1976 for stud books; horses are identified either by HN agents or by vets, on special forms provided by HN. Depending on the value of the foal, registration includes description of whorls and natural markings (coded description used as international lexicon, with new table of coat colours), microchip number (compulsory since 2004) ; DNA type is required for identification and pedigree checking of most of foals registered in 45 stud books; SIRE issues a passport for each, including Universal Equine Life Number, identification, pedigree, stud book registration and all other EC regulations items. The ownership is followed on a separate form re-issued at every ownership change; so the passport remains always with the horse ; SIRE is the basis of an integrated information system linking all other French databases (racing, competition, research center). From 1999 to 2005, nearly 300 000 unknown pedigree horses were registered in addition to 55 000 annual foals . The French living horse population is now evaluated at 900 000 and SIRE holds more than 2 million horses . Les HN are always looking for new modern tools for improving horse identification management and world-wide communication between databases.

Session 3 Theatre 5

Identification of horses in Ireland

Nicholas Finnerty, Irish Horse Board. Block B, Maynooth Business Campus, Maynooth, Co Kildare, Ireland

There are three main approved registration authorities responsible for the registration of horses in Ireland; the Irish Horse Register, Weatherbys Ireland and the Connemara Pony Breeders' Society. The Irish Horse Register is maintained by the Irish Horse Board and Northern Ireland Horse Board and incorporates the Irish Sport Horse Studbook and the Irish Draught Horse Studbook. The main breeds that are integrated into the Irish Sport Horse Studbook are the Irish Sport Horse, Irish Draught Horse, Thoroughbred, Continental Warmbloods and the Connemara Pony. Foals that have been registered in the Irish Horse Register have been DNA tested and micro-chipped since 2001. At present, there are approximately 5,500 foals registered per year in the Irish Horse Register.

Weatherbys Ireland are responsible for the maintenance of the Thoroughbred Studbook in Ireland. There is also a section for Non-Thoroughbred horses in their Studbook. Weatherbys Ireland currently register approximately 10,000 foals per annum. All foals DNA typed since 2000 and micro-chipping foals since 1999.

The Connemara Pony Breeders Society maintains the Connemara Pony Studbook. Approximately 1,000 foals per year are registered with the Connemara Pony Breeders' Society. The Society has been bloodtyping their foals since 1991 and will micro-chip all foals from 2005.

Session 3 Theatre 6

Identification system of horses in Hungary

I. Bodó[1], S. Mihók[1], B. Pataki[2], J. Posta[1], Zs. Tóth[1], [1]Debrecen University 4032 Debrecen Böszörményi út 138, [2]National Institute for Agricultural Quality Control 1024 Budapest Keleti K. 24

There are several methods for the identification of horses

Advantages and disadvantages of chips, hot branding, deep freezed figurees, use of blood groups, other polimorphic blood systems or DNA, branded marks in hoof, the use of description of colour and marks are discussed.

Presently the hot branding is obligatory in Hungary. It can be considered as very useful in the practice, however disliked by the animal protectionist, although it is a thousand year tradition. The problem of stress is disputed.

The everyday system and its use, marks and figures and numbers etc are presented.

An interesting by-product of the hot branding method is the not everywhere known distinction between the grey and roan horse colour.

The Horse-Inform system is presented as the modern electronic system of registration.

Session 3 Theatre 7

Identification of equidae in the European Community

A.-E. Füssel, K.-U. Sprenger, Commission of the European Communities, Health and Consumer Protection Directorate-General, Unit for Animal health and welfare, zootechnics, Rue Froissart 101, 1049 Brussels, Belgium

In accordance with Community animal health and zootechnical legislation laid down in Directives 90/426/EEC and 90/427/EEC respectively, equidae must be identified during their movement. Because equidae are considered food-producing animals, identification is a prerequisite to apply specific medicinal treatment without compromising public health and is part of the food-chain information. In order to improve traceability and to reinforce the requirements set down in Commission Decisions 93/623/EEC and 2000/68/EC for registered equidae and equidae for breeding and production, the Commission proposed a Regulation establishing the method of identification of equidae. The Regulation requires the approved breeding organisations or competent authorities to record in databases the issuing of unique identification documents and the gradual introduction of electronic transponders for active marking of foals. The code for the database in connection with the life number of the animal are the reference point for any modification of identification details of the animal concerned and must not conflict with the Unique Equine Life Number developed by major organisations for equestrian competition and breeding. The Regulation allows Member States to identify equidae moving on national territory by a smart card issued in complement of the passport, to simplify the identification of equidae marked with a transponder and to establish specific regimes for equidae kept under wild or semi-wild conditions.

Session 3 Theatre 8

International horse identification tools

X. Guibert[1], D. De Cadolle[2], Les Haras Nationaux, 19230 Pompadour (France), [1]European and International department, [2]Direction des connaissances, France

Horses are travelling more and more for breeding or competition purposes, and, due to the changes of names and numbers in successive countries and databases, it has become very difficult to maintain the identity of horses throughout the boarders and all along their life. The consequences are numerous: breeders can't promote their stud as they can't know the performances of their offspring abroad; Horses can have several identities and passports, therefore the competition regulations for their qualification can't been applied ; Stolen horses can be re-identified with new identities, can be slaughtered and their meat consumed without any health control... even when the identity is maintained, it is a pity, for the importing stud book, to manage the whole data registration process, although it has already been done in the exporting country! For these reasons, international organisations have been working together since the nineties in order to promote international horse identification tools: first of all bloodtypes, then DNA types, then microchips, natural and artificial permanent markings, now linked with European passports; more recently, tools of automatic communication between databases : the Universal Equine Life Number. Today, they have to set-up altogether standards of communication with universal tools. We hope that in the near future, the identity of each horse will be maintained all along its life and that it will be possible to analyse its entire career through the net!

Session 3 Theatre 9

The Universal Equine Life Number (UELN)

X. Guibert[1], S. Gautier[2], Les Haras Nationaux, 19230 Pompadour (France), [1]European and International department, [2]Direction des connaissances, France

When they started establishing world databases for horses, international federations discovered that the same horse could have as many registration numbers as the different databases in which it had been registered! Added sometimes to several changes of names, to different translations in Latin of Arabic or Cyrillic characters, it was even impossible to establish indexes spotting the same horse in different databases!
For this reason, a first working group was formed by WAHO (arab horses) in 1988 (Pompadour). ISBC for thoroughbreds and WBFSH for sport horses followed the same road, and finally common meetings were organised in Paris with FEI (Olympic competitions), UET (trotters) and ECAHO (shows), which finalized the solution of an Universal Equine Life Number. This UELN is based on the principle that the alpha numeric number given by the original database of birth, or where the horse was registered for the first time, will stay the unchangeable identification number of each equidae for life ; but in order to make it universal, six digits will be added before, three are the ISO code of the country of the database, and the next three are the number of the database in this country. It is the starting point of a world-wide computerised identification system of horses, and it will permit the development of automatic data exchanges. The presentation of UELN and the databases numbers are avalaible on www.UELN.net.

Session 3 Theatre 10

Data exchanges with XML format

S. Gautier, Les Haras Nationaux, Direction des connaissances, 19231 Arnac Pompadour, France

The aim of this project is to allow stud-books to exchange data about registered horses. It is co-financed by the World Breeding For Sport Horses (WBFSH) and by the French organisation Les Haras Nationaux.
The registrar uses a login to connect himself on the website. He enters the name of the horse and he chooses the database where he wants to download information about him. The different kinds of files which can be download on the website for a horse are "the horse application form" / "the pedigree" / "brood mare data" / "approved stallion data" / "progeny" / "performances". He can choose to download the file or to execute it in order to enter data directly in its local database.
The implementation of UELN numbers of horses in each connected database is necessarily because it is the unique key to identify the horse. For the moment, only the French database managed by Les Haras Nationaux and the Belgium database of the BWP are connected with this system. Others stud-books of sport horses has planed to join this project.

Session 3 Theatre 11

FEI new database, link with the original stud books

Fritz Sluyter and Catherine de Coulon, FEI, Av. Mon Repos 24, 1005 Lausanne, Switzerland

The development of a functional database of FEI horses requires that all horses have a permanent registration number and that this number is linked to the identity of the horse, either at FEI level or at NF level. The FEI passport, a document for registration and traceability, provides information on horse identification, movements of the horse and vaccination status which is available at each border crossing . A weak point is that horse owners can apply for a new passport, without mentioning that the horse already had one . Recently, studbook and competition organisations have started new initiatives such as microchip identification or the Unique Equine Life Number (UELN) ; the database attached to these ID-keys should be under the control of the issuing organisation. Clearly, the time is right to have the ID-verification of the horse be linked to a database ; The chip on site would only be used for ID-verification, the database could be consulted at a later stage.

Session 3 Theatre 12

Practical and methodological views on identification and parentage control of horses

S. Mikko and K. Sandberg, Department of Animal Breeding and Genetics, Swedish University of Agricultural Sciences, Animal Genetics Laboratory, Box 7023, S-750 07 Uppsala, Sweden*

Identification of horses and verification of their pedigrees are of great importance when breeding and keeping sport- and race-horses. A false identity of the horse will obstruct the value of performance-based breeding values and will give less security in betting. Upon registration there are many different issues to consider, like coat color and marking sketches, chip/tatoos/freeze markings and sampling for pedigree control.
Today, DNA-typing has replaced the former blood typing techniques as the main method for parentage control of horses. The DNA-typing is routinely performed by PCR of microsatellites and have many advantages compared with blood typing. New methods are also emerging like SNPs (single nucleotide polymorphisms) although they are not yet used routinely for identification and parentage control of horses.
According to Swedish experiences with Standardbreds, North Swedish trotters and riding horses, different aspects of sampling, horse description, marking systems and pedigree analysis will be discussed in relation to the accuracy and efficiency.

Connectedness among five European sport horse populations

E. Thorén, H. Jorjani and J. Philipsson, Department of Animal Breeding and Genetics, Swedish University of Agricultural Sciences, S-750 07 Uppsala, Sweden*

In order to investigate if international genetic evaluation of sport horses is feasible, data on sport horse populations from Denmark (DEN), Netherlands (NLD), Sweden (SWE) and the two German studbooks Hanover (HAN) and Holstein (HOL) were used. Only stallions with progeny tested in at least one of the five populations were used. The material included 2381 stallions which together had 64248 progenies in these five populations. The number of stallions with tested offspring in more than one country was 261, 69, 24 and 5 in 2, 3, 4, and 5 populations, respectively. The number of common stallions between country pairs varied from 32 (HAN and NLD) to 117 (DEN and SWE). Different measurements of genetic connectedness were calculated, including Genetic Similarity (GS). This measures the number of tested horses sired by stallions with progeny in two countries in relation to the total number of tested animals in those two countries. The GS values varied between 0.07 (NLD and SWE) and 0.29 (HAN and NLD). The average GS value among the horse populations was 0.15. This is almost twice as high as has previously been reported for major dairy cattle breeds. Considering these results an international genetic evaluation of sport horses seems feasible. However, a major obstacle is the lack of unique ID numbers of horses across countries.

Genetic correlations between movement and free-jumping traits and performance in show-jumping and dressage competition of Dutch warmblood horses

B.J. Ducro[1], E.P.C. Koenen[2] and J.M.F.M. van Tartwijk[3], [1]Animal Breeding and Genetics Group, Wageningen University, P.O.Box338, 6700AH Wageningen, The Netherlands, [2]NRS, P.O.Box454, 6800AL Arnhem, The Netherlands, [3]Royal Dutch Warmblood Studbook, P.O.Box382, 3700AJ Zeist, The Netherlands*

The objective of this study was to estimate the heritabilities of movement and free-jumping traits of young horses and their genetic correlations with later results in dressage and show-jumping competition. Traits included several detailed judgements of the basic gaits (walk, trot and canter) and free-jumping. In the period 1989-2002, movement scores for walk and trot of 44,840 horses were recorded at studbook entrance. From 1998 canter scores of 12,800 horses and free-jumping scores of 8,700 horses were recorded additionally. Competition data (highest level) were available from 1981 to 2002 and included 30,474 and 33,459 horses for dressage and show-jumping performance, respectively. Movement and free-jumping traits were analysed by an animal model adjusting for the fixed effects of sex*age, thoroughbred percentage, location*date of classification. Estimated heritabilities ranged from 0.16 to 0.47. Performance in dressage showed favourable correlations (r= 0.40 to 0.69) with movement traits. Dressage was unfavourably correlated to free-jumping traits (r= -0.10 to -0.34). Performance in show-jumping had high correlations with free-jumping (r= 0.52 to 0.88). Show-jumping had favourable correlations with canter (r= 0.28 to 0.43). The results indicate that indirect selection for performance using movement and free-jumping traits from studbook entrance can improve selection efficiency.

Session 3 Theatre 15

Efficiency of past selection of French Sport Horse : the Selle Français and suggestions for the future

C. Dubois, A. Ricard, INRA SGQA, Domaine de Vilvert 78352 Jouy en Josas, France

The genetic gain for show jumping and the evolution of genetic variability between 1974 and 2002 for the Selle-Français (SF) population are presented. The three parameters of genetic gain: the intensity of selection (i=1.95 for males, 0.56 for females), the accuracy (r =0.71 for males, 0.57 for females), the generation interval (T=12.9 years sire/sire; 10.4 years broodmare/sire; 12.1 years sire/broodmare; 11.4 years broodmare/broodmare); and the practice of breeders who choose preferentially the best stallions (i=2.2 weighted by the number of covering) but also stallions selected on their progeny (50% of covering) explain the efficiency of the existing breeding scheme for the show jumping and the possibility of improvements. For the males, to increase i, we should use the effective selection rate (1.7%); and to reduce T, we should use the young stallions and reduce the age of reform. For the females, to reduce T, we should help the young mares (4 and 5 years old) to go back to breed. Although the inbreeding rate increases in 30 years (it is now equal to 1.4%), it is not worrying. If we keep the same structure (test of stallion on their own performance), it is possible to add new criteria (conformation, gaits...) in the breeding scheme for SF and to conserve the genetic gain (1.1 points on BSO/year).

Session 3 Theatre 16

Is the freestyle dressage competition a reliable test of the horse's performance?

A. Stachurska[*1], M. Pieta[1], W. Markowski[2] and K. Czyrska[1], [1]Agricultural University, Lublin, Poland, [2]Medical University of Lublin, Poland*

The objective of the study has been to examine if the freestyle dressage competitions are judged reliably and the final results can be used in the horse breeding. Marks given by a judge for single movements of the dressage program were considered. The data contained 9000 marks of ten best ranked horses at seven Grand Prix classes and two Intermediate I classes from six CDI*** and one European Championships. The marks concerned 84 horses in total and were given by 25 judges. The reliability of judging was measured with the disagreement index (α). The index measures the disagreement of ranking by a particular judge relatively to the general ranking based on the marks of five judges.

Only four judges had the mean α index lower than 10%, eight judges had it greater than 20% and the highest mean index amounted to 28.9%. The judging was less reliable in Intermediate than in Grand Prix. The average index ranged from 9.5% to 23.6% at particular classes. In both Intermediate and three Grand Prix classes the average α index of at least one judge exceeded 20%. The study shows that the results of the freestyle dressage competition should be used in the breeding with more caution. The present system of judging should be permanently checked by statistic methods.

Session 3 Theatre 17

Analysis of body shape variation among different horse breeds via Generalised Procrustes Analysis (GPA)

Thomas Druml[1], Ino Curik[2], Johann Sölkner[1]*
[1]BOKU - University of Natural Resources and Applied Life Sciences, Division of Livestock Sciences, Vienna, Austria [2]Department of Animal Science, Faculty of Agriculture, University of Zagreb, Croatia

Characterisation of phenotypes and performance testing are two standard tools in horse breeding. Linear body measurements used to characterise the morphology indicate mostly differences in size or length. The problem of one-dimensionality can be solved by calculating indices from some of those linear measurements. Another approach of body shape analysis, called 'Geometric Morphometrics', considers defined anatomical landmarks. Using measurements and landmarks, torsi can be reconstructed. Torsi of 513 Lipizzans, 48 Kladrubians, 250 Norikers and 32 Posavina horses were reconstructed and analysed by GPA. Lipizzans from 7 studs and Kladrubians cluster together (barocque cluster), whereas Noriker and Posavina horses (draught horse cluster) are clearly distinct. The shape variation, expressed in partial warps, in the barocque group is small, due to stud farm breeding. Shape variation in draught horses is higher, and indicates lower selection pressure on type. The rear quarter is stable across populations, so this body part can be considered conservative. Large deviations in shape are located in the shoulder and elbow regions, again with more variation in the draught horse cluster. The differences between clusters are probably due to selection and different usage of the breeds.

Session 3 Theatre 18

Genetic evaluation of station performance test results of Dutch Friesian horses

E.P.C. Koenen[1] and I. Hellinga[2], [1]NRS, P.O. Box 454, 6800 AL Arnhem, The Netherlands, [2]FPS, P.O. Box 624, 9200 AP Drachten, The Netherlands*

The breeding population of the Dutch Friesian horse includes approximately 100 stallions and 8000 mares. They are selected for conformation and performance. Every year almost 250 3- and 4-yr old horses are scored in a 5-week station performance test (SPT) for their basic gaits, riding ability and driving ability. The aim of this study was to estimate the genetic parameters of these SPT scores and to estimate their relation with conformation and gait scores that are routinely recorded at studbook entrance. The SPT scores of 1282 horses were analysed using an animal model adjusting for the effects of test and sex. Estimated heritability was on average 0.33 and ranged from 0.25 (canter) to 0.38 (driving ability). Genetic correlations among SPT traits were high (0.50-0.97). Genetic correlations between SPT and conformation traits were estimated by including data of 14,011 horses scored at studbook entrance. Especially scores for walk and trot at hand had high genetic correlations (0.55-0.86) with SPT scores. It was concluded that SPT scores can efficiently be used for a genetic selection and that conformation scores at hand have a high predictive value for performance traits. This study has resulted in the implementation of a routine BLUP evaluation for SPT traits combining SPT scores and scores for walk and trot recorded at conformation scoring.

Session 3 Theatre 19

Selection strategies in an endangered Norwegian horse breed

H.F. Olsen[1], G. Klemetsdal[1], M. Holtsmark[1] and J. Ruane[1,2], [1]Norwegian University of Life Sciences (UMB), Department of Animal and Aquacultural Science, N-1432 Ås, Norway, [2]Current address: Food and Agriculture Organization of the United Nations (FAO), Rome, Italy*

Recent studies have shown that the Norwegian Døle horse has low genetic variation and have highlighted the importance of carefully planning future breeding plans for the breed. The main long-term goal is to maintain a healthy population, with a balance between low rate of inbreeding, ensuring sufficient effective size, and genetic gain. In this study, a closed population with 200 animals born per year was stochastically simulated under the following schemes; 1) No selection, 2) Phenotypic selection of sires for one trait (representing current selection practices) and 3) Optimal contribution selection of sires for one trait, a method which maximises the genetic gain with a restriction on inbreeding in the population. Based on the existing distributions in progeny group sizes among 3- and 4-years old sires, the 3-years old were assigned a mating quota for one year, covering approximately 40% of the annual matings, while the 4-years old were assigned a mating quota for 4 years. The selected sires were mated at random to mares, although full sib, sire-daughter or mother-son matings were not allowed. Results from the simulations are presented and the practical implications for the future of this endangered Norwegian horse breed are discussed.

Session 3 Theatre 20

250.

Dry matter production and energy content of a natural pasture grazed by wild horses over a two years period

M. Costantini[1], N. Miraglia[1], M. Polidori[1] and G. Meineri[2], [1]SAVA Department, Molise University, Via E. De Sanctis, 86100, Campobasso, Italy, [2]DIPAEE Department, Torino University, Via L. Da Vinci 44, 10095, Grugliasco (TO),Italy*

In Molise Region (Italy) are bred some wild horses named “Pentro horses” (autochtonous population). The breeding area is 2200 ha extended and includes a broad plane surrounded by wooden hills. This research is part of an important research project concerning the study of the nutritional characteristics of all the area over a two years period. The aim is to improve the management of the herd and to define the stocking rate in relation to the forage production. This paper concerns the dry matter production (May to October) and the energy content over the two years period. The forage samples have been collected from 2 areas for the pasture system and from 3 areas for the grazing meadow system, according with the Corral and Fenlon method. Data are discussed using a one-factor ANOVA test. The results show a low production/ha; nevertheless, the low stocking rate (0,3-0,6 head/ha) largely satisfies the nutrient requirements of the horses for what concerns dry matter and energy and should agree a potential increase of the total number of heads. Anyway, it is necessary to plan a strict monitoring of the environment because of the considerable influence of seasonal climatic conditions and the relative variations from a year to another.

Session 3 Theatre 21

Analysis of the daily biorhythm from horses, measured on movement activity and rest periods, with pedometer in different horse keeping systems

S. Rose[1], U. Brehme[1], U. Stollberg[1], Yvonne Buchor[2], R. v. Niederhäusern[2], [1]Institute of Agricultural Engineering Bornim, 14469 Potsdam, Germany, [2]Haras National - stud-farm, 1580 Avenches, Switzerland*

The growing distribution of the sport and saddle horse keeping throughout the EU requires a customisation of the legal basis and framework conditions for the countries for the horse keeping under the aspects to well-being in different housing systems as stable keeping and grazing. Well-founded statements concerning the daily biorhythm - activity behaviour, rest periods -have to be the new basis, revised guidelines and regulations of the keeping ordinance of horses scientifically for these animals. New development of a pedometer system, named ALT-pedometer, to register **A**ctivity, **L**ying time and **T**emperature for an accurate determination of the daily biorhythm of horses was tested in stud-farms. The main reason for these investigations was the question - is the present dimensions of single boxes, single boxes with paddock and grazing area sufficient and suitable for horse keeping? The first investigation took place in Germany in single boxes with three stallions and one mare over one week. In Switzerland two investigations were carried out, which includes nine stallions and four geldings in single boxes and boxes with run over three month. Through continuous automatic recording with an eligible time interval (5 min) automatic data transmission from pedometer to PC a successful and rapid control of the exact daily biorhythm with detailed results for activity and lying time per day is possible. About first results will be reported.

Session 3 Poster 22

DNS microsatellite test of Hutsul horses in Hungary

Csilla Józsa[1], Beáta Bán[2], Sándor Mihók[3], Imre Bodó[3], Ferenc Husvéth[1], [1]University of Veszprém, Georgikon Faculty of Agricultural Sciences, Department of Physiology and Nutrition, H-8360 Keszthely, Deák Ferenc u. 16, [2]National Institute for Agricultural Quality Control, Immune-Genetic Laboratory, H-1024 Budapest, Keleti Károly u. 24, [3]Debrecen University,H-4032 Debrecen Böszörményi út 13, Hungary

The Hutsul horse breed (synonym Carpathian Pony) is an international breed bred in its mountainous native region in Poland, Ukrain, Slovakia, Hungary and Romania as well as like a hobby horse also in other countries. Kept in primitive conditions it can be considered a primitive horse without the common improver genes of English Thoroughbred or Arabian horses.Since the 1800th years seven genealogic lines have been established and now they are officially registered by the Hutzul International Federation. Therefore its genetic comparison with the improved horse breeds seems to be interesting.

In the course of our investigations 12 microsatellites were used for the analysis.The evaluation was made by means of ABI 310 automatic genetic analyser.

According to our results we found essential differences in number of alleles (VHL20, ASB2, HMS2) and allelfrequences (HTG4, HMS7, HTG7) of some microsatellites in Hutsul compared to Thoroughbred.

Session 3 Poster 23

Assesment of Old Kladruby horse´s constitution

E. Sobotkova, I. Jiskrova, D. Misar and M. Holecova, Mendel University of Agriculture and Forestry in Brno, Department of Animal Breeding, Zemedelska 1, 613 00 Brno, Czech Republic*

The objective of the present study was a detailed analysis of the exterior of the Old Kladruby white horse in the Stud Farm Kladruby. We applied 26 body dimensions, 5 angles of extremity joints and 11 hippo-metrical indices of 83 breeding horses to analyse the population according to the paternal lines (Generale, Favory Generalissimus, Generale Generalissimus, Favory, Sacramoso, Rudolfo), sex (stallions and mares) and age categories (4 classes). The resulting measures were analysed statistically by means of a linear model with fixed effects (GLM). Most of the highly significant differences were differences detected between the stallions and mares. The stallions have a significantly more marked convex profile of the head and circumference of the shank. The mares have a highly significantly larger chest, width of coxae and angles of the shoulder and knee joints. The differences among the lines are minimal. Only line Rudolfo is the least compact.

Session 3 Poster 24

Preliminary analysis of the morphofunctional evaluation in horse-show of Andalusian Horse

M. Valera[1], J.A. Gessa[1], M.D. Gómez[2], A. Horcada[1], C. Medina[2], I. Cervantes[2], F. Goyache[3] and A. Molina[2], [1]Universidad-Sevilla, Spain, [2]Universidad-Córdoba, Spain, [3]SERIDA-Gijón, Spain*

Traditionally, the Andalusian horse performance has been evaluated in morphofunctional horse-shows. In this events, eight type traits (*head-neck, shoulder-withers, chest-thorax, back-loin, croup-tail, forelimbs, hindlimbs, overall forms*) are assessed by means of regional scores and functionality is assessed by scoring the basic movements (walk, trot).

Here, we analyse the morphofunctional performance of 4,366 horses obtained in 58 horse-shows carried out in the last five years. The number of available records and the total number of available scores are, respectively, 8,500 and 236,662.

We observe that, as an average of morphological score (0-80 points), females receive upper than males, in the different group of age. The females ≥4 years old obtain the upper score (59.79) and the males of the same group receive the lower one (53.97).

In the type traits, the *shoulder-withers* in males (7.93) and the *chest-thorax* in females (8.01) are the highest scores. The lowest are *forelimbs* and *hindlimbs* in males (6.94 and 6.91) and females (6.95 and 6.93), standing out the Andalusian horse's defects in leg-stance.

The evaluation of movements (0-20 points) show the highest averages males ≥4 years old (16.68) and males <4 years old obtain the lowest score (15.57).

The fitted model include the following fixed effects: horse-show, year of celebration, stud-farms, show-judge, sex and age. All of them are statistically significant ($p<0.05$).

The estimation of the breeding value of English thoroughbreds in the Czech Republic

Sonia Svobodova[1], Christine Blouin[2], Bertrand Langlois[2], [1]Faculté d,Agronomie, Institut d,élevage des animaux, Université d,Agriculture et de Sylviculture Mendel, 613 00 Brno, République Tchèque, 2Station de Génétique Quantitative et Appliquée, 78352 Jouy-en-Josas Cedex, France [2]Station de Génétique Quantitative et Appliquée, 78352 Jouy-en-Josas Cedex, France*

The aim of the study was to estimate the breeding value of English Thoroughbreds in the Czech Republic using the racing results from a 22-year period (1980-2001). The data includes the performance of 2 and 3-year-old horses That is 30203 racing results of 6 333 horses issued from 762 sires and 2 836 dams. Different criteria were used to measure the performance: Log of earning per race, normalised ranking value, distance if placed, earning and number of starts for the 2, 3, 2+3 year-old horses. After preliminary studies a year effect or a sex by year effect was finally retained. The variance component estimation gave the following values for heritability (±standard errors): 0.14±0.01 and 0.16±0.01 for the Log of earning per race and the ranking value; repeatability was 0.31 and 0.35, respectively. The maternal environment component was evaluated as 0.02±0.004 for the Log of earning per race and 0.03±0.004 for the ranking value. We found that the Log of earning per race and the ranking value were two appropriate criteria for taking into account the racing performance in selection for Thoroughbreds in the Czech Republic. The genetic correlation of the two criteria was 0.98±0.003.

Modelling breeding schemes for two Bavarian horse populations

C. Edel[1,2], L.Dempfle[1], [1]Technical University of Munich, Department of Animal Science, Alte Akademie 12, 85354 Freising/Weihenstephan, Germany, [2]Bayerische Landesanstalt für Landwirtschaft, Institute for Animal Breeding, Prof.-Dürrwächter-Platz 1, 85586 Poing/Grub, Germany

The South German Heavy Horse and the Bavarian population of the Haflinger Horse are currently managed without relying on an estimated aggregated breeding value as criterion for selection. In working out strategies for an optimisation of the breeding process an investigation has been undertaken to compare advantages of different breeding schemes for the two breeds. Deterministic simulations based on previous estimates of demographic and population parameters and on a breeding objective derived from the results of a contingent valuation survey were used. One critical aspect in the modelling turned out to be the realistic definition of the size of the true breeding population, i.e. the identification of animals with non negligible contribution to the next generation. Starting from selection schemes reflecting the actual situation in both breeds (which in that context can be characterized quite well as 'half-sib-schemes'), in a first step necessary adaptations and enhancements for the implementation of a breeding value estimation are suggested. Furthermore, modifications of the schemes are analysed both with respect to genetic progress per unit of time and to effective population size. The question of the practicability of the so found optimal designs is also raised as it was left aside during the optimization process. In conclusion the pros and cons of the current performance tests are discussed in the context of the findings.

Session 3 Poster 27

Genetic analysis of caliber index in Lipizzan horse using random regression

M. Kaps[1], I. Curik[1], and M. Baban[2], [1]University of Zagreb, Faculty of Agriculture, Animal Science Department, Svetosimunska 25, 10000 Zagreb, Croatia, [2]University J.J. Strossmayer, Faculty of Agriculture, Zootechnical Institute, Trg Sv. Trojstva 3, 31000 Osijek, Croatia*

The variability of caliber index in Lipizzan horses was studied. The caliber index was defined as (heart girth / height at withers)*(cannon bone circumference / height at withers)*100. The analysis included measurements (taken at six, 12, 24 and 36 months of age) of 679 horses born between 1931 and 1999. The corresponding pedigree file consisted of 4786 horses. Covariance functions for the caliber index were estimated using a random regression model. Sex, age of dam and stud-year-season interaction were defined as fixed effects while direct additive genetic and permanent environment were defined as random effects. Polynomial functions of order four were adequate to explain additive genetic functions. The average caliber index increased from 94 at birth to 135 at the age of 36 months. The estimated heritabilities ranged from 0.16 to 0.20. Genetic correlations and permanent environment between consecutive measurements ranged from 0.28 to 0.77, and 0.78 to 0.99, respectively. The estimated covariance functions from this study demonstrated heterogeneity of covariances and proved to be an adequate tool in estimating genetic parameters and trends of caliber index in Lipizzan horses.

Session 3 Poster 28

Characterisation of triacylglycerols in donkey milk using HPLC/MS

B. Chiofalo[1], P. Dugo[2], E. Salimei[3], L. Mondello[4], V. Chiofalo[1], [1]Dept. MO.BI.FI.P.A., University of Messina, Italy, [2]Dept. Chimica Organica e Biologica, University of Messina, Italy, [3]Dept. S.T.A.A.M., University of Molise, Italy, [4]Dept. Farmaco-Chimico, University of Messina, Italy.*

In the past, several chromatographic methods were used for the determination of triacylglycerols (TAGs) in human and cow milk but no data have been found about donkey. Aim of this study was to characterize TAG composition in donkey milk for a better knowledge of the mechanism of fat biosynthesis.
The research was carried out on four Ragusana donkeys (stage of lactation = 77±7 days; milk yield = 805±40 mL/milking); milk lipid fraction was extracted using a mixture of methanol:dichloromethane and TAGs were investigated by HPLC/MS.
Fifty-five TAGs were identified with corresponding retention times, partition number, total carbon number, double bonds and fatty acids which were: Caprylic, Capric, Lauric, Myristic, Palmitic, Stearic, Palmitoleic, Oleic, Linoleic and Linolenic. Forty-three TAGs were quantified on the basis of the percentage peak area. The most represented TAGs were: Palmitic-Oleic-Linolenic (6.69%), Palmitic-Oleic-Oleic (6.22%), Palmitic-Palmitic-Oleic (5.25%), Capric-Palmitic-Oleic (4.63%) and Palmitic-Oleic-Linoleic (4.48%). The most significant quali-quantitative differences compared to human and cow milk were: the lower content of Palmitic-Oleic-Oleic and Palmitic-Oleic-Linoleic than human milk and the higher presence of PUFAs in TAG fraction than cow milk; this latter could be correlated with the absence in small intestine of donkey of isomerisation and hydrogenation processes on the fatty acids prior to absorption and esterification.

Session 3 Poster 29

The raising system for Marismeña equine breed in the Natural Park of Doñana (Spain)

M. Luque[1], E. Rodero[1], F. Peña[1], A. Molina[2], F. Goyache[3] and M. Herrera[1], [1]Department of Animal Production, [2]Department of Genetic, University of Córdoba (UCO), Spain, [3]Animal Genetic and Reproduction Area(SERIDA)Gijon, Asturias, Spain

The park has been used since the IXth century for the raising of a breed of horse which is now identified as the ancestor of our present day Spanish breed.

The racial identification of this population and the expression of the special characteristics resulting from them being raised in freedom in a protected area determine the content of this work. The breeding stock is distributed in 4 zones into which the park is divided during the dry season-between March and August. After they go to higher zones in the period of November to March, when the salt marshes are flooded.

The farmers have an average number of 10 mares to which usually they must give extra food during the dry season. On the 26th of June all the mares with their colts are taken from the park to an area with special facilities for this, it is the only time in which they can be studied and indentified.

Session 3 Poster 30

Feeding, riding properties and behaviour

Anders H. Herlin[1], Margareta Rundgren[1] and Mia Lundberg[2], [1]Swedish University of Agricultural Sciences (SLU), Alnarp and Uppsala, Sweden, [2]Flyinge AB, SE-24032 Flyinge, Sweden

The use of the horse is largely related on the behaviour and reliability in the relation between the horse and rider. The horse owner can influence the feeding of the horse and it is in the interest of the rider/trainer to compose a diet that maximizes the performance of the horse including muscle metabolism and reduces unwanted behaviours like excitability, unwanted responses to non-rider stimuli. Inclusion of fat in the diet has earlier been shown to reduce excitability behaviour in horses. In a change-over trial, 13 riding horses were fed two oat varieties differing in the fat content, one high in fat (10%) and one low in fat (4,5%), in two seven week periods each. The horses and the riders were students took part in the equine studies programme. Horses were assed in a number of traits regarding riding properties every week by the riders and by two trainers. The assessment included also a behavioural challenge test. Results indicated some tendencies like the trainers found horses fed high fat oats less susceptible to disturbances ($P<0.1$) and the riders found tendencies of horses being more forward thinking and somewhat more on the aids ($P<0.1$). Most of the differences were attributed to the three participating mares. A discussion on methology and possible explanations for diet induced behavioral changes will be made.

Session 3 Poster 31

Relationship between maintenance conditions and elimination behaviour in horses

C.I. Nedelcu, Mihaela Maria Nedelcu, Gh. Georgescu, A. Marmandiu, University of Agronomical Sciences and Veterinary Medicine Bucharest, Faculty of Animal Sciences, 59 Marasti Blvd., 1 District, Romania

Maintenance conditions in horses into the studs influence elimination behaviour. The observations that were done on 7 individuals of Romanian Trotters for 24 hours (3 stallions and 4 pregnancy mares) pointed out differences between the two categories for some determinations. In 24 hours, pregnancy mares have spent 4.5 hours on the pasture and 8-11 hours in the paddock, fact that determined a faster intestinal transit, which was remarked through higher defecation and urination frequency (8.75 defecation and 7.25 urination as comparison to 6.67 respectively 4.33 urination in stallions). The differences were statistically very significant for the urination frequency. The stallions have been kept into individual boxes (4 x 4 m). Faecal total quantity was higher in mares (9.46 kg as comparison to 8.37 kg), but the difference was non-significant. The stallions produced a higher faecal quantity per defecation (1.249 kg as comparison to 1.082 kg). These indicators were influenced also, by feed structure. Mares were fed by pasture, which stimulates digestion while the stallions were fed only dry feed (alfalfa hay and oats), fact observed also, by higher content in water in mares faecal.

Session 3 Poster 32

Human-mare relationships and behaviour of foals towards humans

S. Henry[1], D. Hemery[1], M.-A. Richard[1], M. Hausberger[1], [1]UMR CNRS 6552, Université de Rennes 1, France*

We investigated the influence of the establishment of positive human-mare relationships on foals' behaviour towards humans. 21 mares were softly brushed and fed by hand during a short period during the first 5 days of their foals' lives (experimental group), whereas 20 other mares were not handled experimentally and their foals received no contact with the experimenter (control group). The reactions of both experimental and control foals were recorded under various conditions (presence of a motionless experimenter, approach test, saddle-pad tolerance test) when they were 15 and 30-35 days old. Approach-stroking tests were also performed, in pasture, by the familiar experimenter and an unfamiliar person when they were one year old. Several observations strongly suggest that mares can influence their foals' behaviour toward humans. 1) During the handling procedure, experimental foals of protective mares were further from the handler than foals of calm mares. 2) Experimental foals remained, at all ages, closer to the experimenter and initiated more physical contacts with the experimenter than control foals. 3) Avoidance and flight responses of experimental foals were considerably reduced during approaches by the experimenter and they accepted saddle-pads on their back more easily and more quickly than control foals. Finally, the consequences of handling mares had effects that lasted at least until foals were one year old and were generalized from experimenter to unfamiliar humans.

Evaluation of horse keeping in Schleswig-Holstein

S. Petersen[1], K.-H. Tölle[1], K. Blobel[2], A. Grabner[3] and J. Krieter[1], [1]Institute of Animal Breeding and Husbandry, Christian-Albrechts-University of Kiel, Germany, [2]HIPPO-Blobel, Ahrensburg, Germany, [3]Clinic for Horses, Surgery and Radiology, Free University, Berlin, Germany*

To get more insight into practical horse keeping, 46 boarding farms in Schleswig-Holstein were inspected once in autumn 2003 and once in spring 2004. On the basis of a questionnaire, 104 individual keeping systems (individual stables) with 1909 horses were evaluated. In total 42 parameters were assessed using a scale from 0 to 4 with regard to animal welfare. The design of function areas, the hygienic conditions and the management were considered and additionally climate data were recorded.
The mean herd size was 53 horses ranging from 18 to 156. 96% of the horses were kept in individual stables. While portion of outside stalls was 35%, only 8% of the keeping systems were group housing systems or stalls having direct external access. Exercise management was significantly different between the farms - 33% offered staying in an outdoor yard for at least 8 hours per day in summer and 22 % for at least 4 hours per day in winter.
On average 106 from maximal 152 points were reached in individual stables, whereas no differences between evaluation in autumn and spring were found. The repeatability of the total score was 93% - repeatabilities of the single partitions were: housing 98%, care 92% and exercise management 84%.

An estimation of reliability of judging the horse dressage competitions

A. Stachurska[1], J. Niewczas[2] and W. Markowski[3], [1]Agricultural University, Lublin, Poland, [2]Polish Academy of Sciences, Lublin, Poland, [3]Medical University of Lublin, Poland*

Results of dressage competitions may be used in horse breeding provided they reliably estimate the horse's performance. It can be assumed that if the marks of five arbiters agree with each other, the horses are judged properly. This agreement should concern not only final sums, which may be coincidental, but also judging successive movements of the program. The problem lies in the agreement of horse rankings made by each judge.
The method proposed in the communication is used to estimate judges with regard to their agreement in ranking horses at dressage competitions. The α index shows how much the horse ranking made by a judge differs from that assumed as proper, i.e. the total ranking in a movement based on the marks of five judges. The fact that top ranking is the most important, whilst further places matter less, is taken into account. The α index defines in per cents the level of disagreement of the judges and so it is called the index of disagreement. The method can be used to assess judging single movements or final results. Usually, only first ten places are considered, or less, if less horses participate in a class. The index may be used to eliminate the marks of a judge who did not agree with the total ranking (mean α >20%).

Which movements of freestyle dressage programs are difficult for horses to execute and for judges to assess?

A. Stachurska[1], W. Markowski[2], M. Pieta[1] and M. Wesotowska-Janczarek[1], [1]Agricultural University, Lublin, Poland, [2]Medical University of Lublin, Poland*

The aim of the study has been to define which movements required in the freestyle dressage programs are assessed low and judged in disagreement. 13600 marks from judges ranking a horse at single movements of the program were considered. 101 horses participated in seven Grand Prix classes and four Intermediate I classes held at seven CDI*** and one European Championships. The significance of differences between the means was examined by the analysis of variance. The reliability of judging was estimated with disagreement index (α). The index measures the disagreement of ranking by a particular judge relatively to the general ranking based on marks of five judges.
The average of marks in Intermediate (6.28) was lower than in Grand Prix (6.69). The lowest marks in Intermediate were scored in pirouettes, flying changes of leg every second stride and collected walk. In Grand Prix they were the lowest in collected and extended walk, pirouette left, flying changes of leg every stride and piaffe. These movements seem to be difficult for horses and they should be exercised more extensively. The average α index of the movements ranged from 9.8% to 26.3% in Intermediate and from 7.1% to 21.5% in Grand Prix. The rules of judging the α>20% movements should be discussed by the judges more precisely.

Evaluation for riding and driving purposes of Bardigiano Horse stallions and mares

A.L. Catalano, F. Martuzzi, S. Filippini and F. Vaccari Simonini, Department of Animal Production, University of Parma, Via del Taglio 8, 43100 Parma, Italy*

Bardigiano Horse is an ancient breed of the mountains in Emilia region, once bred for meat production and work. Recently the breed standard was changed in order to obtain lighter and taller horses as required by the market for riding and light draft use. Aim of this paper is to describe the results of the trials carried out to assess the breed aptitude for these new purposes. 43 stallions and 39 mares, both 3 years old, were evaluated. The horses were housed in a riding centre in the same environmental conditions for feeding, housing and training, except that the trial lasted 3 months for the stallions and 2 months for the mares. After 2 days of adaptation, the horses underwent the first test, including body measurements and examination of limbs, gaits and general health state. Besides behaviour traits were considered. The horses were trained for riding and driving service and once a month assessed for response to training. Processing of the data collected during all trials showed natural predisposition to training and fast improving of all parameters, demonstrating fair aptitude of the breed to saddle service and excellent versatility to draft. Mares even though trained for a shorter period, obtained the same results of the stallions, evidencing therefore more docility and learning easiness.

Session 3 Poster 37

On breeding value of the Tori Horses

Heldur Peterson, Institute of Veterinary Medicine and Animal Sciences of the Estonian Agricultural University, Tartu,51014, Kreutzwaldi 1, Estonia

The Tori Horse breed in the two types is one of the more popular and well-known Estonian cultural breeds. The present investigation focuses on the results of the experiments carried out to estimate the breeding value of the two-year-old offspring of the outstanding stallions of the Tori riding sport type (T_{RSPH}). For Tori universal type (T_{UNH}) we used another test including a tractive power test. The study of Tori $_{RSPH}$ (riding sport type) is based on the results of the experiments conducted in 1998-2004, with 203 tests. Most important qualities were investigated: general index of stallions, indexes of type, body, feet, pace, trot and free jumping. The best general index during 7 years had stallions Casanova 111.17 in 2004, Premium 110.91 in 1998, Gimalai 109.38 in 2002. Impairer was Reval 89.69 in 2000. Last year (in 2004) the best stallion was Casanova 13 581 T, general index 111.17, the second was Vodevil 13 589 T, index 106.33, and the lowest data had Lando 100.83.

Session 3 Poster 38

Evaluation of linear conformation traits of the autochthonous horse breeds in Croatia

A. Ivankovic[1], P. Caput[1], P. Mijic[2] and M. Konjacic[1], [1]Faculty of Agriculture, Svetosimunska 25, Zagreb, Croatia, [2]Faculty of Agriculture, TrgSv. Trojstva 3, Osijek, Croatia

Autochthonous coldblood breeds make the most numerous breed group of horses in Croatia. Their systematic breeding started in the 19th century in Medimurje, Podravina and Posavina. The population of Posavina horse have small variability of conformation features, mostly conditioned by breeding area. In the Croatian Coldblood population two breed types have been noticed (the first type is a bigger body frame, rougher head and stronger bone base, the second type is a bit lighter, smaller head and smaller body frame). The Murinsulaner horse is in the stage of a breeding consolidation (small effective population size).The horse linear conformation traits is of significant importance for its use and breeding value. The linear conformation means the proportionality of built, body measures, movement and axis centres, postures and leg structure, hoof quality and the movement. The research has been 258 horses included (120 Posavina horses, 120 Croatian Coldblood and 18 Murinsulaner horses) for 34 traits. Posavina horse has a significant difference of expressed linear conformation traits in comparison with the Croatian Coldblood and Murinsulaner horse. The difference of expressed linear traits between the Croatian Coldblood and Murinsulaner horse is not significant for most traits.The use of linear evaluation model for the horse conformation as a routine method could help in better gaining knowledge of the autochthonous breeds.

Session 4 Theatre 1

Diversity of livestock farming systems in Europe and prospective impacts of the CAP reform

A. Pflimlin, C. Perrot, Institut de l'Elevage, 149 rue de Bercy, 75595 Paris cedex12, France*

The 2003 CAP reform, which introduces the decoupling payments from production in the First Pillar and gives primacy to market forces, also brings the opportunity for a completely different application in the different EU countries. In this paper, we discuss the potential impact of the CAP reform on the diversity of livestock farming systems (LFS) that still characterises European ruminant husbandry. A simplified geographic classification scheme, based on several land use criteria, allows for the categorisation of the ruminant farming regions into eight areas with specific characteristics, potentials and constraints. In this way, we can quantify their importance and contribution to the regional economy and rural development.

On the basis of this analysis we can conclude that the CAP reform will probably lead to a process of further specialisation and concentration of livestock farming in the medium- and high-productive regions and a severe decline in less-favoured areas. The decoupling scheme can bring important threats for sustainable rural development, especially in extensive grassland and mountain areas, where ruminant farming is an important component of the local economy and also the background for the tourism sector. Such regions include about half of the Arable Area and more than half of the farmers census in EU-15; these proportions being much higher in EU-25. Considering the difficulties observed to allocate significant funds to rural development, due to the EU limited agricultural budget until 2013, there is little hope that the Second Pillar of the CAP can provide sufficient support to restore the balance.

Session 4 Theatre 2

Rapid structural change in Danish dairy production

Susanne Clausen, Danish Agricultural Advisory Centre, The Danish Agricultural Advisory Centre, Udkærsvej 15, Skejby DK-8200 Århus N

Danish dairy production is undergoing a rapid process of structural change. The number of farms is decreasing and the size of the farms is increasing. From 2001 to 2005 the number of dairy farms has decreased from 9,000 to 6,000 with an average herd size of 70 to 90 cows, respectively.

According to our studies, the development of this rapid structural change appears first and foremost caused by the introduction of the quota exchange in 1997, favourable trends in prices of land and milk quota and thus good borrowing facilities, and last but not least, a very low interest rate. Since the mid nineties, almost 3,000 new barns have been built and every second Danish dairy farmer is producing milk in a modern production system. Despite heavy investments and a doubling of farm size, the income from dairy farming has been more or less stable around 26,900 Euros during the last ten years.

The expectation for the next ten years is a further decrease in the number of farms to 2,700 with an average herd size of 160 cows. The general opinion of the farmers is that they would be better off without quota regulation of milk production; in this sense, the spokesman of the Danish Cattle Federation has lately announced that producers will work for an abolishment of milk quotas in the EU in 2015.

Session 4 — Theatre 3

Decoupling: Irish farmers attitudes and planning intentions

W. Dunne and M. Cushion, Rural Economy Research Centre, Teagasc, Malahide Road, Dublin 17, Ireland

The Mid Term Review (MTR) agreement for the CAP in 2003 provided member states with the option of either full or partial decoupling of the animal-based Direct Payments (DPs). For farmers, the decoupling of the DPs is a radical departure from the existing farm income support system. Ireland plumped for full-decoupling post 2004, but most member states have not yet opted for full decoupling based on historical entitlements. Unlike the commodity based DPs, decoupling is a whole farm concept which, within certain broad limits, provides farmers with potential choices to:
- change their enterprise mix of cows, cattle, sheep and cash crops;
- adjust both the method and intensity of their farming activities, particularly for grazing livestock;
- both increase farm revenue and reduce production costs;
- reduce, or even eliminate, the need to keep either grazing or specific types of animals or crops to collect the DPs in the future.

Since the concept is new, there is very little quantitative information available on farmers' attitudes and likely adjustment response. In two surveys of 1,200 farms undertaken in the autumn of 2003 and 2004, Irish farmers were asked a series of questions in relation to their attitudes to decoupling and their farming intentions post full-decoupling. The farmer's responses were analysed by farming system and the main findings will be presented in this conference paper.

Session 4 — Theatre 4

Consequences of the CAP reform for the Bavarian agriculture

A. Heissenhuber and H. Hoffmann, Technical University Munich (TUM), Freising-Weihenstephan, Germany

The 2003 CAP reform represents a logical continuation of the policy started in 1992. Serious effects on the Bavarian agriculture are expected from the decoupling of the direct payments on the one hand and the removal of the price-support schemes (particularly for milk) on the other hand. In grassland areas, even extensive animal husbandry will not be any longer sustainable in economic terms, because adjacent surfaces are not available.

A study carried out at the Institute of Farm Management (TUM), using a land-use model and extrapolation of trends based on accounting data from 2000 till 2002, highlights that dairy systems in Bavaria are doubly threatened by the reform. In addition to the impact of the expected price reduction, the national option to apply an immediate decoupling of the direct payments will induce a substantial acceleration of structural changes. Dairy cattle could completely disappear at some regions in the next years. The results obtained raise an important question with respect to how landscape can be maintained, specifically in those areas where the payments of the second pillar play an important role for sustaining landscape management.

In conclusion, we recommend farmers to improve their competitive strengths during the next years or to re-orientate their production systems (e.g. part-time farming based on low-input systems).

Session 4 Theatre 5

Prospective economic incidence of the CAP reform on Spanish sheep farming systems

E: Manrique and A. Olaizola, Department of Agriculture and Agricultural Economic, University of Zaragoza, Miguel Servet 177, 50013, Zaragoza, Spain

The behaviour and decisions of farmers have a multifactorial origin related to the dynamics of agriculture, to factors relating to the particular farm environment and general environment (technology available, market, policies) and lastly, to the particular sociological and psychological characteristics of the farmers. Economic and agricultural policies have acquired a growing importance, shifting the incidence of market changes.
Over the past 15 years there has been a significant transformation of sheep production systems in Spain, which can be interpreted as adaptive responses. Despite the objectives of the CAP, a process of intensification of production could be observed: increment of herd size, increment of family labour productivity, higher level of farm indebtedness and increment of capital assets per ha. Farm Net Value Added increased notably, derived to a great extent from a large increment of premiums. The last CAP reform decouples subsidies from production and, therefore, is the most radical reform that has been implemented to date. This reform is generating many uncertainties in the sector and is perceived as a factor that will speed up structural changes.
The economic implications of the reform were simulated at the farm level in Aragón (Northeast Spain). Results indicate that the CAP reform will cause a decrease in farms profit, which will mainly depend on percentage of decoupling and herd size.

Session 4 Theatre 6

Sheep breed and system dynamics over the last thirty years: responding to policy and economic changes in the Britain

G. E. Pollott and D. G. Stone, Department of Agricultural Sciences, Imperial College London, Wye campus, Ashford, Kent, TN25 5AH, United Kingdom*

The sheep industry is a large and diverse part of the British agricultural sector. It has been subjected to many policy and economic pressures over the last 30 years, including the UK joining the EU and implementing the EU Sheepmeat Regime. Four sheep breed surveys have been undertaken in the UK to investigate the breeding structure of the sheep industry; in 1971, 1987, 1996 and 2003. These span the period from before joining the EU to the present situation. The results of these four surveys provide an insight in the collective responses of sheep producers in Britain to such policy and economic pressures, as shown by changes in flock numbers, flock size, breed use and sheep system distribution. Sheep producers respond to outside influences in a manner that increases their personal benefit, either by going out of sheep or changing breed and system. This can be seen in the way breeds and crosses have evolved over the last 30 years in Britain. The number of sheep breeds has increased from about 44 (1971) to 90 (2003) and many breeds have been imported during that period. There has been a shift away from pure breeding towards crossbreeding and the dynamics of individual breeds reflects the move towards a more efficient type of ewe.

Session 4 Theatre 7

Short term impact of the EU membership on the Hungarian livestock sector

Zs. Wagenhoffer, Hungarian Animal Breeders Association, H-1134, Budapest, Loportar. u. 16, Hungary

Livestock production in Hungary had long been a significant branch of agriculture given 55% of its output. The political and economic changes of the '90s plunged the sector into a dire position. The sector was one of the transition's victims: animal stock dropped 40 to 60%, total and per animal production both felt dramatically in most of branches. In 2004 Hungary joined the EU causing another upset to the sector. Introduction of the CAP had and will have significant effect on every livestock branches. Hungary chose the SAP system that means 70 /ha direct payment. Land based sectors such as cattle and sheep production can exploit this opportunity while for pig and poultry branch it is disadvantageous. Besides, beef and sheep producers can also have special grants per animal. This is reflected in stock variation of the last 4 years. Beef and sheep population increased slightly whereas dairy cow (-18%), pig (-8%) and poultry (-18%) stock decreased. Animal product consumption seems to move from its nadir of 1995, especially in case of poultry meat (+11%). Trade has been expanded with the EU, in particular with new member states. Beef prices are significantly higher (10 to 30%) while income of dairy producers has been reduced due to low milk prices. Poultry and pig producers are facing stronger competition both in national and in international level.

Session 4 Theatre 8

Economic position of beef sector in Hungary

F. Szabó, Gy. Buzás, Zs. Vincze, M. Török, Zs. Wagenhoffer, University of Veszprém, Georgikon Faculty of Agriculture 8360 Keszthely Deák F. u. 16. Hungary*

EU membership affected the rentability and the reform will also have a great influence on the economic position of cow-calf unit and the finishing cattle production. For more information an economic analyse of input and output data were done in different production systems. Different levels of costs, weaned calf and finishing cattle prices and subsidy were considered during the study. The results show that the profitability of suckler calf production of very extensive system can be better than in less extensive system. Without subsidy at a low calf price profit of extensive system is zero or negative. In case of „good" calf price (2.0-2.5 /kg) and the current EU subsidisation level cow-calf unit with 50 cows can give a reasonable profit that can make a family's living. Cattle fattening and producing finishing bulls in case of approximately 2 /kg production cost, 250 kg weaning weight, 550 kg slaughter weight without subsidization is profitable only in case of low calf price (2.0 /kg or lower) and high slaughter cattle price (1.6 /kg or higher). With special premium fattening can be profitable even in case of high calf price (2.6 /kg) and low slaughter cattle price (1.4 /kg). Due to the decreasing payment by the EU reform the profit, position and competitiveness of beef sector will decrease.

Session 4 Theatre 9

Strategies and management practices of part-time livestock farmers: an example of sheep farming in a French grassland region

C. Fiorelli, J.-Y. Pailleux, B. Dedieu, INRA, 63122 Saint-Genes Champanelle, France*

In the enlarged Europe, part-time farming occurs frequently and its future is worth of interest. In Auvergne, both the regional communities and the sheep production sector are interested in the sustainability of part-time farms. However the diversity of farmers' objectives and management regimes is not well-known, so its specific issues are rarely taken into account by advisors. We present results from a survey (35 part-time farmers) exploring the sheep farming operation, the interactions between sheep farming and off-farm activities in the household and the evolution of the combination of on/off-farm activities. The sample covers a range of flock size (from 46 to 620 ewes) associated with either fixed or unfixed working-hours activities. Sheep farming objectives varied between farms, from husbandry as a leisure pursuit without any economic expectation (26%), to similar systems as specialised farms with equivalent economic return per ewe, but with a smaller number of ewes (23%). Depending on the flexibility of the off-farm activity and the existing skills, farmers implemented practices which enabled exchanges between workers or tasks to be postponed. So sheep farming practices were diverse and led to variable flock productivity (from 0.7 to 1.6 lamb/ewe/year). It is necessary to take into account farmers' own goals, to assess their efficiency and to understand how they will face the CAP reform.

Session 4 Poster 10

The impact of the EU direct payments on beef production systems and feed costs

W. Dunne, Rural Economy Research Centre, Teagasc, Ireland

Grass and grass-silage are joint products from grassland and intermediate products within a grazing livestock production system. They have a multiplicity of "value in use" conditions, and being non-tradable it is difficult to establish either their cost of production, price, or true market value. Land charges are involved both directly and indirectly in estimating the cost of all feed types and these are affected by the EU policy for both beef and cereals. The EU policy switch to the direct payments (DP) income support system, especially extensification, greatly changed the economic circumstances where silage is both produced and used. In cattle farming within the EU, land performs a dual function in that it supplies fodder and provides access to DPs via stocking density compliance criteria. Computer programmes were developed to estimate the net cost of land for silage based on its estimated value for accessing DPs and its opportunity rental charge. This was applied to a cattle production system for a stocking rate range of 1.4 to 2.0 livestock units (LU) per ha. Stocking densities were adjusted by two methods: shedding animals and renting-in land at typical market rates. The aims were to identify the best gross margin for the cattle enterprise and estimate silage costs for the range of stocking densities, and for the different methods of manipulating stocking densities.

Session 4 Poster 11

Beef cattle sector in Hungary as affected by the enlargement

Zs. Wagenhoffer[*1] and D. Mezoszentgyorgyi[2], [1]Georgikon Faculty of Veszprem University, Deak F. u.16., Keszthely, Hungary, [2]National Institute for Agricultural Quality Control, Keleti K. u. 24., Budapest, Hungary*

Beef cattle sector in Hungary has been traditionally export oriented due to the long standing low level of domestic beef consumption and to the relatively high beef production. Cattle stock of Hungary kept on decreasing during the last 4 years (-16%). Within cow population, part of dairy breeds reduced, while stock of dual purpose Simmental and especially of beef breeds increased (75;13;12%). Farm size in dairy sector (300 cow/farm) is much bigger than in beef cattle units, where small herds (5-10 cow/farm) coexist with large holdings (500-1000 cow/farm). The accession seems to shift the beef sector from its nadir. Beef producers have the opportunity to exploit EU and national grants given to the sector. Besides, export has been expanded to EU and 3rd countries as well. Beef prices are significantly higher (10 to 30%) while income of dairy producers has been reduced due to low milk prices (-10%). Only a slight increase of domestic consumption is projected. In spite of the special premium given to cattle finishing, fattening could be profitable only if technological capacity (houses, feed and feeding machinery) is already available and do not require significant further investment. Cow-calf operations with grassland based extensive farming system are recommended to be developed, taking advantage of the country's ecological potential.

Session 4 Poster 12

New grassland management and quality adjustment strategies for dairy farmers in less-favoured areas

V. Thénard, JP. Theau and M. Duru, National Institute for Agricultural Research (INRA), Toulouse, France*

To maintain dairy production in less favoured regions, the value of the milk produced must be increased. The adoption of European quality labels could accomplish this. But the specifications required must change, particularly for the feeding system, to develop high quality cheese production. Any changes in requirements must be compatible with farmers' strategies. It is therefore important to study these practices and their evolution. In this paper, we analyze (i) livestock farming systems and their coherence; (ii) the role of the cheese-making industry in the evolution of farmers' strategies. The evolution of limitation or suppression on the use of maize silage in cows' diets was studied in two mountainous regions in southwest France (Pyrenees and Aubrac).

The methodology was based on interviews and descriptive analysis of farmers' practices (forage and herd management). The coherence of farmer production systems was analyzed using formalized descriptions of livestock farming systems.

It was found that grassland management is tending toward new grazing or hay-making practices that allow increasing grass contribution to the cow diet (100% of the cow feeding with grazing). Also, 1/4 farmers has begun to switch the main milking period to spring so that cows' requirements and grass growth coincide better; despite the fact that for many years the milk industry has been promoting milking in autumn. One case study showed that an increase of milk price was followed by rapid changes (fewer 4 years) in management practices.

Session 4 Poster 13

Cost evaluation of the use of conventional and electronic identification for the national sheep and goat populations in Spain

C. Saa, M.J. Milán, G. Caja and J.J. Ghirardi. Departament de Ciència Animal i dels Aliments, Universitat Autònoma de Barcelona, 08193 Bellaterra, Spain

A cost model was developed to compare different implementation strategies of the Regulation EC 21/2004 for sheep and goat identification and registration in Spain. Strategies were: 1) conventional identification (CID) by two ear tags of all animals (0.15 or 0.3 each for fattening or breeding; 10% annual losses); 2) electronic identification (EID) by one ceramic bolus (2.2 ; 0.3% annual losses) and one ear tag of all animals; and, 3) mixed strategy (MID), consisting of CID for fattening and EID for breeding stock. Total costs per animal identified varied according to complete or simplified implementation option, ranging from 2.48 to 4.64 . The EID was the most expensive strategy (4.47 to 4.64) for all options. Cost of CID and MID strategies ranged from 2.63 to 2.98 and from 2.48 to 3.03 , respectively. Critical values for which MID cost equaled CID depended on the option evaluated, and varied from 7.5 to 11.5% for ear tag losses and from 1.80 to 3.30 for bolus price. In conclusion, the use of a mixed strategy is a currently affordable strategy which fulfills the Regulation requirements for sheep and goats. Reduction of prices and use of EID in performance recording will increase its current benefits for Spanish farmers.

Session 4 Poster 14

Comparison of cattle identification costs using conventional or electronic systems in Spain

C. Saa, M.J. Milán, G. Caja and J.J. Ghirardi. Departament de Ciència Animal i dels Aliments, Universitat Autònoma de Barcelona, 08193 Bellaterra, Spain

A simulation model was developed to evaluate the cost of cattle identification when different strategies are used to fulfill the EU Regulation EC 1760/2000. A decision on the use of electronic identification is foreseen in the regulation but no information is available on the cost impact. Strategies were: 1) CID, conventional identification by two conventional ear tags (1.0 each; 6% annual losses); 2) EID, electronic identification by one electronic ear tag (2.2 each; 3% annual losses) and one conventional ear tag; and, 3) BID, electronic identification by one electronic bolus (2.5 each; 0.3% annual losses) and one conventional ear tag. The model was applied to the Spanish cattle population in 2002 (6,477,900 animals) and for three types farms: dairy (100 cows), suckler (600 cows) and fattening (1,000 calves). Annual costs of identification in Spanish cattle population were 12.20, 15.42 and 15.70 /animal for CID, EID and BID, respectively. Annual values by farm type were (for CID, EID and BID, respectively): dairy (6.91, 6.86 and 6.52 /animal), suckler (6.98, 6.18 and 5.70 /animal), and fattening (1.83, 0.92 and 0.61 /animal). Results indicated that electronic identification of cattle is an affordable option for farmers and cattle industry in Spain. A reduction of device and equipment prices would make the electronic identification the cheapest strategy in the future.

Effect of frequency of manure removal and drying on ammonia, methane, nitrous oxide and carbon dioxide emissions from laying hen houses

C. Piñeiro[1], G. Montalvo[2] and M. Bigeriego[3], [1]PigCHAMP Pro-Europa, Segovia, Spain, [2]Sanidad Animal y Servicios Ganaderos, Madrid, Spain, [3]Spanish Ministry of Agriculture, Fisheries and Food, Spain*

The Best Available Techniques (BAT) for piglets proposed by the "European Reference Document on BAT for intensive rearing of poultry and pigs (BREF)" have been evaluated in two field trials. In these studies, manure removal frequency (once vs twice a week) and drying, was evaluated in a commercial 60,000 laying hens farm. The concentration of NH_3, CH_4, N_2O and CO_2 in the air, the ventilation rate, and the temperature were semi-continuously measured and recorded on site by using infrared photo-acoustic system (Innova 1312 multi-gas monitor, SIR, S.A., Madrid). The emission rate was calculated by multiplying the concentration of gas measured and the ventilation rate of wall exhaust fans. An increase in the frequency of manure removal reduced the emission rate of NH_3 (48%; $P<0.001$), N_2O (29%; $P<0.01$) and CO_2 (13%; $P<0.05$). However, CH_4 emission rate was not affected ($P>0.05$) by manure removal. Manure drying abated the average CH_4 emission rate (31%; $P<0.05$), but the reduction for NH_3 and CO_2 emissions was not significant. Best Available Techniques proposed in the BREF for laying hens were effective to reduce emission rate of gases under Spanish conditions, both frequent manure removal (NH_3, N_2O and CO_2) and manure drying (CH_4).

Best Available Techniques to reduce ammonia, methane and nitrous oxide emissions control from piglets facilities

C. Piñeiro[1], G. Montalvo[2] and M. Bigeriego[3], [1]PigCHAMP Pro-Europa, Segovia, Spain, [2]Tragsega, Madrid, Spain, [3]Spanish Ministry of Agriculture, Fisheries and Food, Spain*

The Best Available Techniques (BAT) for piglets proposed by the "European Reference Document on BAT for intensive rearing of poultry and pigs (BREF)" have been evaluated in field trials. In these studies, lead by the Spanish Ministry of Agriculture, Fisheries and Food, the efficacy of frequent (once a week) slurry removal (FSR), partly-slatted floor (PSF), low protein diet (LPD) and manure channel with sloped side walls (SSW) on emission reduction of noxious gases was compared with the BREF reference system (monthly slurry removal, fully-slated floor, high protein diet, and underlying deep collection). The concentration of NH_3, CH_4 and N_2O were semi-continuously monitored on site with an Innova 1312 multi-gas monitor (SIR, S.A., Madrid). The BAT evaluated reduced ($P<0,05$) the emission rate of NH_3 (24, 51 and 63%), CH_4 (10, 63 and 65%) and N_2O (27, 68 and 51%) for FSR, LPD and SSW method respectively. However, the partly-slatted floor incremented CH_4 and N_2O emissions 7 and 25% respectively ($P<0,05$) whereas NH_3 emission was not affected ($P>0,05$). Frequent slurry removal, low protein diet and manure channel with sloped side walls were effective under Spanish conditions to reduce noxious gas emissions and its practical implication on productive performance of pigs is being currently evaluated. Partly-slated floor is not a recommended technique under Spanish practical conditions.

Session 5 Theatre 1

International trends in genetic evaluation of functional traits in dairy cattle

T. Mark, J.H. Jakobsen, H. Jorjani, W.F. Fikse and J. Philipsson, Interbull Centre, SLU, P.O. Box 7023, S-750 07 Uppsala, Sweden*

Functional traits have become increasingly important in dairy cattle production due to increased costs of production relative to milk prices. This has instigated recording of functional traits to be used for both management and breeding purposes. Although economic values differ between countries, the relative emphasis for functional relative to production traits have increased throughout the western world during past decades. The Scandinavian countries have traditionally recorded and performed genetic evaluations (GE) for a broad range of functional traits, but in recent years many other countries have also implemented GE for udder health, fertility, longevity and calving traits. International GE are available or being investigated for these traits due to the increased trait availability and demand for such services. Across country genetic correlations are high for most trait groups, but further trait harmonizations could lead to better international evaluations. New and more cost effective recording devices give new opportunities of directly or indirectly measuring functional traits, especially in nucleus herd selection schemes, although it is a challenge to maintain emphasis on low heritability traits in such systems. Increasingly sophisticated genetic models (e.g. test-day models) also provide useful information for management purposes. Feed efficiency, metabolic stress and locomotion are important traits that are currently not widely considered for GE, but will likely be so in future.

Session 5 Theatre 2

Functional traits in cattle breeding programs: implementation issues

J. Sölkner[1], A. Willam[1], C. Egger-Danner[2] and H. Schwarzenbacher[1], [1]BOKU - University of Natural Resources and Applied Life Sciences, Department of Sustainable Agricultural Systems, Gregor Mendel Str. 33, A-1180 Vienna, Austria, [2]ZuchtData EDV-Dienstleistungen GmbH, Dresdner Str. 89/19, A-1200 Vienna, Austria*

Functional traits have gained much importance in cattle breeding programs worldwide. This is reflected by many national efforts in breeding value evaluation and inclusion of those traits in total merit indices as well as by the inclusion of udder health, fertility, longevity and calving traits in INTERBULL evaluations.
The change in focus has some practical implications for the implementation of breeding programs. Many of the functional traits have comparatively low heritabilities suggesting that daughter group sizes should be large in order to arrive at reasonable accuracies of sire breeding values. Somewhat surprisingly, calculations show that the differences in genetic gains for different sizes of daughter groups are relatively small. The major step is the inclusion of functional traits in a total merit index, even at conventional daughter group sizes. Another issue is data recording and data quality. Data quality for functional traits tends to be low, either because trait definition is not easy, like for calving ease, or because the data recorded used to be of little consequence, like for number of inseminations. Ways are shown for improvement of data quality through strict monitoring at different levels of resolution. Finally, an outlook is given on how the advent of molecular information is likely to influence selection for functional traits.

Session 5 Theatre 3

US perspective: The importance of functional traits and crossbreeding in dairy cattle

G.W. Rogers, Department of Animal Science, University of Tennessee, Knoxville, Tennessee, USA

Milk and component yields have improved in US dairy populations but performance for many functional traits has declined. Interest in selection for improved functional traits has increased. High rates of transition health problems, poor reproductive performance and high death rates in lactating age cows have become economically important problems. Data from 200 herds in Tennessee indicates that on-farm cow death (8% annually) and culling for diseases other than mastitis or reproduction (8% annually) are the leading causes for cows leaving herds. Other states have similar or higher death rates. Genetic evaluations for daughter pregnancy rate and maternal calving ease are now available in the US and these traits have been added to selection indexes. The feasibility of using disease data collected in progeny test herds has also been established. However, genetic evaluations based on disease data are still unavailable. Use of Lifetime Net Merit for selection should eliminate the undesirable response in reproductive performance that accompanies selection for yield. Many herds are using Holstein, Jersey, Brown Swiss and other European breeds in recently adopted crossbreeding programs. Crossbred cows may represent 10% of the dairy cows in the US by the end of 2005. Crossbreeding will likely continue to increase. Improved selection practices involving more emphasis on functional traits and crossbreeding should enhance the functionality of US dairy cattle.

Session 5 Theatre 4

Threshold *versus* linear model estimates of genetic and environmental correlations between udder health and production traits

E. Negussie, I. Strandén, E.A. Mäntsysaari, MTT Agrifood Research Jokioinen, Finland

Comparison of threshold with Gaussian linear model on estimates of genetic correlations involving categorical data are ambiguous and mostly based on simulation studies. The objective of this study was to compare threshold *versus* linear model estimates of genetic and environmental correlations between clinical mastitis and SCC or production traits based on field data. Data on clinical mastitis (CM), somatic cell count (SCC), milk yield (M), protein (P) and fat (F) from 126,324 first lactation Finnish Ayrshire cows were used. Four bivariate analyses were made using a threshold model and Gibbs sampler. Each analysis fitted one categorical and one continuous trait. Similar bivariate analyses were made using Gaussian linear model. Estimates of heritabilities for CM were 0.068 and 0.021 from threshold and linear models, respectively. Whilst for continuous traits estimates from both models were similar. Genetic correlations between CM and either SCC, M, P, or F from threshold and linear models were 0.64 and 0.62, 0.34 and 0.33, 0.28 and 0.26, 0.31 and 0.29, respectively. Smilarly, estimates for residual correlations were 0.11 and 0.06, -0.12 and -0.02, -0.05 and -0.03, and -0.04 and -0.02, respectively. Comparison between the two models have shown that linear model tends to understate the residual correlation by about 50% whereas a significant underestimation of the genetic correlation has not been found.

Session 5 Theatre 5

Survival analysis for genetic evaluation of mastitis in dairy cattle: a simulation study

E. Carlén[1], U. Emanuelson[2] and E. Strandberg[1], [1]Department of Animal Breeding and Genetics, [2]Department of Clinical Sciences, Swedish University of Agricultural Sciences, P.O. Box 7023, 750 07 Uppsala, Sweden*

A simulation study was performed to compare the traditional linear model methodology for genetic evaluation of mastitis with the method of survival analysis (proportional hazard model). True breeding values for mastitis liability on the underlying scale were simulated for 60,000 first parity cows and 400 sires, and mastitis cases for the cows were created. For the linear model analysis mastitis was defined as a binary trait within 150 d of lactation. For the survival analysis mastitis was defined as a longitudinal trait corresponding to the number of days from calving to either the first case of mastitis (uncensored record) or to the day of next calving or culling because of calving- or fertility related reasons (censored record). The correlation between true breeding values for mastitis liability and estimated breeding values from survival analysis (0.76) was 9 % higher than the corresponding correlation from the linear model (0.70). The higher precision of breeding values estimated by survival analysis can be translated into a higher genetic gain, and therefore survival analysis can be considered a better method to accurately identify bulls carrying genes associated with mastitis resistance.

Session 5 Theatre 6

Genetic analysis of cases of subclinical mastitis

R. Schafberg[1], F. Rosner[1], G. Anacker[2] and H.H. Swalve[1], [1]Institute of Animal Breeding and Animal Husbandry, Martin-Luther-University Halle-Wittenberg, Adam-Kuckhoff-Str. 35, 06108 Halle, Germany, [2]Thüringer Landesanstalt für Landwirtschaft (TLL), Naumburger Str. 98, 07743 Jena, Germany*

Subclinical mastitis most often is undiagnosed resulting in a misleading evaluation of the udder health status of the herd. Infected animals are at high risk for subsequently exhibiting mastitis as well as other dieases. Data for the present study was collected on a single large dairy farm in Thuringia during 1999 to 2001 and comprised a total of 786 cows. On average, eight samples for bacteriological analysis were taken per cow, thus enabling the estimation of environmental effects like parity and season as well as genetic effects.
Heritabilities for binary traits describing the status of healthy vs. infected for infections in general and for infections caused by specific pathogens were estimated using the ASREML package and applying a probit link function for an animal model. Estimates were in the magnitude of around 5 %. Bivariate analyses were used to examine the relationship between the subclinical disease and milk yield as well as somatic cell score (SCS). The joint analysis of subclinical disease and SCS lead to increasing heritabilities. The relevance of an analysis according to pathogen is underlined by the fact that the two most important pathogens in this study, e.g. S. aureus and CNS, are clearly contrasting with respect to the incidence within and across parities.

Session 5 — Theatre 7

Genetic correlations between clinical mastitis, milk fever, ketosis, and retained placenta within and between the first three lactations

B. Heringstad[1,2,], Y.M. Chang[3], D. Gianola[1,3] and G. Klemetsdal[1], [1]Department of Animal and Aquacultural Sciences, Norwegian University of Life Sciences, P.O. Box 5003, N-1432 Ås, Norway, [2]Geno Breeding and A. I. Association, P. O. Box 5003, N-1432 Ås, Norway, [3]Department of Dairy Science, University of Wisconsin, Madison, USA*

Heritability and genetic correlations between clinical mastitis (CM), milk fever (MF), ketosis (KET) and retained placenta (RP), within and between the first three lactations of Norwegian Red were inferred. Records of 372,227 daughters of 2411 sires were analyzed with a 12-variate (4 diseases x 3 lactations) threshold model; CM, MF, KET, and RP were different traits in each lactation. Within lactation absence or presence of each disease was scored as 0 or 1. The model for liability had trait-specific effects of year-season and age of calving (1st lactation) or month-year of calving and calving interval (2nd and 3rd lactation); herd-5-year and sire of cow. Heritability of liability was 0.07-0.08 for CM, 0.09-0.13 for MF, 0.14-0.15 for KET, and 0.08 for RP. Genetic correlations within disease between lactation, ranged from 0.26 to 0.85, and were highest between KET in different lactations. Genetic correlations between diseases were low or moderate (-0.05 to 0.38), within and between lactations. Positive genetic correlations between suggest the existence of some general resistance factor with a genetic component.

Session 5 — Theatre 8

Genetic parameters for predictors of body weight, production traits and somatic cell count in Swiss dairy cows

Y. de Haas[*] and H.N. Kadarmideen, Swiss Federal Institute of Technology Zurich, Institute of Animal Sciences, Statistical Animal Genetics Group, CH-8092 Zurich, Switzerland

Body weight is an important functional trait regulating feed efficiency and energy balance traits in dairy cattle. The main objective of this study was to conduct genetic evaluations and parameter estimations for body measurements and linear conformation traits that are indicators of body weight and investigate their genetic correlations with production traits (milk, fat, protein) and somatic cell count (SCC). Lactation records on 67,839 Holstein heifers, 173,372 Brown Swiss heifers and 106,725 Simmental heifers were used. Heart girth and stature were measured, and linear conformation traits were: body depth, rump width, udder depth and dairy character. Variance components for sires were estimated using ASREML. The statistical model for conformation traits corrected for classifier, season of typing, age, lactation stage, and pregnancy stage at typing and an interaction of herd-year of typing. Heritabilities for stature were high (0.6 - 0.8), and heritabilities for linear conformation traits ranged from 0.3 to 0.5, for all breeds. Genetic correlations with production traits and SCC differed depending on the conformation trait. For all breeds, stature showed moderate positive correlations with all production traits. Udder depth consistently showed negative correlations, indicating that cows with deeper udders have higher milk productions. However, they also have higher SCC, implying that high producing cows are more susceptible to intramammary infections.

Session 5 Theatre 9

Does persistency of lactation influence the disease liability in German Holstein dairy cattle?

B. Harder, J. Bennewitz, D. Hinrichs, E. Kalm, Institute of Animal Breeding and Husbandry, University of Kiel, D-24098 Kiel, Germany,

A lower milk yield peak at the beginning of lactation, i.e. a good persistency, causes less energy imbalance, and thus less metabolic stress for the cow. According to this, the aim of the study was to analyse the relationship between persistency of milk yield (PMY), fat yield (PFY), protein yield (PPY), and persistency of milk energy yield (PEY) and liability to udder diseases (UD), fertility diseases (FD), metabolic diseases (MD), and claw and leg diseases (CLD). Data originated from three commercial milk farms with an overall of 3200 cows. 92.722 medical treatments were recorded from 1998 to 2003. Breeding values for persistency traits were calculated from random regression coefficients obtained from routine sire evaluation. Threshold sire models using Gibbs sampling were fitted for the estimation of heritabilities and subsequently for the estimation of breeding values for the four disease categories. The evaluation of health traits was carried out for the first lactation and for all lactations, respectively. Heritabilities for the health traits ranged between 0.04-0.12 depending on trait and model. Approximate genetic correlations between PMY and CLD were highly significant, while significant correlations could be found between PMY and MD, PMY and FD, and PEY and CLD.

Session 5 Theatre 10

Genetic effects on stillbirth and calving difficulty in Swedish Red and White dairy cattle at first and second calving

L. Steinbock[1], K. Johansson[*2], A. Näsholm[1], B. Berglund[1] and J. Philipsson[1]. Department of Animal Breeding and Genetics, Swedish University of Agricultural Sciences, S-750 07 Uppsala, Sweden, [2]Swedish Dairy Association, P.O. Box 7023, S-750 07 Uppsala, Sweden*

In Swedish Red and White (SRB) dairy cattle, genetic effects on stillbirth and calving difficulty were studied in 804,268 first- and 673,150 second-calvers. Uni- and bivariate linear sire-maternal grandsire models were used to analyse calving data gathered between 1985 and 2000. Mean incidences of stillbirth were low and differed little between first and second parity, 3.6% vs. 2.5%. At first calving, the heritability of stillbirth on the visible scale was 0.7-1.3% for the direct effect and 0.5-0.9% for the maternal effect. For calving difficulty, the heritabilities were around 2.5% and 1.8-2.1% for direct and maternal effects, respectively. Contrasting to previous studies of Holsteins the heritabilities at second calving were similar as at first parity for the two traits in SRB. Genetic correlations between first and second calving results were around 0.8 for direct and maternal effects in stillbirth and around 0.7 for calving difficulty. It was concluded that calving traits at first and second parities could be treated as the same trait, and that bivariate analyses, or a repeatability model, including calving results for both heifers and cows should be preferred in genetic evaluations of SRB bulls as sires and maternal grandsires.

Session 5 Theatre 11

International genetic evaluation of female fertility traits in 11 Holstein populations

H. Jorjani, Interbull Centre, Department of Animal Breeding and Genetics, Swedish University of Agricultural Sciences, Box 7023, S-75007 Uppsala, Sweden*

In spite of the economic importance of fertility traits in the dairy cattle populations, there has not been any international genetic evaluation for them, mainly because national genetic evaluation for these traits doesn't have a long history in a large enough number of countries. Here, preliminary results of a pilot study aiming at an international genetic evaluation are reported. Due to the biological complexity of the fertility traits, and diversity of the measurements used for them, countries participating in this pilot study were asked to provide bull proofs for up to 5 traits. The submitted data were required to signify traits that measure heifer's or cow's ability to become pregnant, traits that measure cow's ability to re-cycle after calving and traits that measure a combination of the above two abilities. Data from 11 populations, each with 1 to 5 traits, for a total of 28 traits and 89715 bulls were available. Despite the low heritability values reported for the female fertility traits, the estimated across country genetic correlations for the similarly designated / defined traits were usually high (ranging from -0.09 to 0.98). The pattern of correlation values followed other traits' correlations, i.e. higher genetic correlations between European and North American populations and slightly lower correlations between New Zealand and Europe.

Session 5 Poster 12

Genotype by environment interaction for udder health traits in Swedish Holstein cows

E. Carlén, K. Jansson and E. Strandberg, Dept. of Animal Breeding & Genetics, Swedish University of Agricultural Sciences (SLU), Uppsala, Sweden*

Genotype by environment interaction for somatic cell count (SCC) and mastitis in the first lactation of Swedish Holstein cows was studied, using a reaction norm model and multiple trait analysis. Data from the Swedish milk-recording scheme containing more than 200 000 cows having their first calving from 1995-2000 were used. Environments were defined as: herd-year averages of number of cows calving, 305-day protein yield, SCC, and mastitis, all measured in first lactation. In the multiple-trait model, the genetic correlation for SCC between low and high quartile SCC environments was 0.8, indicating GxE. For mastitis, GxE was not significant. Significant genetic variation in the slope of reaction norms was however detected for both SCC and mastitis across SCC environments, indicating re-ranking of sires. The environmental scale herd-year average mastitis did not work well, neither for the multiple trait analysis nor for the reaction norm model. This is probably due to that the mastitis frequency is close to zero in low mastitis environments, resulting in a lack of phenotypic variation. There was little or no GxE for the environments herd-year average protein yield or herd size. In practice, the detected GxE for SCC across SCC environments, could affect the selection of bulls in the progeny testing program.

Relationship of somatic cell count and udder conformation traits in Iranian Holstein cattle

M.R. Sanjabi[1], A. Ghoibaighi[2] and R.V.Torshizi[3], [1]Iranian research organization for science and technology (IROST), Tehran, Iran, [2]Department of Animal science, AboRihan higher education Comples,Tehran,Iran, [3]Department of Animal science, Tarbiat Modarres University,Tehran, Iran

Relation ship between somatic cell counts (SCC) and Udder Conformation Traits viz.Udder Height(UH),Udder Weight (UW), Udder Depth(UD)), Fore Udder Attachment(FUA),Udder Cleft(UC), Fore Teat Length (FTL) and Rear Teat Length(RTL) in Holstein breed were obtained from data on 650 cows in 10 herds between years 2002 and 2004 in Tehran's province. The multivariate analyses of DFREML program have been used. The three type traits with negative relationship with SCC were RUH (-.0281), RUW (-.0233) and UC (-.0337).The traits of FTP, RTP and FTL had a positive relationship of .1217, .0572 and .0792 respectively.
The deep udder with weak UC had relatively higher effects on SCC which can be considered in mastities.

Genetic analysis of somatic cell score in Danish dairy cattle using random regression test-day model

R. Elsaid[1], A. Sabry[2], M. Lund[2], P. Madsen[2], [1]Department of Animal Production, Desert Environment Research Institute, Menofia University, Sadat City, Egypt, [2]Department of Genetics and Biotechnology, Danish Institute of Agricultural Science, Research Centre Foulum, PO Box 50, DK-8830 Tjele, Denmark*

Daily SCS records has usually been analyzed as repeated measurements of the same trait, but the correlations between SCS at different stages of lactation are less than 1.0 so a repeatability model is less appropriate. Random regression model (RRM) may be well suited for analysis of test-day SCS, as it induces a covariance structure along a given trajectory. Several sub sets of 1st lactation SCS records from Danish Holstein were analyzed. Each sub set included approximately 15,000 cows with 120,000 test-day records. RRM was fitted using AI-REML. Fixed effects included age at first calving, herd test day and fifth order Legendre polynomials on DIM combined with a Willmink term $e^{-0.09*DIM}$ to describe the mean of SCS along the lactation trajectory. Different orders of orthogonal Legendre polynomials combined with Willmink function were fitted for genetic and permanent environmental effects. Estimates of heritabilities were almost constant over the lactation (~0.15). Genetic correlations between daily SCS were high for adjacent tests (nearly 1) and low between the beginning and the end of lactation. Accounting for the (co)-variance structure along the lactation trajectory should improve the accuracy of genetic evaluations for SCC based on test day records.

Session 5 Poster 15

Comparison of different strategies of quantitative genetic analysis of health traits in dairy cattle

U. Bergfeld[1], R. Fischer[1], C. Kehr[1], M. Klunker[1], [1]Saxon State Institute for Agriculture, Department of Animal Production, Am Park 3, 04886 Köllitsch

A program of recording health traits has been performed in Saxon since 2000. All together 13 different farms are integrated in this program. So far veterinarian diagnosis of 22.301 dairy cattle with 40.975 lactations could be collected. Recording of data was performed on the basis of a complex diagnostic key proposed by Staufenbiel including different traits concerning fertility, udder health, metabolic disorders and the feet and leg complex.
The investigations comprise the comparison of different trait as well as model definitions. As a trait we regarded single diseases or disease complexes. These traits will be handled on the one hand as the total sum of medical treatments or expected days of illness for the whole lactation as well as parts of the lactation and on the other hand as binary trait of the single test day. The expected days of illness were assumed as a certain time span after a diagnosis was detected. For udder health we assumed a time span of 14 days, for metabolic disorders and fertility 21 days and for the feet and leg complex 30 days. An additional half of this time span was added if a new diagnosis was detected between the half of length and the end of this time. Based on these trait definitions, lactation as well as threshold models were applied. Heritabilities and genetic correlation between the health and milk traits were estimated using ASREML. Estimations based on a test day model were very low (from 1 to 5 %). Slightly higher heritabilities were estimated on a lactation model. For nearly all defined traits the results show that the threshold models get the highest estimates (from 8 to 15 %).

Session 5 Poster 16

Longitudinal genetic analyses of functional traits in dairy cows using daily random regression methodology

B. Karacaören[1], F. Jaffrézic[2] and H.N. Kadarmideen[1], [1]Statistical Animal Genetics Group, Institute of Animal Science, Swiss Federal Institute of Technology, ETH Centrum, Zurich CH 8092, Switzerland, [2]INRA Quantitative and Applied Genetics, 78352 Jouy-en-Josas Cedex, France*

This study estimated heritabilities of daily observations on milk yield, milking speed, dry matter intake and live weight as well as genetic correlations among them over time using the random regression methodology. Data were from dairy cattle stationed in an experimental research farm over the past 10 years. All traits were recorded daily using automated machines. Estimated heritabilities were high, as expected from experimental farm datasets, and varied from 0.11 to 0.79 ($\bar{h}^2$=0.53) for milk yield, 0.74 to 0.86 ($\bar{h}^2$=0.82) for milking speed, 0.53 to 0.67 ($\bar{h}^2$=0.55) for live weight. Estimated genetic correlations varied from -0.50 to -0.15 ($\bar{r}_g$=-0.37) between milk yield and milking speed, 0.67 to 0.89 ($\bar{r}_g$=0.81) between milk yield and dry matter intake, -0.82 to 0.56 between milk yield and live weight, 0.24 to 0.85 ($\bar{r}_g$=0.51) between milking speed and dry matter intake, -0.45 to 0.23 between milking speed and live weight and 0.001 to 0.68 ($\bar{r}_g$=0.49) between dry matter intake and live weight. Results could be useful when constructing selection indices for more than one trait, with an emphasis on time-dependent genetic selection.

Session 5 Poster 17

Legendre polynomials for genetic evaluations of persistency of milk production of Holstein cows

J.A. Cobuci[1], C.N. Costa[2], N.M. Teixeira[2], A.F. Freitas[2], [1]CNPq RD Fellow-Embrapa Gado de Leite, Juiz de Fora-MG, Brazil, [2]Embrapa Gado de Leite, Juiz de Fora-MG, Brazil*

First lactation data consisting of 87,045 test day milk yield records of 11,023 Holstein cows calving between 1997 to 2001 from 251 herds were used. Different combinations of Legendre polynomials of order 3 to 5 were used to model fixed and random effects for genetic evaluation of persistency measurements. The random regression models included fixed (herd-year-month of test day and age-season of calving) and random (genetic, permanent environmental and residual) effects. Genetic parameters and rank of animals for persistency changed with the polynomials used for fixed and random effects. Heritability estimates ranged from 0.11 to 0.33 for test day milk yield, 0.33 to 0.36 for 305-day milk yield and 0.00 to 0.36 for persistency. Genetic correlations between persistency and 305-day milk yield ranged from -0.49 to 0.57. Comparison of models based Akaike's Information Criterion and Bayesian Information Criterion confirmed better fit of models with higher order polynomials. Two measurements were appropriate to describe persistency of milk yield for Holstein breed in Brazil.

Session 5 Poster 18

Relationships between milkability traits in Brown Swiss

J. Dodenhoff[1], R. Emmerling[1] and D. Sprengel[2], [1]Bavarian State Research Center for Agriculture, Institute of Animal Breeding, 85586 Poing, Germany, [2]Landeskuratorium der Erzeugerringe für tierische Veredelung in Bayern, 80336 München, Germany*

Milkability is an important functional trait in dairy cattle breeding. It is well documented that milking speed is unfavourably correlated with somatic cell count. Earlier studies in Simmental indicated that milk flow rate also may be unfavourably related with other milkability traits, e.g., bimodality and machine stripping. Aim of this study was to analyse these relationships more in detail. For the analyses 800,000 test-day records from 80,000 Brown Swiss first calf cows from Bavaria were available. Several data sets with approximately 140,000 records each were sampled from the total data set. Estimates of (co)variances for maximum flow rate (MFR) and machine stripping rate (MSR) from six 30-day periods were obtained by REML using an average information method. For MFR, estimates of heritability ranged from .51 to .56 and estimates of genetic correlation between time periods ranged from .91 to .99. For MSR, estimates of heritability ranged from .05 to .09 and estimates of genetic correlation between time periods ranged from .85 to .97. Estimates of the genetic correlation between MFR and MSR were moderate to large in magnitude (.30 to .75). Results suggest that while increased milk flow rates decrease milking time they may require a more thorough milking routine and/or a good milking technology.

Factors affecting days open in Polish Holsteins

W. Jagusiak and A. Zarnecki, Department of Genetics and Animal Breeding, Agricultural University, Al. Mickiewicza 24/28, 30-059 Krakow, Poland*

One useful measure of fertility is days open (DO) consisting of intervals between calving and first insemination (ICI) and between first insemination and last insemination (SP). This study evaluated effects of some environmental and genetic variables on DO, ICI and SP. Data were 185839 Polish Holstein cows that calved for the first time in 1999 - 2003. Herds were divided into quartiles based on size or on average fat production or on average fat breeding value. The GLM procedure of SAS was used to evaluate the effects of year-season subclasses, percentage of HF genes, age at calving and size of herds on each of the dependent variables (DO, ICI and SP). In two other models, herd size was replaced by herd mean fat yield or by herd mean breeding value for fat. All effects were highly significant. Least squares means showed that ICI decreased with increasing herd size. SP increased, while DO displayed small changes. ICI was longest in herds classified in the first quartile according to fat yield. SP showed the opposite tendency, with the shortest intervals in low-producing herds, while the largest least squares mean for DO was in high-producing herds from the fourth quartile. ICI showed no relation to average herd breeding values for fat production. Both SP and DO increased with increasing herd mean breeding value for fat yield.

Phenotypic and genetic variability of reproduction traits of black and white cows

R.. Djedovic[1], V. Bogdanovic[1] and P. Perisic[1]. [1]Institute of Animal Sciences, Faculty of Agriculture, University of Belgrade, Nemanjina 6, 11080 Zemun-Belgrade, Serbia*

Phenotypic and genetic variability and the interrelationship with fertility traits monitored during progeny tests of dairy bulls, were analyzed based on official data from the PK "Beograd" Center for Reproduction and Embryotransfer. Phenotypic variability was analyzed using data on 51 tested Holstein Friesian bulls, while genetic variability and interrelationship with fertility traits were evaluated based on data for 2,664 calves, which were the progeny of 24 sires, as a consequence of the existance of complete data on calving type on only two out of 14 farms. Cows and heifers randomly inseminated with semen from young Holstein Friesian bulls in the progeny test calved normally, without any problems, in 81% of cases. Calves were extracted using additional force in 17.6% of calvings, while only 1.4% of calvings were difficult. Calving type showed highly significant variation ($P<0.001$) under the influence of bulls-sires. At calving, of the total number, 4.0% of calves were stillborn. Average weight at calving for calves of the Black Pied and the Holstein Friesian breed was 37.5 kg, while the average duration of gestation for dams of investigated calves was 277.2 days. Heritability coefficients for investigated fertility traits were relatively low. Evaluated heritability and errors for calving type, number of stillborn calves, birth weight, and gestation duration were: 0.190 ± 0.062, 0.018 ± 0.006, 0.149 ± 0.051, and 0.288 ± 0.086, respectively.

Session 5 Poster 21

Gestation length for genetic evaluation of calving traits in dairy cattle

E. Stamer[1], W. Junge[2], W. Brade[3], E. Kalm[2]. [1]TiDa Tier und Daten GmbH, Bosseer Str. 4c, D-24259 Westensee/Brux, Germany, [2]Institut für Tierzucht und Tierhaltung der Christian-Albrechts-Universität, D-24098 Kiel, Germany, [3]Landwirtschaftskammer Hannover, Johannssenstraße 10, D-30159 Hannover, Germany

Data included 14,095 calvings of German Holstein heifers and cows from three commercial farms. These calvings originated only from single births after normal gestation periods (265 to 295 days). For 12,968 of these calvings birth weights were also available. Calving ease was divided into normal or extreme difficult. Stillbirth levels were alive or dead at birth. Mean frequencies of dystocia and stillbirth were 9.5 % and 7.8 %. Gestation length showed a mean of 279.4 days with a standard deviation of 4.8 days. Variance components were estimated by REML using a multivariate linear model. The model included the fixed effects herd-year-season of calving and the interaction between parity of dam and sex of calf. Random effects were the direct effect of calf and the uncorrelated effect of maternal grandfather. Direct heritabilities are 0.06 for calving ease, 0.02 for stillbirth and 0.67 for gestation length. Direct genetic correlations between gestation length and calving ease and between gestation length and stillbirth are 0.33 and 0.12. The direct genetic correlation between calving ease and stillbirth is 0.26.
Finally, sire evaluations for direct effects of dystocia and stillbirth were carried out. Inclusion of gestation length as an additional information trait leads to both increasing and decreasing mean reliabilities of breeding values.

Session 5 Poster 22

Analysis of genetic and environmental effects on claw disorders diagnosed at hoof trimming

H.H. Swalve[1], R. Pijl[2], M. Bethge[1], F. Rosner[1], M. Wensch-Dorendorf[1], [1]Institute of Animal Genetics and Animal Husbandry, University of Halle, Adam-Kuckhoff-Str. 35, D-06108 Halle(Saale),Germany, [2]Fischershäuser 1, D-26441 Jever, Germany*

Claw disorders can be diagnosed at the time of hoof trimming. In this study, findings were collected at hoof trimming using a personal digital assistant (PDA) with an interface to a data base on a PC and an interface to herd data stemming from the central milk recording computer. A total of around 13,000 records of cows on around 100 farms were evaluated. Data comprised the pathological findings, herd environment information, milk yields and pedigree information.
The most prominent disease of the hoof was laminitis, found in 33% of all cows. Strong environmental effects were age, stage of lactation (or milk yield at trimming) and the herd effect. Only little influence could be attributed to feeding of a total mixed ration, herd size or housing system with the exception of deep-straw free-stalls.
Genetic analysis was carried out using ASREML and fitting a probit link-function for a threshold model. Estimated heritabilities of various disorders ranged from 0.06 to 0.12. Genetic correlations of disorders with first lactation milk yield for most disorders were not significantly different from zero and overall of only moderate size.

Session 5 Poster 23

Phenotypic relationships between type traits and survival in New Zealand dairy cattle

D.P. Berry[1], B.L. Harris[2], A.M. Winkelman[2], and W. Montgomerie[3], [1]Teagasc, Moorepark Production Research Centre, Fermoy, Co. Cork, Ireland, [2]Livestock Improvement, Private Bag 3016, Hamilton, New Zealand, [3]Animal Evaluation Unit, Private Bag 3016, Hamilton, New Zealand*

Data were extracted from the New Zealand national database on all primiparous cows in sire proving herds classified for 16 type traits throughout the years 1987 and 2003. Each type trait was pre-adjusted for age at calving and stage of lactation, normalised within contemporary group and transformed to a qualitative variable. Individual cow type records were merged with longevity records. In total 259,280 cows were included in the analysis. The survival analysis was undertaken using a proportional hazards Cox model stratified by breed. The hazard function was described by the baseline hazard function, herd-year contemporary group, pedigree registration status, age at first calving, heterosis, proportion of breed genes, period of last calving, type score. Production values and milk production explanatory variables were included in the analysis of functional longevity. All type traits had a significant ($P<0.001$) influence on cow longevity. Farmer opinion of the cow was the most influential of the type traits on true and functional longevity. The importance of farmer opinion, relative to other type traits, diminished following the adjustment of longevity for the relative milk production of the cow. Of the individual type traits describing the physical characteristics of the cow, the udder related traits had the largest effect on functional longevity.

Session 5 Poster 24

Relationships between bodyweight, milk yield, and longevity of Estonian test cows

O. Saveli, M. Voore. Institute of Veterinary Medicine and Animal Sciences of the Estonian Agricultural University, Tartu,51014, Kreutzwaldi 1, Estonia

Since 2000 an experiment has been conducted on the Põlula Experiment Farm using the cows of the Estonian Red (ER), Estonian Holstein (EH), Estonian Native (EN), and Red-and-White Holstein (RH) breeds to establish the maximum milk productivity. The animals were kept tethered in a cowshed and were fed energy and protein rich total mixed ration (TMR) *ad libitum,* in summer for half a day on pastures. Pipeline-milking three times per day. Effects of factors (group, reason of culling) on trait (longevity, production per life and per day of productive life) were analyzed with general linear model [PROC GLM of SAS (Statistical Analysis System)].

Milk yield varied between breeds - EH>RH>ER>EN. The optimum bodyweight of Estonian Holsteins was 610...650 kg, that of Estonian Red cows 580...640 kg, respectively, and Red-and-White Holstein cows had linear relationship.

Longevity of test cows depended on culling reasons, a breed had no significant influence. The shortest productive life had the cows culled due to udder, and the longest those culled for feet disorders, therefore the last group had higher milk, -fat and -protein production of life ($P<0.05$). Average milk production (exept fat production) per day of the productive life did not depend on culling reasons, while productivity of cows of Estonian Native breed was lower, compared with the other test groups ($P<0.05...0.001$)

Session 5 Poster 25

Estimation of economic values of longevity and other functional traits in Finnish dairy cattle

M. Toivakka, J.I. Nousiainen and E.A. Mäntysaari, MTT Agrifood Research Finland, Animal Production, 31600 Jokioinen, Finland*

Longevity and functional traits are very important in breeding and management of dairy cattle due to economic and ethical viewpoints. Economic values of production and functional traits were calculated for Finnish Ayrshire and Holstein breeds to evaluate their weights in total merit index. Deterministic and dynamic model at farm level was used in simulating average Finnish milk production system. The model was based on constant herd size and random sampling of replacement cows from a very large base population derived from the Finnish milk recording scheme with real life history data of cows. The model for milk production of cows was based on average solutions from genetic evaluation. The economic return in simulated herds was a function of biological traits and economic parameters.

In Finland, longevity is included in total merit index via its components of health and fertility traits and since 2003 also breeding values for longevity itself have been estimated. For Ayrshire (Holstein) cattle, the relative economic importance of longevity was 16 % (23.2 %). When the economic losses due to culling were accounted only by longevity, the relative weights of production, health and fertility trait groups were 74 % (65.3 %), 6.6 % (7.1 %) and 3.4 % (4.4 %), respectively.

Session 5 Poster 26

Joint genetic evaluation for functional longevity for Pinzgau cattle

C. Egger-Danner[1], O. Kadlecik[2], C. Fuerst[1] and R. Kasarda[2], [1]ZuchtData EDV-Dienstleistungen GmbH, Dresdner Straße 89/19, A-1200 Vienna, Austria, [2]Slovak Agricultural University, Department of Genetics and Biology, Hlohovska 2, SK-94976 Nitra, Slovakia

For the future of the endangered purebred Pinzgau cattle population a close cooperation between the countries Austria, Slovakia and Romania is very important. A common breeding goal and cooperation within the breeding programme could achieve higher genetic response. A joint breeding value estimation is a precondition for a joint selection programme. Because of the comprehensive breeding goal of Pinzgau cattle and the high economic weight of longevity in the total merit index, breeding values for functional length of productive life (fLPL) are important. In Austria breeding values for Pinzgau cattle for fLPL have been routinely estimated for bulls and cows using survival analysis since 1995. Results from a test run show that a joint genetic evaluation for longevity including the Slovakian data is feasible. To increase the reliability of the breeding values and to utilize all the genetic ties between the countries, further emphasis has to be put on the set up of the common pedigree based on the original IDs. A joint breeding value estimation for Pinzgau cattle is an important contribution to increase genetic gain and to secure the future of the Pinzgau population.

Session 5 Poster 27

Study on the longevity of beef cows of different breeds

I. Dákay, D. Márton, Z. Lengyel, M. Török, Zs. Vincze, F. Szabó, University of Veszprém Georgikon Faculty of Agriculture 8360 Keszthely Deák F. u. 16. Hungary*

During the study database of 2115 cows belonging to five breeds (Hungarian Grey, Hereford, Aberdeen Angus, Limousin, Charolais) and two crossbred genotypes (Simmental x Hereford F_1, Simmental x Limousin F_1) born between 1977-1992 was evaluated. Age at first calving (AFC), age at culling (ACU), moreover longevity (LONG) were studied. Longevity was defined from first calving to culling. MS Excel and SPSS for Windows 11.0 were used for statistical analyses. The mean values of AFC, ACU and LONG obtained were 2.71, 9.47 and 6.77, respectively. Breed/genotype and birth year had significant influence (P<0,01) on each evaluated trait, whereas birth month statistically affected only the AFC. Ages at first calving (AFC) of the different breeds and genotypes were: 3.51, 2.08, 2.76, 2.82, 3.02, 2.03, 2.62 years, respectively. Hereford crossbred and purebred cows were the youngest, whereas Hungarian Grey cows were the oldest at first calving. Ages of culling (ACU) of the evaluated breeds and genotypes were as follows: 12.42, 11.09, 11.03, 10.61, 10.89, 12.73, 8.15 years, respectively. The longest life span was reached by Hereford crossbred and Hungarian Grey and the shortest by Limousin crossbred cows. The longevity values (LONG) of the mentioned breeds and genotypes were: 8.59, 9.08, 8.29, 7.81, 7.91, 10.79, 5.55 years, respectively.

Session 5 Poster 28

Multiple breed evaluation for cow survival and fertility in Irish beef cattle

M.H. Pool[1], V.E. Olori[2], A.R. Cromie[2], R.F. Veerkamp[1], [1]ASG WUR, Division Animal Resources Development, P.O. Box 65, 8200 AB Lelystad, The Netherlands [2]Irish Cattle Breeding Federation, Shinagh House, Bandon, Co. Cork, Republic of Ireland*

Cow survival and fertility are together with calving difficulty, gestation length, calf mortality, weaning weight and cull cow carcass weight proposed in a maternal beef cattle index. Beef breeding values for cow survival and fertility are predicted with one multiple trait sire model for reappearance, lifespan and calving interval including beef, dairy and crossbred information. Since information on cow survival or reappearance is scarcely available in historical beef data, reappearance curves are proposed as survival indicators for animals that did not reappear within 300-450 days. Different breed effects were accounted for with a regression on the sire and dam breed percentage.

Average lifespan for beef was with almost 3 years and survival with 0.73 to 0.79 in lact1 to 3 slightly lower than for dairy with the difficulty of having more missing and censored records. Reappearance fractions as survival indicator were included in 8.3% to 2.4% of the censored in lactation 1 to 4. Heritability, first lactations only, was 2% for cow survival and calving interval and 7% for lifespan; correlations were 0.37 and -0.47 for reappearance and lifespan with calving interval and 0.91 for reappearance with lifespan.

Using reappearance fractions and multiple breed data, including dairy and crossbreds, allowed multiple breed evaluation for cow survival and fertility in Irish beef cattle.

Session 6 — Theatre 1

NorFor: the new Nordic feed evaluation system!

A.H. Gustafsson[1], M. Mehlqvist[1], H. Volden[2], M. Larsen[3] and Gunnar Gudmundsson[4], [1]Swedish Dairy Association, Uppsala, Sweden, [2]Norwegian University of Life Sciences (UMB), Norway, [3]Danish Cattle Federation, Århus N, Denmark, [4]Farmers Association of Iceland, Reykjavík, Iceland*

A new feed evaluation system, NorFor, will be introduced in the Nordic countries. NorFor is a cooperative project among the farmers' advisory organizations in Denmark, Iceland, Norway and Sweden and has a successful collaboration with Nordic researchers. The project was initiated in 2002, with the goal to create a common Nordic feed evaluation system for cattle. After evaluating several Nordic and international feed evaluation systems NorFor decided that NorFor Plan, based on a Norwegian model, should be implemented as the new feed evaluation system used for ration formulation and optimisation. NorFor Plan consists of three sub-models; 1) a semi-mechanistic model used for calculation of nutrient supply and nutritional values 2) calculation of feed intake, and 3) calculation of the structural value of the diet based on chewing time. Energy is calculated as net energy for lactation and both ECM and protein yields are predicted. Interactions between animal characteristics, feeding level and feed composition is taken into account when calculating nutrient supply. This implicates that energy and protein values for individual feeds are not constant and additive. The new system makes it possible to better predict the "true" feeding value of a ration, which would lead to a more efficient feed utilisation with economic as well environmental advantages.

Session 6 — Theatre 2

Karoline: the Nordic dairy cow model

A. Danfær[1], P. Huhtanen[2], P. Udén[3], J. Sveinbjörnsson[4] and H. Volden[5], [1]Danish Institute of Agricultural Sciences, Denmark, [2]MTT Agrifood Research, Finland, [3]SLU, Kungsängen Research Centre, Sweden, [4]Agricultural Research Institute, Iceland, [5]Agricultural University of Norway*

We have developed Karoline intended for use as a diet evaluation tool by the advisory services in the Nordic countries. Karoline is a dynamic, deterministic and mechanistic simulation model of lactating dairy cows. It consists of two sub-models: a digestion and a metabolism model. The first sub-model describes digestion in the forestomachs, the small intestine and the hindgut, while the metabolism sub-model comprises portal drained viscera, liver, extracellular fluid, mammary gland as well as muscle, connective and adipose tissues. The digestion part of Karoline is tested against 61 treatment means from studies with cannulated dairy cows and growing cattle, and the full Karoline model is tested against 142 treatment means from production studies with lactating cows carried out in the Nordic countries. It can be concluded from these tests that Karoline is quite accurate in predicting nutrient digestibility and milk production, but the model is less accurate in predicting milk composition. It has been decided by farmer organizations and advisory services in Denmark, Iceland, Norway and Sweden that Karoline shall be included as a diet evaluator in a new feed evaluation system, NorFor, which is to be implemented in these countries during 2005-2006. Karoline will also be used as a diet evaluation tool by the advisory service in Finland.

Session 6 — Theatre 3

Moving from a range of systems, currently assessing practical workloads in equines to a common system

A.D. Ellis, Nottingham Trent University, Southwell, NG25 0LZ United Kingdom

Horses are primarily used for leisure or competitive riding activities. Energy evaluation systems apply a varying range of methods for the practitioner to estimate energy requirements above maintenance for working horses. The energy system currently used in the USA and UK recommends an increase of 1.25, 1.5 or 2 times maintenance requirements for light (pleasure and equitation), medium (ranch work, barrel racing, jumping etc.) and intense work (race training, polo) respectively. The German and Dutch systems are based originally on a calculation from in vivo net energy research which includes weight of rider, tack and horse, duration of exercise and speed. The German system then gives average energy expenditure tables for light, medium, heavy and very heavy work. Both the Dutch and French systems offer similar tables but also give practitioners the option to choose from duration and intensity tables to calculate energy expenditure more accurately. Furthermore, the actual formula which incorporates all above mentioned factors is published in the Dutch system for educated horse owners. Current systems do not take the variety or repetition of exercise periods over an average working week into account. This paper introduces a possible method incorporating the best of current systems with a novel approach for practical application. Energy expenditure for work is derived from the following variables per type of exercise: repetition, duration, intensity and additional effort.

Session 6 — Theatre 4

A model to predict nitrogen excretion from dairy cattle using the PDI protein system

V. Olsson[1,3], J.J. Murphy[1], F.P. O'Mara[2], M.A. O'Donovan[1] and F.J. Mulligan[3], [1]Dairy Production Department, Teagasc, Moorepark Research Centre, Fermoy, Co. Cork, Ireland, [2]Department of Animal Science and Production, Faculty of Agri-Food and the Environment and [3]Department of Animal Husbandry and Production, Faculty of Veterinary Medicine, University College Dublin, D4, Ireland*

A mechanistic model based on the French PDI (protein truly digestible in the small intestine) protein system for ruminants was developed for prediction of nitrogen (N) excretion and it's partitioning between urine and faeces. The model was validated using two independent data sets from eight separate grazing experiments in Teagasc Moorepark and Lyons Research Farm, Ireland. The data set consisted of four all herbage and six supplemented grazing treatments involving 134 dairy cows. Average treatment data range was for milk yield 15.6-29.3 kg/d, dry matter (DM) intake 16.3-19.8 kg/d, diet DM digestibility 0.72-0.87, urine N 213-452 g/d and faecal N 65-184 g/d. Urine N excretion was calculated from measured N intake less N output in milk and faeces. The model successfully predicted urinary N excretion (y), R^2= 0.807, for the combined data set, where y = -3.216 (±9.0974) + 1.159 (±0.0339, p<0.001) x predicted urine N. Faecal N (z) was poorly predicted, R^2= 0.261, where z = -10.977 (±13.8838) + 1.0647 (±0.1071, p<0.001) x predicted faecal N. The model prediction of N excretion in urine was more accurate than for faeces.

Session 6 Poster 5

The effects of urea treatment on in vitro gas production of pomegranate peel

R. Feizi, A. Ghodratnama, M. Zahedifar , M. Danesh Mesgaran and M. Raisianzadeh, Agricultural and Natural Resources Research Center of Khorasan, Mashhad P.O. Box 91735-1148, Iran

The objective of this experiment was to evaluate the effect of different levels of urea on gas production (GP) with and without added polyvinylpolypyrrolidone (PVP) to ensiled pomegranate peel (EPP). In this experiment 4 levels of urea (0, 2.5, 5 and 7.5% of dry matter) were added to pomegranate peel and ensiled for two periods of 30 and 60 days. GP from the samples were measured during 96 h incubation with or without addition of PVP. GP data were analysed in a randomized complete block design using of SAS. The data from gas volume recording were fitted to the exponential equation of the form p=a+b (1-e-ct). The results indicate that addition of urea and then storage decreased (P<0.05) total extractable tannins (TET) content (206, 177, 169 and 170 mg/g respectively). GP after 24 and 48 h was higher for EPP treated with 0% urea (40.37 and 45.20 ml respectively) and lower for EPP treated with 7.5% urea (30.21 and 34.91 ml respectively) (P<0.05). There was a negative correlation between the CP content of EPP and GP at 24 and 48 h incubation. Also non-fiber carbohydrate (NFC) level was positively correlated with GP potential. Tannins have negative effect on in vitro rumen fermentation and PVP could show this effect.

Session 6 Poster 6

Performance prediction using ARC, ME System for different breeds of beef cattle fed with two different feeding periods grown under feedlot conditions

Y. Bozkurt and S. Ozkaya, Suleyman Demirel University, Faculty of Agriculture, Department of Animal Science, Isparta, 32260 Turkey

In this study, data from Holstein, Brown Swiss, Simmental cattle as European type (ET) and Boz and Gak as Indigenous type (IT) grown under feedlot conditions were used to evaluate the ARC, ME system for prediction of beef performance during two different feeding periods.
During growing period, the discrepancies between observed and predicted values of liveweight gains (OLWG and PLWG) were high and significant (P >0.05) for all cattle. OLWGs were underpredicted for IT cattle and those observed values less than 0.9 kg/day. The Mean-Square Prediction Error (MSPE) of the predictions by the model was 0.03 and 0.007 kg/day for ET and IT cattle respectively. During finishing period, the discrepancies between OLWGs and PLWGs were low and were not significant (P <0.05) for all cattle except Gak and there were substantial agreement between observed and predicted values for ET. The model tended to underpredict OLWGs of ET while OLWGs were overpredicted for IT and those OLWGs less than 0.6 kg/day. The MSPE was 0.006 and 0.02 kg/day for ET and IT respectively.
The accuracy of measurement predictions were within the acceptable range only for OLWGs obtained for ET at finishing period. The results indicated that the model does not provide very close agreement with reality for prediction of LWG for IT cattle.

Prediction of nutritional content of silages by analyses of green chop

J. Bertilsson, Swedish University of Agricultural Sciences (SLU), Uppsala, Sweden

Knowing the nutritional content of silage in advance give good possibilities to plan the winter-feeding. Taking samples from the green chop when loading the silos give good prerequisites to get representative samples. But will the results correlate with the corresponding silage? Calculations have been made for silages from 54 silos during a 12-year period at Kungsängen Research Centre, SLU, Uppsala in order to elucidate this. Representative samples were taken from the green chop and from the silage during the emptying period. Calculations are based on average per silo. Analyses were made for dry matter, ash, crude protein (CP) and in vitro digestibility (VOS). Metabolisable energy was calculated from VOS. In general there were good correlations between the analyses. For CP the difference between green chop and silage was 0.15 percent units and this difference was the same independent of CP content. For the other analyses the values for silage was in general higher than for the green chop, especially at low levels. The conclusion is that nutritional analyses of the green chop give a good overview of the silages to feed. It can be used as a tool to plan the coming feeding season. Complementary analyses of silage quality and dry matter content is however essential.

Session 6 Poster 8

The effect of grain type and processing on chewing activity in horses

C. Brøkner[1], P. Nørgaard[1], L. Eriksen[2] and T. M. Søland[2], [1]Department of Basic Animal and Veterinary Sciences, [2]Department of Large Animal Sciences, KVL, Grønnegårdsvej 2, DK-1870 Frederiksberg C, Denmark*

The aims of the present experiment were to study the effect of grain type and processing on chewing activity in horses. Three adult trotters (Exp.I) and 3 adult Icelandic horses (Exp.II) were fed 3 daily meals during 3 consecutive days in two 3 x 3 completely randomized block design experiments. Meals of (Exp.I: 1.0 kg) and (Exp.II: 0.5 kg) oats, barley and wheat were fed whole, rolled and ground at 10 am, 12 pm, 2 pm. Jaw movements (JM) were identified from pressure oscillations in a tube around the mouth. The efficient chewing time (EPTIME, min/kg DM), corrected for pauses, the basic chewing rate (PBCR, JM/s) and the standard deviation of time interval between JM (SPDDT), which indicate chewing regularity, were estimated. The mean EPTIME was 24 and 15 min for the Icelandic and trotter horses, respectively. The EPTIME for whole grain was 20 min and shorter than for ground grain ($P = 0.02$). EPTIME for oats was 25 min and longer than for wheat ($P=0.05$). The SPDDT value for whole grain was low (0.06) and therefore more regularly chewed than ground grain ($P<0.01$). PBCR value was unaffected ($P>0.50$). In conclusion, grain processing increased chewing time and affected chewing regularity in horses perhaps due to decreased palatability.

Investigations on protein requirements of fattening bulls of the German Holstein breed

U.Meyer, P. Lebzien, G. Flachowsky and H. Boehme, Federal Agricultural Research Centre, Institute of Animal Nutrition, Bundesallee 50, D-38 116 Braunschweig, Germany

The recommendations of protein supply for bulls of the German Holstein breed, as laid down by the GfE (1995), were tested in a feeding trial with 62 animals (194 to 550 kg live weight, LW) allotted to four groups with a different crude protein supply. The rations were based on maize silage, which was fed ad libitum and two concentrates on wheat, sugar beet pulp, soybean oil as energy source and soybean meal or alternatively peas as protein source. The concentrates and the protein feedstuffs were combined in the way that the XP/ME ratio met, exceeded or remained below the GfE-recommendations. The concentrates were fed according to requirements, i.e. increasing amounts in the course of the trial. Feed intake and LW gain were daily registered, data of slaughter performance at the end of the experiment.
Dry matter intake showed differences between groups only up to 300 kg LW. The statistical analysis of XP-intake and LW-gain, for which the broken-line-model was applied, resulted in crude protein requirements of 10.7, 11.7 and 11.5 g/MJ ME for the fattening phases 200 - 300 - 400 - 500 kg LW. The result obtained for the phase 300 - 400 kg is in agreement with the GfE - recommendations, whereas the value for the preceding phase was lower by 1.8 gXP/MJ ME. In the final phase the value exceeded the GfE - recommendations by 0.7 gXP/MJ ME.

Evaluation of buffering capacity for ruminant ration formulation and its effects on rumen fluid

A. Moharrery, Animal Science Department, Agricultural College, Shahrekord University, Shahrekord, Iran

The usual ruminant's feed samples were divided to seven groups as: forage, silage, straw, protein concentrate, grain, by-product material, and feed supplements. Buffering capacity (BC) and buffer value index (BVI) were determined on these samples.
Six totals mixed rations for dairy cows formulated using the above feedstuffs with known BC or BVI. The study on rumen fluid pH was conducted using 5 simple rations on 5 adult sheep. The rumen fluid sample was collected after 21 days adaptation period. Correlation was calculated between rumen fluid pH and fiber or ash content of the diets.
Results showed that higher BC for forage comparing to grain material but the pattern of each feed in reaction to acid or base addition was different. The analyzed BC appeared to be lower than calculated values for each diet ($P< 0.05$). The analyzed BVI also was two times higher than the calculated BVI for each diet ($P< 0.05$). In spit of this, the correlation between calculated and analyzed values were 0.933 ($P=0.0066$) and -0.107 ($P=0.840$) for BC and BVI, respectively.
Rumen fluid pH decreased as starch increased in the ration, but rumen degradable protein in the ration did not revealed any significant correlation with rumen fluid pH ($p> 0.05$).
In conclusion, BC and BVI of rations must be considered before feeding of any ration to dairy cows.

Effect on milk production when maize silage from two hybrids harvested at two stages of maturity is partly substituted with grass silage

L. Hymøller and M.R.Weisbjerg, Danish Institute of Agricultural Sciences, Research Centre Foulum, PO Box 50, DK-8830 Tjele, Denmark*

Sixty-four Danish Holstein cows (32 in first lactation) were allocated to eight different treatments, arranged in a 2x2x2 factorial design, after a two-week covariate period. Mean days in milk were 58.6 ± 24.3 at the start of the covariate period. Cows were fed total mixed rations with 60% forage on DM basis. Treatments were: two maize hybrids Bangay and Pretti (high and low NDF digestibility, respectively), two maturities (28 and 35 % dry matter (DM)) and +/- substitution (grass silage 0 and 1/3 of forage DM). Daily yield of energy corrected milk (ECM) was on average 2.3 kg higher in cows fed Bangay compared to Pretti ($P<0.0001$). Substitution with grass silage increased daily ECM on average from 32.0 kg to 34.6 kg ($P<0.0001$). However, grass silage with Bangay only increased ECM yield from 33.8 to 35.1 kg whereas grass silage with Pretti increased yield from 30.2 to 34.1 kg. Maturity did not alter ECM yield significantly. Influence of treatments on milk yield reflected mainly differences in DM intake and forage digestibility. Treatment differences increased linearly during the 12-week period wherefore treatment differences were much larger by the end of the experiment than the above mentioned LS-means indicate.

Session 6 Poster 12

Relative energy requirements of beef bull breeds

M.J. Drennan[1], M. McGee[1] and A. Grogan[2], [1]Teagasc, Grange Research Centre, Dunsany, Co. Meath, Ireland, [2]Irish Cattle Breeding Federation, Tully, Co. Kildare, Ireland

The objective was to estimate the energy requirements of pure bred bulls that exited the Irish bull performance test station between January 1998 and June 2004. A total of 28 batches, comprising of 112 Aberdeen Angus (A), 87 Hereford (H), 255 Charolais (C), 432 Limousin (L) and 251 Simmental (S) were used. Feed energy requirements (UFV/day) were calculated for each breed by regression of daily energy intake on mean liveweight and daily weight gain during the final period (varied from 98 to 168 days) of each test. Estimated energy requirements (relative) of A, H, C, L and S bulls at 500 kg liveweight and gaining 1.5 kg daily were 10.2 (100), 10.1 (99), 9.3 (92), 9.1 (90) and 9.8 (97) UFV per day, respectively. Relative values reported by the Agricultural Research Council were 100, 88 and 78 for bulls of small, medium and large mature size, respectively. Relative values reported by INRA, France for Holstein/Friesian and large beef breeds (C, L) were 100 and 83, respectively. The present results show similar values for the early-maturing small breeds (H, A) which were approximately 0.10 greater than the C and L (late-maturing breeds) which were also similar. The higher value for the late-maturing S reflects the influence of higher milk potential. The data indicate that tables of energy requirements should specify breed.

Session 6 Poster 13

In vitro gas production profile of non-washable, insoluble washable and soluble washable fractions in some concentrate ingredients

A. Azarfar, S. Tamminga and B. A. Williams, Wageningen University, Animal Nutrition Group, P.O. Box 338, 6700 AA Wageningen, The Netherlands*

It is assumed that rumen degradation of material washed out of nylon bags is instantaneous and complete. Using a standardised laboratory procedure that mimics the results of washing in the *in situ* technique (Azarfar et al., 2004) along with an *in vitro* gas production technique seems to be a promising method to verify this assumption. In a 6×4 factorial arrangement of treatments with three replicates, samples of 6 concentrate ingredients (maize, barley, milo, peas, lupins and faba beans) and 4 fractions (whole: WHO, non-washout fraction: NWF, insoluble washout fraction: ISWF and soluble washout fraction: SWF) were subjected in two runs to an *in vitro* incubation technique, which measures gas production continuously in an automated system (APES) for 72 h. The Gas production profiles, obtained with APES, were fitted to a multi-phasic model (Groot et al., 1996). The fermentation characteristics of ISWF were more similar to those of WHO and NWF than to those of SWF. Dividing the gas production profile of SWF into two phases, revealed a very rapidly degradable fraction in the first phase of degradation. Fitting the gas production profiles of WHO with a tri-phasic model revealed that the calculated sub-curve of the first phase did not match with the gas production profile of SWF.

Session 6 Poster 14

Degradability characteristics of dry matter of some feedstuffs using _in vitro_ technique

A. Taghizadeh, H. Abdoli and A. Tahmasbi, Dep of Animal Science, Tabriz University, Tabriz, Iran

The degradability characteristics of test feeds were determined using In Vitro Technique. The feeds were barley grain (BG), soybean meal (SBM) wheat bran (WB). The In Vitro studies were conducted consecutively over a period of two weeks using rumen fluid obtained pre-feeding from two sheep (38±4kg, fed a diet containing (as fed) 600 g kg^{-1} a 16 % cp concentrate and 400 g kg^{-1} alfalfa hay). Dry matter degradation (DMD) were estimated 0.0, 2, 12, 24 and 48 h of incubation. Degradabilities data were fitted to a equation of $p=a+b(1-e^{-ct})$; where (p) is DM degradability at time, t, (a) is intercept and ideally reflects the soluble fraction, (b) is the degradable of insoluble fraction, (c) is the fractional rate at which b is fermented per hour. The DM soluble fraction (a) for BG, SBM and WB was 29.7, 39.6 and 29.6, respectively. The insoluble (but with time fermentable) fraction (b) was 57.9, 50.2 and 47.3, respectively. The fractional rate of fermentation (c) was 0.13, 0.05 and 0.05, respectively. The results showed that the soluble fraction (a) in SBM was more than the other feedstuffs ($P<0.05$), while DM insoluble fraction (b) and fractional rate (c) in BG were more than the other feedstuffs ($P<0.05$). These variability in DM fermentation parameters can be resulted from the differences in fractions of constitute of DM in feeds.

Session 6 Poster 15

Monitoring the fate of untreated and micronized rapeseed meal proteins in the rumen using SDS-PAGE

A.A. Sadeghi and P. Shawrang, Department of Animal Science, Faculty of Agriculture, Science & Research Campus, Islamic Azad University, Tehran, Iran

This study was carried out to monitor the fate of untreated and micronized rapeseed meal (RSM) proteins in the rumen using nylon bags and SDS-PAGE techniques. Nylon bags of untreated and treated RSM (burner temperature: 200°C for 20, 40 and 60s) were suspended into the rumen of three Holstein steers from 0 to 48-h, and data was fitted to non-linear degradation characteristics to calculate effective rumen degradation (ERD). Electrophoresis results showed that RSM proteins are composed of napin (2S albumin) and cruciferin (7S globulin). The molecular weight of 10.3 and 8.2 kDa for two sub-units of napin and 31.2, 26.8, 21.1 and 20.5 kDa for four sub-units of cruciferin were estimated. Napin sub-units of untreated, 20, 40 and 60s micronized RSM were disappeared completely within 0, 2, 4 and 8-h incubation in the rumen, whereas cruciferin sub-units were not degraded completely until 48-h incubation. ERD of CP decreased ($p<0.05$) as processing time increased. *In vitro* digestibility of ruminally undegraded CP for 20 and 40s micronized RSM was higher ($p<0.05$), but for 60s was lower ($p<0.05$) than untreated RSM. In conclusion, RSM proteins can be effectively protected from ruminal degradation by micronization for 40s.

Session 6 Poster 16

Ruminal starch and protein degradation kinetics of untreated and microwave treated barley grain

A.A. Sadeghi, P. Shawrang and A. Nikkhah, Department of Animal Science, Faculty of Agriculture, Science & Research Campus, Islamic Azad University, Tehran, Iran*

The present study was designed to evaluate the effects of 800W microwave treatment for 180, 300 and 420s on ruminal crude protein (CP) and starch degradation characteristics of barley grain (BG). Nylon bags of untreated and microwave treated BG were suspended into the rumen of three Holstein steers from 0 to 48-h, and data was fitted to non-linear degradation characteristics to calculate effective rumen degradation (ERD). ERD of CP decreased ($P<0.05$) as processing time increased. Microwave processing for 5-min increased ($P<0.05$), but for 420s decreased *in vitro* crude protein digestibility of BG. Microwave heating for 180 and 300s increased ($p<0.05$), but for 420s decreased ($p<0.05$) ERD of starch. From electrophoretic analysis, three major protein components contain B-Hordein, C-Hordein and D-Hordein fractions were observed. B, C and D-Hordein subunits of untreated BG were disappeared completely within 0, 6 and 2-h incubation, whereas in microwave treated BG were not degraded completely until 4, 12 and 6-h incubation, respectively. Microwave processing decreased degradation of C-Hordein effectively. The bulk of the rumen undegradable protein at 12-h incubation for microwave treated CG were C-Hordein subunits. In conclusion, the best processing time with 800W microwave power is 300s.

Session 6 Poster 17

Chemical composition, in vitro DM and OM digestibility of ten pasture species

P. Shawrang[1] and A. Nikkhah[1], [1]Dep. of Animal Science, Tehran University, Tehran, Iran

This study was carried out to determine telly and terry *in vitro* DM and OM digestibility of ten pasture species (Vicia villosa, Bromus tomentellus, Hordeum bulbusum, Festuca ovina, Agropyron tauri, Agropyron trichophorum, Prangus ferulacea, Ferula orientalis, Lathyrus odoratus, Taeniatherum caput-medusae). CP content of these species were 197, 143, 94, 54, 123, 72, 122, 102, 212 and 124 g/kg DM. NDF content of these species were 398, 451, 611, 626, 604, 638, 251, 239, 307 and 512 g/kg DM. Significant ($P<0.05$) differences between DM and OM digestibility were observed in most of the pasture species. The lowest DM digestibility was seen in Festuca ovina (301 g/kg) and highest in Prangus ferulacea (703 g/kg). DM digestibility of ten pasture species were 566, 639, 328, 301, 547, 465, 703, 666, 567 and 648 g/kg, respectively. The lowest OM digestibility was seen in Festuca ovina (303 g/kg) and highest in Prangus ferulacea (720 g/kg). OM digestibility of ten pasture species were 574, 651, 334, 303, 553, 477, 720, 688, 580 and 667 g/kg, respectively. In conclusion, chemical composition, DM and OM digestibility were different among pasture species. These characteristics must be considered as main parameters in developing models for physical fill, intake capacity and ration formulation of grazing ruminants.

Session 6 Poster 18

Manure evaluation in dairy cows

K. Steen[1], T. Eriksson[2] and M. Emanuelson[1], Kungsängen Research Center, S-753 23 Uppsala, Sweden, [1]Swedish Dairy Association, [2]Swedish University of Agricultural Sciences SLU, Department of Animal Nutrition and Management, Sweden

Manure analysis has been suggested as a management tool. A study was therefore conducted to investigate various characteristics of manure under Swedish conditions to gather background information. Manure was collected from 16 cows during 3 periods at the University Research herd and from 9 commercial herds (5 cows/herd at one occasion). The manure samples were analysed for dry matter, ash, starch, pH-value, particle size (wet sieved) and consistency. The dietary composition was analyzed and Penn State Forage Particle Separator was used for determining the particle size of the roughage.

The starch content in the faeces averaged 1.2% for all cows and the highest individual value observed was 6 %. These values are low, indicating a high utilisation of starch from the Swedish diets. The pH in faeces was for most diets 7 or higher (variation 6.2 - 8.0) but decreased with lower roughage:concentrate ratio in the diet. There was a significant, negative correlation between the pH-value and the starch content in the faeces. Mucine casts were occasionally found in faeces during the study and the occurrence increased during a boost of diarrhoea.

Very few particles in the faeces were greater than 2 cm which indicates enough structure in the different diets. Penn State Forage Particle Separator was not suitable for evaluating the particle size of the grass-clover silage based diets.

Session 7 Theatre 1

Economic changes under the CAP: the implications for animal science and animal welfare

M.F. Seabrook, School of Biosciences, University of Nottingham, LE12 5RD, UK*

Research into the intensification of animal production in an attempt to reduce the unit cost of production, has run concurrently with research into ethical and sustainable production (e.g. the psychological interaction of animals and animal caretakers and the designing of production systems related to the animals' 'natural environment'). The consumer is a crucial player and some of this research is consumer and market driven, with product differentiation and "welfare friendly" food as an effective marketing image. These developments have sometimes been reflected by changes in the legislation for animal welfare (e.g. transport and housing).

The changes in the CAP, the breaking of the link of EU subsidies from production, has in theory removed the incentive to intensify production. It will certainly give some farmers the opportunity to consider what the market requires and develop products for this, often based on less intensive or innovative production systems. The challenge is to ensure that these systems fully meet the needs of the animals' welfare. However, since much food is merely a commodity which cannot readily be differentiated it is also likely to lead to some producers perceiving that economic returns can best be enhanced by specialisation, increased scale and size of units and intensification in order to be a cost effective commodity producer. These developments challenge all with a research interest in animal welfare and health.

Session 7 Theatre 2

The challenge of competing goals: animal welfare, the environment, human health and the profitability of livestock production

C.E. Milne[1], A.W. Stott[1], G.E. Dalton[2], [1]Animal Health Economics Team, SAC, Aberdeen, UK,[2]Dalton Associates, Aberdeen, UK*

Producing livestock for meat, milk, or other good requires the skillful combination of a wide variety of resources, such as land, labour, equipment, veterinary medicines. From an infinite number of potential combinations the challenge for farmers and farm managers is to select the 'best' solution, which is dynamic and may change in light of evolving governmental policies and market conditions. Governmental policies include regulation which establishes accepted minimum standards for issues that are of concern to the public, such as the welfare of farmed animals, environmental care and the protection of human health. Farmers must comply with these standards or face prosecution and any associated penalties. However, regulation is a blunt instrument and given practical constraints presented by operating in a natural environment some conflicts now exist between individual regulations. Furthermore conflict can occur between regulatory compliance and achieving an acceptable level of profitability. Regulations in combination may thus require farmers to risk breaching a minimum standard, where no course of action is expected to assure compliance and acceptable profit simultaneously or led to unexpected changes in livestock production where unacceptable levels of profitability are derived.

Using the example of sheep ectoparasite control in Scotland this paper describes some conflicts that currently arise and considers potential production consequences, highlighting challenges for producers and researchers.

Session 7 Theatre 3

Integration of animal welfare in the food quality chain: from public concern to improved welfare and transparent quality

H.J. Blokhuis[1], R.B. Jones[2], R. Geers[3], M. Miele[4] and I. Veissier[5], [1]Animal Sciences Group, Wageningen UR, Lelystad, The Netherlands, [2]Roslin Institute (Edinburgh), Roslin, Scotland, [3]Katholieke Universiteit Leuven, Leuven, Belgium, [4]University of Pisa, Pisa, Italy, [5]INRA, Theix, France*

European citizens expect their food to be produced with greater respect for the welfare of farm animals. Indeed, their perception of food quality is determined not only by the nature of the product but also by the welfare status of the animals from which it was produced. This means that efforts to inform the consumers and to address their concerns need to be at the forefront of the future agricultural policy agenda. Furthermore, worldwide marketing strategies confirm that producers and retailers today are ready to apply new criteria so as to provide consumers with extra value. Transparency of the product quality chain requires reliable on-farm monitoring of welfare status and the standardised conversion of welfare measures into accessible and understandable information, thereby addressing consumer and stakeholder concerns and allowing clear marketing and profiling of products.
Within the European project Welfare Quality, leading European groups are integrated to: a) analyse and address the perceptions and concerns of consumers, retailers and producers about animal welfare, b) develop pan-European standards for on-farm welfare assessment and product information systems, and c) identify practical strategies for improving animal welfare.

Session 7 Theatre 4

Assessing the costs of infection by Bovine Viral Diarrheoa Virus (BVDV) in dairy herds

C. Fourichon, F. Beaudeau, N. Bareille and H. Seegers, Veterinary School - INRA, Unit of Animal Health Management, BP 40706, 44307 Nantes cedex 03, France*

New infection by the BVDV may induce dramatic signs in totally naive herds (abortions, youngstock mortality,...), whereas it often happens not to be detected when resulting only in decrease in herd performance. Both situations have to be considered when assessing costs consecutive to infection at a regional level. Such results can be combined with incidence of clinically diseased herds and prevalence of non-clinical infected herds, in order to assess potential benefits that could be expected from regional qualification or control programmes. Production losses associated with BVDV infection were calculated by partial budgeting based on disease and production data collected in dairy farms in Brittany. Source data were issued from (1) all farms detected as very likely to be infected based on bulk-milk antibody testing, through a systematic regional surveillance scheme and (2) a few newly infected farms with severe clinical forms of the disease. The year when BVDV infection was detected in the herd, overall costs (losses plus expenditures) resulted in a decreased gross margin of 11 per 1000 litres of milk for an average non-clinical-but-likely-to-be-infected farm up to 39 per 1000 litres of milk in a severe-clinical-case farm. Partial budgeting may result in underestimating costs of disease in farms where the farming system is temporarily but highly disrupted.

Session 7 Theatre 5

Economic value of mastitis incidence in dairy herds in the Czech Republic

M. Wolfová, M. Stípková and J. Wolf, Research Institute of Animal Production, P.O.Box 1, CZ 10401 Praha-Uhríneves, Czech Republic*

The calculations were based on data collected from 1996 to 2003 on five Holstein farms in the Czech Republic. Clinical mastitis (CM) incidences (number of CM cases) per cow-year at risk in the whole data set were 0.68, 1.00 and 1.27 for the 1st, 2nd and 3rd and subsequent lactations, respectively. The CM incidences per cow and year averaged over lactations on the individual farms ranged from 0.53 to 1.56. For the whole data set, a value of 0.94 was calculated for this trait. The economic value was defined as change of the total profit per cow and year when increasing the CM incidence by 0.1 cases. The part of the profit function influenced by CM included losses from discarded milk, costs for drugs, veterinary service and herdman´s time, cost for extra milking machine and cost for antibiotic drying of cows. The economic value of CM incidence ranged from -169 to -267 CZK (1 ≈ 30 CZK) per 0.1 case of CM per cow and year. The economic value for the total data set was -186 CZK. The milk production level and the level of CM incidence had the largest impact on the economic weight for CM incidence.

The financial support from the project MZE 0002701401 is acknowledged.

Session 7 Theatre 6

Variation in milk yield associated with the cow-status to *Mycobacterium avium* subspecies *paratuberculosis* (Map) infection in French dairy herds

F. Beaudeau[1], M. Belliard[1], A. Joly[2], H. Seegers[1], [1]Unit of Animal Health Management, Veterinary School-INRA, BP 40706, F44307 Nantes cedex 03, [2]UBGDS, BP 110, F 56000 Vannes cedex*

This study aimed at quantifying the variation in test-day milk yield (TDMY) of dairy cows according to *Map*-infection status of cows. The cow-status was determined combining (i) her testing(s)-result(s) (ELISA, faecal culture (FC), PCR, Ziehl staining), (ii) the *Map*-status of her herd, and (iii) her possible vaccination against *Map*. 14 cow-statuses were defined. A total of 23,219 cows in 569 herds located in Bretagne was considered. The effect on TDMY of the cow-status to *Map*, adjusted for herd-year (random), lactation number, days in milk and breed was assessed using mixed linear models. The average TDMY was significantly lower in cows from herds with at least one *Map*-infected cow (defined as positive herds). Individual TDMY showed a reduction of 1.6, 2.5, 2.1 and 6.2 kg/day ($P<0.001$) for cows tested negative in a positive herd, not vaccinated and ELISA-positive, PCR- or FC-positive, and Ziehl-positive, respectively, in comparison with cows in *Map*-negative herds. ELISA-tested positive but vaccinated cows had a smaller loss in TDMY than those not vaccinated, suggesting the coexistence of infected and not-infected animals among ELISA-positive cows. These estimates can be used to further assess the economic impact associated with *Map*-infection in dairy herds or to enlighten the culling decision making process regarding infected cows.

Session 7 Theatre 7

Economic analysis of foot-and-mouth disease in Turkey-I: acquisition of required data via Delphi expert opinion survey

B.Senturk[1], C. Yalcin[2], [1]The Ministry of Agriculture, General Directorate of Agricultural Researches, Department of Animal Health Research, PO Box:78, 06171 Yenimahalle, Ankara, Turkey, [2]Ankara University, Faculty of Veterinary Medicine, Department of Livestock Economics, 06110 Diskapi, Ankara, Turkey*

The main obstacle in assessing the impact of Foot and Mouth Disease, which is considered to be economically most important disease both in Turkey and the other countries in the world, on Turkish Economy is unavailability of reliable data. Considering this issue, this study aimed at using a Delphi Expert Opinion Survey Method to obtain data required form economic analysis of FMD in Turkey. This study concluded that although there were problems in obtaining some information from the experts, in general the Delphi technique is a promising way of obtaining animal health data which is missing and/or not regularly recorded in developing countries.

Session 7 Theatre 8

Economic analysis of foot-and-mouth disease in Turkey- II: an assessment of financial losses in infected animals and cost of disease at national level

C. Yalcin[1], B.Senturk[2], [1]Ankara University, Faculty of Veterinary Medicine, Department of Livestock Economics, 06110 Diskapi, Ankara, Turkey [2]The Ministry of Agriculture, General Directorate of Agricultural Researches, Department of Animal Health Research, PO Box:78, 06171 Yenimahalle, Ankara, Turkey*

This research paper, the second of this study series, aimed at estimating cost of FMD in Turkey in 1999, considering financial values of FMD induced production losses in different livestock species and breed in Turkish field conditions, and expenditures for FMD outbreak management and annual vaccination programs, estimated by using data from Delphi Expert Opinion Survey. The average financial losses per infected animal were estimated to be US$294 for milking cow, US$152 for dairy heifer, US$197 for beef cattle, US$69 for sheep and US$64 for goat. However, the amount of the financial losses were considerably varied amongst the animals depending on species, breeds, age and sex. The overall cost of FMD in 1999 was estimated to be US$51.3 million of which financial losses, disease control expenditures at the outbreaks and rutin disease prevention expenditures accounted for %33.6, %2,6 and %63,8 respectively.

Economics of sub-clinical helminthosis control through anthelmintics and supplementation in Menz and Awassi-Menz crossbred sheep in Ethiopia

M. Tibbo[1,2], K. Aragaw[2], J. Philipsson[3], B. Malmfors[3], A. Näsholm[3], W. Ayalew[1], J.E.O. Rege[1], [1]International Livestock Research Institute (ILRI), Animal Genetic Resources, PO Box 5689, Addis Ababa, Ethiopia, [2]Debre Berhan Agricultural Research Centre, Debre Berhan, Ethiopia, [3]Swedish University of Agricultural Sciences (SLU), Animal Breeding and Genetics, Uppsala, Sweden

We evaluated the profitability of anthelmintic treatment and nutritional supplements using partial budget analysis on 108 weaned Menz (n=39), 50% Awassi-Menz (n=38) and 75% Awassi-Menz (n=31) crossbred sheep genotypes kept on-station under sub-clinical helminthosis at Debre Berhan, Ethiopia. Data were collected on feed intake, live weight, EPG, PCV, fleece weight, slaughter weight, carcass weight, dressing percentage, and adult worm burden counts. Input and output prices were recorded. Anthelmintic treatment reduced EPG and worm burden, which was consistent with significantly higher PCV, slaughter and carcass weights for treated than for non-treated sheep ($P<0.05$). Supplemented lambs had higher ($P<0.0001$) weight gain, carcass weight and dressing percentage. Anthelmintic treatment resulted in a marginal profit (MP) of 13.46 ETB per sheep whilst supplementation resulted in MP of 17.13 ETB per sheep. MP per sheep was higher for Menz than the crosses due to the higher market price of its skin. MP from anthelmintic treatment increased as exotic blood increases from indigenous Menz to 50% and 75% Awassi-Menz, indicating higher dependence of the crossbreds on anthelmintics. Sheep breeding objectives were discussed in relation to low input system.

Solid dosage medicine for poultry

A.E. Ustyanich[1], M.A. Ustyanich[1] and E.P. Ustyanich[2], [1]24 Kilburn Place, Toronto, M9R 2X5, Canada, [2]Lviv-79000, Slovatskogo 1, P.O.Box 50, Ukraine*

Taking into consideration the specifics of the intestine and masculine stomach in poultry, a solid dosage preparation with prolonged release action utilizing the liquid medicine substances was developed. The capillary-porous granules of silica gel, porous glass and other porous materials, which are indifferent to the poultry organism, can serve as a matrix for such drugs.

The method of preparation of these drugs is based on the localized saturation of the capillary-porous granules with active liquid medicine with post encapsulation of these granules into a polymer, wax and/or solid fat. These granules may be fed to the poultry along with the fodder. When entering the stomach these granules perform as milling agents and while gradually wearing release the drug substance over the extended period of time, from a few days to several weeks.

The drug granules can be prepared along with various vitamins, growth inhibitors and other biologically active substances.

Session 7 Poster 11

Clinicopathological studies on *Theileria Annulata* infection in Siwa Oasis in Egypt

T.A. Abdou[1], T.R. Abou-El-naga [2]and Mona A Mahmoud[2], [1]Faculty of Veterinary Medicine, Cairo University [2]Animal Health Dep.-Desert Research Center

One hundred and twenty five (125) cross and native cattle breeds were examined to estimate the prevalence of *T. annulata* infection for the first time in Siwa Oasis and evaluate its effect on some blood constituents before and after treatment with buparavaquone. The prevalence of tropical theileriosis was 40.3% and 29.4% in cross and native breed respectively using blood smear examination. IFAT could identify *T. annulata* in 80.7% of cross breed and 70.5% of native cattle. Also there was seasonal prevalence variation. The tick species *H. a. anatolicum* was recovered from 65.6% of examined cattle. Clinically infected with *T. annulata* had significantly low levels of total proteins, albumin, magnesium, potassium and iron concentrations ($P \leq 0.05$) but AST, L γ glutamyl transferase activities, total, direct and indirect bilirubin, creatinine levels were significantly high ($P \leq 0.05$).Buparvaquone was effective against both stages of *T. annulata* and succeed to control fever and temperature returns to normal range by 7th day post treatment. Also, some serum elements to its normal values especially in native but not in cross breed post treatment. In brief our data showed that tropical theileriosis was prevalent in Siwa Oasis especially among cross breed cattle and had effect on hepatic and renal functions. There is a need for the use of immunization methods to reduce the losses of the disease.

Session 7 Poster 12

Economic aspects of cystic ovarian disease treatment

M. Farhoodi[1] and K. Valipoor, [1]Department of clinical sciences, Faculty of Veterinary Medicine, University of Azad, Karaj, Iran*

Cystic ovarian disease (COD) is a major reproductive disorder of dairy cattle. The objectives of the present study were to evaluate that it is profitable to treat cows early in postpartum period.
The study was conducted in a large dairy herd of 720 Holstein cows in Tehran. In this herd all cows were examined 30 days postpartum (clean test) and cows which had fluid-filled structures greater than 2.5 centimeters in diameter determined as cystic cows. Cystic cows were divided in two treatment groups. Some were treated in the same time as diagnosed (day 30 postpartum), (early treated group), (no =86) and the other was treated 12 days later if they did not recovered (late treated group), (no=51) with GnRH and $PGF_{2\alpha}$ (dicloprostenol) 10 days later. In each group days open, days to first service, service per conception and direct costs (vet, drugs and labor) were estimated. Then difference of economic losses has been estimated.
Days open and days to first service of early treated group (75.74±43.36 and 53.86±14.28) were better than late treated group (96.14±42.82 and 65.78±19.12) ($P<0.05$). Service per conception of early treated group (1.86) was lower than late treated group (2.1). The early treated group had 102.66 $ lesser cost than late treated group that indicates the advantage of treating cystic ovary in 30days postpartum over treatment in 42 days postpartum.

Duration effect of propylene glycol oral drenching on animal health and production in transition dairy cows

H.A. Malek Mohammadi, H.R. Rahmani, G.R. Ghorbani, Department of Animal Sciences, Isfahan University of Technology, Isfahan, Iran.

Sixty-four multiparous Holstein cows were used to determine the duration effect of propylene glycol (PG) oral drenching during 3 weeks pre-partum (prep) and 3 days pos-tpartum (pstp) on metabolism, animal health, and milk production. Treatments were: 1) water (Control), 2) 1 wk prep and 3 d pstp, 3) 2 wk prep and 3 d pstp, and 4) 3 wk prep and 3 d pstp of PG. Milk yield recorded from d 1 to 260. Blood samples collected 1, 2, and 3 d prep and d 1, 2, 3, 6, 10 and 17 pstp. PG Drenching decreased milk production for Trt 4 from d 1-140 ($p < 0.05$). Milk production from d 140-260 did not differ between the treatments. Rumen pH and fermentation pattern was not affected by PG drenching. Insulin and glucagon, and inslin to glucagon ratio, 2 d prep, were not affected by PG draenching. Insulin was higher for control group ($p < 0.05$) on d 2 pstp, but the ratio did not differ between treatments. Serum glucose was higher in control group in d 1-3 and 1-17 pstp ($p < 0.05$). Other blood metabolites; NEFA, total keton bodies, TG, BUN, Ca and P were not different statistically. Data do not support any benefit from PG drenching in any of prepartum periods.

Effect of two type of stress on the plasmatic levels of cortisol and some haematic parameters

M.C. Mura[1], V. Carcangiu[1], G.M Vacca[1], A. Parmeggiani[2] and P.P.Bini[1], [1]Dipartimento di Biologia Animale, via Vienna 2, 07100 Sassari, Italia, [2]Dipartimento di Morfofisologia Veterinaria e Produzioni Animali, via Tolara di Sopra 50, 40064 Ozzano dell'Emilia (Bo), Italia*

The aim of the study was to determine the effect of two type of stress by measuring plasma concentrations of cortisol, glucose and triglycerides. Three blood samples were taken from 30 sheep at 10 minute intervals, beginning at isolation. The animals have been subdivided in three groups, everyone of 10 animals: Group A separated from the flock, Group B separated and tied down, Group C control. Plasma cortisol concentrations were evaluated by RIA and the haematochemical parameters were measured by colorimetric assay. The data were subjected to analysis of variance. Cortisol and glucose levels in the second and third samples showed a higher increase ($P<0.01$) in animals that had been isolated and tied down than in the others ($P<0.05$). A comparison between the isolation and tied down groups revealed significant differences ($P<0,01$) in the third sample in levels of cortisol and glucose. All the two types of handling caused a certain amount of stress, shown by an increase in blood levels of cortisol and glucose, but tying down had a greater effect on the response of the organism. In order to respect animal welfare it would be right to reduce to the necessary the time of handling.

Session 7 Poster 15

Dairy cattle show preferences between different types of cow-brushes

F. Tuyttens[1], K. Van den Bossche[2], L. Lens[2], J. Mertens[2] and B. Sonck[1], [1]Agricultural Research Centre (CLO-DVL), Merelbeke, Belgium, [2]Department of Biology, Ghent University, Belgium*

Cattle have the urge to rub their skin against walls, trees and other objects. Although the function of this behaviour is not entirely clear, there are cow-brushes on the market that offer cows the possibility to better satisfy this need. Based on the percentage of animals that make use of them and on the frequency and duration of usage in time, cow-brushes have been reported to be a suitable environmental enrichment for cattle. Because cattle farmers have little information to compare the variety of cow-brushes that are on the market, we tested whether cows show a preference between different cow-brushes in an open-choice test. Four cow-brushes were installed in a loose house (cubicles) for dairy cattle. The position of the brushes was rotated so that each brush was tested on each position. After a 3-week habituation period, 234 h of digital images were recorded during October - December 2002 to quantify the usage (5-min interval) and preference (percentage of times a cow chose a certain brush when all four brushes were vacant) for each brush by every cow. Adjusting for technical defects and position, cattle showed a difference in preference between the four cow-brushes ($P<0.0001$). The order of this preference was: Heido-VPG3 (Heitmann) > E-brush (Brouwers) > Borstelgarnituur (Brouwers) > HappyCow (Mayer).

Session 7 Poster 16

Yield losses associated with clinical mastitis in Swedish dairy cows

C. Hagnestam[1], U. Emanuelson[2], H. Andersson[3], J. Philipsson[1] and B. Berglund[1], [1]Dept. of Animal Breeding and Genetics, P.O. Box 7023, [2]Dept. of Clinical Sciences, P.O. Box 7019, [3]Dept. of Economics, P.O. Box 7013, Swedish University of Agricultural Sciences (SLU), SE-750 07 Uppsala, Sweden*

With increasing production levels, the risk of clinical mastitis increases. In order to take appropriate mastitis preventive measures, in breeding programs as well as in herds, it is of outmost importance to know the magnitude of the yield losses caused by the disease. Yield losses resulting from the first lactational incident of clinical mastitis were assessed in the Swedish Red and White (SRB) and the Swedish Holstein (SLB) breeds. The data set consisted of 38 569 weekly production records from 1193 lactations (506 cows), sampled during 1987-2004 in the university's experimental herd. Furthermore, detailed information on health status, fertility and calving ability was available. Analyses of yield losses were carried out using a mixed model correcting for systematic environmental effects of clinical mastitis, week in milk, breed, other diseases, reproductive status and year/season of calving. First calvers and older cows were analysed separately. The production decreased several months before the mastitis incident. Yield losses were considerable for two months after the case and mastitic cows never returned to their initial production level.

Session 7 Poster 17

Teat closure and condition after dry-off in high producing dairy cows

M.O. Odensten[1,], K. Holtenius[1], and K. Persson Waller[2]*
[1]Department of Animal Nutrition and Management, Swedish University of Agricultural Sciences, SE-753 23 Uppsala, Sweden, [2]Department of ruminant and porcine diseases, National Veterinary Institute, SE-751 89 Uppsala, Sweden

Prompt teat closure and good teat condition is important for good udder health in dairy cows. The aim was to examine these factors after dry-off in thirty-seven primi- and multiparous cows. They were dried off 8 wk prior to expected parturition, and milked twice during dry-off, in the mornings of day 2 and 5. Teat closure was examined once a week after dry-off in two teats from each cow by applying light pressure to the teat in a gentle milking action. If the pressure resulted in a drop of secretion at the teat orifice, the quarter was classified as "open". The cows were divided into two groups depending on milk yield prior to dry-off, e.g. low (< 15 kg energy corrected milk (ECM); n=40) and high (>15 kg ECM; n=34). The groups had a mean milk yield of 11.7±0.6 and 20.8±0.8 kg ECM, respectively prior to dry-off. One week after dry-off, the proportions of open teats were 41% and 20% in high and low groups, respectively. Five weeks after dry-off the corresponding numbers were 15% and 5%. Teat condition did not differ between groups. These preliminary results indicate that the ability to create a keratin plug in the teat-end was decreased in high producing dairy cows.

Session 7 Poster 18

The map of hip dysplasia of the Hungarian police dog population

E. Hegedüs and M. Horvai Szabó, St István University, Faculty of Agricultural and Environmental Sciences, Gödöllö, H-2103, Hungary

Hip dysplasia is the most common heritable disease of dog. The connected data include different information about the age of the dogs at the exact time of screening, the sex and the breed of them, hip-joint scored as normal, slight, moderate hip dysplasia. We measured the Norberg angle and the dimension of trochanter minor with the aim of looking for one measurable anatomical parameter. Realising a conditional relation between those parameters the diagnosis can be more precise. Data of 856 police dogs screened from 1997 to 2002 were analyzed using correlation and regression analysis.

By our evaluation the effect of breed and sex were statistically significant. Hip dysplasia was the most frequent disease in the German Shephard dog, and its females had slightly worse hip-joint than males. The mean of dimension of trochanter minor decreased year by year, that it proved positive correlation with the degree of dysplasia. There is a positive correlation (0.54) between the left and right hip-joint angles as well as the left and the right trochanter minor (0.78). We found a negative correlation between the right hip-joint angle and the dimension of the left trochanter minor in severe cases of dysplasia. It always was characteristical that the mean of the left hip-joint angle was better than the right one, and the dimension of the left trochanter minor was smaller as well.

Session 8 Theatre 1

Biological robustness of pigs

J. ten Napel, Animal Sciences Group of Wageningen UR, P.O. Box 65, 8200 AB Lelystad, The Netherlands

Ideally, pigs in commercial production systems are able to effectively deal with the dynamics of their environment. There are basically two strategies to achieve this: minimise occurrence of disturbance, or make the pig robust.
A pig that is robust to normally occurring challenges has the genetic potential to cope, has learned by experience how to use its ability effectively, and has the means at the time of exposure to mount an appropriate response.
Unbalanced breeding goals and insufficient natural selection may pose a risk to the genetic potential of the animal to cope with common conditions, so avoiding these risks should be an integral part of the selection strategy. The main risk of genetic selection seems to be overstretching animals in more demanding environments, rather than increased susceptibility to disease in general.
A pig also needs to learn to cope with disturbance by doing. This requires natural or controlled exposure with some form of support or protection, such as maternal antibodies. Experience is particularly relevant to the immune system and the central nervous system.
The production system and management need to facilitate the pig in its attempts to adapt to challenges. This may be in the form of providing various micro-climates, allowing time to adapt, avoiding large and stressful changes, etc. Problems that people associate with lack of robustness often arise from frustration from fruitless attempts to adapt.

Session 8 Theatre 2

Traits associated with sow stayability

M. Knauer[1], T. Serenius[1], K.J. Stalder[1], T.J. Baas[1], J.W. Mabry[1] and R.N. Goodwin[2], [1]Iowa State University, Ames, USA, [2]National Pork Board, Des Moines, USA

The purpose of this study was to analyze the complex interactions of factors that influence stayability to parities 1, 3, and 5. The study was carried out by analyzing data from the 1996 NPPC Maternal Line Evaluation Program, which included six genetic lines and consisted of 2,935 female pigs. Genetic lines evaluated included Newsham, National Swine Registry, American Diamond Swine Genetics, Danbred, and two Monsanto (DK44 and GPK347) genetic lines. The affects of genetic line, gilt backfat, average daily gain, loin depth, and age at puberty on stayability were included in the statistical model for all stayabilities. In addition to these effects, total number born, incoming pre-farrow backfat, backfat loss during lactation prior to removal, and feed intake during lactation prior to removal were also included in the statistical models for stayabilities 3 and 5. The analysis was carried out by a modified chi-squared automated interaction detection algorithm. The results showed that age at puberty, lactation feed intake, and backfat loss during lactation were the factors most significantly associated with stayabilities 1, 3, and 5, respectively. High age at puberty, low lactaion feed feed intake, and high backfat loss during the lactation were all associated with low stayability. The results showed further that the associations between stayabilities and the factors mentioned above are not linear, extremes being most detrimental on stayability.

Session 8 Theatre 3

Analysis of true sow longevity

K.J. Stalder[1], T. Serenius[1], T.J. Baas[1], J.W. Mabry[1] and R.N. Goodwin[2], [1]Iowa State University, Ames, USA, [2]Goodwin Family Farms, Ames, Iowa, USA

Data from the National Pork Producers Council Maternal Line National Genetic Evaluation Program were used to compare the sow longevity of six different genetic lines, and to estimate the associations of gilt backfat thickness, age at first farrowing, litter size at first farrowing, litter weight at first farrowing, average feed intake during lactation, and average backfat loss during lactation on sow longevity. The lines evaluated were American Diamond Genetics, Danbred North America, Dekalb-Monsanto DK44, Dekalb-Monsanto GPK347, Newsham Hybrids, and National Swine Registry. In the trial, no culling due to poor production was allowed until a sow had reached her fourth parity. The line comparison was carried out by analyzing all the lines simultaneously, and including genetic line in the statistical model. Because of the survival distribution functions differ between the genetic lines, the association analysis were carried out separately for all the genetic lines. All the analysis were based on nonparametric proportional hazard (Cox model). The results showed that the sows of Dekalb-Monsanto GPK347 had a clearly lower risk of being culled than the sows of other five lines. The results showed further that sows with poor feed intake and greater backfat loss during lactation had the shortest productive lifetime in most of the genetic lines. As a conclusion, the between line differences indicate that it is possible to select for sow longevity.

Session 8 Theatre 4

Estimates of additive and dominance genetic effects for sow longevity

T. Serenius and K.J. Stalder, Iowa State University, Ames, Iowa, USA

The purpose of current study was to estimate variance components, especially dominance genetic variation, for overall leg action, length of productive life, and sow stayability until third and fifth parity. That was carried out by analyzing the data from Finnish litter recording scheme. The variance components were estimated in two purebred (Landrace and Large White) and crossbred (Landrace x Large White) datasets. There were information on 38 941, 36 907, and 26 777 sows in the Landrace, Large White, and crossbred datasets, respectively. The fixed effect of herd-year, and random effects of additive sire, parental dominance, and litter were included in the statistical model of all the traits. Moreover, the fixed effect of breeding consultant, and linear regression of test weight were also included in the statistical model for overall leg action. The estimated heritabilities ranged between 0.04 and 0.06, being very similar between the different breeds. Similarly, the estimates for ratio of dominance variance to phenotypic variance (d^2) varied between 0.01 and 0.17, being highest in crossbred dataset. Moreover, all the d^2 estimates in crossbred population were higher than the corresponding heritability estimates. Similarly in purebred populations, 5 out of 8 d^2 estimates were higher than the corresponding h^2 estimates. Based on current results, it seems that the effect of dominance should be accounted in the breeding value estimation of sow longevity, especially in crossbred model.

Session 8 Theatre 5

Breeding against osteochondrosis in swine: the Swedish experience

N. Lundeheim, Swedish University of Agricultural Sciences, Dept. of Animal Breeding and Genetics, P.O. Box 7023, S-750 07 Uppsala, Sweden

Osteochondrosis is defined as a disturbance of cell differentiation in metaphyseal growth plates and joint cartilage, and is regarded to cause leg weakness symptoms and pain. Already in 1982, recording of the presence and severity of osteochondrosis started within the national Swedish pig breeding programme. The medial condyle of one elbow joint and one knee joint were scored after slaughter for all pigs from the progeny- /sib-testing stations. A 6-point scoring was used (0=no lesions; 5=severe lesions). This information has since 1988 been included in the breeding evaluation, with the primary intention to balance the unfavourable genetic correlations between osteochondrosis and production (growth rate and carcass leanness). The traditional Swedish pig progeny-/sib-testing at testing stations ended in year 2000. On the basis of data on female pigs entering test in the period 1987 to 2000, genetic analyses (multitrait animal model) of osteochondrosis in relation to weight and age at slaughter were performed. In total, records on 6200 Landrace and 6400 Yorkshire female pigs were included. The estimated heritabilities for the two osteochondrosis scores were in the range 0.15 to 0.25. Genetic correlations estimated indicate that high weight and low age at slaughter (both being components of fast growth) are genetically linked to high osteochondrosis burden.

Session 8 Theatre 6

Removal of Swedish sows

L. Englblom[1], N. Lundeheim[1], A.-M. Dalin[2] and K. Andersson[1], Swedish University of Agricultural Sciences (SLU), [1]Department of Animal Breeding and Genetics, P.O. Box 7023, SE-75007 Uppsala, Sweden, [2]Department of Clinical Sciences, P.O. Box 7039, SE-75007 Uppsala, Sweden*

Worldwide, approximately 50 % of all commercial sows are replaced every year. To generate knowledge on sow removal, data was collected from 21 commercial piglet-producing herds in the south-central part of Sweden. This study was based on Landrace*Yorkshire sows with at least one farrowing in the period 2001-2003. The statistical analyses were restricted to data on the first five parities (51 412 farrowings; 21 134 sows). A removal code describing if the sow got a next parity (=0), or if the sow was removed (=1) was assessed each farrowing. This information was analysed using the SAS PROC MIXED and the GLIMMIX macro.

All effects included in the statistical model were highly significant ($p<0.001$): herd, farrowing year within herd, farrowing date (two-month periods), parity number and the interaction between herd and parity number. The proportion of farrowings followed by sow removal varied between herds (range: 10.4-28.6 %). Farrowings in parities 4 and 5 were to a significantly higher proportion (18.1 % and 22.8 %) followed by sow removal, compared with farrowings in lower parities (15.2-15.9 %) ($p<0.01$). Furthermore, farrowings during May-June and July-August were to a significantly higher proportion (18.7 % and 18.6 %) followed by sow removal, compared with farrowings during the rest of the year (16.5-17.1 %) ($p<0.01$).

Session 8 — Theatre 7

Environmental sensitivity and robustness

E. Strandberg, Dept. of Animal Breeding & Genetics, Swedish University of Agricultural Sciences (SLU), Uppsala, Sweden

The ability of an individual to change the phenotype in response to changes in the environment is called phenotypic plasticity or environmental sensitivity. Plasticity has a genetic basis can be observed at the biochemical, physiological, behavioural, and other levels of the organism. One possible definition of robustness are animals that are not environmentally sensitive for the traits of importance (e.g. production, health, fertility). However, in practice we would also like them to have a high level of these traits. Differences in plasticity between individuals (genotypes) will result in genotype by environment interaction (GxE), i.e. the difference between the expected phenotypic values of two genotypes is not the same in two environments. If the difference changes sign, we have reranking of genotypes.
In this presentation I will describe some methods for studying the genetic background of environmental sensitivity, both multiple trait analysis, covariance functions and the reaction norm approach. I will also discuss what type of data materials are necessary for such analyses, and the consequences for breeding programs if GxE exists. A point of discussion is also whether we want environmentally insensitive animals, and if so, for what traits. Are there any ethical or animal welfare problems with environmentally insensitive animals?

Session 8 — Theatre 8

Developing the breeding goal for Norwegian Landrace: aiming at a robust and superior sow

B. Holm, E. Gjerlaug-Enger and D. Olsen. Norsvin, P.O. Box 504, NO-2304 Hamar, Norway

Since 1992, the breeding goal for Norwegian Landrace has included number of live born piglets, analysed by a repeatability model. As s result, the genetic and phenotypic trends show an increase of 1.2 live born piglets/litter since 1998. To ensure a sustainable and robust overall genetic progress, the following additional recordings were required from the nucleus and multiplying herds in 2001: oestrus symptoms, birth assistance, parturition length and individual piglet weight at three weeks of age. Genetic analyses including correlations to important production traits were conducted. Genetic parameters revealed that litter size in different parities should be treated as separate traits, and that substantial genetic correlations between litter and production traits exist. Especially growth, litter size and weight were highly correlated. Several reproductive traits contributing to number of weaned piglets per sow/year revealed genetic variation to such an extent that selection would be possible. Multiple trait breeding value estimation should be conducted, utilizing the genetic correlations between production and reproduction traits. In 2004 the breeding goal of Norwegian Landrace was altered to include the following traits; number of live born piglets and litter weight at three weeks in first, second and third parity, age at first service, weaning to first service interval. In addition, important production and efficiency traits, as well as number of functional teats, conformation and osteochondrosis already were included.

Session 8 Theatre 9

Investigations on the impact of genetic resistance to oedema disease on performance traits and its relation to stress susceptibility in pigs of different breeds

S. Binder[1], K.-U. Götz[2], G. Thaller[1] and R. Fries[1], [1]Lehrstuhl für Tierzucht, Techn. Universität München, Freising-Weihenstephan, [2]Bayerische Landesanstalt für Landwirtschaft, Inst. für Tierzucht, Grub, Germany*

Oedema disease is caused by the adherence of specific *E. coli* strains to F18 receptors in the porcine intestinal tract. The α-(1,2)-fucosyltransferase gene (*FUT1*) was identified as a candidate for the expression of these receptors. *FUT1* is located on SSC6, 2 cM away from the *RYR1*, which is responsible for the porcine stress syndrome (MHS). Breeders are eager to utilize genetic resistance to oedema disease in pigs by direct gene tests. Prior to such an application it has to be investigated, whether *FUT1*-variants affect economically relevant traits. We analysed the impact of different *FUT1*-genotypes on traits of fattening performance, carcass composition, meat quality and fertility for the two breeds Pietrain and German Landrace. The results showed that there are no negative effects of *FUT1* on any of the performance traits. However, we found a highly significant linkage disequilibrium between *FUT1* and *RYR1* in the Pietrain breed. Moreover, there were no animals in the Pietrain breed that carried the preferable haplotype A-N which simultaneously confers resistance to stress and to oedema disease in an homozygous state. Prospects for a selection strategy that facilitates both, resistance to stress and resistance to oedema disease at a time, while maintaining genetic progress on production traits are discussed.

Session 8 Theatre 10

Enzymes, proteins, *Escherichia coli* and Ryanodin receptor gene polymorphisms and their association with osteochondrosis in pigs

H.N. Kadarmideen and P. Voegeli, Swiss Federal Institute of Technology Zurich, Institute of Animal Sciences, ETH Zentrum (UNS D8), CH-8092 Zurich, Switzerland*

This study investigated polymorphisms of enzyme (EZ), serum protein (SP) and blood group (BG) systems as well as of Ryanodin receptor (RYR) and *Escherichia coli* F18 receptor (ECF18R) locus and estimated their association with development of six different osteochondrosis (OC) lesions in station tested pigs (data from SUISAG). Genotype information on six BG systems, four EZ (Adenosine deaminase, ADA; Glucose phosphate isomerase; 6-phosphogluconate dehydrogenase, 6PGD; Phosphoglucomutase 2, PGM), two SP (Postalbumin 1A, PO1A; Alpha-1-B glycoprotein) and RYR1 and ECF18R were available. All pigs with OC data with genotypes for above systems (n=707) were used. 'Allele substitution effects' were estimated by multiple regressions of OC lesions on each allele of each locus separately (14x6=84 analyses). Estimated allele substitution effects for OC scores were significant for some combinations but not all; e.g. 0.03-0.18 OC score for BG-A; 0.03-0.11 OC score for ECF18R; 0.03-0.05 OC scores for PGD1 as well as ADA and 0.03 to 0.09 OC score for PGM. Both PO1A and RYR1C/T had significant effects on 2 of 6 lesions. Complete mixed model analyses as well as association with other production traits were also conducted. Preliminary results show that some polymorphisms may be related to biochemical pathways in the development of osteochondrosis, which in turn could help in the identification of candidate genes for osteochondrosis.

Session 8 Poster 11

Parity and production in sows

M. Neil, Swedish University of Agricultural Sciences, Box 7024, S-750 07 Uppsala, Sweden

Number of piglets born per litter increases with parity in sows, but how are other traits affected? Data from 1924 farrowings in Swedish Yorkshire sows belonging to the University pig herd were used to study the influence of sow parity on performance. Sows were group-housed with a maximum of 16 sows per group during pregnancy, and housed individually in farrowing pens without crates during lactation. Piglets were not cross fostered, and were weaned at 5 weeks of age. Sows were weighed and their backfat was measured ultrasonically at service, after farrowing, and at weaning. Piglets were weighed at birth, at 3, 5 and 9 weeks of age. Data from sows in parity 7 to 10 were pooled to parity 7+ due to few observations. Number of piglets born totally and alive was smallest in litters of gilts, and increased with parity to parity 4-5 and 4, respectively. The number of surviving piglets increased from parity 1 to parity 3, and was smallest in parity 7+. Piglets of gilts were lightest and had the slowest growth rate. Piglets were heaviest and grew fastest in parity 2. Piglet growth rate decreased with sow parity from 2 to 7+. In summary, piglet production in terms of litter size, weight and growth rate was inferior in gilts, superior in parity 2-4, and decreased with parity thereafter.

Session 9 Theatre 1

Genetic control of environmental variability: evidence from snails, pigs and rabbits

D. Sorensen, Danish Institute of Agricultural Sciences, Department of Genetics and Biotechnology, Research Center Foulum, PB50, DK-8830 Tjele, Denmark

The classical model of quantitative genetics assumes that genotypes differ in mean performance but not in environmental (or residual) variance. An extension postulates that environmental variance is partly under genetic control, and that these genes may be correlated with those acting at the level of the mean. The extended model is relevant in studies of maintenance of variation, in the interpretation of results from selection experiments, and from an animal breeding perspective, it opens the possibility for controlling variation through selection. Several versions of the extended and classical model were fitted to litter size data of pigs and rabbits, and to mature weight data of snails, using a Bayesian, McMC approach. Model testing using posterior predictive distributions, and criteria of overall model fit and complexity such as Bayes factors and deviance information criteria, favour in all cases versions of the model postulating genetic control of environmental variation. Further, the means of posterior distribution of the correlation coefficient between genes affecting the mean and those affecting environmental variance was -0.60, -0.74 and 0.81, in pigs, rabbits and snails, respectively. The support of the posterior distributions were shifted far away from zero in all cases. There is need for more work in this area, including data from well designed selection experiments to further test the validity of the extended model.

Session 9 Theatre 2

Genetic homogenization of birth weight in rabbits: evolution of the characteristics of the genital tract after two generations of selection

G. Bolet[1], H. Garreau[1], T. Joly[2], M. Theau-Clement[1], J. Hurtaud[3], L. Bodin[1], [1]INRA, SAGA, BP 52627, 31326 Castanet-Tolosan, France, [2]ISARA, 31 place Bellecour, 69002, Lyon, France, [3]GRIMAUD FRERES Sélection, La Corbière 49450 Roussay, France

In order to study the possibility of selection to reduce the variability of a production trait, a divergent selection for the homogeneity of rabbit birth weight was initiated. After two generations of divergent selection, the difference in birth weight standard deviation between lines was highly significant, and a large indirect correlated response was observed for mortality at birth and later at weaning in favour of the homogeneous line. An experiment was conducted in order to determine whether the divergence between lines was due to an evolution of the morphological and physiological characteristics of the genital tract. After their third kindling, 33 females from the heterogeneous line and 31 from the homogeneous line were superovulated and slaughtered 72 hours after the insemination. Embryos were collected in the oviduct to be frozen; the number of ova shed, the weight of the uterine horns and there length and elasticity were measured. The irrigation of the uterine horns was evaluated by the shape and number of blood vessels. Results of this experiment are presented in the paper.

Session 9 Theatre 3

Possibilities for selection for uniformity in pig carcasses

E.F. Knol[1], P.R.T. Bonekamp[2] and P.E. Zetteler[1], [1]Institute for Pig Genetics (IPG), Beuningen, [2]Wageningen University, Wageningen, The Netherlands, P.O. Box 43, 6640 AA Beuningen*

Retail product standardisation is easier to realise in uniform carcasses. Slaughter plants react by improving their carcass evaluation systems, by decreasing the boundaries of acceptance and by increasing penalties for out of specification carcases. Our research question was whether or not genetics can play a role in increasing uniformity of pork. Traits of interest were HGP backfat and loindepth. First dataset came from a farrow to finish research farm with main emphasis on product comparisons (6 sire lines, 3 sow crosses). Both traits were analysed for variation. Weight, sex, sire line and dam cross explained respectively 23%, 18%, 5%,and 1% for backfat, and 31%, 2%, 8% and 0% for loin depth. Within line variation for sires and dams for backfat (loindepth) explained 3% (3%) and 6% (4%).
Within litter variation for backfat and loindepth was genetically analysed using ASREML, with fixed effects for trait litter level, HYS and litter size at slaughter (525 litters analysed). Heritabilities were 0.17± 0.08 for backfat and 0.04± 0.06 for loin depth. Data from three other farms with similar data were added (+ 2500 records, total of 3025 litters) to arrive at heritabilities of 0.18± 0.04 for backfat and 0.04± 0.03 for loin depth. Correlated reponses of uniform low backfat in dissection results were favourable for level of backfat, loindepth and deboned, defatted ham weight.

Session 9 Theatre 4

A weighted regression approach for the detection of QTL effects on within-subject variability

Dörte Wittenburg, Volker Guiard and Norbert Reinsch, Forschungsbereich Genetik und Biometrie, Forschungsinstitut für die Biologie landwirtschaftlicher Nutztiere FBN, Wilhelm-Stahl-Allee 2, 18196 Dummerstorf, Germany*

Quantitative trait loci may not only affect the mean of a trait but also its variability. A special aspect is the variability between multiple measurements of genotyped animals, for example the variance of somatic cell count around a lactation curve. A weighted regression approach was developed, assuming a normally distributed trait and taking the transformed sample variance s^2 between repeated measurements as observation for every genotyped individual. The weight for every observation was derived by a first-order approximation of the variance of the observations. For a daughter-design this weighted regression approach was evaluated in terms of precision of the estimate for the QTL-position, statistical power and compliance with the desired error probability under the null hypothesis.

Session 9 Theatre 5

Using the reaction norm approach to investigate genotype-by-environment interactions in the UK Suffolk Sire referencing scheme

G.E. Pollott, Department of Agricultural Sciences, Imperial College London, Wye campus, Ashford, Kent, TN25 5AH, United Kingdom

Genotype-by-environment interactions can be a significant source of variation in breeding schemes involving many flocks genetically linked by common sires. The use of the reaction norm approach to explain the systematic effect of different environments on genetic merit is becoming more widely used and has led to some interesting results on the way heritability changes with environment. This study investigates how bodyweight, fat and muscle depth in lambs are inherited in a range of UK environments using the reaction norm approach with an animal model. It also discusses the use of different measures of the environment and their effect on the genetics of the three traits. The analyses undertaken indicate that there are systematic effects of the environment, as measured by a range of lamb factors on scanning weight, fat depth and muscle depth. However, models including a random regression of sire on the environmental factor do not explain as much of the variation in scanning weight as the random effect of sire/flock/year. The use of a random regression model in place of the sire/flock/year results in a correlation between direct and maternal genetic effects more in line with the expected relationship between growth and milk yield, derived from other studies of these factors separately. The heritability of the recorded traits varied over the different environmental ranges found.

Session 9 Theatre 6

Evidence of genotype by time interaction for protein production in dairy cattle

P. Madsen[*1], *J. Pedersen*[2], *U.S. Nielsen*[2] *and J. Jensen*[1], [1]*Danish Institute of Agricultural Sciences, DK-8830 Tjele, Denmark,* [2]*The Danish Agricultural Advisory Service, DK-8200 Aarhus N, Denmark*

A sample of 212.806 first lactation protein records of Danish Holsteins, with at least 280 d in milk, was selected for this analysis. The data spanned the period from 1990 to 2002 and were divided into 5 three year periods. Each period was considered a different trait. Sire and residual variances were estimated for each trait in a multivariate model. Genetic ties across time periods were used to estimate genetic covariances between time periods.
Genetic variance increased slightly, but insignificantly, over the 15 year period. However, the environmental variance increased significantly with 50.1%, resulting in a decrease in heritability from .33 in early years to .25 in late years. Genetic correlations between time-periods were all significantly lower than 1.0 and were .92-.95 between consecutive periods and .75-.85, .63, and .32 as time between periods increased.
Considerable changes have taken place in the Danish dairy production system during the period. Production have increased by more than 20%, feeding have changed and herd sizes have doubled. Which factors contribute to the interactions found is presently unknown.
Clearly a time by genotype interaction for protein production was found in Danish Holsteins. This needs to be reflected in systems for genetic evaluation. Covariance functions could probably be used to model the development in covariance structure.

Session 9 Theatre 7

A test of quantitative genetic theory using *Drosophila*: effects of inbreeding and rate of inbreeding on heritabilities and variance components

T.N. Kristensen[1,2], *A.C. Sørensen*[2,3]*, *D. Sorensen*[2], *K.S. Pedersen*[1], *J.G. Sørensen*[1], *and V. Loeschcke*[1], [1]*University of Aarhus, Denmark,* [2]*Danish Institute of Agricultural Sciences,* [3]*The Royal Veterinary and Agricultural University, Denmark*

We investigated the effects of inbreeding treatments (fast (effective population size, N_e=2) and slowly (N_e=8) inbred to the same absolute level of inbreeding - and a control) on heritability, phenotypic, genetic and environmental variances of sternopleural bristle number in *Drosophila melanogaster* in ten replicate lines within each treatment. Standard least squares regression and Bayesian methods were used in the analysis. Heritability and additive genetic variance within lines were highest in the controls and higher in slowly than in fast inbred lines. Thus, slowly inbred lines retain more evolutionary potential. Between line additive genetic variance was larger with inbreeding and more than twice as large among fast than among slowly inbred lines. The different patterns of redistribution of genetic variance within and between lines in the two inbred treatments cannot be explained by invoking selective neutrality and additive gene action. Inbreeding and the rate of inbreeding affect environmental sensitivity, as environmental variances were higher with inbreeding, and more so with fast inbreeding. Phenotypic variance decreased with inbreeding, but was not affected by the rate of inbreeding. No inbreeding depression was observed. These results are important for understanding quantitative genetic and phenotypic variation within and between large and small populations.

Session 9 Theatre 8

Estimation of ancestral inbreeding coefficients

S. Suwanlee[1], R. Baumung[1], J. Sölkner[1] and I. Curik[2], [1]Division of Livestock Sciences, University of Natural Resources and Applied Life Sciences Vienna, Gregor Mendel-Strasse 33, A-1180 Vienna, Austria, [2]Department of Animal Science, Faculty of Agriculture, University of Zagreb, Zagreb, Croatia*

Ballou (1997) proposed ancestral inbreeding coefficients as a means for evaluating the purging of deleterious alleles through inbreeding. A computer simulation was conducted to examine the validity of the formula which aims at estimating the proportion of alleles within a genome that has undergone inbreeding in the past. Different breeding population sizes with a sex ratio 1:2 were considered. Population size was kept constant over 50 discrete generations of random mating. Three genetic models (neutral, detrimental and lethal alleles) influencing individual fitness were investigated. Further, two initial allele frequencies, 0.01 and 0.005 for lethal and detrimental alleles were simulated. In all scenarios investigated, the ancestral inbreeding coefficient is overestimated by Ballou's formula, because the probability of an allele being identical by descent (*F*) and having undergone inbreeding in the past (*Fa*) are not independent. Overestimation decreases with increasing breeding population size but initial allele frequency and genetic model have only small effects on estimation. The gene dropping method is proposed as an alternative providing unbiased estimates of ancestral inbreeding coefficients.

Session 9 Theatre 9

Modelling of the inbreeding depression as a function of the age of the inbreeding

D. Hinrichs, M. Holt, J. Ødegård, O. Vangen, J.A. Woolliams and T.H.E. Meuwissen, Department of Animal and Aquacultural Sciences, Norwegian University of Life Sciences, N-1432 Ås, Norway*

Due to the fact that slow inbreeding causes less inbreeding depression than fast inbreeding several studies analysed the effect of new versus old inbreeding on inbreeding depression.
The aim of this study was to fit a model for the inbreeding depression, which takes the age of the inbreeding into account. A pedigree, including 74,630 animals and 125 generations was analysed. The data contained information on 29,635 litter sizes. The increase of inbreeding per generation looking further back in time was estimated and summarised into blocks.
The estimated inbreeding depression for the total level of inbreeding was -9.72 pups. Depending on block size and age of the inbreeding the estimates for the depression ranged from -26.25 to 11.15. The estimated heritability of litter size was 0.17. The analyses of the inbreeding depression of the blocks of generations suggested an exponential function to model the effect of the age of inbreeding.
The main conclusion from fitting an exponential function was that inbreeding had an effect over 32 generations and thereafter it disappears. The results suggest that it is possible to model the inbreeding depression as a function of the age of the inbreeding, and that inbreeding should stay low for 32 generations in order to avoid the negative aspects of inbreeding depression.

Evaluation of the epistatic kinship based on genotyping fullsib pairs of three Göttingen Minipig populations

C. Flury and H. Simianer, Institute of Animal Breeding and Genetics, Albrecht-Thaer-Weg 3, 37075 Göttingen, Germany*

The epistatic kinship describes the probability that chromosomal segments of length are identical by descent. It is an extension from the single locus consideration of the kinship coefficient (Malécot, 1948) to chromosomal segments. The parameter reflects the number of meioses separating individuals or populations. Hence it is suggested as measure to quantify the genetic distance of sub-populations that have been separated only few generations ago.
The properties of the approach were investigated in a Monte Carlo simulation. Based on theoretical results seven segments < 0.15 Morgan on six porcine chromosomes were chosen. The length of those segments was roughly determined by the choice of six microsatellites from the genetic maps USDA-MARC_v1 and USDA-MARC_v2.
For the epistatic kinship haplotypes are relevant. Therefore an easy and fast method that derives haplotypes without pedigree information was sought. We present a method for the reconstruction of haplotypes based on fullsib information. The power of the method decreases with increasing segment length and with decreasing number of alleles per locus. The ideal segment length seems to lie between 0.01 and 0.10 M. The insights from genotyping 55 fullsib pairs of each of the three Göttingen Minipig populations for the seven chromosomal segments are expected to lead to a further optimization of the epistatic kinship as a new measure for the genetic similarity between populations.

Session 9 Theatre 11

The role of gene banks as a safe guard in scrapie genotype eradication schemes

T. Roughsedge[1], B. Villanueva[1] and J.A. Woolliams[2], [1]Sustainable Livestock Systems, SAC, West Mains Road, Edinburgh EH9 3JG, [2]Roslin Institute, Roslin, Midlothian EH2 59PS, United Kingdom*

Breeding plans have been adopted in many European countries to eradicate scrapie from sheep populations. The plans, through selection for particular alleles at the PrP locus, aim to increase the frequency of the resistant ARR allele and to remove the most scrapie susceptible allele (VRQ). Adopting a genotype eradication breeding programme incurs a number of risks including those associated with TSEs that are as yet unknown, but may emerge in the future. A new TSE may appear, more threatening than scrapie, for which the currently favoured allele may not confer resistance and the currently disfavoured allele may be found to be resistant. Also, favourable attributes of particular breeds may be lost as a consequence of selecting for specific PrP alleles. Gene banks provide a useful mechanism to alleviate the risks identified. The role of semen gene banks in future reintroduction of removed alleles was investigated. Computer simulations were run for various breed types to quantify the capacity of the bank required for reintroducing a particular allele. The effect of the differing levels of fertility, fecundity and mating ratio between breed types on the speed of reintroducing banked alleles was quantified. Terminal breeds required two and a half times as many semen straws as hill breeds to produce the same frequency of ewes homozygous for the desired allele in the population after a ten year reintroduction period.

Session 9 Theatre 12

Individual-based assessment of population structure and admixture levels among Austrian, Croatian and German draught horses

I. Curik[1], T. Druml[2], R. Baumung[2], K. Aberle[3], O. Distl[3], and J. Sölkner[2], [1]University of Zagreb, Faculty of Agriculture, Animal Science Department, Zagreb, Croatia, [2]BOKU - University of Natural Resources and Applied Life Sciences, Vienna, Austria, [3]University of Veterinary Medicine Hannover, Institute of Animal Breeding and Genetics, Hannover, Germany*

Population structure and admixture levels among Austrian (Noriker Horse Salzburg-S, Noriker Horse Carinthia-K), Croatian (Posavina Horse-P, Croatian Draught Horse-H) and German (South German Draught-G, Black Forest Horse-B, Schleswig Draught Horse-C, Rhenish Draught Horse-R, Mecklenburg Draught-M, Saxon Thuringa Draught-T, Altmaerkish Draught-A) draught horses were assessed using an individual based approach to genotypic information of 31 microsatellite loci and 434 horses. When a priori population assignment was neglected the presence of five genetically different groups was suggested: a) S, K and G populations, b) B population, c) P and H populations, d) C population and e) A, R, M and T populations. Classical distance analyses, multivariate analyses and graphical presentations showed that group a) and e) are the most distinctive groups. Molecular information revealed recent migration histories with surprisingly high accuracy. Nonetheless, microsatellites were powerful in making correct assignments (more than 90%) to the originally defined populations. The information obtained should aid conservation strategies and policies.

Session 9 Theatre 13

The estimation of extinction probabilities of five German cattle breeds

J. Bennewitz[1], T. H. E. Meuwissen[2], [1]Institute of Animal Breeding and Husbandry, Christian-Albrechts-Universität, D-24098 Kiel, [2]Department of Animal Science, Agriculture University of Norway, Box 5052, 1432 Aas, Norway*

The estimation of the expected loss of genetic diversity and of the breed marginal diversities in a set of breeds within a defined future time horizon requires estimates of breed extinction probabilities. In this study the extinction probabilities of five German dual purpose dairy cattle breeds were estimated by population viability analysis. A regression was used to estimate the infinitesimal mean and variance of the population growth and this is based on the diffusion approximation of the density independent population growth (also known as the Dennis regression model). Annual number of milk recorded cows was used as census data. Based on the regression results, the extinction probabilities and their confidence intervals were estimated for a wide variety of future time horizons using Monte Carlo time series simulations.

The obtained extinction probabilities were reasonable but in two cases they depended heavily on the time horizon considered. Additionally, the confidence intervals became very wide with an increased time horizon.

We recommend the use of extinction probabilities for a set of future time horizons rather than for a single future time and the largest time horizon should not be too large. The validity of the use of milk recorded cows as census data and of the model assumptions are discussed.

Estimation of genetic parameters for fertility and hatchability in the two laying cycle (four periods) of three White Leghorn strains (repeated measurement)

M. Tazari, I. McMillan, J.W. Wilton, V.M. Quinton, Centre for Genetic Improvement of Livestock, Department of Animal and Poultry Science, University of Guelph, Guelph, Ontario N1G 2W1, Canada*

This research involves a multiple-trait analysis using repeated measurements to estimate the genetic parameters for fertility and hatchability in three White Leghorn Strains, strain11 and 44 with long-term multiple-trait selection and strain77, an unselected control strain. Production was synchronized according to physiological age.

The traits were measured in four periods of egg production (pre-peak, peak, post-peak and second cycle). An animal model (multiple-trait) with restricted maximum likelihood method (REML) was used in the analysis. The data were analyzed using SAS and DMU packages.

The 4 periods had a significant effect on age at first egg, egg weight and mature body weight ($P<0.05$). The estimates of genetic correlation among periods for fertility and hatchability were highest, between the last periods of the first cycle and the second cycle. The highest heritability accurated in the second period of the first cycle (peak).

In conclusion, keeping birds for better fertility and hatchability performance in the second cycle of production is best performed by using the phenotypic value in the last period of the first cycle. Also, the optimal selection strategy for improving performance in laying birds involves using the fertility and hatchability data from period 2 when both are at their peak.

The evaluation of genetic variability of wild ancestors, founders and autochthonous cattle breeds in Serbia

M. Savic, S. Jovanovic, R. Trailovic, Faculty of Veterinary Medicine, University of Belgrade, Department of Animal Breeding and Genetics*

The genetic characterization of endangered autochthonous breeds has been performed to allow the reconstruction of biodiversity in Serbia. The identification of genetic variability, allelic genes and breed specific gene distribution of endangered cattle breeds, such are Podolian and Busha, are proposed topics of the research. The evaluation of genetic variability in autochthonous populations, based on paleozoological ancestor remnants, was performed, in order to establish characteristics of domesticated and transitional animals and wild ancestral forms inhabiting Balkan peninsula during Neolithic period, as well as anthropogenic influence on their evolution. Radiocarbon dating of Neolithic animal remnants in settlement Laole, 5500-4800 years B.C., containing 3487 animal bone fragments, has been performed. The morphological analysis showed that 42% of the bone remnants were of bovine origin. According to morphometric analysis of 18 bovine horns, the two different ancestors of domesticated cattle were established : Bos primigenius and Bos brachyceros. Zoomorphic figurines from Laole are among the best preserved animal Neolithic figurines. Lateral holes on the rostral scull part are clearly visible on ox figurines. These findings point to possible cattle restraining by humans in Laole. Considering that migrations from Middle/Far East to Europe have passed through Balkan, these data can give further information about wild ancestors, early domesticated animals and the development of local animal husbandry.

Session 9 Poster 16

Estimation of inbreeding in Iberian pigs using microsatellites

E. Alves, C. Barragán and M.A. Toro, Departamento de Mejora Genética Animal, INIA, Carretera La Coruña km., 28040 Madrid, Spain

Genetic markers have been proposed as a useful tool to estimate inbreeding in the absence of pedigree. In the present work 62 pigs from two related strains of Iberian breed, Guadyerbas and Torbiscal, belonging to a conservation programme with completely known pedigree since 1945, have been genotyped for 49 microsatellites distributed over all the genome and the molecular inbreeding calculated as the proportion of loci homozygous. These values were compared with the genealogical inbreeding that ranges from 0.13-0.37 (all animals),(0.33-0.37 (Guadyerbas) and 0.13-0.21 (Torbiscal). The correlation between the molecular and the genealogical inbreeding was 0.69 when all animals were considered, but it decreases remarkably to -0.32 and 0.19 for the Guadyerbas and Torbiscal populations respectively. Simulated results using the actual pedigree and the same number of markers resulted in correlation values of 0.90 for all animals and 0.15 and 0.27 for the Guadyerbas and Torbiscal respectively. The relevance of these findings is discussed.

Session 9 Poster 17

Sequence variation of the PRKAG3 gene in two cattle breeds: a phylogenetic analysis

E. Ciani[1], M. Roux[2], R. Ciampolini[1], E. Mazzanti[1], F. Cecchi[1], H. Leveziel[2], S. Presciuttini[1], V. Amarger[2], [1]Dept of Animal Production, University of Pisa, Italy, [2]Animal Molecular Genetics Unit, INRA, University of Limoges, France

Thirty-three SNPs were characterized in a region of 6.8 kb of the bovine PRKAG3 gene, a member of a 5'-AMP-activated protein kinase (AMPK) family. Here, we show results of haplotype and network analysis carried out in a sample of 97 Chianina and 100 Holstein-Friesian animals typed for 11 markers. Only 10 different haplotypes were present in the total sample (out of more than a thousand theoretically possible), two at high frequencies (53% and 32%), three at low frequencies (3% to 8%), and five present in a single or two instances only ($p < 0.005\%$). Haplotype distribution was significantly different between breeds, and haplotype diversity (heterozygosity) was significantly lower in Holstein-Friesian. Based on haplotype sequence comparison, a phylogenetic network was constructed, leading to an unambiguous topology. A high-frequency haplotype was located at centre, and two main branches radiated from it, connecting the more distant haplotypes through four or five sequential base substitutions. Three SNPs were associated to amino acid change; considering only these substitutions, the total sample was subdivided in four groups of 255, 126, 12 and 1 different haplotypes, respectively. This work is preliminary to the analysis of possible associations of specific haplotypes with phenotypic traits of commercial interest.

Session 9 Poster 18

Genetic relationships between populations of Andalusian bovine local breeds in danger of extinction from the polymorphism of microsatellites of ADN

E. Rodero[1], P.J. Azor[2], M. Luque[1], A. Molina[2], M. Herrera[1], M. Valera[3] and A.Rodero[2], [1]Department of Animal Production, University of Cordoba (UCO), [2]Deparment of Genetic, University of Cordoba (UCO), [3]Department of Sciences Agroforestry (EUITA), Univerity of Seville, Spain

The Berrenda en Negro, Berrenda en Colorado, Pajuna and Cárdena Bovine breeds have been studied in a complete program of characterization and evaluation in order to conserve and improve them. We have genotyped a total of 437 animals using 31 microsatellites. They are breeds that according to our census, are in danger of extinction and are raised in extensive meadow systems (mediterranean wood Quercus) in Andalusia, Extremadura, Castilla-La Mancha and Castilla-Leon areas. They are reared for meat production, although also they are used in the handling of the fight cattles and in annual pilgrimages to religious shrines.

All loci have been polymorphic in all the breeds and although in some of them we could not to work with a high number of animals, the drift had not an important paper.

The differences between the results seem to show that the Wahlund effect has been of little intensity.

The molecular variance analysis (AMOVA) considered for the four breeds showed that 2,56 % of the total genetic variability were due to differences between breeds and the rest due to individual differences.

Session 9 Poster 19

Preliminary study on microsatellite variation of Baluchi sheep breeding flock

Saber Qanbari[1#], Morad Pasha Eskandari Nasab[1], Saeed Esmaeel Khanian[2], Rahim Osfoori[3], [1]Department of Animal Science,- Zanjan University, P.O. Box 313, Zanjan, Iran, [2]National Research Institute of Animal Science, Karaj, Iran, [3]Research Center of Agricultural-Jihad, Zanjan Province, Iran

Genetic variation at 10 microsatellite loci was investigated in Baluchi sheep breeding flock. The study was conducted on overall 150 sheep. Microsatellite markers were amplified by polymerase chain reaction (PCR), followed by 8% denatured polyacrylamide gel electrophoresis. A part of two monomorphic loci microsatellites showed extensive polymorphism with allele numbers ranging from 2-9 and polymorphism information content (PIC) values in the range of 0.09-0.83. Heterozygosity as an intrapopulation variation criterion was calculated and Hardy-Weinberg proportions were also considered. The average heterozygosity calculated from data on polymorphic and monomorphic loci was estimated as 0.48 (SE= 0.11). An estimate of average heterozygosity excluding the monomorphic data was also calculated as 0.63 (SE= 0.09). The results of this study indicate, despite the historical selective breeding and closed flock system a relatively high level of heterozygosity still exists in the representative sheep flock.

Session 9 Poster 20

The pool of erythrocyte antigenes in Yakutsky cattle

E.V. Kamaldinov, V.L. Petukhov, O.S. Korotkevich, A.I. Zheltikov, Research Institute of Veterinary Genetics and Selection of Novosibirsk State Agrarian Universit, Russia

For centuries Yakutsky cattle (YC) have been adapted to severe conditions of northern Russia. YC have high growth rate and fat content in milk in early postnatal period of ontogenesis. They were imported about 6 generations ago to Novosibirsk region (NR). Unfortunately, there are only two small populations of purebred cattle mentioned above in Yakutia and NR. The erythrocyte antigens of seven blood groups of Yakutsky cattle were studied. High frequencies of antigens G_2, G_3, Y_2, those of C_1, C_2, X_2 and H' in blood groups B (0.75-0.83), C and S-U, respectively, were typical of YC. Yakutsky cattle differ from other Siberian breeds in high frequency of antigene O_1 of the B system. The index of genetic similarity of YC to Black-and-White cattle in NR was equal to 0.8423 and higher than that between two YC populations in NR and Yakutia. At present the question about YC gene pool preservation is under discussion.

Session 9 Poster 21

Analysis of bovine PIT1 gene polymorphism in Iranian Sarabi cattle (*Bos taurus*) using PCR based RFLP

J. Tavakolian[1], S. Zenali[2], B. Azimifar[2], N. Asadzadeh[1], A. Javanmard[3]. [1]Animal Science Research Institute, Karaj. [2]Human Genetic laboratory, Tehran. [3]Agriculture Biotechnology Research Institute, Tabriz, IRAN

The part of the bovine genome showing a superior action and explaining the major part of variation of the economical production traits were known as QTL. Pit1, which is also termed hormone factor-1, is a pituitary-specific transcription factor is responsible for pituitary development and hormone expression in mammals. The main factions of Pit1 are binding and trans-activity the promoters of both growth hormone (GH) and prolactin (PRL) genes and polymorphism in this gene had significant relationships with both milk and meat production traits. This gene was subjected to different molecular studies as key for genetic variation in dairy cattle. This study carried out to analysis of Hinfl polymorphism in Pit1 in Iranian Sarabi cattle. DNA was extracted from blood or sperm samples collected from 82 Sarab bulls and cows and submitted for polymerase chain reaction (PCR) followed by digestion with Hinfl restriction enzyme. The frequency of the A and B alleles of This gene was 73.2 and 26.8 percent in Sarabi cattle breeds respectively.

Session 9 Poster 22

Molecular analysis of the bovine leptin gene in Iranian Sarabi cattle (Iranian Bos taurus)

A. Javanmard[1], G. Elyasi-Zarringabayi[1], A.A. Gharadaghi[2], M.R Nassiry[3], A. Javadmanash[3]
[1]Biotechnology Laboratory, East Azarbejan Agriculture and Animal Research Center, Tabriz, Iran. [2]Animal Science Research Institute, karaj. Iran· [3]Department of Animal Science, Faculty of Agriculture, Ferdowsi University of Mashhad, Mashhad, Iran.

Progress in animal breeding may be improved by combining traditional performance data and molecular genetic information on quantitative trait loci in selection index. Candidate genes are chosen for study on the basis of known relationships between biochemical and physiological process. Leptin, a 16 KD-protein producing by obesity gene in white adipose tissues, is involved in regulation of feed intake, energy balance, fertility and immune function. In cattle, the leptin gene is located on chromosome 4. It consists of 3 exons and 2 introns that only 2 exons translate into protein. Totally 66 animals from Sarab and Shabestar stations were genotyped for this project. A strategy employing standard polymerase chain reaction was used to amplify a 422-bp fragment of intron 2 from blood DNA. Digestion of amplicons with restriction enzyme Sau3AI revealed two alleles. Allele A had 2 fragments, 390 and 32 bp and allele B had 3 fragments, 303, 88 and 32 bp (fragments 88 and 32 have not detected on the gel). Frequencies of genotypes were 0.32, 0.43 and 0.25 for AA, AB and BB respectively. The populations were in Hardy-Weinberg equilibrium. This polymorphism could be evaluated for marker assisted selection and the developed PCR method would expedite screening for large number of animals.

Session 9 Poster 23

Genetic polymorphism of ovine calpastatain locus in Iranian kordi sheep by PCR-RFLP

M. R..Nassiry[1], A. Javadmanesh[1], M. Nosrati[1], A. Mohamadi[1], A. Javanmard2. [1]Ferdowsi University of Mashad, Faculty of agriculture, Center of Excellence for Animal Science, P.O. Box: 91775-1163,[2] Biotechnology Laboratory, East Azerbaijan Agriculture and Animal Research Center

Marker-assisted selection is one of the new methods that improve accuracy and progress of selection in animal breeding. The calpastatin gene is a candidate gene in order to meat tenderness. It is located on ovine chromosome five and plays important roles in formation of muscles, degradation and meat tenderness after slaughter. This study was designed to investigate the calpastatin genotype's using PCR-RFLP analysis. Blood samples were collected from 100 pure bred Kordi sheep. The extraction of DNA was based on Boom R. method. A 622 bp fragment from Exon and intron I of the ovine CAST I gene was amplified. The digestion of PCR product by MspI restriction endonuclease generated two alleles, M (presence of cutting site) and N (absence of cutting site) with frequencies of 88% and 12% respectively. Three patterns were observed which frequencies were 76.0%, 24.0% and 0.00% for MM, MN, and NN respectively. G and $\chi 2$ test confirmed the Hardy-Weinberg equilibrium in this population.

Session 9 Poster 24

Polymorphism of growth hormone and growth hormone receptor genes in two Iranian Sarabi (*Bos taurus*) and sistani (*Bos indicus*) cattle breeds

A. Javanmard[1], R. Miriaei-Ashtiani[2], A. Torkamanzehi[3], M. Moradi Sharbabak[2], M. Esmailzadeh[4], M.R. Nassiry[5], G. Elyasi Zarringabayi[1]. [1]Agriculture Biotechnology Research Institute, Tabriz· [2]Department of Animal Science, Faculty of Agriculture, University of Tehran, Karaj .[3] Department of Animal Science, Faculty of Agriculture, University of Zabol· [4] Razi Serum and Vaccine Research, Institute·[5]Department of Animal Science Ferdowsi University Mashad. IRAN

The use of molecular techniques could help to solve of the limitation of the traditionally Animal breeding methods. Candidate genes are chose for study on basis known relationship in biochemical or physiological processes. Genetic variation in GH/GHR genes have been shown to be responsible for traits such as milk production, feed intake, immune function, spermatogenesis, and steredogensis, wool growth. The present study investigated genetic variability of one polymorphism to the GH and GHR genes between Iranian Sarabi and Sistani cattle breeds. DNA was extracted from blood or sperm samples collected from 50 Sistani and Sarab bulls and cows and submitted for polymerase chain reaction (PCR) followed by digestion with AluI restriction enzyme. The frequency of the AluI (+) allele of GHR and Leu allele of GH was 0.56,0.20 and 0.45,0.46 for Sarabi and Sistani cattle breeds respectively. The result of the study may demonstrate uniquely different between taurin and indicince BGH gene is consistent of the two races. This polymorphism could be evaluated for marker assisted selection and the developed PCR method would expedite screening for large number of animals.

Session 9 Poster 25

Polymorphism of bovine lymphocyte antigen DRB3.2 alleles in Iranian Holstein cattle

M. Pashmi[1], A.R. Salehi[1], A. Ghorasi[2], M.R. Mollasallalehi[3], A. Javanmard[4], S. Qanbari[5], R. Salehi-Taba[2], [1]Department of Animal Science, University of Tehran.[2] National Research Center for Genetic Engineering and Biotechnology, Tehran. [3] Animal Breeding Center, Karaj. [4] Animal Research Institute, Tabriz. [5] Department of Animal Science, University of Zanjan

Breeding goals for dairy cattle have focused mainly on increasing productivity and have ignored disease resistance. In this study polymorphism of the second exon of BoLA-DRB3 gene of 96 Iranian cattle was investigated. Genomic DNA extracted from whole blood samples and two-step polymerase chain reaction (PCR) was carried out in order to amplify a 284 bp fragment of target gene. Nested- PCR products were digested with three restriction endonucleases RsaI, BstYI and HaeIII. Digested fragments were analyzed by Polyacrylamid gel electrophoresis. Twenty-two BoLA-DRB3 alleles were distinguished with frequencies ranging from 0.5 to 19.3%. Identified alleles include: BoLA-DRB3.2*3, *6, *7, *8, *9, *10, *11, *12, *13, *14, *15, *16, *20, *21, *22, *23, *24, *25, *27, *28, *32 and *51. Their frequencies found to be 2.6, 2.6, 1.0, 13.5, 0.5, .05, 13.0, 1.6, 1.6, 0.5, 2.6, 14.1, 0.5, 1.6, 7.3, 5.2, 19.3, 2.1, 2.1, 1.6, 0.5 and 5.7% respectively. The most frequent alleles (BoLA-DRB3*8, *11, *16, *22 and *24) accounted for 67.2% of the observed alleles. Results of this study indicate that BoLA-DRB3 locus is highly polymorphic among tested animals.

Session 9 Poster 26

Genetics parameters of variability for litter size and litter weight at birth in *Mus musculus*

J.P. Gutiérrez, B. Nieto, P. Piqueras and C. Salgado, Departamento de Producción Animal, Facultad de Veterinaria, U.C.M. Av. Puerta de Hierro s/n, 28040 Madrid, Spain*

Genetic parameters of variability of litter size, and litter and mean individual weight at birth in mice, have been explored from a population designed to conduct an experiment aiming to decrease genetic variability of litter size.
The population consisted of 875 females daughters of 43 males with two consecutive parities, Genealogical data was traced back 18 generations and the whole data set consisted of 4129 records of litter size and 1796 of mean individual and total litter weight records considering up to 3997 pedigree records.
The methodology used in the treatment of data was that proposed by SanCristobal-Gaudy *et al.*, 1998. Based on this, the logarithm of the squared of the estimated residuals resulting from a BLUP evaluation, were used to estimate variance components via REML.
The heritabilities of the variability under an univariate model were 0.0040, 0.0072 and 0.0139 for litter size, litter weight and mean individual weight at birth respectively. Heritabilities were slightly higher when bivariate models for mean and variability traits where analysed together and genetic correlations between them were highly negative. These results suggest a low genetic base for variability of the evaluated traits. A selection generation can be carried on in order to have an estimate of realized heritability.

Session 9 Poster 27

Study of canalization in an experiment of divergent selection for uterine capacity in rabbit

N. Ibáñez[1], D. Sorensen[2], R. Waagepetersen[3], A. Blasco[1], [1]Departamento de Ciencia Animal, Universidad Politécnica de Valencia, P.O. Box 22012. 46071 Valencia, Spain, [2]Departament of Genetics and Biotechnology, Danish Institute of Agricultural Sciences, PB 50, DK-8830 Tjele, Denmark, [3]Departament of Mathematical Sciences, Aalborg University, 9229 Aalborg, Denmark*

One of the most important traits in domestic prolific species is litter size. In recent years, homogeneity of litter size has become important, because it may influence economic efficiency. A 10 generation divergent selection experiment for uterine capacity in rabbits was carried out at the Universidad Politécnica de Valencia. The objectives are first, to evaluate the response to selection for uterine capacity with two models: the classical infinitesimal model (additive genetic model with homogeneous variance) and an additive genetic model with heterogeneous residual variance (San Cristobal-Gaudy et al., 1998). Secondly, to investigate whether there are additive genes controlling residual variation and whether these genes are correlated with those affecting mean uterine capacity. The inferred response to selection based on either model was similar. Studies of both, model comparison based on the deviance information criterion, and model fit, give higher relative credibility to the heterogeneous variance model. The statistical analysis provides evidence for the existence of additive genes affecting residual variation and for a negative correlation between these genes and those affecting mean uterine capacity. This opens the possibility for controlling residual variation of litter size through selection.

Session 9 Poster 28

Distinct genotypes from different origins for black coat color in pigs

S.H. Han[1], Y.L. Choi[1], M.S. Ko[1], M.Y. Oh[2] and I.C. Cho[1], [1] National Institute of Subtropical Agriculture, Rural Development Administration, Jeju 690-150, Korea, [2] Department of Life Science, College of Natural Sciences, Cheju National University, Jeju 690-756, Korea*

Melanocortin receptor 1 gene plays a key role in development of black coat colour. A total of six *MC1R* alleles has been documented in various pig breeds including wild boars. Of those, there are well-known two different genotypes responsible for porcine black coat colour, which were defined as *MC1R**2 and *MC1R**3, respectively. The present study investigated the polymorphisms of *MC1R* gene in Korean wild boars and native pigs. SSCP analyses divided those of Korean wild boars into five haplotypes corresponding to the *MC1R*1* for the wild type and those of Korean native pigs did into two to *MC1R*2* and *MC1R*3* for the black. The phylogenetic tree showed two major clusters; one is comprised of those from European wild boars and black pig breeds, and another did those from the Asian wild boars and native black pigs. Clustering patterns of the *MC1R* gene sequences indicate that the black coat color of the present pigs originated from wild boars but did not by mutations during domestication. Consequently, this study suggests that two representative genotypes for black coat color in pigs originated probably from two kinds of ancient black wild boars different continents, Asia and Europe.

Session 9 Poster 29

Genetic diversity and maternal origins of Jeju (Korea) native pigs

I.C. Cho[1], Y.L. Choi[1], M.S. Ko[1], J.T. Jeon[2], S.J. Kang[1] and S.H. Han[1], [1] National Institute of Subtropical Agriculture, Rural Development Administration, Jeju 690-150, Korea, [2] Division of Animal Science, College of Agriculture and Life Sciences, Gyeongsang National University, Jinju 660-701, Korea*

The Jeju native pigs which are small-sized, black, and raised only on Jeju Island in South Korea, were studied to elucidate the genetic diversity and evolutionary maternal origins. Molecular genetic analyses for the nucleotide sequences of mitochondrial genome and *MC1R* gene regulates the coat color development, were carried out for five domestic pig breeds, Korean wild boar, and Jeju native pig. The results of DNA sequencing and subsequent analyses showed breed-specific characteristics, and divided the pig populations into several mtDNA haplotypes. We found two mtDNA haplotypes in the population of Jeju native pigs; one of both is common with several domestic pig breeds has been introduced from Europe in twentieth century, and another does specific for Jeju pig population. The coat color of Jeju native pigs are uniform-black so that the *MC1R* genotypes (*MC1R*2* and *MC1R*3*) found in this study support genetically their black coat color in spite of several silent mutations in some animals. However, mtDNA haplotypes and *MC1R* genotypes found in the population of Jeju native pig suggests a hypothesis on the multiple maternal origins from domestic European pig lineages and indigenous Asian pig lineages.

Session 9 Poster 30

Allele frequencies of Stearoyl CoA desaturase genetic variants in various cattle breeds

B. Moioli, Fr. Napolitano, G. Congiu, L. Orrù and G. Catillo, Istituto Sperimentale per la Zootecnia, via Salaria 31, 00016 Monterotondo, Italy*

We genotyped 23 Chianina, 48 Friesian, 26 Maremmana, 24 Piedmontese, 26 Podolica, 23 Simmental animals at exon 5 of Stearoyl-CoA desaturase (SCD) gene, which codes for the rate-limiting enzyme in the biosynthesis of monounsaturated fatty acids. A Single Nucleotide Polymorphism in this exon gives rise to a different amino-acid codon, the Alanine residue becoming a Valine residue. Percentage of heterozygous animals was 51.19 %, with a maximum of 65.22 % in the Piedmontese and a minimum of 36.36% in the Chianina. Alanine homozygous were 40.48 %, with a maximum of 59.09 % in the Chianina and a minimum of 21.74 in the Simmental. Valine homozygous were only 8.33 %, with a maximum in the Simmental (17.39 %) while in the Friesian no homozygous was detected. Genotype frequencies of the Friesian resulted significantly different from those of Maremmama, Podolica, and Simmental. Because the Valine at this position is referred to be the ancestral amino-acid of SCD, our results confirm that frequency of this variant is higher in the ancestral ones (Maremmama, Podolica, Simmental). Because it has been referred that the Alanine variant is associated to higher content of monounsaturated fatty acid in carcasses, the low frequency of Valine homozygous in the Chianina and Piedmontese makes to suppose that selection for better flavour of meat has been performed.

Session 9 Poster 31

Ascertainment of evolutionary processes from the genetic variation associated to each geographic point in a map

J.R. Quevedo[1], E. Fernández-Combarro[1], L.J. Royo[2], I. Álvarez[2], A. Beja-Pereira[3], I. Fernández[2], A. Bahamonde[1] and F. Goyache[2], [1]Centro de Inteligencia Artificial de la Universidad de Oviedo, Gijón, Spain, [2]SERIDA, Gijón, Spain, [3]CIBIO-UP, Universidade do Porto, Vairão (VCD), Portugal*

Available techniques suited for spatial analysis of genetic variation are scarce, preventing the ascertainment of the source and direction of evolutionary processes. Here we propose a new approach using actual information at population level (size, allelic frequencies) to infer a distribution of the individuals forming said population in an area around a node considered as its center of the distribution. We subsequently define an area around each geographical point on our map to sample the individuals (belonging to any 'real' population) contributing to the genetic variation at this given geographical point and compute the genetic variables associated with it. The genetic information associated with each geographical point will be graphically projected onto the map to allow visual inspection of possible patterns of variation. The methodology is demonstrated by its application to published microsatellite data sets for two livestock species and for Eurasian otter (*Lutra lutra*) in Scotland, using two single parameters (variability and uniqueness) that allow us to take into account both the highly stochastic nature of gene variation and the distribution of rare alleles. The methodology described here has been implemented in the freely downloadable program GeoGen (http://www.aic.uniovi.es/GeoGen/).

Session 9 Poster 32

Combined use of genealogy and microsatellites in the endangered Xalda sheep

I. Álvarez[1], J.P. Gutiérrez[2], L.J Royo[1], I. Fernández[1], E. Gómez[1] and F. Goyache[1], [1]SERIDA, Gijón, Spain, [2]Universidad Complutense de Madrid, Madrid, Spain

In small populations with shallow pedigree the combined use of genealogical and molecular markers information is recommendable to assess losses in genetic variability. Here we compare the 'real' molecular coancestry in the base population (Mol_0) with different estimations of this parameter in the rare Xalda sheep breed of Asturias. The Xalda individuals were classified in 4 different groups according to their pedigree knowledge using the 'equivalent generations' parameter: founder generation (0) when no parent is known, and generations 1th, 2nd and 3th. Up to 160 Xalda individuals corresponding to the 0 (46), 1th (37), 2nd (38) and 3th (39) generations were genotyped with 14 microsatellites. Molecular and genealogical coancestry coefficients have been computed for each pair of the 160 genotyped individuals. Average molecular coancestry was of 0.338, 0.357, 0.468 and 0.416 for, respectively, generations 0, 1th, 2nd and 3th. Estimations of the average molecular coancestry in the base population (M_0) were obtained as $M_0 = (M_g - f_g) / (1 - f_g)$, where is the genealogical coancestry for the generation g. Difference between the estimates of M_0 and Mol_0 were 0.006, 0.102 and 0.023 for generations 1th, 2nd and 3th, respectively. When non-founders are considered as a whole this difference was of 0.019. This methodology tends to overestimate the genetic variability in the base population, at least in the earlier generations.

Session 9 Poster 33

Molecular analysis of MC1R gene in the Italian pig breed Mora Romagnola

M. Marilli, F. Fornarelli, E. Delmonte, P. Crepaldi, Istituto di Zootecnia Generale, Facoltà di Agraria di Milano, Italy

The MelanoCortin-1 Receptor (MC1R) gene, known as Extension locus, is involved in mammals coat pigmentation. So far, several mutations have been identified in the porcine MC1R gene. We have investigated the presence of three of them, the MC1R*2, MC1R*3 and MC1R*4, by three independent PCR-RFLP in an Italian autochthonous pig breed, the Mora Romagnola (n.13), characterized by a solid phaeomelanic coat colour. The same alleles have been analysed in the wild boar (n.5) and in two cosmopolitan breeds, the Duroc (n.4) and the Large White (n.7). All the Mora Romagnola animals are homozygous for the MC1R*4 (e) allele and don't carry the MC1R*2 and MC1R*3 alleles. The 4 Duroc analysed present the same polymorphism as the Mora Romagnola, while the Large White only presents the MC1R*3 at the homozygous state. The wild boars don't carry any of these polymorphisms. The results on the two cosmopolitan breeds and on the wild boar confirm previous studies on this gene, while this is the first characterization of the MC1R gene in the Mora Romagnola breed, suggesting the fixation of the MC1R*4 allele. It could be interesting to also analyse the other known alleles, in order to complete the MC1R haplotype. Anyway these data confirm the useful role of pigment gene polymorphisms in the identification of breed specific markers for pig products traceability.

Genetic characterization of indigenous Southern African sheep breeds using DNA markers

P. Buduram[1], J.B. van Wyk[2] and A. Kotze[1,3], [1]Animal Genetics, Business Livestock Division, ARC Animal Improvement Institute, Private Bag X2, Irene, 0062, [2]Department Animal, Wildlife and Grassland Sciences, University of the Free State Bloemfontein, [2]Department Plant Sciences: Genetics, University of the Free State, Bloemfontein, 9300, South Africa*

Indigenous and locally developed sheep breeds of Southern Africa have developed unique combinations of adaptive traits to best respond to pressures of the local environment. Little is known about the genetic relationships within and between these breeds. The aim of this study was to compare genetic variability, genetic relationships and genetic differentiation within and between these breeds. A total of 640 blood samples comprising 12 breeds were collected from different regions. The applicability of 24 microsatellite loci from the FAO and ISAG recommended list for sheep molecular analysis was investigated. Results indicated an allele range of 4 to 7. The genetic distance estimate ranged from 0.223 (Blackhead Speckled Persian and Redhead Speckled Persian) to 1.103 (Ronderib Afrikaner and Namaqua Afrikaner). Clustering in a dendogram based on these genetic distances revealed no specific link with geographic distance. Although two of the breeds were phenotypically very different, results indicated close genetic similarity. These results will contribute to a routine DNA typing service for individual and parentage verification to the advantage of the sheep industry and for forensic application to the South African Police Servic.

Genetic variation among six Iranian goat breeds using RAPD markers

A. Javanrouh Aliabad[1], S. Esmeeelkhanian[2], N. Dinparast[3] and R. Vaez Torshizi[1], [1]Dept. of Animal Science, Faculty of Agriculture, University of Tarbiat Modares, Tehran, Iran, [2]Dept. of Biotechnology, Iranian Institution of Animal Science Research, Karaj, Iran, [3]Dept. of Biotechnology, Pastour Institution of Iran, Tehran, Iran

In order to evaluate genetic variation of six Iranian goat breeds including: Markhoz (MR), Korki of South Khorasan (KK), Black Lori (BL), Najdi (NJ) and Tali (TL), blood sample were collected from various locations of these breeds. DNA extraction was done by modified Salting-Out method (Miller, 1988). Polymorphims were surveyed for 16 RAPD primers and 10 primers were selected for high polymorphism among all of breeds RAPD-PCR were done on 20 individuals per breeds. The length of PCR products were varied from 220bp to 2310bp. The percent of polymorphism were obtained 53.9%. On the basis of polymorphic bands, the highest and lowest genetic distance was obtained for Black Lori and Korki of Khorasan (./237) and Tali and Korki of Raeini (./081), respectively. The phylogenetic tree was reconstructed on Neighbor-Joining method and showed two main separated groups, including: Korki of South Khorasan, Tali, Korki of Raeini and Najdi together and Black Lori and Markhoz. This research showed that RAPD-techniques is an useful tool for evaluation of genetic variation among of domesticated animals.

Session 9 Poster 36

Effects of different mating systems on genetic variance and the average of breeding value in dairy cattle

M. Aminafshar[1], A.A. Sadeghi[1] and P. Shawrang[2], [1]Department of Animal Science, Faculty of Agriculture, Science & Research Campus, Islamic Azad University, Tehran, Iran, [2]Department of Animal Science, Tehran University, Iran, maryamsa1381@yahoo.com*

Computer simulation was developed to compare the effect of three different mating systems on genetic variance and the average of breeding value in dairy cattle during five generations. Visual basic computer program was used for simulating breeding value of all individuals, average and variance of breeding value of milk yield traits in every generation. In all schemes animals were sorted according to their breeding value, then the best bulls are selected to mating to herd. In scheme A, B and C, random mating, positive assortive mating and negative assortive mating were used respectively. In base population the genetic variance and the average of breeding value were equal to 237818.8 and 2.8, respectively. After five generation the genetic variance of schemes A, B and C were equal to 131777.0, 163922.2 and 111169.0, respectively. The averages of breeding value of above schemes were equal to 2330.4, 2548.6 and 2171.8, respectively. Results showed that, the genetic variation was the best in scheme B and was the worst in scheme C. the genetic variation in scheme A was intermediate. According to genetic response, scheme B was better than scheme A and scheme A was better than scheme C due to their genetic variation.

Session 9 Poster 37

Pairwise comparison of mtDNA sequences in two Croatian sheep populations

Martina Bradic[1] B. Mioc[1] Vesna Pavic[1] Z. Barac[2], [1]Department of Animal science, University of Zagreb, Faculty of Agriculture, Svetosimunska 25, 10000 Zagreb,Croatia, [2]Croatian livestock center, Croatia

The sequencing analysis of the proximal part of the D loop region at mtDNA was performed to assess the genetic divergence and population structure of two sheep breeds in Croatia (Krk island sheep and Dubrovnik sheep). The mtDNA sequences of approximately 482 base pairs were obtained. Sequencing of the this part of the region from 20 unrelated animals revealed 23 polymorphic sites characterizing 17 haplotypes. Of the 17 haplotypes, 14 (88.88 %) were not shared. The average haplotype diversity was Hd: 0,984 and the average nucleotid diversity was Pi: 0,01444, and it indicates higher level of genetic diversity in Krk sheep.

The topology according to the neighbor-joining and maximum parsimony methods showed mosaic composition of the 17 haplotypes, suggesting that the populations were not completely divergent (the samples comprised one significant cluster). One type of sequinces were similar to European Merinolandschaf type of sheep mtDNA and the other 3 out of 20 sequences were different. Our data support the hypothesis that the first haplotype represents a trace of Merinisation of the this populations. The another 3 sequences (from Krk island sheep) belongs to the original primitive sheep, that have inhabited the island of Krk for thousands of years. These findings suggest that modern Krk island sheep were derived from at least two maternal lineages, the autochthonous and imported one.

Session 9 Poster 38

Characterisation of the founder matrilines in Asturcón pony *via* mitochondrial DNA

L.J. Royo, I. Álvarez, I. Fernández, E. Gómez and F. Goyache, SERIDA, Gijón, Spain*

The recovery of the Asturcón pony breed, in the 70's, involved 50 founder mares. They were recovered basically in various management units in the Sueve's range. The aim of this work is to ascertain possible differences among founder matrilines in the Asturcón pony *via* mtDNA and compare the obtained sequences with those from the western-asturian pony population (CCo), which is not included in the Asturcón pony stud-book. Here we analyse a 361 bp D-loop fragment in 32 individuals representing the following founder geographical areas: *Borines* (7), *Cerecea* (5), *Raicéu* (5), *La Vita* (4), *La Goleta* (6) and *Potes* (5). Most of them are common land of the Sueve's mountains, whilst *La Goleta* is representative of the Sueve's lowlands and *Potes* includes a little number of individuals recovered out of the Sueve. The obtained sequences were compared with 14 samples of the CCo population. Up to 13 different haplotypes (10 in Asturcón samples) defined by 32 variable sites were identified, seven of them being unique. Haplotypic diversity was of 0.31 and 0.36 for Asturcón and CCo samples. Two haplotypes are shared by 27 of the 46 samples. No significant among-populations Φ_{ST} distance was found. Differentiation among the Asturcón geographical populations and with the CCo population do not present a maternal genetic support. This work is founded by INIA-RZ03-011.

Session 9 Poster 39

Genetic variability in pigs assessed by pedigree analysis: the case of Belgian Landrace NN and Pietrain in Flanders

S. Janssens[1], J. Depuydt[2], S. Serlet[2], Vandepitte W.[1], [1]Centre for Animal Genetics and Selection, Department of Animal Production, K.U.Leuven, Kasteelpark Arenberg 30, 3001 Heverlee, Belgium, [2] Vlaams Varkensstamboek, Van Thorenburghlaan 20, 9860 Scheldewindeke, Belgium*

Pure breeding of pigs by independent, pig book breeders is under increasing pressure in Flanders. As a consequence, populations of pure breeds are decreasing which will eventually lead to higher rates of inbreeding and a loss of genetic variability. The homozygous stress-resistant Belgian Landrace (BL-NN) is considered most at risk and its current genetic variability was evaluated by pedigree analysis. The Belgian Pietrain (P), used as a terminal sire, was included as the second breed.

The variance of family size of sires was markedly different between P and BL-NN, and the high variance reduces the effective population size to approx. 80 in the BL-NN. The rate of inbreeding is increasing in the BL-NN whereas it remains stable in Pietrain.

Genetic variability measured by ancestors, effective founders and founder genomes in the BL-NN is about half of the value computed for P. However, compared to larger foreign populations, levels of genetic variability are still satisfactory.

Inbreeding control at farm level was introduced using an inbreeding matrix between all available boars and the sows in each herd. At breed level, a maximum number of pure breed matings per boar was set to reduce the variance in family size.

Session 9 Poster 40

Survey of milk protein polymorphism in the "Rossa Siciliana" dairy cattle

A. Zumbo[1], R. Finocchiaro[2], M.T. Sardina[2], J.B.C.H.M. van Kaam[3], A. Rundo Sotera[1], E. Budelli[2], B. Portolano[2], [1]Dipartimento MO.BI.FI.PA.-Sez. Zootecnica e Nutrizione Animale, Polo Universitario Annunziata, Università di Messina, [2]Dipartimento S.En.Fi.Mi.Zo.-Sez. Produzioni Animali, Viale delle Scienze, 90128 Palermo, Università di Palermo, [3]Istituto Zooprofilattico Sperimentale della Sicilia "A. Mirri", Via Gino Marinuzzi 3, 90129 Palermo.*

"Rossa Siciliana" is a dairy cattle population traditionally linked to cheese production. Protein polymorphisms from 62 individual milk samples from 19 farms were analyzed by isoelectrofocusing. High variability was found at the CSN1S1, CSN2, CSN3, LGB and LALBA loci, while the CSN1S2 locus was monomorphic for the A allele. The casein haplotype frequencies and the occurrence of the linkage disequilibrium, taking into account the association among loci were investigated. The population was not at Hardy-Weinberg equilibrium (Chi-square=19.0, p-value=0.0407), probably due to genetic drift. Frequencies of the three alleles detected at CSN1S1 locus were B (0.734), C (0.258) and D (0.008). High frequencies were found for $CSN2^{*}A^{2}$ (0.573) and $CSN3^{*}B$ (0.637). Whey proteins showed high frequencies for $LALBA^{*}B$ (0.903) and $LGB^{*}B$ (0.815). Strong linkage disequilibrium was detected for the polymorphic casein loci CSN1S1-CSN2-CSN3. Out of the 27 possible combinations from the 3×3×3 alleles considered, only four haplotypes had a frequency above 0.10. The most frequent haplotype frequency was BA^2B (0.236) followed by CA^2B (0.214), BA^1A (0.168), BA^1B (0.131), and BA^2A (0.094). The CSN1S1*B variant showed an allele frequency of 0.734 and occurred in almost all haplotype combinations.

Session 9 Poster 41

Development of variation in a random bred mouse population

W. Schlote[1], A. Wolc[2], T.A. Schmidt[3] and T. Szwaczkowski[2], [1]Populationsgenetik, Institut für Nutztierwissenschaften, Humboldt-Universität zu Berlin, 10099 Berlin, Germany, [2]Department of Animal Genetics and Breeding, August Cieszkowski Agricultural University of Poznan, 60627 Poznan, Poland, [3]Institute for Animal Breeding (FAL), 31535 Neustadt-Mariensee, Germany*

A population of mice randomly bred for 42 generations with constant family size was used to analyze variation for weights and growth characteristics. The data comprised 24 363 individual records for weight at 21 (W21) and 42 days of age (W42) and average daily gain from birth to weaning at 21 days (G21) and from weaning to 42 days (G42). The mice were kept under non-sterile conditions in macrolon cages at 22^0C and fed with standard feed ad lib. Population size varied between 30 and 40 pairs. Fixed effects generation, sex, litter size and coat colour were highly significant ($P<.0003$) and included in the individual animal model analysis. The means were 12.4, 24.9, .515 and .593 g in 42nd generation for W21, W42, G21 and G42, resp., and showed a slight positive trend. The phenotypic standard deviations varied quite considerably, the coefficients of variation were 15.2, 11.7, 16.1 and 19.1 percent for the 4 traits. Overall heritability was, 0.56 (±.012), 0.59 (±.014), 0.55 (±0.13) and 0.50 (±.016) for W21, W42, G21 and G42, resp. The analysis showed that differences in variation between generations were relatively large despite of a considerable data base and variances were not systematically reduced.

Session 9 Poster 42

Investigation of the Russian sheep breeds using DNA microsatellites

E. Gladyr[1], N. Zinovieva[1], M. Müller[2], G. Brem[3], [1]All-Russian state research institute of animal breeding of the Russian academy of agricultural science, Dubrovitzy, Podolsk district, Moscow region, 142132 Russia, [2] Institute of animal breeding and genetics of the Vienna veterinary university, Vienna, Austria, [3]Ludvig Boltzman institute of immune- cyto- and molecular genetic research, Vienna, Austria*

In this study, 13 Russian sheep breeds (romanov, romney, soviet merino, kujbyshevskaja, karakul, volgogradskaja, tzygaj, edilbaj, groznenskaja, stavropolskaja, kuchugurovskaja, Russian longhair, kalmytskaja) were compared for DNA-microsatellite markers. The samples were collected from a total of 1421 animals using the TypiFix sample collection system and the DNA bank was created. The DNA samples were genotyped for 11 microsatellite markers (McM42, OarFCB20, MAF65, McM527, OarCP49, OarAE119, HSC, MAF214, TGLA53, INRA49) using the Mega BACE 1000 Genetic Analyzer. The number of allelic variants significantly differed between the breeds and was in average of 13.15±0.87. The new alleles were identified. The expected levels of heterozygosity differed between the breeds and varied of 56.7% in stavropolskaja breed to 74.3% in romney sheep. The heterozygote deficiency was observed in all breeds analyzed. The genetic distances between the investigated Russian sheep breeds were calculated.

This work was performed under the financial support of the Russian foundation of basic research, projects 00-04-02002, 03-04-20016.

Session 9 Poster 43

A genetic linkage map of the blue fox (*Alopex lagopus*)

S. Keski-Nisula, K. Elo, J. Tähtinen and M. Ojala, Department of Animal Science, P.O.Box 28, University of Helsinki, 00014 Helsinki, Finland

A linkage map was constructed for the blue fox using a three-generation mapping population. The population was developed by crossing 20 females and 4 males in parental generation and F1 matings were made using 33 females and 10 males. Most F1 animals had two parities. The number of F2 animals was 392 and the total number of animals in mapping population was 455. We tested 130 previously published canine microsatellite markers to find highly informative markers for the blue fox. We were able to genotype 102 markers in mapping population, 28 primer pairs tested were either monomorphic or did not amplify DNA fragments with interpretable banding patterns. Percentage of informative meioses per marker varied from 30 to 99 % (mean 71%). The sex-averaged map consisted of 89 markers assigned to 17 linkage groups and the total lenght of the map was 1658 cM. Lengths of sex-averaged linkage groups ranged from 4.9 cM to 322.2 cM. Thirteen markers remained unlinked with any other marker. Our study suggests that most linkage groups are conserved between dog and blue fox. The order of microsatellites in the linkage map of the blue fox is mostly the same as in canine linkage map.

Session 9 Poster 44

Canalising selection on ultimate pH in pig muscle: consequences on meat quality.

C. Larzul[1], P. Le Roy[1], T. Tribout[1], J. Gogué[2], M. SanCristobal[3], [1]SGQA, Jouy-en-Josas, [2]Domaine de Galles, Bourges, [3]LGC, Toulouse, INRA, France*

The purpose of the project was to judge feasibility of a canalising selection to improve meat quality by reducing the heterogeneity of ultimate pH, measured in a muscle of the ham (*semimembranosus* muscle). Two divergent lines (Low and High, for homogeneity and heterogeneity, respectively) of Large White pigs were created made up from a selection of the boars from Artificial Insemination Centers, indexed on the pH values of their offspring. Four generations of selection were conducted. The tests on the differences in variance in G3 showed a lower variance of the ultimate pH in the Low line compared to the High line for the *semimenbranosus* ultimate pH, but not in the LD muscle. For the other measurements of meat quality, no significant differences in variance were observed. For growth and carcass traits, a significant evolution of the variances was observed, especially for backfat thickness. On the level of the phenotypical means, there was a significant evolution in backfat thickness measured on the half carcass: the animals of the High line were less fatty than the animals of the Low line. In addition, the Low line pigs were shorter than High line pigs in G3.

Session 9 Poster 45

Phylogenetic relationship and divergence time between two Iranian Kordi sheep populations using microsatellite markers

M.H. Banabazi[1,2], S.R. Miraei Ashtiani[1] and S. Esmaeelkhanian[2], [1]Dept. of Animal Science, Faculty of Agriculture, University of Tehran, Iran, [2]Dept. of Biotechnology, Animal Science Research Institution, Iran

Based on historical evidences, two Iranian Kordi populations have common origin due to obligate migration Kord people to Khorasan province at 400 years ago. This study has evaluated phylogenetic relationship, divergence time and gene diversity for Iranian Kordi sheep populations (Khorasan and Kordistan) using five microsatellites (McMA2, McMA26, MAF64, OarCP26 and OarFCB304). Three populations including Sanjabi, Mehraban and Moghani were considered in study. D_A genetic distance between five population were estimated and dendrogram drawn. using neighbor-joining (NJ) method. With unbiased Nei`s standard genetic distance (D), mutation rate (α) equal $4.5*10^{-4}$ and 3.5 years for generation interval; divergence time between two Kordi were calculated according to = $2\alpha t$ formula and then compared with historical evidences. The lowest D_A genetic distance was obtained between Kordi Kordistan and Kordi Khorasan populations (0.234). Two Kordi populations and then Sanjabi were at one cluster. The lowest average heterozygosity (H_e) were estimated in Kordi Khorasan (0.506). Small population size and short time for variability may be a possible reason for the low H_e in Kordi Khorasan. The estimated time of divergence for two Kordi populations was 445 years that has accordance with historical evidences. These findings confirm probable common origin for Kordi Kordistan and Kordi Khorasan sheep populations.

Session 9 Poster 46

Breed and typological composition of cattle in townships Prozor-Rama and Konjic

S. Ivankovic[1], A. Zelenika[1] and J. Pavlovic[2], [1]Faculty of Agriculture, University of Mostar, Mostar, BiH, [2]Selection Service of Herzegovina-Neretva Canton, Bosnia and Herzegovina*

The townships Prozor-Rama and Konjic are located in mountain region of North Herzegovina (BiH). Extensive and semi-intensive cattle breeding and sheep-farming characterize animal production. This region is also characterized by traditional production of cheese, milk and soft pickled cheese. Breed composition has devastated during the war and now is lasting after-war renewal of cattle fund. The breed composition of cattle is various and numerous of autochthonous Busa and its bustards are in permanent decreasing on this region. In township Konjic on sample of 326 milk-cows is considered share of 23,24% grey breed, 12,88 % of brown, 37,42 % of Busa and its types. In township Prozor-Rama on sample of 94 milk-cows grey breed is represented with 20,21%, Simmental breed with 11,70% and Busa and its types with 22,34%. Comparing considered results with earlier results of breeding composition is noticed significant share of more productive breeds of cattle that had not represented before.

Session 9 Poster 47

Different types of sheep breeds in Algeria: further molecular characterizations

S. Gaouar[1], M. Aouissat[2], L. Dhimi[3], A. Routel[4], B. Kouar[5], N. Saidi-Mehtar[1], [1]Département de Biotechnologie, Université d'Oran, Es-Senia, Algeria, [2]ITElv de Aïn El-Hadjar, Saïda, Algeria, [3]ITElv de Aïn M'Lila, Constantine, Algeria, [4]ITElv de Kssar chellala, Tiaret, Algeria, [5]Laboratoire de Génétique Moléculaire et Cellulaire, USTO, Oran, Algeria

In order to get a better management of domestic animal production and to define a global strategy for conservation of Algerian sheep breeds, the first step is to know their genetic variability by direct analysis of their genome at the molecular level. The objective of this study is to evaluate biodiversity and the relationships between these breeds which are well adapted to arid areas.

In this study, we have realized the phenotypical identification of the following local breeds: Hamra, Ouled-Djellal, Rembi, Taâdmit, Sidaou, D'men, Berbere, and obtained their present relative proportions. These data allowed to establish a new geographic distribution for these breeds. Currently, we are collecting bloods samples from unrelated sheeps of each breed and extracting their DNAs. Their molecular polymorphism will be studied using microsatellites listed by FAO and PCR-SSCP method which had been previously used in our laboratory in the framework of a preliminary study on DNAs from two breeds: Hamra and Ouled-Djellal.

Session 10 Theatre 1

Unravelling stress-induced subfertility in ruminants

H. Dobson, S.P. Ghuman, S. Prabhaker, R. Morris, S. Walker, V. Gandotra and R.F. Smith, Faculty of Veterinary Science, University of Liverpool, United Kingdom*

Many cows under the stress of production diseases (e.g., lameness, mastitis, dystocia) have smaller follicles and are less fertile. How and why? Microscopy reveals new locations for CRH/AVP neurones as well as novel connections between the brain stem and hypothalamus to influence GnRH and CRH/AVP secretion. Noradrenergic and GABA systems are involved. Differences in neurotransmitters between normal and stressed animals may alter neuroendocrine and behavioural centres in the brain. Stress or ACTH also reduces the amount of LH released during GnRH self-priming via the pituitary. After ovarian stimulation by FSH in normal ruminants, pulses of LH drive dominant follicles to produce oestradiol, which causes oestrous behaviour and pre-ovulatory surges of LH. Experimental stressors (transport, restraint/isolation or insulin-induced hypoglycaemia) decrease the amplitude and frequency of LH pulses. A stressor in the follicular phase before an expected LH surge in cattle or sheep lowers oestradiol and delays the onset and/or lowers, or even blocks, LH surge secretion. Follicular phases are extended by 2-20 days resulting in 'delayed ovulation'. Once formed, persistent follicles produce oestradiol for as long as 20 days exerting a pathological effect back on the hypothalamus. Another consequence of low oestradiol could be altered expression of oestrus. Our clinical investigations emphasise the importance of our experiments. Many clinical cases with persistent follicles do not have an LH surge in response to oestradiol injections, have abnormal patterns of oestradiol secretion and display abnormal behaviour. Do clinical cases of persistent follicles occur as a direct consequence of stressful situations?

Session 10 Theatre 2

Studies on stress and reproduction in the mare: effect of ACTH on adrenal steroid hormone levels in mares

Y. Hedberg, A.-M. Dalin and H. Kindahl, Department of Clinical Sciences, Swedish University of Agricultural Sciences, Uppsala, Sweden*

The aim of this presentation is to review the knowledge of "stress and reproduction" in the mare and to present our studies using ACTH (adrenocorticotrophic hormone). Activation of the hypothalamus-pituitary-adrenal axis is one of the many events occurring when an animal is exposed to stress. The use of exogenous ACTH is one way to mimic part of this pathway. Studying the effect of ACTH in mares is of particular interest, since oestrous behaviour in the mare is thought to be affected by adrenal steroid hormones. In our studies, the effect of ACTH on adrenal steroid hormone levels in intact mares with and without disturbed oestrous signs, as well as ovariectomized mares before and after oestradiol treatment was investigated. ACTH treatment led to significant increases in plasma levels of cortisol, progesterone and androstenedione in all mares. However, no significant differences between mares with normal and disturbed behaviour were found. In ovariectomized and oestradiol treated mares, a significantly lower increase in cortisol was found compared with only ovariectomized mares. This is in contrast to several other species studied. Three out of five ovariectomized mares showed behavioural oestrus during the breeding season, whereas only one mare displayed oestrous signs out of season. Such oestrous behaviour may be due to adrenal production of steroid hormones, such as androgens.

Session 10 Theatre 3

Effects of stress during pregnancy on endocrine and immune responses in pigs

E. Kanitz[1], W. Otten[1], K.P. Bruessow[2], M. Tuchscherer[1] and F. Schneider[2], Departments of [1]Behavioural Physiology and [2]Reproductive Biology, Research Institute for the Biology of Farm Animals, Dummerstorf, Germany

Stress impacts on the reproductive axis at the hypothalamus, pituitary or gonads by activation of endocrine, paracrine and neural pathways. Maternal stress during pregnancy in pigs may influence the function of the placenta and corpora lutea and may also exert lasting effects on gonadal function, behaviour, regulation of hypothalamic-pituitary-adrenal (HPA) axis and immunocompetence of the offspring. It was shown that repeated ACTH applications to pregnant sows are able to influence the release of reproductive hormones in the mother and offspring, depending on the stage of pregnancy. Furthermore, transiently enhanced maternal cortisol levels increase the cortisol exposure at the foetuses and induce changes in gene expression of specific markers for the HPA axis and neuronal activity in foetuses and neonatal pigs. Besides the effects on the ontogeny of the HPA feedback system increased maternal cortisol affected the activity of the noradrenergic system in a gender-specific manner. Restraint stress during late pregnancy also caused alterations in neuroendocrine stress regulation, impaired humoral and cellular immune function and increased the morbidity and mortality in suckling piglets. In summary, the studies showed that maternal stress during pregnancy in pigs can influence the offspring's physiological development with possible consequences on animal health and welfare as well as on economic aspects of production.

Session 10 Theatre 4

Effect of social stress on embryonic survival in indoor and outdoor housed gilts and sows

W. Hazeleger and N.M. Soede, Dept. Of Animal Sciences, Wageningen University, Wageningen, the Netherlands*

Outdoor housing challenges robustness of animals due to environmental variability. In pigs the 2nd week of pregnancy is considered to be sensitive for stress resulting in increased embryonic mortality. To test if outdoor housed animals are more resistant to social stress (robust), indoor and outdoor housed sows and gilts were socially stressed by mixing unfamiliar groups of animals in week 2 after insemination. In Experiment 1, 17 indoor and 16 outdoor housed 1st parity sows were inseminated following oestrus synchronisation. In Exp. 2, same numbers of gilts were treated similarly. In each experiment, two batches of indoor and outdoor housed animals were mixed at about day 8 of pregnancy. Animals in Exp. 2 were additionally stressed around Day 10, 12, 15 and 20 by several other social stressors (space restriction etc.). Around Day 35 (Exp. 1) or Day 48 (Exp. 2) of pregnancy fetal survival was evaluated by slaughter. Pregnancy rate was 100% and 85% in Exp. 1 and 2 respectively, with no differences between indoor and outdoor housed animals. Embryonic survival was 87.4±16.2 % and 86.5±9.5 % in Exp. 1 and 64.0±18.8 and 71.5±12.8% in Exp. 2 for respectively indoor and outdoor housed animals (P>.05). Outdoor housing (robustness) did not affect embryonic survival induced by social stress. Possibly the applied stressors were not severe enough to affect embryonic survival.

Session 10 Theatre 5

Effect of summer stress on quantity and quality of semen of Holstein-Friesian and Jersey bulls under subtropical environments

M. Fiaz[1] and R.H. Usmani[2], [1]Semen Production Unit Kherimurat,Tehsil Fatahjang, Attock, Pakistan,[2]Animal Sciences Division, Pakistan Agricultural Research Council, P.O. Box-1031, Islamabad, Pakistan*

Effect of summer stress on semen producing ability of Holsein-Friesian and Jersey bulls maintained at the Semen Production Unit Kherimurat, Pakistan from 2001 to 2003 was examined. The calendar year was divided into stressful season (April to October) and stress free season (November to March) on the basis of daily ambient temperature. Based on the relative humidity percentage, the stressful season was further sub-divided into dry summer (April to June) and wet summer (July to October). Friesian and Jersey bulls produced highest ejaculate volume during stress free season (4.15 ml) and wet summer (3.02 ml), respectively. The seasonal pattern of mass motility and individual motility of sperms in terms of maximum and minimum values was similar in both breeds i.e. maximum during dry summer and minimum during wet summer. Both Holstein-Friesian and Jersey bulls had smallest number of semen doses acceptable for freezing per bull per month during the wet summer as compared to other seasons ($P<0.05$). It is concluded from the present study that semen quality of exotic bulls of H.Friesian and Jersey breeds gets deteriorated during hot and humid season (wet summer) under subtropical environmental conditions in Pakistan.

Session 10 Theatre 6

Factors describing stress of dairy cows around insemination and their effect on non-return rate

C. Schrooten[1,] and W.A.J. Veldman[2], [1]HG, Arnhem, The Netherlands, [2]CR Delta Zuid-West, Rijen, The Netherlands*

Prior to insemination, dairy cows are often tethered in a cow cubicle or to a fence. In some cases, the cow still needs to be tethered on arrival of the AI-technician, resulting in additional labour and time for the technician. Besides, handling of animals just before insemination may cause stress and therefore may have a negative impact on pregnancy results. The hypothesis to be tested is whether there is a positive influence on non-return rates when handling just before insemination is restricted as much as possible. In this experiment, routine inseminations (n > 100.000) carried out by AI service organisation CR Delta Zuid-West during three months (October 2004 - January 2005) were included. Data were collected on the behaviour of the cow around time of insemination (quiet vs. restless), environment of the cow (herd-mates present or not), and the moment the cow was tethered before insemination (already tethered, tethered just before insemination, not tethered at all). In order to minimize labour for the technician, scores were collected by optical reading of barcode score charts in combination with PDAs. Non-return data for the inseminations will be analyzed with a model containing the fixed effects scored in this experiment, and other factors related to these inseminations and available in the national database.

Session 10 Theatre 7

Plasma cortisol level in relationship to welfare conditions in dairy farms

E. Trevisi, R. Lombardelli, M. Bionaz, G. Bertoni, Istituto di Zootecnica, Facoltà di Agraria, Università Cattolica del Sacro Cuore, 29100, Piacenza, Italy*

The cortisol is an index proposed to evaluate the stress status in dairy cows, but its interpretation remains puzzling. In particular, the real basal level remains difficult to establish, as any common cow/man interactions could raise it. To deepening the knowledge about this potential index, cortisol level and some other blood parameters, were monitored in 8 commercial herds, characterized by good (GW) or poor (PW) welfare status. Using any care to avoid restless during cow capture, 8-10 lactating multiparous cows for each herd were bleed. The bleeding was repeated 15 and 30 minutes later, leaving animals restrained in the rack. For statistical evaluation, cows of each herd were grouped according to their basal value of cortisol (<3, 3-10, >10 ng/ml) and cortisol increase (<3, 3-10, >10 ng/ml) after 1st bleeding.

At 1st bleeding (basal), the cortisol showed the lowest values, but only 37% of cows had values lower than 3 ng/ml. Interestingly, GW herds exhibited higher frequency of cows with cortisol under 3 ng/ml (63 vs 12% of PW, $P<0.05$). During the following 30' the cortisol raised, usually peaking at 15', but about 40% of cows was not responsive. The responsiveness at bleeding challenge was not different between groups.

Basal plasma cortisol seems related to aggregate factors of discomfort of cows and less to the acute stressors.

Session 10 Theatre 8

Glutathione peroxidase activity and its relathionship with cortisol and oppiod responses to training in trotters

S. Diverio[1], A. Barone[1], G. Tami[1], D. Beghelli[2] and C. Pelliccia[1], [1]Dpt. Scienze Biopatologiche, Igiene delle Produzioni Animali e Alimentari, Perugia University, Via San Costanzo 4, 06126 Perugia, Italy, [2]Dpt. Scienze Veterinarie, Camerino University, Macerata, Italy*

Aim of the study was to assess how physical exercise influenced Glutathione Peroxidase (GPx) activity in athletic horses. The experiment was carried out on October-November 2004, on a group of 7 trotters, aged 2-3 years, after the beginning of their training period. Each trotter was subjected to a standard training (ST)(20 minutes): after a warm up of 3 laps horses went at steady trot for two laps. Blood samples were taken just before beginning the ST (T1), soon after (T2), and after 60 (T3), 120 (T4), 180 (T5) and 240 minutes (T6) the end of the ST. Blood β-endorphin, cortisol, PCV and GPx were determined. Data were statistically analysed using GLM and correlation analysis. Training induced significant increases ($P<0.001$) of cortisol, PCV and HR at T2, with values returning to baseline within 1hour after the ST (T3). Plasma β-endorphin concentrations showed a similar trend but differences were not significant. GPx activity also significantly ($P<0.001$) increased after the ST, but peak values were observed at T4 and values remained still elevated to T6. No significant variation of GPx activities were recorded for age, sex and individuals. No correlations were found.

Session 10 Poster 9

Physiological and hematological responses of Baladi goats to tree-sheltering in summer

Y.M. Shaker[1], S.A. Kandil[2] and A.A. Azamel[1], [1]Animal and Poultry Physiology Department, Animal and Poultry Division, Desert Research Center, 1 Mathaf ElMataryia St., Cairo, [2]Animal Production Department, Cairo University, Egypt.

This study was carried out to study the effect of providing tree-sheltering on the thermorespiratory-responses and some hemobiochemical parameters of Baladi goats during summer. Thirty-six adult Baladi goat does in two groups (one was tree-sheltered and the other left unsheltered) were used. Thermorespiratory parameters of goats (rectal, skin and coat temperatures and respiration rate) and Radiant and ambient temperatures (AT and RAT) and relative humidity were recorded weekly twice a day (06.00 and 14.00 hr.), some hemobiochemical parameters (PCV, Hb, TP, AL, GL and A/G ratio) were measured weekly.

Providing sheltering at the hottest period of the day reduced AT and RAT values by 4.7 and 11.43 °C, respectively. Tree-sheltered goat does had significantly lower values of rectal, skin and coat temperatures and respiration rate as compared to the unsheltered ones. At 14.00 hr., the heat gradient from body core to the surrounding environment was wider for the sheltered goats than that of their counterparts (7.62 vs. 3.28 °C). Sheltering was found to affect significantly PCV, Hb, AL and GL while TP and A/G were reduced insignificantly.

The study revealed that providing the natural tree-sheltering alleviated the heat load falling on the Baladi goats during summer.

Session 10 Poster 10

Calving season affects reproductive performance of high yielding but not low yielding Jersey cows

E. Soydan[1], E. Sirin[1], Z. Ulutas[2], M. Kuran[1], Department of Animal Sciences, Universities of [1]Ondokuz Mayis, Samsun, Turkey and [2]Gaziosmanpasa, Tokat, Turkey*

Days from calving to first insemination, days open and calving interval in Jersey cows with low (1522-2476 kg) and high (3917-7116) lactation milk yield were determined to evaluate the effect of calving season on the reproductive performance of Jersey cows. A total of 1269 lactation milk yield record of 462 Jersey cows were analysed and data were classified according to calving season into four different climatic seasons (winter, spring, summer and autumn). Days from calving to first insemination in high producing cows (104.0±3.4 d) was 44 days longer ($P<0.001$) compared to low producing cows (59.5±3.3 d), while this period in high producing cows calving in summer months (128.4±7.4 d) was found to be 42 days longer ($P<0.001$) than those calving in winter months (86.0±5.7 d). Days open in high producing cows calving in summer months (151.2±8.7 d) were 35 days longer ($P<0.001$) than those calving in winter months (116.4±6.7 d). Calving interval was 18 days longer ($P<0.05$) in high producing cows calving in spring (388.1±7.7 d) compared to high producing cows calving in winter months (367.9±6.9 d). These results show that reproductive performance of high producing Jersey cows are adversely affected by the high environmental temperature and/or vegetation in summer months compared to the low producing Jersey cows.

Session 10 — Poster 11

Application of eustress and distress to pigs and the effect on meat quality

U. Küchenmeister, I. Fiedler, B. Puppe, K. Ernst, G. Manteuffel and K. Ender, Research Institute for the Biology of Farm Animals, Wilhelm-Stahl-Allee 2, D-18196 Dummerstorf, Germany*

Stress influences health, growth and meat quality. However, current views on stress emphasise that the consequences can be negative and harmful (distress) or positive and beneficial (eustress). In one experiment with Landrace pigs negative stress just before slaughter was applied (nose snare, electrical goad). In a different experiment also with Landrace pigs, eustress (a computer controlled "call-feeding-station" trained the animals to recognize a specific acoustic signal as a summons for food) was applied to pigs from the 7th to the 20th week of age.

Compared to control animals the nose snare (5 min) distress was of no significant influence on meat quality, not even a tendency could be detected. However, the electrical goad distress, applied for 5 min, significantly deteriorated the meat quality (lower pH45, higher conductivity 45 min p.m. and 24 h p.m., paler colour, and higher drip loss). Biochemical values (R-value, lactate, glycogen, carcass temperature) indicated an increased p.m. metabolism.

The applied eustress was of no influence on growth and the effect on meat quality was only partly significant. The drip loss was significantly reduced as was the intramuscular fat content. However, even if colour and conductivity had favourable trends, the differences did not reach statistical significance. In conclusion, stress (distress or eustress) has to have a high level to be of effect on meat quality.

Session 10 — Poster 12

Impact of stress during oestrus on the sperm reservoir and on progesterone concentrations in the sow

Y. Brandt[1], A. Lang[1], A. Madej[2], H. Rodriguez-Martinez[1] and S. Einarsson[1]
[1]Department of Clinical Sciences, SLU, Sweden; [2] Department of Anatomy and Physiology, SLU, Sweden

Due to increased consumer awareness of pig production and legal requirements, several countries seek alternatives to keeping sows crated or tethered. However, group-penning is not universally considered an improvement in sow welfare. While the sows can interact, they can also hurt each other and this stressful experience may influence their reproductive capacity. This study investigated whether simulated stress in sows during standing oestrus has an impact on semen transport in the female genital tract and progesterone concentrations in blood plasma. Fourteen sows were surgically fitted with jugular catheters and blood samples were taken every 2h. Seven sows (ACTH-group) were given synthetic ACTH intravenously every 2h for 48h starting from the onset of standing oestrus, while the remaining seven sows were given NaCl-solution (C-group). The sows were artificially inseminated once before ovulation, and were euthanised 6h after ovulation (detected using ultrasonography). The oviducts and their sperm reservoirs were flushed and the number of spermatozoa retrieved determined. There was a tendency towards a greater number of spermatozoa among sows in the ACTH-group. During treatment, concentrations of progesterone in jugular blood were significantly higher in the ACTH-group ($p<0.001$). Conclusion: simulated stress in sows results in elevated concentrations of progesterone and may disturb semen transport through the female genital tract. Supported by Formas.

Session 10 Poster 13

Search for regulatory DNA variation in genes related to stress response in pigs

E. Murani[1], S. Ponsuksili[1], K. Schellander[2] and K. Wimmers[1], [1]Research Institute for the Biology of Farm Animals (FBN), Wilhelm-Stahl-Allee 2, 18196 Dummerstorf, Germany, [2]Institute of Animal Breeding Science, University of Bonn, 53115 Bonn, Germany*

Stress hormones glucocorticoids and catecholamines play important role in the metabolism of fat, proteins and glucose. Their levels have been shown to correlate with carcass and meat quality traits in pigs and to vary between porcine breeds. Hence genes regulating synthesis/secretion of stress hormones as well as genes mediating and modulating their effects represent candidate genes for carcass and meat quality. We assume, that a large proportion of the variation in the activity of the neuroendocrine stress systems is attributable to regulatory DNA variation affecting expression of the stress responsive genes. Corticotropin releasing hormone acts as a central coordinator for neuroendocrine and behavioral responses to stress. To identify regulatory SNPs in the porcine CRH gene we isolated and comparatively sequenced ca 500bp of the proximal promoter. To prove the in vivo effect of a promoter SNP on the CRH gene transcription we designed an allele specific RT-PCR on the basis of a transcribed polymorphism in the exon2. In addition we analysed the association of the putative regulatory SNP with several biochemical indicators of stress in the German Landrace. Currently we are using this approach to identify regulatory variation in additional genes related to stress.

Session 10 Poster 14

Assessment of Fas and FasL immunoreactivity in ejaculated bull spermatozoa

D. Meggiolaro[1], F. Porcelli[2], A. Carnevali[2], P. Crepaldi[1], M. Marilli[1], B. Ferrandi[2], [1]Istituto di Zootecnia Generale, Facoltà di Agraria, Milano, [2]Istituto di Anatomia degli Animali Domestici, Facoltà di Veterinaria, Milano, Italy*

The Fas receptor is abundantly expressed in various tissues, particularly in activated T and B cells, thymocytes, hepatocytes and heart, whereas its ligand, FasL, is more restricted and tightly regulated in its expression. Fas/FasL signalling has a key role in apoptosis of target cells during cytotoxic immune response, activation-induced cell death of lymphocytes, elimination of self-reactive lymphocytes and escape of tumor cells from immune surveillance. Apoptosis also occurs during spermatogenesis and different authors suggested that Fas/FasL might be involved in it. In this research the expression of Fas/FasL molecules was detected in ejaculated spermatozoa of fertile bulls, to identify membrane determinants that could influence sperm survival or the fertilization processes. Spermatozoa were tested by immunoperoxidase staining, using anti-Fas and anti-FasL antibodies. Staining intensity was evaluated by microdensitometry in individual cells to get quantitative data. The results show that both Fas and FasL are present in mature spermatozoa and they are predominantly co-localized in the pericrosomal region. These data confirm what is already known about the marked regionalization of the spermatozoon plasma membrane. Furthermore, as there is increasing evidence that Fas activation can also result in non apoptotic responses, it is possible that, in the mature spermatozoa, Fas/FasL complex provides resistence to apoptosis, performing immunoprotection in the female genital tract.

Session 10 — Poster 15

Fertility and productive postpartum traits in crossbreds from Pelibuey sheep

L. Avendaño[1], F.D. Alvarez[1], L. Molina[2], R. Santos[3], A. Correa[1], and A. Pérez[1], [1]Instituto de Ciencias Agrícolas, Universidad Autónoma de Baja California, [2]Centro de Bachillerato Tecnológico Agropecuario No. 41, [3]Departamento de Zootecnia, Universidad Autónoma de Chapingo, México*

The objective of this study was to evaluate the fertility, prolificacy, birth weight, and weight at 30 d in two crossbreds hair sheep obtained from artificial insemination in an arid region of Mexico. Forty three Pelibuey ewes were distributed in three groups to be inseminated by laparoscopic method involving three breeds (Katahdin, n=14; Dorper, n=15; and Pelibuey, n=14). Data were analyzed using analysis of variance and chi-square test. There were no significant differences (P > 0.05) in fertility for Katahdin X Pelibuey (57%), Dorper X Pelibuey (53%), and Pelibuey (50%) groups. The average number of lambs per ewe was also similar (P > 0.05) in Katahdin X Pelibuey (2.38±0.24), Dorper X Pelibuey (2.27±0.28), and Pelibuey (2.0±0.27) groups. Even though birth weights were similar (P > 0.05) in all groups (Dorper X Pelibuey 2.72±0.27 kg; Pelibuey 2.57±0.29 kg; Katahdin X Pelibuey 2.47±0.29 kg), the average weight at 30 d old was higher (P < 0.05) in Dorper X Pelibuey (7.5±0.8 kg) than Pelibuey (6.2±0.9 kg), while Katahdin X Pelibuey lambs had similar (P > 0.05) weight (6.95±0.8 kg) than the previous two groups. Results show that meat production may be improved using Dorper X Pelibuey crossbreed under arid conditions of México.

Session 10 — Poster 16

Effect of a spray and fans cooling system on productive and physiological response of Holstein steers under heat stress

A. Correa[1], M. Morales[1], L. Avendaño[1], C. Leyva[1], A. Pérez[1], R. Díaz[2] and F. Rivera[1], [1]ICA, [2]Facultad de Medicina, Universidad Autónoma de Baja California, Mexicali, México*

The purpose of this study was to evaluate the effects of a spray and fans cooling system on daily weight gain, respiration rate and surface temperature (shoulder) of Holstein steers during summer heat stress. Sixty Holstein steers were randomly allotted in two treatments. One treatment consisted of only shade in the center of the pen (CONTROL, n=30). Other treatment was with shade similar to control plus spray and fans under the shade (COOLED, n=30). Daily gain was estimated biweekly, while respiration rate measured as breaths per minute (bpm) and steer surface temperature established via infrared gun were determined three times per week. Ambient temperature and humidity were recorded hourly to calculate the temperature-humidity index (THI). Data were statistically analyzed using a repeated measurements design. The weekly THI average during the trial ranged from 78 to 84. The daily gain was 1.37 ± 0.05 kg/d in the control and 1.46 ± 0.05 kg/d for the cooled (P>.05). Respiration rate and surface temperature were higher (P<.05) in the control (104 ± 13.9 bpm, 38.4 ± 0.96 °C) compared to cooled (77 ± 9.8 bpm, 35.9 ± 1.41 °C). The cooling improved steer comfort over control, but did not result in a higher daily weight gain.

Session 10 Poster 17

The effect of hormone treatment on reproductive performance of pigs

A. Siukscius, Institute of Animal Science of LVA, R. Zebenkos 12, Baisogala, LT-82317 Radviliskis distr., Lithuania

Gilts of Lithuanian White pig breed were used in a trial to determinate the effects of follitropin (FSH) and surfagon on the development of ovaries, conception rate and fertility of pigs, weight of newborn piglets and their viability. The trials indicated that the average number of ovarian follicles maturing per pig was 17.33±1.45 after a double FSH injection for pig oestrus stimulation. This index was by 8.8% (P>0.5) lower compared with the control group, in witch pigs were treated with pregnant mare blood serum (PMS). Replacement of PMS with single FSH injection for stimulation of oestrus reduced the conception rate and fertility by, respectively, 33.4 and 27.1% in comparison with the control group (P<0,001). However, if FSH (300 i.u.) was injected twice at on interval of 12 h, the efficiency of ovulation, conception rate and fertility were similar to those at PMS treatment (P>0.5). In synchronization of pig oestrus and ovulation, replacement of choriogonin (500 i.u.) with 2 ml (10μg) of surfagon had no significant effect on follicle growth, ovulation, conception rate and litter size. To sum it up, the trial results indicated that two FSH (300 i.u.) injections at an interval of 12 hours may be used for stimulation of oestrus, and surfagon injections (2 ml, 10 μg) may be used for stimulation of ovulation.

Session 10 Poster 18

The effect of early wearing on the reproductive performances on Romanian sows

M. Parvu[1], C. Dinu[1], A. Marmandiu[2], [1]Faculty of Veterinary Medicine Spiru Hare Bucharest, Romania, [2]USAMV Bucharest, Faculty of Veterinary Medicine, Romania*

It was study the effect of early weaning on fertility, prolificacy and annual productivity of 44 sows of Synthetic Line Peris 345. The animals were distributed into 4 groups according to lactation length 21, 35, 42 and 56 days. The sows had free access to standard diets and to water; they were followed during three parturitions. The interval between weaning and fertile mating was decreased from 25 days (too early weaning) to 16 days (too late weaning). The prolificacy and the annual productivity were not influenced by the early weaning (p≥0.05). It has no economic effect as compared to the increase with 4.7% annual piglets at too early weaning. Viewpoint of the reproductive performances, to the Romanian sows it was recommended the weaning at 35 days.

Session 10 Poster 19

The influence of some transport associated factors on heart rate response in sheep

A. Barone[1], S. Diverio[1], R. Cavallina[2] and N. Falocci[3], [1] Dpt. Scienze Biopatologiche, Igiene Produzioni Animali e Alimentari, Perugia University, Via San Costanzo 4, 06126 Perugia, Italy, [2]Istituto Zooprofilattico Sperimentale Regioni Lazio e Toscana, Rome, [3] Dpt. Scienze Statistiche, Faculty of Economy, Perugia University, Italy*

This study is part of a multidisciplinary project aimed to assess transport stress in sheep, by evaluating several physiological, immunological and behavioural parameters. The aim was to study the effect of stationary or moving vehicle and road type on the heart rate (HR) response of transported sheep. At this purpose, just before loading on the truck 5 subjects, within a group of 20, were equipped with a Polar Accurex Plus HR monitor. HR was recorded at 1 minute interval and road conditions video-recorded during a 4.30 hours journey. Videos were analysed by Instantaneous Scan Sampling at 1 minute interval, to record vehicle status (Stationary *vs.* Movement) and to score road type (3-point rating scale for cornering; 4-point rating scale for slope). Data were analysed by ANOVA and linear regression. HR was significantly influenced by truck motion ($P<0.001$) and type of road (cornering and slope: $P<0.001$). In particular, HR proportionally increased of 5,3 bpm for each cornering point and of 14,77 bpm for increasing slope. In conclusion, within the transport associated factors considered in this study, vehicle movement itself and increasing number of climbs seemed to be the most relevant.

Session 10 Poster 20

Reproductive performance evaluation of different prostaglandins for repeated synchronization program in postpartum dairy cows: Preliminary results

C.C. Pérez, J.M. Sánchez, L. Molina, M. Luque y J. Perea, Veterinary School, University of Cordoba (UCO)

This trial evaluates reproductive performance in a early-routine estrus synchronization program using two different PGF2α. Cows were administered D(+)cloprostenol sodium (n=88; Group A) or dinoprost (n=78; Group B) between 35-42d postpartum. Pedometers were used to detect estrus. Evaluation of secondary signs of estrus and evaluation of vaginal mucus quality was made prior to AI. When estrus was not detected by day14 following PGF2α, the treatment was repeated, up to a maximum of 3 time. Progesterone were analysed by RIA in blood samples collected on the day of treatment. There was no difference in the estrus detection to both PGF2α treatments (A=79.7 vs. B=75.6%). No treatment-related variation was recorded in respect of estrus signs (A=90.2 vs. B=94.9%) nor good mucus quality (A=95.08 vs. B=88.13%). First-service conception rate and conception rate at 200d were 45.9 and 65.2% in group A, and 22.0 and 67.9% in group B, respectively. There was no difference for the interval from parturition to first-AI; however, group A showed a significantly shorter interval from parturition to conception and services than group B (101.5 vs 122.6; 1.9 vs 2.6). Progesterone levels were lower in cows that needed 2 or 3 repeated treatment. In those individuals estrous response and conception rates were also lower. D(+)cloprostenol proved to be as efficacious for estrus response as dinoprost, but better fertility results were reported with D(+)cloprostenol.

Session 10 Poster 21

The effect of training and competition on the endocrine-metabolic response to stress in trotters

S. Diverio[1], A. Barone[1], C., Tami G.[1] , Beghelli D. [2] and N. Falocci[3], [1]Dpt. Scienze Biopatologiche, Igiene delle Produzioni Animali e Alimentari, Perugia University, Via San Costanzo 4, 06126 Perugia, Italy, [2]Dpt. Scienze Veterinarie, Camerino University, Macerata, [3]Dpt. Scienze Statistiche, Faculty of Economy, Perugia, Italy*

Since it is possible that additional stressors may exacerbate the agonistic related stress response in athetic horses, we investigated and compared the endocrine-metabolic changes elicited by training and sporting competition in trotters. A group of 8 trotters, aged 2-3 years, already trained and participating to competitions was used. Animals were randomly subjected to two different treatments: a standard training (ST) and a sporting competition (C). Blood samples were taken soon after each treatment (T1), and after 60 (T2) and 120 minutes (T3) and 24 hours (T4). Plasma were assayed for β-endorphin, cortisol, PCV, total proteins, BUN, glucose, Ca, P, Fe, Mg, Cl and LDH. Data were statistically analysed using GLM. Higher plasma β-endorphin, cortisol, PCV, total proteins, LDH and Ca concentrations were recorded after C compared to ST ($P<0.001$). In both treatment, plasma β-endorphin, cortisol, PCV, glucose, total proteins and P concentrations significantly decreased after exercise, to return their normal values within T3.

Session 10 Poster 22

Effects of two different exercises on physiological stress responses of training trotters

S. Diverio[1], A. Barone[1], C. Pelliccia[1], Moscati L.[2] and N. Falocci[3], [1]Dpt. Scienze Biopatologiche, Igiene delle Produzioni Animali e Alimentari, Perugia University, Via San Costanzo 4, 06126 Perugia, Italy, [2]Istituto Zooprofilattico Sperimentale Regioni Umbria e Marche, Perugia, [3]Dpt. Scienze Statistiche, Faculty of Economy, Perugia*

Aim of the study was to assess the influence of intensity and duration of exercise on the magnitude of the physiological stress response in trotters. The experiment was carried out on March-July 2004, on a group of 12 trotters, aged 2-3 years, after the beginning of their training period. Horses were randomly subjected, with at least one-month interval, to two exercises of different intensities: a standard training (ST) and a high-speed training (HST). Blood samples were taken just before beginning exercise (T0), soon after (T1), and after 60 (T2) and 120 minutes the end of training. At T0 and T1 heart rate (HR) were also recorded. Plasma were assayed for β-endorphin, cortisolo, PCV, total proteins, BUN, glucose, Ca, P, Fe, Mg, Cl and LDH. Data were statistically analysed using GLM and correlation analysis. A significant effect of sampling time ($P<0.001$) on plasma β-endorphin, cortisol, PCV, glucose, total proteins and HR was recorded. A large interindividual variation ($P<0.001$) in electrolytes concentrations was observed. Increasing intensity of exercise significantly affected ($P<0.001$) plasma cortisol, glucose and LDH concentrations. A positive correlation between adrenal cortex and oppioid responses was found.

Session 10 Poster 23

Evaluation of ovarian follicular growth patterns between the left and the right ovary in control line gilts and selected for high ovulation rate gilts

G. Vatzias[1], G. Maglaras[1], E. Asmini[2], R.V. Knox[3], C.H. Naber[4] and D.R. Zimmerman[4,5], [1]Technological Educational Institute of Epirus, Arta, Greece, [2]Technological Educational Institute of Larisa, Larisa, Greece, [3]University of Illinois, Urbana, U.S.A., [4]University of Nebraska, Lincoln, USA, [5]deceased*

Gilts previously selected for high ovulation rate (RS, n=76) and randomly selected control line (C, n=56) gilts from the University of Nebraska Gene Pool population were used. Gilts from each genetic line were randomly assigned for ovary recovery on day 1, 2, 3, 4, 5 or 6 following administration of prostaglandin $F_{2\alpha}$ (Lutalyse) on day 13 of the estrous cycle. Ovaries were recovered and individual follicles were measured, counted and classified by size into small (S: 2-2.9 mm) medium (M-1, 3-4.9 mm; M-2, 5-6.9 mm) and large (L: L-1, 7-8.9 mm; L-2, 9-10.9 mm) follicle classes. The total number of follicles across days for both lines was 7145. The follicular population in the left and the right ovaries was 3667 and 3478 respectively. The analysis showed that there was no evidence of an interaction between line and ovary (P>0.1). Similarly, there were no differences in follicular development (P>0.05) between the two ovaries within each line, except on day 5, where the L-2 follicular population of the left ovary in the RS line gilts was larger (P<0.05) compared to the right ovary.

Session 11 Theatre 1

Local products for genetic resources sustainability in Southern Europe: a solution or a problem?

François Casabianca, INRA LRDE Quartier Grossetti 20250 Corte, France

In many marginal regions from Southern Europe, particularly the mountain areas, producers have built strong connections between the local genetic resources from low productivity breeding systems and the protection of origin labelled products. Obviously, this connection has kept some threatened resources from disappearance, their productivity potential being too far from intensive standardized areas. So we can observe many positive impacts which contribute to a sustainability of the resources as well as the associated know-how.

However, in the light of few remarkable situations, we must ask ourselves about the consequences of such a dependent system. By the way, the local product, as well with its expected characteristics as the technical rules set in order to protect it, becomes a world of obligations for the local breed evolution. We have to be careful about the selection objectives and the problems of compatibility between quality of the products and potential productivity. This is more evident when the technical rules take measures of limitation of production that in fact reduce the possibility of expression of the animal potential.

All this is complicated because the real interests of the social groups who manage these two territorial resources are not always convergent. Therefore, it is essential to ask the right questions upon the long-term evolution of the local animal populations which future is directly dependent on the local products protection.

Session 11 Theatre 2

Examples of successful commercialisation of sheep and goat products in Alpine regions

F. Ringdorfer, Agricultural Research & Education Centre, Raumberg - Gumpenstein, Austria

The area of the Alps includes 8 nations and covers a surface of about 191.000 km^2 with more than 13 million inhabitants. It is an important living, recreation and economic space. Sheep and goat keeping significantly contributes to the preservation and sustainable development of this unique area. Many steep grassland areas can be used only by sheep or goats.

In Austria, most sheep and goat farms are very small (average flock size 20.4 sheep and 5.2 goats) and their production conditions vary considerably. This causes very variable quality of products, specially in lamb meat. For the commercialisation of large amounts of lamb, the quality differences are a disadvantage. However, most products are sold by the producer directly. In this case, they get a better price but they have more work.

Another way is to group several small farms into local production and commercialisation associations. The farmers produce according to defined instructions, resulting in better and standardised quality of products and better opportunities for selling. In most cases the name of the products include the name of the region where it is produced, e.g., Ennstaler Quality Lamb, Kärntner Lamb, Tiroler Alp Lamb, Salzburger Mountain Lamb, Steirisches Kid Meat, Wolfgangsee Sheep Cheese.

With the name of the product, the consumer connects, for example remembrance of a holiday. He knows, where the product is coming from and will buy it.

Session 11 Theatre 3

Specialised small ruminant products and genetic resources in the Middle and Eastern European countries

S. Kukovics[1], K. Kume[2], D. Dimov[3], E. Gyarmathy[4], J. Dubravska[4], E. Martyniuk[5], S. Stojanovic[6], D. Kompan[7], V. Matlova[8] and I. Padeanu[9], [1]Research Institute for Animal Breeding and Nutrition, Gesztenyés u. 1.Herceghalom, 2053 Hungary; [2]Institut de Recherches Zootechniques, Tirana, Albania; [3]University of Plovdiv, Bulgaria; [4]University of Agricultural Science, Nitra, Slovakia; [5]National Research Institute of Animal Production, Warsaw, Poland; [6]Ministry of Agriculture and Water Management, Belgrade, Serbia and Montenegro; [7]University of Ljubjana, Slovenia; [8]Research Institute of Animal Production, Uhrineves, Prague, Czech Republic; [9]University of Timisoara, Romania

In order to evaluate the small ruminant genetic resources and the specialised products in the region a questionneries were developed and the data were collected. The questionneries was covering the native breeds, the number of animals, the production-, reproduction and farming systems used, the presence of subsidy and its effect, the economic values of the breeds, breeding plans for these breeds, traditional goods/foods made using the products of these breeds, the labelling situations of the goods/foods, the income and costs of the traditional products, the organisations helping the breeding as well as preparing and selling the traditional products, characteristics of the selling and marketing of these products, the presence of industry in producing these products, presence of special prices of these products, bio and or organic system in producing these products and its effect on profitability, and expected future of these breeds and products/foods/goods.

Session 11 — Theatre 4

Genetic relationships between growth and pelt quality traits in the Gotland sheep breed

A. Näsholm, Department of Animal Breeding and Genetics, Swedish University of Agricultural Sciences, P.O. Box 7023, SE-750 07 Uppsala, Sweden

The Gotland breed is an indigenous Swedish sheep breed used for production of meat and pelts, which are characterized by natural shades of grey and a lustrous curly fleece. Besides functional traits of the ewe and traits that are aiming to increase amount and quality of the products the breeding objectives also include traits that make the animals adapted to the marginal grassland resources in Sweden. During the last ten years there has been a genetic progress both in 4-month weight (about 0.5 % per year) and in fleece quality (about 0.3 %). The aim of the present investigation was to study the genetic relationships between pelt quality traits and traits of importance for meat production. Data from the Swedish Sheep Recordings Scheme included observations on 4-month weight and pelt quality traits for 51 400 lambs and on commercial carcass traits for 12 000 lambs. The lambs were born during the period 1991-2003. The genetic correlations between pelt quality traits and 4-month weight and carcass traits, respectively, were generally low and breeding for improved growth and meat quality would not influence the pelt quality negatively.

Session 11 — Theatre 5

Fatty acid composition of lamb's meat from two different genotypes

G.M. Vacca[1], M.L. Dettori[1], L. Cengarle[2], G. Tillocca and V. Carcangiu[1], [1]Dipartimento di Biologia Animale, via Vienna 2, 07100 Sassari, Italy; [2]Dipartimento di Scienze Fisiologiche Biochimiche e Cellulari, via Muroni 25, 07100 Sassari, Italy*

Milk lamb is the most appreciated meat product of Italian ovine breeding. It is naturally fed and slaughtered at the age of about one month; lamb's meat value is recognized for its excellent organoleptic characteristics. It is obtained from dairy breeds and, sometimes, by crossbreeding with meat breed rams, set up to improve productivity and quality of its meats. The present work, carried out on 10 Sarda breed lambs and 10 Sarda x Mouflon (Ovis g. musimon) crossbred lambs, at 40 days of age, has the aim of verifying fatty acid composition of longissimus dorsi muscle and pelvic adipose tissue. Fatty acid composition was determined by capillary gas chromatography of fatty acid methyl esters. The data were submitted to Student's t test. As concerns longissimus dorsi muscle, crossbreds showed higher percent values ($P<0.05$) of C12:0, C14:0 and lower ones of C20:5 and C22:4 and ($P<0.01$) of C18:1 and C22:5. Sarda lambs showed higher levels of MUFA ($P<0.01$), lower SFA ($P<0.05$) and a better n-6/n-3 ratio. The fatty acid profile of adipose tissue was similar in the two groups, with the exception of C18:1, which was higher in Sarda lambs. The values found indicate, on the whole, a good meat's nutritional quality for the two compared genotypes.

Session 11 Theatre 6

Relationship between tissue thicknesses measured on live Pinzgau bulls by ultrasound and weight of hot carcass

P. Polák, E.N. Blanco Roa, E. Krupa and J. Huba, D. Peskovicová, M. Oravcová, Research Institute for Animal Production, Hlohovská 2, 949 92 Nitra, Slovakia*

Pinzgau breed is a dual-purpose breed categorized as a rare breed. Breeding for beef production was not recognized as a priority during last decades. For this reason our work was aimed to investigate relationships between tissue thicknesses measured by ultrasound on live Pinzgau bulls and slaughter parameters, and to analyse their prediction ability for hot carcass weight estimated in vivo. Tissue thicknesses at five positions, slaughter and carcass traits of 25 fattened bulls of Pinzgau cattle were analysed. The equipment comprised Aloka SSD 500 echo camera, 3.5 MHz probe UST 5813 and ECM-9 echo coupler. The highest tissue thickness (93.4 mm) was found on the back behind the shoulder blade. Highest correlations were observed among hot carcass weight and tissue thickness on the os ischia (r= 0.67) and on the back behind the shoulder blade (r= 0.39). Linear regression model where tissue thickness behind the shoulder blade measured by ultrasound were combined with weight at ultrasound measurement fitted best (R^2= 0,85). Weight at ultrasound measurement showed highest influence on estimation of hot carcass weight.

Session 12 Theatre 1

Use of genomic information for genetic improvement of livestock

T.H.E. Meuwissen, Norwegian University of Life Sciences, Ås, Norway.*

Genetic gain equals *intensity* times *accuracy of selection* times the *genetic standard deviation* divided by the *generation interval*. The new genomic information mainly affects the *accuracy of selection*. This implies that MAS is especially useful in situations where conventional *accuracy of selection* is low, ie. when heritability is low; or trait recording is limited (expensive; late in life; intrusive; not on selection candidates). Alternatively, MAS can be used to reduce the generation interval, while avoiding a reduction in accuracy of selection as much as possible. Current MAS schemes consist: 1) find the biggest, statistically significant QTL(s); 2) set up IBD matrices at the QTL positions given marker data (difficult due to massive amount of missing marker genotyped in practice); 3) calculate MA-EBV by fitting these IBD matrices next to the usual relationship matrix in animal model evaluations. Step 1) will only find the biggest QTL, which explain a limited amount if the total genetic variance, and hence limit the value of the MAS scheme. Genomic Selection avoids this limitation by using the estimated effects of individual chromosomal segments, without testing for their significance, and using genome wide high-density marker maps to identify the segments. Accuracies of selection were up to 0.84 for animals without performance records, which makes the accurate selection of animals without performance records possible, and thus opens opportunities for new traits and new breeding schemes with short generation intervals.

Session 12 Theatre 2

An approximate interval mapping procedure for selective DNA pooling

M. Dolezal[1], H. Schwarzenbacher[1], J. Sölkner[1], P.M. Visscher[2], [1]Department of Sustainable Agricultural Systems, BOKU - University of Natural Resources and Applied Life Sciences Vienna, Austria, [2]Institute of Cell, Animal and Population Biology, School of Biological Sciences, University of Edinburgh, West Mains, Road, United Kingdom

Selective DNA pooling is a very powerful method for QTL mapping. It considerably reduces genotyping costs while maintaining high statistical power. Applied to a Daughter Design, milk samples of offspring with extreme phenotypic values for a trait of interest are assigned to high and low pools, respectively, and within each pool the DNA is used for genotyping. A single marker test for linkage between marker and QTL considers marker allele frequency differences between high and low pools. Single marker across sire test statistics are more or less strongly affected by the number and QTL status of the specific sires that are heterozygous for a given marker. Given the single marker test statistics an approximate multiple marker method was developed to predict test statistics for markers for which a sire was homozygous or at any other location on the chromosome, to extract maximum information on linkage. A simple selection index analogy was used to make multipoint predictions, making use of the fact that the prediction of a test statistic at location j, given an observed test statistic on location i, is only a function of the genetic distance and hence the recombination rate. Power and map resolution of the proposed method are assessed by simulation.

Session 12 Theatre 3

Background bias on cDNA micro-arrays

M.H. Pool, B. Hulsegge, L.L.G. Janss, Division Animal Resources Development, P.O. Box 65, 8200 AB Lelystad, The Netherlands

cDNA microarray experiments measure differences in mRNA expression of many genes exposed to specific conditions. Unfortunately measures are not simple reflections of differences in gene-expression but are biased by several steps in the procedure. Since micro-arrays are applied as a quantitative tool it should find genes with any effect. Especially for genes with smaller effects, correct prediction of deviations from the background noise becomes important. Correlations were found between foreground and local background measures ranging from 0.2 to 0.5, also across channels. To fix the foreground-local background correlations blank array spots were used, working from the assumption that the average local background should equal the average blank spot intensity. Regression of local background by color on the blank spots confirmed background bias on all of a set of 8 slides analyzed and on three levels: a constant (base line) difference, representing a systematic bias by color, a more than proportionate increase of local background vs. blank spot intensities (biased by a factor 1.5 to 2), and a dependency of local background on spot foreground intensity. By fitting a local background, or preferable a surface, on the blank spots we were able to correct for the methodologically introduced background bias. This correction resulted in less negative background corrected spots, smaller error ratio's and less noise by lower spot variation.

Session 12 Theatre 4

Modelling liveweight change over a lactation in Irish dairy cows

N. Quinn[1,2], L. Killen[1] and F. Buckley[2], [1] School of Computing, Dublin City University, Dublin 9, Ireland, [2] Dairy Production Research Centre, Teagasc, Moorepark, Fermoy, Co.Cork, Ireland

The objective of this study is to derive an equation, which is biologically interpretable, to model the liveweight of Irish dairy cows over the lactation period. The dataset consists of liveweight recordings on weekly or monthly intervals throughout 12,501 lactations of spring calving cows from 83 herds. As the data used in this study is time series in nature, initial work focussed on time series techniques. Splines were also examined to ascertain the number of dimensions necessary to fit the dataset involved in this study. As the liveweight curve is similar to an inverted milk yield curve the models that have been used to predict milk yield were also tested. Ultimately liveweight changes between two calvings were modelled as a function of age, lactation and pregnancy. As multicollinearity was a severe problem with this function, the variance inflation factor was examined and principal component analysis was carried out on the variables responsible for severe multicollinearity. The new model has an R^2 of 0.76; the effect of multicollinearity is weak and the residuals are normal, homoskedastic and independent. This new model therefore provides an acceptable level of accuracy in representing the shape of the liveweight curve for Irish dairy cows.

Session 12 Theatre 5

Genetic selection against cannibalism related mortality in layer chicken

E. D. Ellen[1], W. M. Muir[2] and P. Bijma[1], [1]Animal Breeding and Genetics Group, Wageningen University, The Netherlands, [2]Department of Animal Science, Purdue University, W. Lafayette, USA*

In contrast to selection for production traits, selection for behavioural traits has been difficult, and populations have occasionally responded in the opposite direction. For example, selection against cannibalism-related mortality in chicken has sometimes increased mortality. Artificial selection against cannibalism-related mortality in commercial chicken is difficult for two reasons, *i*) cannibalism-related mortality is a trait of multiple individuals, the actors and the recipient of the behaviour, and *ii*) commercial breeding populations are kept individually under good conditions, to reduce loss of animals and enable individual recording of production traits. Though between-group selection has been proposed to overcome the first obstacle, opportunities to decrease mortality using individually housed selection candidates have not been investigated to our knowledge.

Our results show that selection of individually kept animals, based on mean phenotype of their relatives kept in groups, yields similar or even higher response than classical between-group selection. Response to classical group selection is proportional to $[(n\text{-}1)r+1]r_{pc}$, whereas response to selection of individuals based on group-kept relatives is proportional to nr, where n is group size, r relatedness between candidate and its relatives, and r_{pc} the purebred-crossbred genetic correlation. Consequently, the use of group-kept full sibs of selection candidates may enable efficient breeding against cannibalism. A selection experiment applying this method in chicken line is currently executed

Session 12 — Theatre 6

Selection for intramuscular fat in Duroc pigs using real-time ultrasound

T.J. Baas, C.R. Schwab, and K.J. Stalder, Iowa State University, Ames, IA, USA*

Intramuscular fat (IMF) is an important meat quality trait that has been shown to positively impact consumer acceptance of pork. The objective of this study was to evaluate the effectiveness of selection for IMF in pigs using real-time ultrasound. A selection project to increase IMF in purebred Duroc pigs was initiated at Iowa State University. Intramuscular fat is estimated using longitudinal images collected 7 cm off-midline across the 10th-13th ribs with an Aloka 500V ultrasound machine. Control and select lines are maintained and selection is based on estimated breeding values (EBV) for IMF from ultrasound estimates on the live pigs and from littermate carcass data. After three generations of selection, the average EBV for select line pigs is 0.83% greater than for control line pigs. Selection for IMF has, however, resulted in slightly more backfat and less loin muscle area, and a trend toward more days to 113.5 kg in the select line compared to the control line. Carcass evaluation of a sample of pigs from each litter indicated a similar increase in IMF, increase in backfat, and reduction in loin muscle area for select line pigs. No differences between lines were found for Hunter L color, Minolta reflectance, and ultimate pH. Ultrasound technology offers seedstock producers the opportunity to select for improved IMF in live animals and speed genetic progress for improvement of this trait.

Session 12 — Theatre 7

From single loci to chromosome segments: a different quantitative genetic perspective

H. Simianer, C. Flury, M. Tietze and H. Täubert, Institute of Animal Breeding and Genetics, Georg-August-University, Göttingen, Germany*

Although epistatic effects are well defined and, in principle, can be exploited in quantitative-genetic selection, they often are ignored or even treated as nuisance parameters in practical applications. Traditionally, epistasis is considered as an interaction between genes at unspecified loci. Inspired by the observation that functional genes are often organised in physical clusters, we developed a model to combine additive effects and addititve x additive interactions in gene clusters of defined length. Extending Malécot's kinship concept to identity by descent probabilities of chromosome segments of a given length in Morgan units, the basic concept called epistatic kinship is introduced and epistatic relationship and inbreeding coefficients are defined. Simple rules are given to set up the epistatice relationship matrix and its inverse directly. The method was used to obtain REML estimates of the epistatic variance from a data set on litter size in a commercial pig population. The results illustrate that with the suggested model correlations between relatives are differentiated according to the amount of epistatic covariance. Given that a substantial variance component is attributed to this effect, the expected genetic gain can be increased on the short term, but will gradually diminish through recombination in subsequent generations. Despite some practical problems yet to be solved, the suggested model and algorithms open new perspectives to use a higher proportion of genetic variability in selection and breeding.

Session 12 Theatre 8

Genetic analyses of traits affected by interaction among individuals

P. Bijma[1]* *and W.M. Muir*[2]*, *[1]*Animal Breeding and Genetics Group, Wageningen University, The Netherlands, *[2]*Department of Animal Science, Purdue University, W. Lafayette, USA*

Many traits of interest to animal breeders are affected by interaction among individuals. Examples include juvenile growth in pigs, calving ease in cattle and aggression related mortality in mink and chicken. Statistical analyses of such traits frequently yield strongly negative estimates for the genetic correlation between direct and indirect (*e.g.* maternal) effects, both in animal breeding and evolutionary genetics. Those negative correlations have raised scientific debate, because an obvious biological reason is lacking. Our results show that strongly negative estimated genetic correlations may be a statistical artefact resulting from inconsistency of the statistical and biological model. For example, with maternal effects, the biological model is $P_{\text{off}} = A_{D,\text{off}} + E_{D,\text{off}} + A_{M,\text{mother}} + E_{M,\text{mother}}$, whereas the common statistical model is $y_{\text{off}} = \textit{fixed} + A_{D,\text{off}} + A_{M,\text{mother}} + e_{off}$, with $\text{Cov}(e_i,e_j) = 0$. The biological model, however, reveals that covariances between relatives may contain an environmental term, *e.g.* $\text{Cov}(y_{\text{mother}}, y_{\text{off}}) = 1/2V_{A,D} + 11/4C_{A,DM} + 1/2V_{A,M} + C_{E,DM}$. The last term is neglected in the usual statistical model. Simulations showed that, when $C_{E,DM}$ was neglected, (absolute) genetic correlations were overestimated by a factor of ~2, whereas allowing $\text{Cov}(e_i,e_j) \neq 0$ yielded unbiased results. Genetic correlations between direct and maternal effects may therefore be less negative than currently believed. Similar results were obtained for other types of interaction.

Session 12 Theatre 9

Validation of an approximate approach to compute genetic correlations between longevity and linear traits

J. Tarres[1]*, *J. Piedrafita*[1] *and V. Ducrocq*[2]*, *[1]*Department of Animal Science, Universitat Autonoma, 08193 Bellaterra (Barcelona), Spain, *[2]*Station de Génétique Quantitative et Appliquée, Institut National de la Recherche Agronomique, 78352 Jouy en Josas, France*

The genetic correlation between longevity and linear traits is computationally difficult on large datasets. A two step approach is proposed and was checked via simulation: first, univariate analyses are performed to get genetic variance estimates and to compute "pseudo-records" and their associated weights. These pseudo-records are virtual performances free of all environmental effects that can be used in a BLUP animal model to get the same breeding values as in the (possibly nonlinear) initial analyses. Combining these pseudo-records in a multiple trait animal model and fixing the genetic and residual variances, one can get correlation estimates by AI-REML and approximate MTBLUP breeding values that blend direct and indirect information on longevity. Mean genetic correlations and reliabilities obtained under a sire and an animal model on simulated data confirmed the suitability of this approach in a wide range of situations (different genetic correlations, heritabilities and levels of censoring). When nonzero residual correlations exist between traits, the sire model gave nearly unbiased estimates of genetic correlations, while animal model estimates were biased upwards. Finally, when an incorrect genetic trend was simulated to lead to biased pseudo-records, the joint analysis could adequately correct for this bias through the inclusion of a time (year or generation) effect.

Session 12 — Theatre 10

Different selection strategies for improving lactation milk yield and persistency

C.Y. Lin[*1] and K. Togashi[2], [1]Dairy and Swine Research and Development Centre, Agriculture and Agri-Food Canada, Dept. of Animal and Poultry Science, University of Guelph, Canada, [2]National Agricultural Research Centre for Hokkaido Region, Sapporo, Japan*

This study compared 6 different selection strategies for improving lactation milk and persistency (P) defined as EBV_{280}/EBV_{60}: (1) Selection on lactation EBV (EBV_L) used a basis for comparison; (2) Index I_1 subject to restriction of equal genetic gains at DIM 60 and 280; (3) I_2 subject to restriction of zero genetic gain at DIM 60; (4) I_d designed to improve lactation milk without altering the lactation curve; (5) I_u=EBV_L+P; and (6) I_w=EBV_L/σ_L+P/σ_P where σ_L and σ_P are SD of EBV_L and P, respectively. Index I_u and selection on EBV_L exhibited the same response pattern, achieving the greatest response in milk coupled with the worst response in persistency. I_1 yielded the second largest response in milk while maintaining constant persistency. I_2 achieved the greatest persistency coupled with the smallest response in milk. I_d maintained the shape of the lactation curve at the expense of lactation milk. None of the 6 methods compared excelled in both lactation milk and persistency. I_w ranked second in both traits. The results suggest that I_1 is the method of choice for maximizing lactation milk while holding persistency constant, and I_w is a viable strategy for simultaneously improving the two target traits. The index procedure developed provides a useful tool for impcsing different restrictions on different days of the lactation to modify the lactation curve.

Session 12 — Theatre 11

Estimation of realised genetic trends in French Large White pigs from 1977 to 1998 using frozen semen: farrowing and early lactation periods

L. Canario[1], T.Tribout[1], J. Gogué[2], J.P. Bidanel[1], [1]INRA, Station de Génétique Quantitative et Appliquée, Jouy-en-Josas, France, [2]INRA, Unité porcine du domaine expérimental de Galle, Osmoy, France

An experiment was set up in order to estimate realized genetic trends in the French Large White (LW) pig breed. Two groups of pigs (G77 and G98) were produced by inseminating LW sows with semen of LW boars born in 1977 or in 1998. Three generations of G77 and G98 pigs were produced by inter se matings. Farrowing and early lactation periods were thoroughly investigated on a total of 137 first and second parity litters from second generation G77 and G98 sows (66 and 70 litters, respectively). The data were analysed using mixed and generalized linear models methodologies (for binary data). The total number of piglets born per litter and farrowing length did not significantly differ between G77 (11.9±0.5 and 244±23mn, respectively) and G98 (12.7±0.5 and 275±25mn, respectively) sows. G98 piglets were heavier at birth than G77 (+0.11±0.05 kg when adjusted for TNB), but had a higher probability of being stillborn (odds ratio of 1.5). More detailed analyses of the farrowing process (kinetics of birth, ...), of sow behaviour and of piglet vitality at birth (time to first suckling, ...) are also presented.

Session 12 Theatre 12

Graphic explanation of response prediction in long-term selection program

A. Nishida, K. Suzuki and Y. Ohtomo, Graduate School of Agricultural Science, Faculty of Agriculture, Tohoku University, 981-8555 Sendai, Japan*

It can be graphically explained that in a Cartesian coordinate system defined by the x-axis of environmental value E and y-axis of breeding value G, the axis for phenotypic value P is in the middle between the G and E axis. Then the methods to estimate the bi-variate distribution of G and E in the base population is proposed and an example of the distribution is shown in three-dimensional (3d) graph drown by "gnuplot". Truncation selection based on P is shown in the 3d graph, and the way to estimate the marginal distribution in G is given for the selected portion. The effect of gene segregation in meiosis and of re-coupling by mating on the distribution of G is explained using the change in the graph of the probability density functions. The change in the distribution of G, and in the simultaneous distribution of G and E with the generation of selection are shown as the 2d and 3d animation of the graphs. Finally, the fact should be stressed that all the information necessary to predict the long-term selection response accurately is in the two-variate distribution of G and E in the previous generation of population.

Session 12 Poster 13

The construction of indexes with constant restrictions by iterative procedure

C.Y. Lin, Dairy and Swine Research and Development Centre, Agriculture and Agri-Food Canada, Correspondence: Department of Animal and Poultry Science, University of Guelph, Ontario, Canada N1G 2W1

The objective of this study was to present an iterative procedure for deriving selection indexes with constant restrictions. By constant restriction, it means that the genetic responses of the restricted traits are preset to actual amounts for a specific level of selection intensity ($\bar{i}$). The resulting index equations possess three distinctive properties: 1) the coefficient matrix of the index equations is non-symmetric and nonlinear; 2) the coefficient matrix contains unknown $\bar{i}$, pointing to the necessity of predetermining the value of $\bar{i}$ intended for a selection program in order to derive the corresponding index coefficients (*b*); and 3) the coefficient matrix contains unknown *b*, thus requiring an iterative approach to solve the index equations. As a result of these unique characteristics, the index coefficients and selection responses of the index with constant restriction change nonlinearly with varying levels of $\bar{i}$. Thus, an index with constant restriction has no meaning unless it is associated with a specific value of $\bar{i}$. Animal breeders should prudently determine realistic levels of both selection intensity and constant restriction before constructing the constant restricted indexes. These unique properties contrast sharply with conventional restricted selection indexes which are developed and compared independently of selection intensity. Numerical examples were given to illustrate the methods for constructing the indexes with constant restriction alone or with both constant and proportional restrictions.

Multiple traits Bayesian analysis of birth, weaning and yearling body weights of Egyptian Zaraibi goat

I. Shaat[1], L. Varona[2] and W. Mekkawy[3], [1]Animal Production Research Institute, Agriculture Research Centre, Ministry of Agriculture, Cairo, Egypt, [2]Area de produccio Animal, Centre UdL-IRTA. 25198, Lleida. Spain, [3]Departamento de Ciencia Animal, Universidad Politecnica de Valencia. 46071, Valencia, Spain*

Data from 6620 Egyptian Zaraibi kids were collected during 1990 - 2003 to estimate genetic parameters for birth, weaning and yearling body weights using a multiple traits Bayesian animal model. Systematic effects included in the model were, year-season of birth, sex, type of birth and the age of doe. Gibbs Sampling was used to calculate the marginal posterior distributions of the variance components, heritabilities, genetic and residual correlations. The posterior mean and 95% HPD interval of the heritabilities were 0.37 (0.30-0.44), 0.18 (0.14-0.23) and 0.45 (0.36-0.53) for birth, weaning and yearling weights respectively. Genetic correlations were moderate between birth weight and weaning weight 0.41 (0.27-0.56) and between birth weight and yearling weight 0.33 (0.20-0.49). A substantially greater genetic correlation was obtained between weaning weight and yearling weight 0.79 (0.68-0.88). The residual correlations were 0.21 (0.17-0.26), 0.15 (0.07-0.21) and 0.47 (0.42-0.52) for birth-weaning, birth-yearling and weaning-yearling weights.

Session 12 Poster 15

Development of epistatic variance components in the two locus bi-allelic model

I. Curik[1], J. Sölkner[2], and M Kaps[1], [1]University of Zagreb, Faculty of Agriculture, Animal Science Department, Zagreb, Croatia, [2]BOKU - University of Natural Resources and Applied Life Sciences, Vienna, Austria*

Estimation of variance components for a quantitative trait is very difficult when genetic interactions between loci can not be neglected. The presence of interactions, even among few loci, very soon becomes extremely complicated due to the large number of possible combinations. Here, we presented development of variances components in the simplest model that includes epistatic effects (two locus bi-allelic model) for a population that is in Hardy-Weinberg equilibrium. The presented formulas enable three dimension graphical presentations of all variance components (including additive by additive, additive by dominance and dominance by dominance variances) under different models for all possible gene frequencies and can be used in scholastic purposes. Further, these formulas are also suitable for Monte Carlo based studies related to deviations of the observed from theoretical values (for example the effects of inbreeding on epistatic variances) as well as for further developments in segregation analyses related to two loci models and epistatic effects.

Session 12 Poster 16

LDLA, a package to compute IBD matrices for QTL fine mapping by variance component methods

L.L.G. Janss[1] and H.C.M. Heuven[1], [1]Animal Sciences Group, Wageningen UR, P.O. Box 65, 8200AB Lelystad, The Netherlands*

QTL detection and fine mapping has been performed in experiments using regression interval mapping methods. These methods rely entirely on linkage (LA) information, require pre-correction of the phenotypic data for fixed effects, and in most cases take only the segregation of the sire marker haplotypes into account. LDLA offers a more flexible approach, allowing for complex pedigrees, multi QTL, multi-trait analyses where QTL effects and other fixed and random effects are simultaneously estimated. It also allows for a variety of QTL models such as Mendelian and (partly) Imprinted modes of inheritance.

Analyses exist of three separate steps:-1- reconstruction of haplotypes using SIMWALK2, LOKI or other software, -2- calculation of IBD-matrices at a range of genomic locations using LD and/or LA information -3- variance component estimation using programs such as ASREML. The LDLA module is based on subroutines developed by Meuwissen and it is written in FORTRAN. The IBD matrix may not be positive definite due to pair wise IBD calculations, and in that case will be bended to allow its use in variance component analysis. The final analysis using ASREML allows large modelling flexibility.

LDLA is prepared to integrate well with ASREML, and additional tools are provided to summarize ASREML output into table and/or graphs. A license can be obtained.

Session 12 Poster 17

Genetic evaluation using markers completely linked to QTLs

Y. Liu and P. K. Mathur, Canadian Centre for Swine Improvement, Ottawa, K1A 0C6 Canada*

The markers that are completely linked to QTLs are valuable information for tracking QTL segregations in marker-assisted selection. However, these markers result in linear dependency in the gametic relationship matrix (G-matrix) that makes the marker information difficult to use in the gametic model method, currently used for marker-assisted genetic evaluation. The number of marker alleles at a locus is usually small and the number of animals to be evaluated is usually very large. Therefore, the rank of G-matrix could be very low in practical situations, and genetic evaluation of the gametic model method could become quite complicated in comparison with its usual procedure. In this study, a method based on mixture model approach was proposed as an alternative method for marker-assisted genetic evaluation using the completely linked markers. Probabilistic analyses of marker-QTL cosegregation were developed for using the marker information and handling the uncertainty of QTL segregation. The approach and its statistical properties have been discussed. Numerical example data were adopted to demonstrate the approach.

Session 12 Poster 18

Estimate of genetic parameters for competition effect in selected line of Duroc pigs

T. Oikawa[1], A. Nagata[1], M. Tomiyama[1], K. Suzuki[2], H. Kadowaki[3] and T. Shibata[3], [1]Faculty of Agriculture, Okayama University, Okayama, [2]Faculty of Agriculture, Tohoku University, Sendai, [3]Miyagi Prefecture Animal Industry Experiment Station, Miyagi,Japan*

Competition effect is additive genetic effect of an animal affecting performance of the other animals in the same pen, which was reported to be significant on growth of poultry, but no result has been reported for mammalian species. This study was aimed to reveal significance of the competition effect in Duroc pigs. A total of 1262 pigs (394 males and 868 females) selected for growth and meat quality, were measured at the average weight of 105 kg. Traits studied were body weight (BWF), age (AGF), daily gain (ADG), ultrasound backfat thickness (BFT) and loin area (LMA). The statistical model included fixed effects, pen effect, direct genetic effect and competition effect. (Co)variance components were estimated by REML using maximizing engine of "remlf90". Heritabilities of direct effect for BWF, AGE, ADG, BFT and LMA were 0.24, 0.44, 0.37, 0.75 and 0.44, respectively in candidate female animals, and 0.04, 0.68, 0.57, 0.70 and 0.56, respectively in slaughtered animals. Heritabilities of competition effect for the traits were low, ranging from 0.002 to 0.012. Proportion of pen variance was also low (c^2<0.1) except for ADG (0.20). Genetic correlations between direct and competition effect largely varied. Extra analyses indicated that pen size didn't affect the low heritabilities of the competition effect.

Session 13 Theatre 1

A protective role of seminal plasma in sperm-induced endometritis

M.H.T. Troedsson[1], A.S. Alghamdi[2] and A. Desvousges[1], [1]University of Florida, Gainesville, FL USA, [2]University of Minnesota, St. Paul, MN, USA*

Breeding-induced endometrits in horses is characterized by an influx of polymorphonuclear neutrophils (PMNs) into the uterine lumen. This facilitates elimination of excess semen from the uterus and provides a compatible environment for the conceptus. While non-viable spermatozoa must be eliminated from the uterus, viable spermatozoa need to be protected from binding and phagocytosis of PMNs in order to safely reach the oviduct for fertilization. We hypothesized that seminal plasma (SP) protects viable spermatozoa from binding and phagocytosis by PMNs. *In vitro* assays were used to test this hypothesis in a series of experiments. SP reduced binding and phagocytosis of spermatozoa by PMNs. Protection of spermatozoa was removed by heat inactivation of SP, but not by charcoal treatment. Ammonium sulfate precipitation of SP retained the biological activity. Addition of precipitated SP protein (SPP) reduced binding between PMNs and spermatozoa dose-dependently. Preliminary data suggest that specific SPP selectively prevents binding between viable spermatozoa and PMNs. This function was reduced in the absence of protease inhibitors. Non-viable spermatozoa were not protected from binding to PMNs in the presence of SPP. In a subsequent *in vivo* experiment, mares were inseminated with washed spermatozoa (w/o SP) 24 hours after induction of a uterine inflammation. Insemination of washed viable spermatozoa into an inflammatory environment reduced fertility almost completely. Fertility was restored if SP was added to the inseminated semen.

Session 13 Theatre 2

The cow in endocrine focus before and after calving

H. Kindahl[1], B. Kornmatitsuk[1] and H. Gustafsson[2], [1]Department of Clinical Sciences, Swedish University of Agricultural Sciences, SE-750 07 Uppsala and [2]Swedish Dairy Association, SE-631 84 Eskilstuna, Sweden*

During the period one month before and one month after parturition in the cow, several events have to take place. The dam has to be prepared for the impending parturition and the uterus and ovaries must return to a certain stage to be prepared for a new pregnancy. Most of these processes are due to or reflected in endocrine changes. A special interest is of course the status of the foetus - 'foetal well being'. The processes could either be considered as normal in a clinical perspective or as impaired (dystocia, small calves, stillbirth, retained foetal membranes, etc.). The main question for this presentation is if normal and impaired performance could be mirrored in endocrine parameters. Many studies have been performed to follow endocrine changes during the periparturient period in the cow. The following parameters have been shown to be the most important and seem to be the most suitable for an endocrine supervision:
Progesterone; corpus luteum, maternal adrenals, placenta
Prostaglandin (PG) metabolite; placenta, uterus, inflammation
Cortisol; regulator of prostaglandin synthesis, stress
Free oestrogens; placenta, ovaries
Oestrone sulphate; placenta, calf weight
Pregnancy associated glycoproteins (PAG); placenta

Session 13 Theatre 3

Current physiological aspects of pregnancy between implantation and partus in the pig

M. Wähner, Anhalt University of Applied Sciences, Strenzfelder Allee 26, 06406 Bernburg, Germany

Pregnancy is the central occurence of all reproductive endocrinological functions. The endocrinilogy after implantation is the central aspect for viviparity. The process of implantation is the beginning of a wide range of morphological - functional relations between endometrium and chorion. It is the time between apposition (13th to 14th day of gestation) and adherence (15th to 16th day of gestation). The pregnancy research has contributed to a good knowledge about sensitive physiological processes in fertilisation, blastulation, nidation, implantation and time after implantation until partus. With regard to an effective piglet production the influences of genetic, endocrinological and physiological aspects on embryo and foetus survival are very complex. In comparison to less-prolific European pig breeds the high fertility of Meishan pigs is the result of a markedly decreased placental size and an increased pig weight : placental weight ratio (placental efficiency). Crossbred gilts have a higher uterine capacity than purebred gilts. The prenatal growth correlates with surface, number of villuses and weight of placenta, circulation and nutrient supplies (glycogen). Progesteron levels on or near day 13 of gestation may be associated with early gestation uterine capacity.
Changes in progesteron - dependent on the secretion of the uterus are major caurses for embryonic and foetal losses. The various concentration of estrogen is caused by secretion in embryo and also in placenta and endometrium.

Session 13 Poster 4

Arachidonic acid activates the matrix metalloproteinase of placental fibroblast cells

H. Kamada[1], Y. Ueda[1] and M. Murai[1], [1]National Agriculture Research Organization for Hokkaido Region, Sapporo, Japan

The mechanism of fetus discharge at delivery is well understood, however, there are few information about the process of placenta discharge. Some reports have shown evidence of a relationship between the matrix metalloproteinase-2 (MMP-2) activities and placenta discharge at delivery. MMP-2 is primarily expressed in fibroblasts, so placental fibroblast cell culture was used to clarify the mechanism of placenta discharge after delivery.
$2x10^5$ viable cells were seeded per well in medium 199 containing 10% fetal calf serum (FCS). On the following day, FCS was removed and the culture medium was replaced with fresh medium containing cortisol. Last day, arachidonic acid (Ara) was added to the culture medium. Data were subjected to ANOVA using the general linear models procedure (SAS).
Ara addition induced a rapid exfoliation of fibroblast cell, which was inhibited by EDTA (general inhibitor of MMP) and phenylpropionamide (MMP2,9-inhibitor). However, any protease inhibitors (Pefabloc, PMSF, leupeptin) did not inhibit it. And this cell exfoliation was not inhibited by cyclooxygenase inhibitor, and did not reappear by prostaglandins (PGs). While lipoxygenase inhibitor inhibited the exfoliation of fibroblast cell. It is known that selenium decrease the incidence of retained placenta. Fibroblast cells pre-treated with selenium showed higher sensitivity to Ara addition. These results suggested that Ara metabolite by lipoxygenase, not PG, might activate placental MMP at delivery.

Session 14 Theatre 1

High health pig systems: the Danish approach

P. Bækbo, The National Committee for Pig Production, Danish Bacon & Meat Council, Denmark

Healthy pigs and safe pork products are in focus in the Danish pig production. Since 1968 Denmark has maintained an SPF program (Specific Pathogen Free) that has gained increasing national importance. All SPF pigs originate from, but are not necessarily born by sows derived from caesarean sections. New herds are established by total depopulation-repopulation. Apart from the monitoring schemes to document health status, all farms under the SPF program have to follow high standards of biosecurity. On top of the high biosecurity on farm level, all transportation of pigs between farms takes place in specially equipped trucks run by the SPF Company. Further more many so-called conventional Danish sow herds have eliminated diseases, typically by a combination of partial depopulation and medical programs. Today, reliable programs are in operation for elimination of several diseases. To insure an optimal location of high health farms a geographical information system (GIS) that contains information of the herd size and the health status of all Danish herds, is used. A national program was started in 1995 to eliminate pork as an important source of human salmonellosis. The program operates at all stages of the production chain. On farm level all breeding and multiplier herds are blood tested each month for Salmonella antibodies. All finishing herds that produce >200 finishers per year are tested for Salmonella antibodies in meat juice at slaughter and put into three categories according to their infection level.

Session 14 Theatre 2

Porcine embryo vitrification and transfer: a way to maintain high health status

C. Cuello, J.M. Vázquez, J. Roca and E.A. Martínez, Animal Medicine & Surgery, University of Murcia. Spain

Health is a key factor for efficient pig performance, since disease is expressed by weight loss, lowered rate of weight gain and reduced feed conversion efficiency. Therefore, a disease security system is needed mainly in herds with high health status. The greatest risk of disease introduction in a herd occurs from the incoming pigs. The use of safe methods for introducing new genetic material in a farm, such as caesarian derived piglets, artificial insemination or embryo transfer (ET), greatly reduce the possibility of introducing disease. Probably, ET is the safest technique, its use allow the movement of genetic resources with enhanced animal welfare, minimal risk of disease transmission and reduced transportation costs in comparison with transport of animals. Despite the advantages of ET, its application is still limited due to the high cost of surgical embryo transfer procedures, and to the difficulties for the long-term storage of porcine embryos. Recently, new non-surgical ET methods in pigs are being developed, which may be appropriate for field use. Additionally, new perspectives for pig embryo preservation have been occurred with the development of vitrification, specifically of the Open Pulled Straw (OPS) technology. Recent results show that non-surgical ET and vitrification are promising procedures to be used for safe transport and introduction of genetic material. However, these technologies have to be optimized in practice and several parameters, which may affect embryo vitrification and transfer, have to be discussed and studied in order to enhance their efficiency.

Session 14 Theatre 3

Mortality pattern and causes in crossbred pigs

D.Srinivasa Rao, All India Coordinated Research Project On Pigs, College of Veterinary Science, ANGRAU, Tirupati-2, India

An attempt has been made to study the mortality pattern and causes in crossbred pigs at the All India coordinated Research Project On Pigs,Tirupati, during the period 1994-2004. The piglets were weaned at 56 days after birth. Mortality in different age groups of pigs i.e., pre-weaning, post-weaning and adults (above 1 year) was recorded and causes were elucidated based on the post mortem findings.

A total of 2676 piglets were born from 432 farrowings. The mortality during pre-weaning , post-weaning and in adults was 11.66, 3.76 and 2.04% respectively and was well below the 15, 5 and 5% levels prescribed by the AICRP. Major causes of mortality were gastroenteritis (44.2%), pnuemonia (14.8%), agalactia (12.4%) and haemorrhagic septicemia (10.4%). Majority of the pre-weaning deaths (59%) were occured during 0-14 days of age. So, special attention should be given to the nutrition of gestating sow and good husbandry practices in early life i.e., colostrum feeding, proper protection from cold, good stockmanship, proper medication and health care etc., to lower the mortality rate.

Session 14 Theatre 4

The relationship between birth weight, the condition of the umbilical cord and the time interval at birth

K. Fischer, M. Wähner, Anhalt University of Applied Sciences, Strenzfelder Allee 26, 06406 Bernburg, Germany

The birthweight of a piglet is considered to be a substantial criterion for the evaluation of its vitality. A weight of 1600g is aimed. Underweight piglets - with a birth mass of less than 1000g - need to struggle a lot of disadvantages. They need obviously more time to get the first colostrum. The relationship between the birth weight, the shape of the umbilical cord as well as the birth interval was examined in 3 pig breeding enterprises. The evaluation includes the data of sows of the breed German Landrace, Large White and cross bred sows (German Landrace x Large White).

In these investigations the birthweight of the piglets does not have a significant influence on the condition of the umbilical cord. The results of those three enterprises point out the tendency that an extended birth interval accompanies with the possibilty of an already interupted umbilical cord. In that case it is of importance as heavier piglets show a longer birth interval. Evaluations of one enterprise approve that fact. If the birth interval is less than 5 minutes the average birth weight is about 1.35kg. A longer time interval comes along with heavier piglets in this case about 1.57kg. Despite that the average time interval is not influenced by the situation of the piglet, if it is alive or already dead.

Session 14 Theatre 5

Relation between (the breeding value for) weaning survival and periparturient sow behaviour

K.A. Uitdehaag[1], E.D. Ekkel[2], E. Kanis[1], E.F. Knol[3], T. van der Lende[4], [1]Animal Breeding and Genetics Group, [2]Ethology Group, Wageningen University, PO Box 338, 6700 AH Wageningen, The Netherlands, [3]IPG, Institute for Pig Genetics B.V., Schoenaker 6, 6641 SZ Beuningen, The Netherlands, [4]Animal Sciences Group, Division Animal Resources Development, PO Box 65, 8200 AB Lelystad, The Netherlands*

The sow's breeding value for mothering ability (EBV_{ma}) is estimated as the genetic, maternal effect of the foster sow on piglet survival at weaning. Litters from sows with high EBV_{ma} have shorter mean interval from birth until first colostrum intake than litters from sows with low EBV_{ma}. This study investigated whether there is a relation between EBV_{ma} and periparturient sow behaviour, and therefore possibly with this interval length. Piglet survival at weaning and periparturient behaviour of 25 sows with known EBV_{ma} were recorded. EBV_{ma} was not related to periparturient behaviour, but sows with high piglet survival at weaning showed longer durations of sitting and shorter durations of standing during parturition than sows with low piglet survival at weaning. These results indicate that periparturient behaviour has an effect on piglet survival at weaning. It remains unclear whether behavioural differences can explain shorter mean litter intervals from birth until first colostrum intake. More insight in periparturient behaviour beneficial to piglet survival at weaning can provide breeding companies with tools to accurately select for sow behaviour.

Session 14 — Theatre 6

Development of a protocol to record functional traits and inherited disorders affecting welfare in pigs

X. Fernàndez[1], J. Tibau[1], J. Piedrafita[2], E. Fàbrega[1], [1]Institut de Recerca i Tecnologia Agroalimentaries, Monells, Spain, [2]Universitat Autònoma de Barcelona, Bellaterra, Spain*

Genetic selection in most pig breeds has been mainly focused on performance traits. Recently, animal welfare awareness has increased and new concepts, like "functional traits" (e.g. leg weakness or teat functionality), have emerged. These new traits have not received much attention by pig breeders yet, mainly because little information is available on how to evaluate them. The first objective of this study was to prepare a protocol to record leg weakness and teat functionality using three pig pure breeds (Duroc, Large White and Landrace), which will be applied in 2 nucleus and 2 multiplier farms per breed. This protocol is based on the assessment of leg and teat morphology, using a numerical scoring system from 0 to 3 (0= worst and 3=best), indicating also the reason of the assigned category.

An important number of inherited disorders in pigs affecting mortality or welfare have been reported. An effective detection of these disorders could reduce their incidence in production farms. The same farms mentioned before will collect the most common inherited disorders (hernias and splay leg), which appear in the farm, according to a standard recording sheet.

The final objective will be to define a practical on farm protocol to record functional traits and inherited disorders at nucleus herd level to help pig breeders to include these traits in their selection strategies.

Session 14 — Theatre 7

Effects of rearing system on performance, animal welfare and meat quality in two pig genotypes

B. Lebret[1], M.C. Meunier-Salaün[1], A. Foury[2], E. Dransfield[3] and J.Y. Dourmad[1], [1]INRA-SENAH, 35590 Saint-Gilles, [2]INRA-INSERM Neurogenetics and Stress, 33077 Bordeaux, [3]INRA-SRV, Theix, 63122 Saint-Genès-Champanelle, France*

The effects of an alternative rearing system (O) for growing-finishing pigs (sawdust-shave bedding with free outdoor access, 2.4 m^2/pig) compared to a conventional (C) one (slatted floor, 0.65 m^2/pig) were evaluated for performance, animal welfare and meat quality in two (Duroc or synthetic line crossbreds) genotypes. Trials were conducted in spring and winter, each involving one pen of 10 pigs/genotype/system (a total of 40 pigs/season).

No significant interactions between rearing system and genotype were observed on any of the traits evaluated. In the whole, the O pigs spent 40% more time on exploratory activities, in particular towards the bedding, suggesting an improved animal welfare with the O system. Urine levels of cortisol and catecholamines in the O were similar with those in C pigs at 70 kg. The O pigs exhibited a higher growth rate (+ 6%) and were heavier (+ 5kg) at slaughter at the same age. Back fat depth and lean meat content, as well as plasma ACTH and cortisol, and urine cortisol and catecholamines levels at slaughter were not significantly affected by the rearing system. The O pigs exhibited similar pH_1 and pHu values, higher drip losses, but also higher intramuscular fat contents. The O system improved loin juiciness, but did not influence other eating quality traits.

Session 14 Theatre 8

The influence of rearing conditions on meat and back fat quality of large white breed

G. Holló, J. Seregi, J. Csapó, E. Varga-Visi and I. Holló, University of Kaposvár, Faculty of Animal Science, Kaposvár, H-7400 Guba Sándor str. 40., Hungary*

The purpose of this trial was to investigate the meat quality traits and fatty acid composition of backfat of large white (n=28) under intensive vs. extensive conditions. Half of the breed was kept in the outdoor system low with pasture and the other group under intensive condition. The weight of hot carcass, pH_1 and the lean meat content estimated by Fat-o-meater differed not significantly between groups, means for intensive and extensive were 106.50±14.21kg, 5.92+0.28 and 49.08+4.66% vs. 109.77±17.33kg, 5.85+0.29 and 50.30+4.62% respectively. The moisture and crude protein content of longissimus were higher in the extensive group, however the differences between groups were not significant. The intramuscular fat content of longissimus in the intensive group was about 1 % more, than that of in extensive. The higher values for MUFA of backfat and the lower ones for SFA in the extensive group (E:45.33%, 41.75% vs. I.:43.08%, 43.49%) are advantageous from human-nutrition point of view. Higher concentrations of eicosatrienoic and arachidonic acids ($P<0,01$) in backfat were recorded in the extensive group. Marked statistical differences ($P<0,001$) were represented between groups in the ratio of linoleic and linolenic acid of backfat. Outdoor rearing modified the fatty acid composition of backfat, particularly the level of n-3 fatty acids leading to a decline in the n6/n-3 ratio.

Session 14 Theatre 9

Effects of reduced phosphorus levels during growth and pregnancy on leg weakness, osteochondrosis and longevity in sows

*B. Jørgensen * and H.D. Poulsen*
Danish Institute of Agricultural Sciences, P.O.Box 50, DK-8830 Tjele, Denmark

The existing Danish standards of phosphorus were compared with reduced levels (10% to 25%) during growth and pregnancy in a 2x2 factorial design. The study included 152 female pigs of which 63 were slaughtered at 100 kg and 89 continued as sows. After two pregnancies the sows were given standard feed and were culled according to normal practice. The animals were examined for leg weakness at six months of age as well as at the end of each pregnancy, and the slaughtered pigs were examined for osteochondrosis. Furthermore, the longevity of the sows was analysed by means of survival analyses. A reduced phosphorus level *during growth* resulted in more weak pasterns in forelegs ($P<0.05$) (which is a positive trait) as well as more stiff locomotion in rear ($P<0.05$) at six months of age, less out turning legs ($P<0.05$) and upright pasterns in hind legs ($P=0.09$) in the first pregnancy, but no effect on longevity. A reduction of phosphorus *during pregnancy* resulted in more standing under position ($P<0.05$) in the first pregnancy, but less stiffness in front ($P<0.05$) and rear ($P=0.07$) in the second pregnancy. Furthermore, a significantly better longevity was found ($P<0.01$). It is concluded that in general a reduction of phosphorus allocation during growth and in particular during pregnancy has positive effects on leg weakness and longevity.

Session 14 Poster 10

Effect of zeolite clinoptilolite on biochemical and hematological parameters in weaned piglets fed with increased zearalenone level

M. Speranda[1], B. Liker[2], T. Speranda[3], V. Seric[4], Z. Antunovic[1], D. Sencic[1], Z. Grabarevic[5], Z. Steiner[1], [1]Faculty of Agriculture, Osijek, [2]Agricultural Faculty, Zagreb, [3]Zito,d.d. Osijek, [4]Clinical Hospital in Osijek, [5]Veterinary Faculty, Zagreb, Croatia*

The effect of zeolit clinoptilolite (CLIN) on some metabolic parameters in blood serum (total protein, albumine, glucose, triacylglycerols (TRY), cholesterol (CHOL), bilirubin, urea, creatinine, Fe, ALT, AST, GGT, ALP, CK, LDH) and hematological values (RBC, MCW, MCH, MCHC, Hb, hematocrit, thrombocyte, WBC) in weaned piglets fed with increased level of zearalenone (ZEN) was investigated in 14 day period. The trial was conducted with 3 groups, each with 10 female piglets (fed with starter mixture 20% CP; 29,95 MJ ME/kg). Group C was supplied with food containing 0,2 g/kg CLIN (ZEN level< 5,1 ng/g), group E1 3 mg/kg ZEN and 0,2 g/kg CLIN and group E2 3 mg/kg ZEN. Blood samples were collected et 8th and 14 thday of the trial. Group C was characterized by the highest body weight and daily gain (P<0,01). The values of CHOL were the lowest on 8. day and the values of TRI the highest on 14th day (E1:C, P=0,058; E1:E2, P=0,0006) in E1 and the lowest on 14th day in E2 (E2:C, P=0,0096). Serum Fe was the lowest on 8th day (P=0,002) and 14th day (P=0,15) in E2. The highest activity of AST and CK was on 8th day in E1 (P<0,05) and 14th day in E2 (P<0,05). MCV was the lowest on 8th day (P=0,02) and 14. day (C:E1, P=0,02) in E1. Antiestogenic metabolic effect of ZEN determened to be in E2, but with added CLIN, due to its reduced apsorption, ZEN mostly acted as estrogen agonist.

Session 15 Theatre 1

Milk urea content as affected by roughage type

S. De Campeneere, D.L. De Brabander and J.M. Vanacker, Agricultural Research Centre, Department Animal Nutrition and Husbandry, Scheldeweg 68, 9090 Melle, Belgium

Milk urea content (MUC) is used to manage protein nutrition and predict nitrogen excretion of dairy cows. However, MUC might depend on the roughage type offered. To evaluate that, three diets were compared using 18 lactating Holstein cows: 100% maize silage (MS), 50%/50% MS/grass silage (GS) and 100% GS. For all treatments, cows were fed to supply 105% of their net energy and digestible protein requirements. For the 100% groups, N balance was determined.
MS and MS/GS treatments had higher DM-intake (19.5 and 19.6 vs. 18.7 kg/d; P<0.001) and improved milking performance: 26.7 and 26.5 vs. 24.3 kg milk/d; 1163 and 1138 vs. 1050 g milk fat/d and 854 and 845 vs. 752 g milk protein/d (P<0.001). MS and GS fed cows ingested the same amount of degraded protein balance (73 g/d), while the MS/GS group ingested 106 g/d. MUC of MS (230 mg/l) and MS/GS (214 mg/l) was significantly (P<0.001) different from GS (171 mg/l). N balances indicated that for the MS diet, 33.2, 32.8 and 34.0% of the excreted N (392 g/d), was excreted with the faeces, urine and milk respectively, while for the GS diet this was 40.0, 30.1 and 29.9% (total excretion: 389 g/d). These results suggest that MUC is roughage dependent and that a system to predict N excretion can not merely be based on MUC.

Session 15 Theatre 2

Effect of rumen escape starch in maize silage based diets for dairy cattle

D.L. De Brabander, S. De Campeneere, J.M. Vanacker and N.E. Geerts, Agricultural Research Centre, Department Animal Nutrition and Husbandry, Scheldeweg 68, 9090 Melle, Belgium*

The effect of rumen escape starch (RES) level in maize silage (MS) based diets on milk yield and composition was investigated in two trials (T1, T2) carried out in early lactation. MS was given as the sole roughage in T1 or with prewilted grass silage in T2. In T1 three RES-levels, 40, 52 and 59 g.kg DM^{-1}, originating from two MS-cultivars and two concentrates (C), were compared in a latin square design with 18 Holstein cows: MSlCl, MShCl, MShCh (l, h: low, high RES-content). The two MS-cultivars had a similar S-content (315 and 318 g.kg DM^{-1}), but a different S-fraction escaping the rumen (18.5 and 24.4 %). The two concentrates only differed in starch degradability. In T2 two RES-levels, 28 and 35 g.kg DM^{-1}, originating from the previous two MS-cultivars, were compared in a cross-over design with 16 Holstein cows: MSlCl, MShCl.

Dry matter intake amounted to 19.1, 20.0 and 20.1 kg for MSlCl, MShCl and MShCh, in T1, and to 20.7 and 21.3 kg for MSlCl and MShCl, in T2. Milk yield tended (P>0.05) to increase with increasing RES-level and amounted to 26.1, 26.3 and 26.6 kg in T1 and to 28.8 and 29.7 kg in T2. Milk composition was almost unaffected. It is not clear whether the higher milk yield may be directly attributed to the RES-content and/or indirectly to the higher DM-intake.

Session 15 Theatre 3

Nutritional value and effect of Vicia ervilia seed on Holstein dairy cow performance

M. Moeini[1], H. Amanlo[2], M. Azari[1] and M. Souri[1], [1]Razi university,Kermanshah, [2]Zanjan university, Zanjan, Iran

Vicia ervilia is one of the leguminous crops, which has a potential and good adaptability in low rainfall area. Vicia ervilia seed (V.E.s) can be used as a protein supplement in animal nutrition (Arabi 1997, Enneking et al. 2000). Fifteen Holstein cows (1st lactation, 554 ± 57 Kg) randomly divided into three groups feeding V.E.s at the rate of %0, %7 and %14,DM of dietary, which replaced with Soya been meal and barely from 7th milking day for 60 days (NRC 2001). Milk records, changes in milk compositions and body weight were mesureaed. Blood samples taken at days; 30 and 60 monitoring serum constituent's changes.

Data statistically were analyzed using randomized completed design and Duncan's multiple range tests. V.E.s contain approximately: %93.9 DM, %22.8 CP, %5.95 Ash, %3.8 EE, %63.8NFE, % 15.01 NDF, and %6.13 ADF.

The result indicated that the effective ruminal degradability of CP, DM and CF of V.E.s amounted to 74.8, 70.6 and 42.9% respectively. Milk production and the average body weight of cows in %14 group increased significantly compare to other groups. The values of serum parameters were higher in cows in %14 group but were not significantly difference. It can be concluded V.E.s can be used up to %14 DM of dietary in dairy cattle.

Session 15 Theatre 4

Effect of Ca-soap of linseed oil on rumen fermentation pattern and on the characteristics of goat milk

É. Cenkvári[1], S. Fekete[1], H. Fébel[2], T. Veresegyházi[1] and E. Andrásofszky[1], [1]Szent István University, Faculty of Veterinary Sciences, Institute of Animal Breeding, Animal Nutrition and Laboratory Animal Sciences, Str. István 2, H-1078 Budapest, [2]Research Institute of Animal Breeding and Animal Nutrition, Str. Gesztenyés 1, H-2035 Herceghalom, Hungary*

A model experiment was performed with 6 rumen-cannulated rams to investigate the effects of Ca-soap of linseed oil on VFA composition and cell wall degradation in the rumen fluid. 14 days after the supplementation by Ca-SLO (5.4% of DM), pH stayed stabile (6.07 vs. 6.06), levels of C2 decreased (88.33 vs. 84.82 mmol/l) and that of C3 increased (21.06 vs. 22.32 mmol/l). After 24 hours of incubation of control and Ca-SLO-containing rumen liquid, ADF was less digested in the feed samples including Ca-SLO (2.10 vs. 2.05 g in the control diet and 0.15 vs. 0.16 g in the Ca-SLO-diet; $P<0.01$ and $P<0.001$, respectively).

A farm trial including 14 Saanen goats in early lactation was carried out to determine the effects of Ca-SLO-supplementation (4.6% in DM) on the milk ingredients. Milk fat and milk protein increased in the control goats by 7.9 and 10.4%, respectively. Ca-SLO-supplementation decreased levels of saturated fatty acids (from 53.6 to 48.3%), but C18:0 was higher by 43.7%. Amounts of C16:0 and C18:1 decreased by 6.8 and 6.7% and levels of C18:2 and C18:3 increased by 7.9 and 5.9%, respectively.

Session 15 Theatre 5

Increasing amounts of sunflower seeds increase CLA and vaccenic acid content in milk fat from dairy cows

T.S. Nielsen[1], E.M. Straarup[2], M.T. Sørensen[1] and K. Sejrsen[1], [1]Danish Institute of Agricultural Sciences, Tjele, Denmark,[2] BioCentrum, Technical University of Denmark, Lyngby, Denmark

The objective was to determine the effect of increasing levels of sunflower seeds high in linoleic acid on milk fat conjugated linoleic acid (*cis*-9, *trans*-11, C18:2; CLA) and vaccenic acid (*trans*-11, C18:1; VA) content. Twenty-four cows were randomized into four treatment groups (I, II, III, IV) receiving a grass silage based diet including sunflower seeds at 0, 5, 10 or 16% of dry matter for 35 days. Feed intake was lower in groups III and IV compared with groups I and II (13.5 vs. 16.4kg DM/d) ($P=0.007$). Energy corrected milk yield was numerically lower in groups III and IV; 26.3, 26.7, 22.6 and 23.1kg/d in groups I through IV, respectively ($P=0.32$). Milk fat CLA and VA content increased as dietary supplement of sunflower seeds increased, i.e. 0.5, 0.8, 1.2 and 1.8g CLA/100g fatty acids ($P<0.001$) and 1.1, 2.1, 3.5 and 4.8g VA/100g fatty acids ($P<0.001$) in groups I through IV, respectively. Surprisingly, milk fat percentage was higher in group IV (4.6%) compared with group I-III (3.8%) ($P=0.008$). Overall, milk fat CLA can be increased more than 3.5 times by adding high levels of sunflower seeds to the ration without a decrease in milk fat percentage, but feed intake and perhaps milk yield may be negatively affected.

Session 15 Theatre 6

Kinetics of *trans* and conjugated fatty acids (FA) concentrations in cow milk after addition of plant oils to different basal diets

A. Roy, A. Ferlay, A. Ollier, Y. Chilliard, Herbivore Research Unit, INRA-Theix, 63122-Ceyrat, France*

Twenty-four lactating Holstein cows received one of 4 diets: a grass Hay-rich diet with Linseed oil (5.0% of diet DM) (HL, Forage/Concentrate ratio F/C=60/40), two Maize silage + grass hay-based diets with Sunflower oil (5.2% or 6.5-9.0%) (M_1S or M_2S^+, F/C=52/48) and a high Concentrate + maize silage-rich diet with Sunflower oil (5.0%) (CS, F/C=25/75). Milk FA composition was determined 2 d before and every second day until 20 d (D20) after oil introduction in the diets. The kinetics of vaccenic acid (VA) and rumenic acid (RA) in milk were similar and varied according to the diet: either a peak of concentration was attained at D4-D6 (VA: 7.0, 5.5, 5.4%; RA: 3.1, 1.7, 2.9% of total FA) for M_2S^+, M_1S, CS, respectively, or a gradual enhancement of both FA, reaching a maximum at D11-D15 (VA: 8.7, RA: 3.0%), and then a plateau until D20 for HL diet. The kinetics of *trans*10-18:1 varied among diets, in a way opposite to VA changes: *trans*10-C18:1 remained at basal value until D20 (0.2-0.6%) with the HL diet, while it rose to a maximum at D11 (11.2%, M_2S^+) or D18 (18.6%, CS) and then remained at high levels until D20. *In conclusion*, concentrations of RA, VA and *trans*10-C18:1 are extremely variable in cow milk fat according to the type of diet and the duration of oil supplementation, as shown by the successive "waves" of these FA.

Session 15 Theatre 7

Lamb vigour is affected by DHA supplementation of ewe diets during late pregnancy

R.M. Pickard, A.P. Beard, C.J. Seal, S.A. Edwards, University of Newcastle upon Tyne, School of Agriculture Food and Rural Development, Newcastle NE2 7RU, United Kingdom

This study explored the effects of feeding an algal biomass supplement, rich in docosahexaenoic acid (DHA), to pregnant ewes on measures of lamb viability. 48 twin-bearing English mule ewes were allocated between 4 treatment groups fed, during different time periods, either a control diet based on silage and a commercial ewe concentrate, or a similar diet containing algal biomass (AB) to provide 12g DHA/ewe/day. Treatments were: ewes fed solely on the control diet for 9 weeks prior to lambing (C); ewes fed the AB diet for the first 3 weeks of the trial (3wk) then returned to control diet; ewes fed the AB diet for the first 6 weeks of the trial (6wk); and those receiving AB diet for 9 weeks up to parturition (9wk). AB supplementation tended to increase gestation length (P=0.08). Lambs born from ewes in the 6 and 9 week groups stood significantly sooner after birth than lambs born from ewes in the C group ($P<0.05$). Lamb birth weights and subsequent weight gain did not differ between groups. Ewe plasma and colostrum EPA and DHA levels at parturition showed a graded increase with supplementation ($P<0.001$), with lamb plasma at birth showing a similar but less pronounced response. This study shows that lamb vigour, reflected by reduced latency to stand, can be improved by supplementing gestational diets with omega-3s. There appears to be a threshold level of supplementation for this effect.

Session 15 — Theatre 8

Influence of feed withdrawal on plasma leptin concentrations in lambs of different carcass composition

E. von Borell[1], H. Sauerwein[2] and M. Altmann[1], [1]Institute of Animal Breeding and Husbandry with Veterinary Clinic, Martin-Luther-University, Halle, Germany,[2] Institute of Physiology, Biochemistry and Animal Hygiene, Bonn University, Bonn, Germany*

From the literature it is evident that feed restriction or starvation leads to decreasing plasma leptin concentrations and that leptin highly correlates with body fat tissue in many species. Until now, only little is known about the influence of feed withdrawal on the relationship between leptin and body fat tissue. Plasma leptin concentrations were measured in 30 ad libitum fed lambs of 40 kg live weight and after a feed withdrawal period of 24 h. The lambs were slaughtered immediately following the last blood collection. Leptin was analysed with a specific enzyme immunoassay. Visceral fat was weighed and the left carcass half was dissected into lean, subcutaneous and intermuscular fat and bones. The feed withdrawal leads to a reduction of the correlations between leptin and various fat tissues (before feed withdrawal: $r = 0.36 - 0.56$; after feed withdrawal: $r = 0.10 - 0.33$). The decrease in leptin concentration was lower in lean lambs (1.36 ± 0.60 ng/ml) than in fatty lambs (2.50 ± 0.82; $P < 0.01$). It can be concluded, that plasma leptin concentration after feed withdrawal reflects more the actual metabolic situation rather than the extent of body fat reserves.

Session 15 — Theatre 9

Central effects of histamine on food intake, and kind of histamine receptors in sheep brain

H.R. Rahmani[1], M. Mohammadalipour[1] and C.D. Ingram[2], [1]Dept. of Animal Sciences, Isfahan University of Technology (IUT), Isfahan, Iran, [2]Institute of Neuroscience, University of Newcastle, Newcastle-Upon-Tyne, NE1 4LP, UK*

Histamine as a central amine has several functions in the brain and feeding behaviour as well, via specific H_1, H_2 and H_3 membranous receptors. Ruminants such as sheep are under the influence of fermented nutrients which enhance histamine release seriously. Four intra-cerebroventricularly cannulated ram were used in a Latin square designed experiment which received 0 (control), 100, 400 and 800 nM dissolved histamine chloride in 100 µl of PBS into their ventricle after 12 hr fasting, and had access to water and TMR food after 15 minutes. Each sheep received four doses of histamine at least four times on subsequent days, and the recorded data (n) for each dose was sixteen. Results of this stage revealed that 400 and 800 nM of histamine suppressed food intake significantly (P<0.01) for three hours. In the next stage, the experiment was repeated with the same conditions, but 10 minutes earlier the sheep were pre-treated with 300-400 nM of ventricular injections of three specific histamine antagonists; chloropheniramine, ranitidine, and thioperamide, (H_1, H_2 and H_3 antagonists, respectively). Results of this stage showed H_1 antagonist significantly (P<0.01) blocked the histamine effect, concluding histamine as a central feeding neuromodulator and H_1 as a candidate receptor in this relation in this species.

Session 15 Theatre 10

Selenium status around peripartum in beef cows and calves offered grass silage and barley produced with selenium enriched fertilizers

J.F. Cabaraux[1], J.L. Hornick[1], N. Schoonheere[1], L. Istasse[1], I. Dufrasne[2], [1]Nutrition Unit, [2]Experimental Station, Faculty of Veterinary Medicine, Liège University, Liège, Belgium

Selenium (Se) is a trace element of importance in animals and in humans owing to its implication in many metabolisms. Grass silage and barley, both produced with fertilizers enriched or not in Se, were included in a diet of pregnant double-muscled Belgian Blue cows. Blood samples were obtained at calving and at days 4 and 15 post calving from cows and calves. Plasma Se concentrations were significantly higher in the Se cows groups than in the control (27.1 vs. 15.9µg/l, $P<0.001$). Both the concentrations of fibrinogen and haptoglobin, markers of late and acute inflammation processes were higher in the control cows. There were no day effects on the plasma Se content while fibrinogen and haptoglobin concentrations were the highest on day 4. In the Se group, the Se content was also increased in the colostrum (50.3 vs. 39.8µg/l) and in the milk obtained at day 4 (26.6 vs. 16.0µg/l). The feeding of Se enriched feedstuffs to the dams improved the plasma Se content in their calves (17.5 vs. 13.1µg/l, $P<0.05$) but did not affect the inflammation markers. It could be concluded from the present trial that the use of Se enriched fertilizer to produce grass and barley can improve the Se status in beef cows and their calves.

Session 15 Theatre 11

Analysis of n-alkanes in kidney fat for tracing feeding systems in meat producing animals

S. De Smet[1], K. Raes[1], E. Claeys[1], M.J. Petron[2], K. Vervaele[1], [1]Laboratory for Animal Nutrition and Animal Product Quality, Department of Animal Production, Ghent University, Proefhoevestraat 10, 9090 Melle, Belgium, [2]Food Technology and Biochemistry, Universidad de Extremadura, Carretera de Cáceres s/n 06071, Badajoz, Spain*

There is a need for chemical markers that allow distinguishing animal products derived from different production systems. We examined the potential of n-alkane analysis by gas chromatography in kidney fat. N-alkanes are components of vegetable wax and are absorbed and deposited in fat depots to a small extent. In trial 1, kidney fat was analysed from three groups of bulls that had been fattened on diets differing in the content of grass and grass silage versus maize silage and concentrate. In trial 2, samples from lambs that had been fattened for two or three months on different concentrate/hay diets were compared with samples from lambs slaughtered at the beginning of the trial. All lambs had been similarly reared with their mother on exclusive pasture feeding before the trial. In both trials, there were differences between feeding groups in the profile of 21 linear (odd and even) alkanes. However, the largest differences were seen for the content of 4 unknown peaks (near C18), that significantly increased with increased proportions of grass or grass silage in the diet. After identification, these compounds could be potential markers for grass feeding.

Session 15 Poster 12

Effect of starting time of feeding milk replacer on the performance of Holstein calves

Sh.J. Ghassemi[1], Y. Rouzbehan[1], A. Nikkhah[2], [1]Animal Science Dept., Faculty of Agriculture, Tarbiat Modarres University, Tehran, P.O. Box 14115-336, [2]Animal Science Dept., Faculty of Agriculture, University of Tehran, Karaj, Iran

To assess the effect of starting time of feeding milk replacer (MR), thirty two newborn Holestein calves were used. The animals were divided into four groups (n=8) in which the control group1 were fed cow milk. The other groups, were offered MR either on 5 (group 2), 12 (group 3) or 19 days (group 4) after birth. The average daily gain (ADG), starter intake (SI) and fecal scores (FS) were measured throughout the trial which lasted for two months.The data were statistically analysed using completely randomised design. Mean valus of animals performance for the groups 1, 2, 3 and 4 were as follows: ADG (from 5 to15 days after birth) 454, 223, 409, 477 g/day (s.e.m. 31.8, $P<0.05$); ADG (from 15 to 30 days after birth) 475, 387, 325, 275 g/day (s.e.m. 24.3, $P<0.05$); ADG (from 30 to 45 days after birth) 467, 475, 500, 448 g/day (s.e.m. 32.7, $P>0.05$); ADG (from 45 to 60 days after birth) 950, 804, 871, 856 g/day (s.e.m. 64.3, $P>0.05$); During the first month of age, SI 69.6, 103, 95.8, 65.2 g/day (s.e.m. 2.33, $P>0.05$); During the second month of age SI 703, 692, 837. 694 g/day (s.e.m. 13.01., $P>0.05$); FS 1.28, 1.35, 1.37, 1.33 (s.e.m. 0.028., $P>0.05$). Apart from the control group, the performance of calves fed MR on 12 day after birth was the highest.

Session 15 Poster 13

The influence of feeding level and milk replacer protein content on growth and blood protein levels of Holstein-Friesian calves

H.C.F. Wicks[1], R.J. Fallon[2], J. Twigge[3], L.E.R. Dawson[1] and M.A.McCoy[4], [1]Agricultural Research Institute of Northern Ireland, Hillsborough, Co. Down, BT26 6DR. [2] Teagasc Grange Research, Dunsany, Co.Meath, Ireland. [3]Nutreco Ruminant Research Centre, 5830 AE Boxmeer, The Netherlands. [4]Veterinary Sciences Division, Stoney Road, Belfast, BT4 3SD*

Data from the USA suggests that the current UK industry recommendations for feeding the neonatal calf (~500g milk replacer/d, at ~230 g crude protein per kg fresh milk powder) are inadequate to sustain high growth rates in early life. At the Agricultural Research Institute of NI a study to evaluate the effects level of milk replacer intake and milk replacer crude protein content on live weight gain, skeletal size, body condition and blood protein profiles, has been initiated. Holstein-Friesian calves were fed two levels of milk replacer (5 and 10 l/d at 120g/l)) and two milk replacer crude protein contents (230 and 300g CP/kg) in a 2x2 factorial design experiment. Increasing level of milk replacer feeding significantly increased live weight at weaning ($P<0.001$). Calves fed milk replacer with 300 g CP/kg had lower growth rates compared with calves fed the lower (230 g CP/kg) protein milk replacer ($P<0.01$) and had significantly higher total blood protein ($P<0.05$) and blood urea ($P<0.001$). The results from this study do not support increasing CP content of milk replacers to 300g/kg in order to enhance calf growth rate.

Session 15 Poster 14

The effect of by-pass methionine supplement served before and after calving on milk yield and physiological parameters in dairy cows

V. Kudrna[1], J. Illek[2], P. Lang[1,], P. Mlázovská[1], [1]Research Inst. of Anim. Prod., Přátelství 815, 10401, Prague-Uhříneves, [2]VFU Brno, Palackého 1/3,61242,The Czech Republic*

33 high-yielding dairy cows were used in an experiment to find out an effect of by-pass methionine (MET), which was/wasn't served up before and/or after calving. Before calving cows were allocated into one of two groups: with (M) or without (0) a MET supplement in their feeding ration and then after calving each from two sets was divided into two groups - with (M) or without (0) the MET supplement. A large scale of miscellaneous parameters was followed: intake of feeds and nutrients, milk yield, contents and production of milk components, milk FA profile, and other parameters. The highest effects of MET were found out in contents of milk proteins and daily milk and FCM production, while all other parameters followed were uninfluenced by MET-delivery. The highest FCM production was recorded in cows with MET both before and after calving (M/M - 32.44; 0/0 - 30.13 kg/head/day). Production of FCM was higher in the groups M/0+M/M (31.71 kg/head/day) comparing the groups 0/0+0/M (30.82 kg/head/day). Daily milk yield in animals with MET after calving was statistically higher, than in the animals without MET (34.27 vs. 32.88 kg/head/day). The MET supplement served during whatever stage of the experiment increased milk protein concentration (the lowest in group 0/0 - 1.02 %).

This project No. 1G46086 was supported by NAZV (National Agency for Agricultural Research of the Czech Republic).

Session 15 Poster 15

Effect of L-glutamine-containing oral rehydratation solution on the absorptive function of small intestine in diarrhoeal calves infected with *Cryptosporidium parvum*

P. Klein[1], H. Lelkova[2], J. Lastovkova[1], M. Soch[2], [1]Research Institute of Animal Production, Praha - Uhrineves, CZ-10401,Czech Republic, [2]Faculty of Agriculture, University of South Bohemia, CZ-38005, Czech Republic*

L-glutamine is a preferable energy source for enterocytes. Effect of its peroral application in rehydratation solution (RS) was assessed in 18 calves with cryptosporidium-associated gut inflammation and profuse diarrhoea. Calves were experimentally infected with $10x10^6$ oocysts and allocated into 3 groups, six in each. Five days p.i. animals were switched for 7 feedings (3.5 days) from their diet on RS (55 mL/kg of BW). Group 1 received glutamine-free RS ("WHO type"), RS for group 2 contained L-glutamine (11.6 mg/mL) and calves in group 3 received their diet continually without break. Intestinal absorptive capacity was assessed at 10th day p.i. by oral tolerance test with 0.5 g of D-xylose/kg of BW. Plasmatic concentration of xylose was recorded for consecutive 4 h in 30'-intervals. Clinical pattern of the infection was similar in all groups. Maximal concentration of xylose was recorded for time 2.5 h in all groups. Average values at this time were 3.4^a, 1.6^b and 2.4^c mmol/L in groups 1, 2 and 3, respectivelly (a,b: $p<0.001$; a,c: $p<0.01$). L-glutamine in oral RS had positive effect on absorptive function of the infected gut epithelia and was found as effective component of RS for calves with intestinal infection. (Project MZe-000 270 1403).

Session 15 Poster 16

The effect of addition of selenium to a milk diet of calves on the meat quality

V. Skrivanova[1], Y. Tyrolova[1], M. Marounek[1,], M.Houska[2], [1]Research Institute of Animal Production, Pratelstvi 815, 104 01 Prague-Uhrineves, [2]Food Research Institute Prague, Radiova 7, 102 31 Praha 10, Czech Republic*

The aim of our work was to study the effect of addition of selenium (selenium yeast) to a milk replacer on quality the calves meat.The experiment was conducted on 18 Holstein bulls at average age of 20 days and weight of 50 kg. All animals were fed by milk replacer (0.4 kg with 3 l water) twice a day and a concentrate mixture *ad libitum*. The animals were divided into three groups: 1) addition of 1 mg Se per day (1st group), 2) without Se (2nd group), 3) addition of 0.5 mg Se per day (3rd group). Daily weight gains of calves were not significantly different. The calves were slaughtered at the age of 4 months. The average slaughter weight was 161 kg. The *m. longissimus dorsi,* kidney and liver were sampled. The contents of Se was investigated. The differences in Se contents of the *m. longissimus dorsi* between the groups (0.38, 0.23 and 0.31 mg/kg) were significant ($P < 0.05$). Contents of Se in the kidney and in the liver were not significantly different. The storage stability of the meat (0 day, after 3 days, after 6 days) using the malonaldehyde parameter was investigated as well. The differences between the groups were not significant. Thus, the selenium addition to the milk replacer had positive effect on the contents of selenium in the meat, but not on other parameters investigated.

(This work was funded by the project MZE 0002701403).

Session 15 Poster 17

Effect of feed blocks on the growth of grazing heifers

A.S. Chaudhry[1], C.J. Lister[2] and W. Taylor[1], [1]School of Agriculture, Food and Rural Development, University of Newcastle-upon-Tyne NE1 7RU, UK, [2]Caltech, Solway Mills, Silloth. CA7 4AJ, United Kingdom*

We examined the impact of *ad libitum* access by grazing heifers to feed blocks (Booster) on their growth. Thirty Holstein-Friesian heifers in two balanced groups (Control; Booster) were housed separately in fields of similar ryegrass swards where only the Booster heifers received blocks containing protein, sugars, oils and vitamin-minerals. The study lasted two periods (P1, P2) with 10-days interval without blocks. Sward height (SH) was measured and grass was chemically analysed. When grass availability declined during P2, barley straw (straw) was offered in ring feeders to both groups. Straw and block intake and heifer LW were recorded and statistical effects of group, period and group/period interaction on LW and SH were studied. The grass quantity remained good for both groups in P1 but declined in P2. However, the block intakes in P1 and P2 remained consistent at 330g/day/heifer. In P2, the Booster heifers consumed 2.2 times more straw than the Control heifers suggesting stimulatory effects of Booster on the utilisation of poor quality forage. All heifers gained LW in both periods but this gain (LWG) was significantly greater in P1 than P2 ($P<0.001$). Booster heifers showed greater LWG than Control heifers in both periods but showed significance only for P1 ($P<0.06$). When averaged over periods, LWG was significantly greater for Booster than Control ($P<0.06$) but the group/period interaction was non-significant ($P>0.06$). The heifers accepted Booster well and grew more than the Controls consuming grass alone.

Session 15 — Poster 18

Effect of offering two levels of crude protein and two feeding levels of milk replacer on calf performance

R.J. Fallon[1], H.C.F. Wicks[2] and J. Twigge[3], [1]Teagasc, Grange Research, Dunsany, Co. Meath, Ireland, [2]The Agricultural Research Institute of Northern Ireland, Hillsborough, Co. Down, BT26 6DR, UK, [3]Nutreco Ruminant Research Centre, 5830 AE Boxmeer, The Netherlands

The objective was to evaluate a calf milk replacer (CMR) with two protein levels and two feeding levels on calf performance. Sixty-four 2 to 3 week-old purchased Holstein/Friesian calves with an initial weight of 50 kg (+/- 1.8) were allocated to (1) 23% crude protein CMR at 600 g/d (LL), (2) 23% crude protein CMR at 1200 g/d (LH), (3) 30% crude protein CMR at 600 g/d (HL) or (4) 30% crude protein CMR at 1200 g/d (HH). The milk replacer was offered warm by bucket for 56 days. All calves had *ad libitum* access to a concentrate diet throughout the 112 day experimental period. Liveweight gains (g/d) were 690, 760, 720 and 870 (sem 48)1 to 56 days for treatments LL, LH, HL and HH, respectively. The corresponding values from day 1 to 112 for liveweight gains were 860, 950, 900 and 990 (sem 48), for concentrate DM intake were 215, 218, 209 and 234 kg, for CMR DM intake were 30, 57, 30 and 57 kg. Increasing the daily CMR allowances from 600 g to 1200 g increased LWG in the period 1 to 56 day and there was no response to increasing the level of crude protein from 23 to 30%.

Session 15 — Poster 19

Bioelectrical impedance analysis for the prediction of saleable products in buffalo

A. De Lorenzo[1], F. Sarubbi[2], F. Polimeno[2], R.Baculo[2], M. Servidio[1], P. Abrescia[3], L. Ferrara[2], [1]Dipartimento Neuroscienze, Università Tor Vergata, Roma, Italy,[2]ISPAAM-CNR, Via Argine 1085, 80154, Napoli, Italy,[3]Dipartimento Scienze Biologiche- Università di Napoli Federico II, Napoli, Italy*

Many studies defined the buffalo aptitude for meat production, and nutritional characteristics of buffalo meat were appreciated. Various techniques and procedures for estimating body composition of live animals have been used under specific conditions and different species. Bioelectrical impedance (BIA) technology, previously developed for humans because of its accuracy and simplicity, has been applied conveniently in many animal species but in buffalo. The objective of this study was to develop prediction equations of saleable cuts from buffaloes carcass. Twenty buffaloes were fed *ad libitum* with a concentrate feed and vitamin-mineral integration, for 14 months. Seven days before slaughtering, the animals were weighted and somatic and BIA measurements (Rs, Xc) were collected. Carcasses were kept at 4 °C for seven days and then measured for BIA, carcass half weight, yield grade, longissimus muscle area, and fat mass. The best subset regression procedure was used to select a group of likely models including BIA and measurements in live buffalo. Then the adjusted R^2 selection and residual analysis were used as criteria to select the independent variables in the models and individuate optimum values of frequency (500 - 1000 khz). Our study suggests that BIA can be used conveniently to predict the yield of saleable cuts.

Intestinal digestibility of rumen undegraded protein determined by mobile bag method in rapeseed, rapeseed meal and extracted rapeseed meal

P. Homolka, V. Koukolová, Research Institute of Animal Production, Uhríneves, 104 00 Prague, Czech Republic

In this study, nutritive value of rapesead, rapesead meal and extracted rapesead meal were compared. The experiments were performed usin the mobile bag technique with three dry cows (Black Pied), fitted with a large ruminal cannula and a T-piece cannula in the proximal duodenum. The procedure involves three steps:
1. Incubation of feed samples for 16 hours in the rumen of cattle to obtain the undegraded residues.
2. Incubation of the residues for 2.5 hours in an artificial stomach (abomasum).
3. Estimation of protein digestibilities of residues in the intestine using mobile bags.
The cows were fed twice a day (at 6 a.m. and 4 p.m.) and their daily rations consisted of 4 kg alfalfa hay, 10 kg maize silage and 1 kg barley meal with a vitamin and mineral supplement.
Intestinal digestibility of rumen undegraded protein was 30 % for rapeseed, 15 % for rapesead meal and 65 % for extracted rapesead meal. There were statistically significant difference among the feeds ($P<0.05$). This work was supported by the Ministry of Agriculture of the Czech Republic (MZE 0002701403).

Ruminal degradability and mobile bag intestinal digestibility of individual amino acids of pasture forage

P. Homolka[1], J. Trinácy[2], A. Skeríková[1], [1]Research Institute of Animal Production, 10400 Prague, Czech Republic, [2]Research Institute for Animal Breeding, Ltd., Department Pohorelice, Czech Republic*

The mobile bag technique was performed with two dry cows, each fitted with a large ruminal cannula and a T-piece cannula in the proximal duodenum. The procedure involves three steps:
1. Incubation of feed samples for 16 hours in the rumen of cattle to obtain the undegraded residues.
2. Incubation of the residues for 2.5 hours in an artificial stomach.
3. Estimation of amino acids digestibilities of residues in the intestine using mobile bags.
Degradation of selected amino acids after 16 hours incubation in the rumen was following: Lys 52 %, Met 54 %, Thr 56 %, Phe 76 %, Leu 47 %, Val 61 %, Ile 50 %, Cys 61 %, Tyr 53 %, Arg 56 %, His 51 %, Ala 64 %, Gly 54 %, Ser 63 % and Pro 80 %, respectively. Intestinal digestibility of selected amino acids from pasture forage was following: Lys 86 %, Met 55 %, Thr 79 %, Phe 76 %, Leu 80 %, Val 79 %, Ile 80 %, Cys 63 %, Tyr 81 %, Arg 84 %, His 80 %, Ala 82 %, Gly 77 %, Ser 76 % and Pro 24 %, respectively. This project was supported by the Ministry of Agriculture of the Czech Republic (NAZV No.1B44037 and MZE0002701403).

Session 15 Poster 22

Characteristic size dimensions of washed faeces particles from dairy cows fed different concentrate/forage ratios

P. Nørgaard[1], M. R. Weisbjerg[2], K. F. Jørgensen[2] and D. Bossen[2], [1]Department of Basic Animal and Veterinary Sciences, KVL, Grønnegårdsvej 2, DK-1870 Frederiksberg, [2] The Danish Institute of Agricultural Sciences, Department of Animal Heath, Welfare and Nutrition, P.O. Box 50, DK-8830 Tjele, Denmark

Our aim was to study effect of concentrate/forage ratio on the size dimensions of washed faecal particles from dairy cows. Ten cows were fed on high (45:55) (H) and 9 on low (26:74) concentrate/forage ratio (L) in mixed rations. Cows were individually supplemented with 3 kg concentrate. Forage consisted of 64% maize and 36% grass/clover silage on dry matter basis. Concentrate consisted of barley, rapeseed meal and sugar beet pulp. Faecal samples were washed in nylon bags before drying and sieving into 4 fractions: >2.8 mm, 1-2.8 mm, 0.5-1 mm and bottom bowl. Samples of particles from the sieving fractions were scanned and the area, length and width of individual particles were identified using image analysis. The overall mean, mode (most frequent) and 95 percentile values were estimated from a composite function. Higher concentrate/forage ratio significantly reduced the proportion of faecal particles retained in 2.8 mm sieve fraction ($P=0.05$) whereas the overall mean, mode or 95 percentile length and width values were not significantly affected ($P > 0.1$). The overall mode, mean and 95 percentile length values were 0.52±0.06, 3.10±0.67, 10.5±2.4 mm, respectively, and width values were 0.15±0.02, 0.76±0.17, 2.8±0.7 mm, respectively.

Session 15 Poster 23

Effect of plant phenolic compounds on growth of some rumen bacteria

R. Rullo[1], A. Tava[2], L. Ferrara[1] and G. Maglione[1], [1]ISPAAM - CNR Via Argine 1085, 80147 Neaples, Italy, [2]C.R.A. Istituto Sperimentale per le Colture Foraggere, viale Piacenza 29, 26900 Lodi., Italy*

The increasing demand for sustainable livestock production systems implies the need to focus on a better utilization of new grazing lands. The aim of this study was to investigate the effects of condensed tannins (CT) and flavonoids extracted from leaves of some typical Mediterranean forage crops to be used as an alternative feedstuff. For this purpose, two different strains of rumen bacteria were treated with CT purified from *Hedysarum coronarium*, *Lotus ornithopodioides* and flavonoids from *Medicago sativa* and *Cicorium Hyntibus*. The rumen bacteria studied were *Streptococcus bovis* JB1 and *Ruminococcus albus* (DSM 20455). Their growth rates were monitored *in vitro* by measuring a change in optical density at 600nm under anaerobic conditions. Purified CT were dissolved into anaerobic phosphate buffer pH 6.8 to yield a final concentration ranging from 100 to 400 µg/ml in the medium. Both strains reached their maximum optical density after 7 - 8 hours in the absence of phenolic extracts. At 400 µg/ml of flavonoids from *C. hytibus* and *M. sativa,* both strains continued to grow. Only 100 µg/ml of CT from *H. coronarium* or 400 µg/ml of *L. ornithopodioides* reduced significantly the growth of both strains ($P<0.001$).

Session 15 Poster 24

Comparing metabolic traits glucose and insulin in their relationship to milk production

L. Panicke[1], G. Freyer[1], R. Staufenbiel[2] and E. Fischer[3], FBN 18196 Dummerstorf[1], Freie Universität 14163 Berlin[2], Universität 18059 Rostock[3], Germany*

Early information on the evaluation of growing young dairy bulls is of interest to breeders. Our investigation is aimed at metabolic traits that can be observed before the individual estimated breeding values (EBV) for milk production is known. Insulin has a central position within the energy metabolism and the reaction can be measured by glucose challenge (GTT). Heritability coefficients of insulin area (Ia) were 0.37 and 0.33 for logarithmic Ia, respectively. Glucose half life time (Ghwz) and glucose area (Ga) resulted in heritability coefficients of 0.49, 0.33, 0.59 and 0.37 for original and logarithmic values. The correlation coefficients between insulin and GTT traits was negative, ranging from r = -0.10 to -0.44. Based on these metabolic traits and ancestral breeding values (PBV) we calculated breeding values (CBV) for young bulls via linear and quadratic regression. These CBV were evaluated by means of EBV of 83 bulls. Based on Ghwz, Ga and PBV, the correlation coefficient was r = 0.39. The best evaluation was obtained by combining Ga and Ia/Ga with PBV (r = 0.42). When prediction based on PBV alone, the correlation coefficients were much lower (r = 0.16 to 0.23). These results have been confirmed by investigations on different groups of sires. It is concluded that metabolic traits are suited for an early evaluation of young dairy bulls.

Session 15 Poster 25

Effect of feeding pistachio hulls on performance of lactating dairy cows

P. Vahmani[*], A.A. Naserian, J. Arshami and H. Nasirimoghadam, Animal Science Department of Ferdowsi University of Mashhad, khorasan, P.O.Box:91775-1163, Iran

Pistachio hulls (Pericarp) are a by-product of de-hulling of pistachio nuts soon after harvest. This experiment was carried out to evaluate the effect of dried pistachio hulls (DPH) as a feed ingredient for lactating dairy cows. Eight multiparous Holstein dairy cows (Body Weight = 606±24kg and 160±18 days in milk) were used in a replicated 4×4 latin square design (4 periods & 4 treatments). The treatments were 0 (control diet), 2, 4 or 6% DPH in dietary dry matter. The DPH substituted for beet pulp in the control diet. Cows were offered total mixed ration (TMR). Each experimental period was for 21 days (adaptation, 14d; sample collection, 7d). Chemical analysis of DPH indicated that this by-product contained 12%crud protein, 5% etter extract, 45% NDF, 34% ADF, 5.20% ash and 8% tannin of dry matter. Increasing levels of DPH in diets had no significant effect on DMI, ruminal PH, blood urea nitrogen (BUN), milk yield, milk fat, protein, lactose and SNF(P>0.05), but there was a tend for decreased DMI(23.24, 23.19, 23.01 and 22.70 kg/day, respectively)as dietary DPH content increased. The decline of the DMI may be attributed the tannin content of DPH in diets. Results from the chemical composition and production responses indicate that the use of the DPH did not adversely affect animal performance.

Session 15 Poster 26

Nutritive evaluation of processed cottonseed fed Holstein dairy cows

A.R. Foroughi[1], R. Valizadeh[2], A. A. Naserian[2] and M. Danesh mesgaran[2], Education centre of khorasan Jihad-Agriculture, Animal Science Department,Mashhad, [2]Ferdowsi University of Mashhad , Animal Science Department, Iran

The objective of this study was to evaluate the effect of processing (grinding and moist heat) of whole cottonseed (WCS) on its crude protein(CP) degradability and digestibility of diets containing processed cottonseed fed to Holstein lactating cows. Eight multiparious dairy cows averaging 84.50±10.34 days in milk and 36.10±4.46 milk yield were used in a 4x4 Latin Square design. Cows were divided into four dietary treatment groups. Dietary treatments were 1) WCS; 2) ground cottonseed(GCS); 3)GCS heated in 140°C and steeped for 2.5 minute (GHCS1); or 4) GCS heated in 140°C and steeped for 20 minute (GHCS2). Moist heat and steeping of WCS decreased CP degradability of HGCS1 and HGCS2 by 11 and 18 %, respectively. The mean DMI was significantly (P<0.01) affected by diets and in treatments of 1,2,3 and 4 were 25.97, 27.24, 27.63, and 27.63 (kg/d), respectively. DM digestibility was significantly (P<0.01) affected by the diets and was greatest for HGCS1 (64.14%) and the lowest for HGCS2 (60.10%). Moist heat treatment and steeping reduce rumen protein degradability and affect digestibility of diets containing processed cottonseed.

Session 15 Poster 27

Evaluation of processed cottonseed and ruminally protected lysine and methionine for lactating dairy cows

A.R. Foroughi[1,], A. A. Naserian[2], R. Valizadeh[2] and M. Danesh mesgaran[2] , Education centre of khorasan Jihad-Agriculture, Animal Science Department,Mashhad, [2] Ferdowsi University of Mashhad , Animal Science Department,Iran*

The objective of this study was to evaluate the effect of processing (grinding and moist heat) of whole cottonseed (WCS) and ruminally protected lysine(Lys) and methionine(Met) on milk composition and production of Holestein lactating cows during early lactation. Multiparious cows (n=12) averaging 24.50±11.05 (DIM) and 38.66±4.27 milk yield (MY) were used in a 4x4 Latin square design. Cows were fed one of the following treatments: 1) WCS; 2) WCS + 16gr Met&20gr Lys(WCS2); 3) ground cottonseed (GCS) heated in 140°C and steeped for 20 minute (GHCS1); or 4) GCS heated in 140°C and steeped for 20 minute +20gr Met&30gr Lys(GHCS2). The mean DMI was significantly (P<0.01) affected by diets and in treatments of 1,2,3 and 4 were 21.08, 21.19, 22.57 and 27.63 (kg/d), respectively. MY was significantly (P<0.01) affected by the diets and was greatest for HGCS2 (35.78 kg/d) and the lowest for WCS (33.07kg/d). Milk fat percentage and yield were unaffected by diets. Milk protein percent was progressively increased, averaging 3.21%, 3.30%, 3.28% and 3.48% for 1,2,3 and 4 treatments, respectively. Results indicated that when cows were fed WCS and processed cottonseed associated ruminally protected lysine and methionine ,milk yield and composition were improved.

Session 15 Poster 28

The effect of application a dry feed additive on the fermentation characteristics of lupin silage

P. Dolezal, L. Zeman, J. Dolezal, Pyrochta, V., Department of Animal Nutrition, Mendel University of Agriculture and Forestry Brno, Zemedelska 1, 613 00 Brno, Czech Republic

A fresh green matter of lupine plants (Lupinus L. variete Juno, 197.85 g/kg DM) in full waxy stage of maturity was cut to size ca 25 mm. The crop was wilted for a periody 24 hours and ensiled for 98 days in in laboratory silos with capacity 4 L. There was supplemented feed additive in following amount: 5, 10, 30, 40. 50 or 70 kg/tone of forage. The composition of additive was dry whey (30 %), maize meal (40 %) and dry molasses (30 %). The silages fermented rapidly and changes in volatile fatty acids (VFA) production ($P<0.01$) and in sum of acids were noted. All treated silages were well fermented with low levels of ammonia and pH. The experimental silages with higher (50 and 70 kg) feed supplementation was of better quality (significantly higher ratio LA/sum acids, higher content of lactic acid, lower NH_3 content and pH value) than the control silage, or silage with lower concentration. In experiment feed additive-treated silages had significantly higher alcohol content and dry matter than control untreated silage. It was concluded that feed additive used as a silage additive improved fermentation of lupine, reduced acetic acid production and increased silage nutritive value.

This study was supported by Project MSM 432100001.

Session 15 Poster 29

Development of a quick method of evaluating flavour preferences in concentrates for lactating cows

E. Roura[1], C. Ossensi[2], R. Mantovani[2] and L. Bailoni[2], [1]Lucta SA, Barcelona, Spain, [2]Department of Animal Science, University of Padova, Italy

A 50-day experiment with 32 Friesian lactating cows was designed as a quick method to determine flavour preferences in concentrate feeds. During the 12 double choice tests, each cow was offered simultaneously with two identical buckets one with a low palatable concentrate (ConcB) and the other one with the same concentrate flavoured with one of six different flavours (Bx were x=1,2,3,4,5,6). ConcB was formulated like a standard concentrate (ConcS) supplemented with urea and calcium salts at 0.5%. Tests lasted 5 minutes and only one flavour at a time was tested (each flavour was tested twice). ConcS was fed during an adaptation period (10 d) and for 2/3 days between each test. The amount of feed consumed from each bucket was recorded and divided by the time of exposure (ingestion rate: IR, g/s). On the overall average, flavoured concentrates Bx showed a significantly ($P<0.01$) higher IR than unflavoured ConcB (6.43 vs. 7.14 g/s). Differences of IR associated with individual flavours B1, B2, B3, and B4 were not significant. On the contrary, IR was significantly higher with addition of B5 (6.39 vs. 7.83 g/s; $P=0.02$) and B6 (6.71 vs. 7.77 g/s; $P=0.10$). In conclusion, the method appeared very useful to evaluate flavour preferences in concentrates for lactating cows.

Session 15 Poster 30

Milk production and composition as affected by feeding supplemental fat in Sahiwal cattle

A. Iqbal[*1], J. Akbar[1], M. Abdullah[2] and M. Sarwar[3], [1]Faculty of Animal Husbandry, University of Agriculture, Faisalabad.Pakistan [3]Institute of Animal Nutrition and Feed Technology. University of Agriculture, Faisalabad. [2]University of Veterinary and Animal Sciences, Lahore, Pakistan*

The present project was conducted to study the effect of feeding supplemental fat on the milk production and composition in four primiparious Sahiwal cows. The animals were offered four diets viz. A, B, C and D, allotted to them at random, having 0, 2, 4 and 6 % levels of animal fat (tallow), respectively. All the diets were made nitrogenous and isocalonic and fed for four periods, each of 21 days duration in 4x 4 Latin Square Design. The cows were housed individually in a tie stall barn and fed *ad-libitum* a total mixed ration (TMR). The daily milk production of cows increased by feeding tallow supplemental diets (8.08-9.50 lit/day), being statistically significant ($P<0.05$). Similar trend was noticed in its composition in terms of fat, total solids (TS) and solids-not-fat (SNF). Contrary to this, protein contents showed a declining trend with increased levels of dietary fat. The specific gravity, however did not differ significantly ($P>0.05$) on different treatments. Various milk components were determined by the standard methods, ranged: fat 3.89-5.08; TS 12.25-13.52; SNF 8.23-9.18; and protein 3.00-3.20 %. The study concluded that tallow is an economical source of energy for supplementation up to 4 % level of the diet dry matter in Sahiwal cows.

Session 15 Poster 31

Degradation of dry matter and fiber of five feeds by rumen anaerobic fungi of sheep

T. Ghoorchi[1], S. Rahimi[2], M. Rezaeian[3] and G.R. Ghorbani[4], [1]Iran, Gorgan, Faculty of Animal Science, Gorgan University of Agricultural Sciences and Natural Resources. [2]Tarbiat Modarres Univ., Tehran, Iran. [3]Tehran Univ., Tehran, Iran, [4]Isfahan Univ. Tech., Isfahan, Iran.

An experiment was carried out to estimate the potential activity of rumen anaerobic fungi in the degradation of dry matter and fiber of feeds. Samples of wheat bran, bagasse, cotton seed, alfalfa and corn silage were used as the substrates to culture rumen fungi which were isolated from a fistulated Shal sheep. Loss percentages of dry matter (DML), neutral detergent fiber (NDF), acid detergent fiber(ADF),acid detergent lignin(ADL), cellulose, and hemicellulose of samples were measured after 0,3,6 and 9 days of incubation. Dry matter and NDF loss of substrates varied from 10.6% to 29.4% an 11.7% to 48.7% after 9 days of fungi growth. The highest and lowest DML and NDF were related to alfalfa and bagasses, respectively. The highest values for the ADF loss (39%),hemicellulose loss (65.6%) and cellulos loss (55.6%) were measured from alfalfa. The results indicated that rumen anaerobic fungi have the ability of degrading dry matter and fiber from different types of feed.

Session 15 Poster 32

The effect of Polyethylen Glycol (PEG) addition on in vitro organic matter digestibility (IVOMD) of grape pomace

D. Alipour[1] and Y. Rouzbehan[1], Department of Animal Science, Faculty of Agriculture, Tarbiat Modarres University, P.O. Box 14115-336, Tehran, Iran.*

Grape pomace is an agro-industrial waste, which has numerous environmental adverse effects because of its high tannin content. To evaluate the nutritive value of grape pomace three samples of this by-product were collected from two provinces in Iran. The chemical composition includes: NDF, ADF, ADL, crude protein (CP), ether extract (EE), total phenolics(TP), total tannins(TT), condensed tannin(CT) and protein precipitation capacity(PPC) and IVOMD with and without PEG were evaluated. TP and TT and CT ranged from 13 to 140, 7.6 to 91.1 and 8.76 to 245.9 g/Kg DM, respectively and PPC was from 0 to 41.94% . Addition of PEG to grape pomace significantly ($P < 0.05$) increased the IVOMD.

Session 15 Poster 33

In vitro enzymatic proteolysis of different protein sources

A.S. Chaudhry, School of Agriculture, Food and Rural Development, University of Newcastle-upon-Tyne NE1 7RU, UK

The suitability of enzymes to estimate *in vitro* proteolysis was tested. Experiment 1 compared the proteolytic activity over various times of Protease (*Streptomyces griseus)* with Papain (*Papaya latex)* by using 1.33 (high) or 0.4 (low) units (U, amount) of enzyme/mg protein (CP) of purified proteins (bovine-albumin; egg-albumin). Experiment 2 used 0.66U of each enzyme/mg CP of semi-purified proteins (casein; wheat gluten; maize gluten, MG). Incubations were terminated and the supernatant analysed to estimate total amino acids as the measure of proteolysis of each protein over time. The data were used to derive constants for solubility *(a)* and rate *(c)* and extents ($a+b$) of proteolysis of each food by each amount of each enzyme. Significant differences were observed between foods, enzymes and enzyme amounts (experiment 1) for proteolysis in both experiments ($P<0.001$). While Protease gave about 4X more proteolysis than Papain ($P<0.001$), the high amount gave only 2X greater proteolysis than the low amount ($P<0.001$). On average, the proteolysis by protease was over 3X faster *(c)* than papain ($P<0.001$) and 2X faster for high versus low amount of enzyme ($P<0.001$). While purified foods were similar in solubility ($P>0.05$), they differed in rate and extent of proteolysis ($P<0.001$). Amongst semi-pure proteins, casein gave the highest but MG the lowest proteolysis at each time ($P<0.001$). However, the extent of proteolysis depended upon the enzyme, food and incubation time. Clearly, the protease and not papain can be used to estimate *in vitro* proteolysis of different feed proteins for ruminants.

Session 15 Poster 34

Nutritional assessment of genetically modified rape seed and potatoes, differing in their output traits

H. Boehme[1], B. Hommel[2], E. Rudloff[3] and L. Huether[1], [1]Federal Agricultural Research Centre (FAL) - Institute of Animal Nutrition, Bundesallee 50, D-38116 Braunschweig, [2]Federal Biological Research Centre for Agriculture and Forestry, Kleinmachnow, [3]Federal Centre for Breeding Research on Cultivated Plants, Gross Luesewitz, Germany

Rape seed with modifications in the fatty acid pattern (increased portions of myristic and palmitic acids at costs of oleic acid) and potatoes with modifications in the carbohydrate fraction were analysed for their nutrient content and their feeding value for growing pigs. Both GM-crops were tested against their non-modified counterpart.
The chemical analysis of the GM-rape seed showed only marginal differences in the nutrients, but the glycosinolate content was increased (20 vs. 13 μmol/g dry matter). The digestibility of the diets containing 15% isogenic or transgenic rape seed dry matter remained unaffected . Correspondingly, production responses of the pigs were only marginal due to the replacement.
The genetic modification of potatoes to synthesize high molecular weight inulin resulted in a decreased starch content (68% vs. 74% in the dry matter) and an increased content of sugars (7.7% vs. 2.5% in the dry matter). But the silages of both potato varieties did not show differences in their energetic feeding value for pigs and effects on growth and slaughter performance were not observed, when the ensilaged potatoes were incorporated at a level of 40% in the diets (on a dry matter basis).

Session 15 Poster 35

Investigation of some preparation procedures of fatty acid methyl esters for capillary gas-liquid chromatographic analysis of conjugated linoleic acid in feed

L. Wágner, K. Dublecz, F. Husvéth, L. Pál, Á. Bartos, G. Kovács, Z. Garádi, Sz. Stiller, J. Karnóth, Department of Animal Physiology and Animal Nutrition, University of Veszprém, Georgikon Faculty of Agriculture Keszthely, H-8360 Keszthely, Deák Street 16. Hungary

The five most widely accepted procedures for preparing fatty acid methyl esters in feed lipids were investigated for their suitability in capillary gas- liquid chromatographic analysis of cis-9, trans-11 conjugated linoleic acid (c-9, t-11 CLA) in feed. A modified procedure of fatty acid methyl esterification was developed and the method was applied to determine c-9, t-11 CLA content in some feeds. This method involved hydrolysis of lipids with 0.5 M KOH in methanol at 100°C for 5 min, followed by esterification with aqueous HCL (35%)/methanol (1:1 v/v) at 100°C for 5 min. The resulting c-9, t-11 CLA methyl ester was separated by gas-liquid chromatograph equipped with a capillary column and determined using nonadecanoic (C19:0) as an internal standard.

Session 15 Poster 36

Effect of vitamin B_6 on lysine synthesis from 2,6-diaminopimelic acid by mixed rumen protozoa and bacteria

A.M. El-Waziry, Department of Animal Production, Faculty of Agriculture, Alexandria University, El-Shatby, Alexandria, Egypt

It has generally been thought that vitamin B complex (B vitamins) were produced in the rumen of ruminants and the B vitamins were unnecessary to be given to the ruminant animals. In the earlier papers, however, vitamin B_6 was shown to effectively stimulate the cellulose digestion by washed cell suspensions of mixed rumen bacteria and the growth of cellulolytic rumen bacteria such as *Ruminococcus flavefaciens* and *Fibrobacter succinogenes*. An *in vitro* study was conducted to examine the effect of vitamin B_6 (pyridoxine hydrochloride) (B_6) on the production of lysine from 2,6-diaminopimelic acid (DAP) by mixed rumen protozoa (P) and mixed rumen bacteria (B). P and B, were isolated from the rumen of goats given a concentrate mixture and lucerne cubes and separately incubated for 12 h with and without DAP (5 mM) as a substrate and B_6 (10 µg/ml) as additives. In P suspensions, B_6 increased the disappearance of DAP by 19.9% ($P < 0.05$), and also increased the production of lysine by 26.8% ($P < 0.05$), during 12 h incubation. In B suspensions, the increase of the disappearance of DAP with B_6 was 2.7%, and increased lysine production by 32.9 % ($P < 0.001$), during 12h incubation. B_6 was demonstrated to enhance the production of lysine from DAP by mixed rumen protozoa and bacteria by the present study.

Session 15 Poster 37

Pollen composition of six plant species in some nutrients, microminerals and aromatic substances and potential use in animal nutrition

D. Liamadis[1], T. Zisis[1], Ch. Milis[2], A. Thrasivoulou[1], [1]Aristotle University of Thessaloniki. [2]Ministry of Agriculture, Greece

A preliminary study was conducted in purpose to evaluate the nutritive value of pollen. Pollen samples of six plant species (Papaver, Taraxacum, Sinapis, Lamium, Pinus, and Prunus) were analyzed according to Weender procedure. Crude protein (CP) content was 23.1; 17.8; 19.0; 25.9; 14.5 and 22.9 % DM, respectively, revealing a great variation in CP content. Some pollen species could be used as protein supplements in animal nutrition. The fat content was 3.0; 6.6; 8.1; 2.1; 0.7 and 1.1 % DM, respectively. Additionally, pollen samples were analyzed at four crucial microminerals (Fe, Cu, Zn and Mn). This analysis have showed high Fe and Zn levels in all species under study (38.2; 34.6; 42.9; 86.2; 45.6 and 39.6 p.p.m Fe, respectively) and (40.7; 24.1; 24.1; 47.7; 38.1 and 52.0 p.p.m Zn, respectively). The high content in those minerals indicates that pollen could be used as a mineral additive especially in high yielding ruminants. This has been observed previously in human nutrition. Finally, 18 aromatic substances were determined by using gas chromatography. The substances which were present in all species were a-hexanol, 3,5-octadien-one nonanal, p-xylone, toluene and nonanal, in quantitative order. Some of these substances have antibiotic properties. Moreover, above results suggest that pollen may be able to increase diet's palatability and in consequence DM intake.

Session 15 Poster 38

Apparent ileal amino acid digestibility in diets supplemented with phytase and pancreatine for pigs

F. Copado[1], M. Cervantes[2], J. Yánez[3], JL. Figueroa[4], and W. Sauer[5], [1]Zootecnia-UACH, [2]ICA-Universidad Autónoma Baja-California, México; [3]UAT; [4]Colegio-Postgraduados, [5]University of Alberta, Canada*

An experiment was conducted to determine the effect of phytase and pancreatine supplementation to sorghum diets on the apparent ileal digestibility (AID) of AA. Eight pigs (BW 22.1 kg) were used. Treatments (T) were: T1) basal, sorghum-soybean meal diet, T2) basal plus 1,050 units of phytase activity (FTU), T3) basal plus 591 mg pancreatine/kg, T4) basal plus 1,050 FTU and 591 mg pancreatine/kg. Diets were added with vitamins and minerals to meet the requirements. Chromic oxide was the digestibility marker. The AID (%) of AA for T1, T2, T3 and T4 were: arg, 83.0, 82.6, 83.0, 82.2; his 75.6, 75.7, 75.1, 73.9; ile, 73.5, 73.0, 72.8, 72.2; leu, 73.2, 73.3, 72.3, 71.5; lys, 77.8, 77.7, 77.0, 76.3; met, 67.5, 66.3, 62.9, 66.0; phe, 74.6, 74.5, 74.0, 73.3; thr, 64.0, 63.6, 62.2, 61.8; val, 70.0, 70.0, 69.5, 68.4, respectively. There was no effect of phytase supplementation on the AID of AA ($P>0.10$). Except for met ($P = 0.07$), there was no effect of pancreatine supplementation on the AID of AA. There was no phytase x pancreatine interaction ($P>0.10$). The AID of arg was highest whereas the AID of threonine was lowest. These results indicate that the supplementation of either phytase or pancreatine does not affect the supply of digestible AA to growing pigs.

Session 15 Poster 39

Mustard seed (*Sinapis Alba*): Nutrient content and digestibility in swine

T. Ács, A. Hermán, M. Szelényi, A. Regius and J. Gundel, Research Institute for Animal Breeding and Nutrition, 2053 Herceghalom, Hungary*

Authors studied mustard seed samples of different genotypes treated in different ways. The experiments included untreated, enzyme activated and reduced eruca acid containing variants of mustard seed. Mineral matter and nutrient content, amino acid and fatty acid composition, and digestibility of nutrients (in metabolic experiments with pigs) were measured. N metabolism and digestibility of nutrients of feed mixtures with different mustard seed and cereal content and amino acid composition were also analysed.

It was concluded that crude protein content of mustard seed was 300-357 g/kg, the crude fat content was 252-303 g/kg and the crude fiber content was 63-86 g/kg. Lysine content was 1.3-2.1 g/kg, the phosphorus level was 6.3-9.9 g/kg which hardly exceeds the similar values of soybean. Eruca acid level in crude fat was 2.6% in samples with reduced eruca acid content and 28-40% in other samples. Digestibility of nutrients in mustard seed hardly changed (crude protein: 84-86%, crude fat: 75-77%, crude fiber: 50-54%) irrespectively of quantity of diet (3, 6, 9, 10, 20, 30%). Apparent digestibility of crude protein content of mustard seed containing diet increased by 6% and N retention by 7% in case of amino acid supplementation. Productive protein utilisation showed a similar improving tendency.

Influence of dietary fibre on the gut morphology and pancreatic and intestinal enzyme activities in the weaned piglet

P. Trevisi[1], J.P. Lallès[2], I. Luron[2] and B. Sève[2], [1]DIPROVAL, University of Bologna, 42100 Reggio Emilia, Italy,[2]UMR Systèmes d'Elevage et Nutrition Animale et Humaine, INRA, 35590 Saint-Gilles, France*

Thirty four 28-d old piglets were used to determine the effects of fibre sources on the development of gastro-intestinal tract and on intestinal and pancreatic enzyme activities in weaned piglets. The pigs were fed four different diets: control (C), low fibre (LF), high insoluble fibre (HIF), high soluble fibre (HSF). Soluble and insoluble fibres were provided by sugar beet pulp and wheat bran, respectively. On d-14 post-weaning the piglets were sacrificed and gut segments weighed and sampled. Data were statistically analysed as incomplete block design by the GLM procedure of SAS including gender, diet and pair in the model. Villus morphology was not affected by the diets but colonic crypts were wider with the fibre diets and jejunal crypts area were higher with the HSF diet. The high fibre diets reduced jejunal maltase (-36%), sucrase (-41%) and aminopeptidase-N (-40%) activities. Moreover, the HSF diet tended to decrease the specific activity of pancreatic trypsin by 23%. It also tended to reduced caecal pH. In conclusion, the addition of a high level of fibre in diets for weaned piglets reduced digestive enzyme activities in the mid-intestine. This may decrease digestion in the small intestine and consequently increase the load of undigested material fermentable in the large intestine.

Mannan oligosaccharide enhances absorption of colostral Ig G in newborn calves and piglets

Miodrag Lazarevic, Faculty of Veterinary Medicine, University of Belgrade, 11000 Belgrade, Serbia and Montenegro

The objective of the present study was to evaluate the effect of Clinoptilolite Minazel (CM) and mannan oligosaccharide (MOS) on plasma Ig G concentrations.
36 Holstein calves were randomly assigned to 3 groups (Control (no additive); CM: 5 g/L of CM; MOS 5 g/L of Bio-Mos(r) (Alltech Inc.)) and fed colostrum 3 times during the first 24 hours of life (mean intake 3 L in 24 hrs). After day 1 all calves received the same diet. Blood samples were taken 6, 12, 24, and 48 hrs and 4, 7, 14 and 21 days following parturition. The piglet trial was a complete randomised block design with 6 Swedish land race sows and their litters. Four piglets per sow served as control, and four piglets received 0.75 g of MOS suspended in 10 ml of water (Bio-Mos(r), Alltech Inc.) each 0 and 24 h after birth. Blood samples were taken 48 hrs after parturition. Sera Ig G was estimated by radial immuno-diffusion and data analysed by Students T-test.
In calves, CM tended to improve, whilst MOS significantly improved Ig G concentrations compared to controls. MOS also improved plasma Ig G concentrations in piglets (Control: 42.04 mg/ml; MOS: 44.41 mg/ml) significantly ($P < 0.01$).

Session 15 Poster 42

Growth promoting effects of Rare Earth Elements in piglets

U. Wehr, C. Knebel and W.A. Rambeck, Institute for Physiology, Physiological Chemistry and Animal Nutrition, University of Munich, Schoenleutnerstr. 8, D-85764 Oberschleissheim, Germany*

For decades Rare Earth Elements (REE), such as lanthanum, cerium and praseodymium are used in Chinese agriculture and farming for yield increase and in parts tremendous growth promoting effects are reported.
Several feeding trials with mixtures of REE under western feeding and housing conditions were carried out in our working group resulting in positive effects concerning growth performance. So far, REE chlorides were used in previous studies. In the present study we tested for the first time the effects of REE citrate in growing pigs.
A feeding study with 28 piglets was carried out. They were divided into 4 groups (n=7). The animals received a regular diet, supplemented with a mixture of REE citrate in concentrations of 0, 50, 100, and 200 mg/kg feed for a 6 week period. A positive effect of the REE citrate on growth performance parameters was observed. In the trial period the daily body weight gain of the two high supplemented groups increased by 8.6 % respective 22.6 % compared to the control group. An improvement of the feed conversion rate from 2 % to 6 % was shown in all REE supplemented groups. In summary, REE citrate might be of interest in animal production as a new alternative feed additive. However, the mechanism of the action of REE is still unknown.

Session 15 Poster 43

An hydrolyzed protein concentrate (Palbio 62(r)) increases feed intake and villus height in early weaning pigs

E. Borda[1], D. Martinez-Puig[1] and F. Perez[2], [1]Bioiberica S.A., Barcelona, Spain. [2]SYBA S.L., Barcelona, Spain.*

The transition from suckling to eating solid food is associated with a critical period of underfeeding during which the piglet is adapted to digest a dry vegetable diet. The temporal low feed intake after weaning is associated with suboptimal growth rate and increased risk of diarrhoea (Nabuurs, 1991). The aim of the present study was to test the palatability, growth performance and histological measurements of the intestine of early weaned pigs fed on 3 different diets. Diets contained a soy protein concentrate (SPC; 10%), spray dried porcine plasma (SDPP; 4%) or Palbio 62(r), an hydrolyzed protein concentrate obtained from pig intestinal mucosa (PB; 5%). 60 piglets (n=20) were fed during 15 days with the experimental diets. At the end of the trial, 4 animals per treatment were euthanized and samples of ileum were collected for histological measurements. An increased average daily feed intake ($P<0.05$) was observed with PB (938 g/kg) respect SDPP (603 g/kg) and SPC (562 g/kg). However no differences were found among treatments on average daily gain due to the reduced number of replicates. Villus height was higher ($P<0.05$) for PB (569 µm) than for SDPP (391 µm) and SPC (269 µm) showing a positive correlation with feed intake. The overall results suggest that PB prevents villus atrophy through an increase in feed intake.

Effects of supplemented dietary L-tryptophan on growth performance of 15- to 30-kilogram pigs

L. Buraczewska[1], E. Swiech[1] and L. Le Bellego[2], D. Melchior[2], [1]The Kielanowski Institute of Animal Physiology and Nutrition, PL 05-110 Jablonna, Poland, [2]Ajinomoto Eurolysine, 153, rue de Courcelles, 75817 Paris Cedex 17, France*

To determine the optimal tryptophan:lysine (Trp:Lys) ratio, four diets (without antibiotics) were designed to contain 13.3 g Lys and 14 MJ ME per kg and were supplemented with L-tryptophan to contain Trp:Lys ratios of 0.14, 0.17, 0.20, and 0.23. The diets were fed *ad libitum* during 25 days to pigs with a high lean gain potential (synthetic line 990, Poland). Each diet was fed to 16 pigs housed in 8 pens with 2 pigs in each (1 barrow + 1 gilt). In the groups fed diets with increasing Trp:Lys ratio of 0.14, 0.17, 0.20, and 0.23, ADG from d 0 to 14 was 675, 741, 775, and 730g and from d 0 to 25 was 725, 759, 783, and 753g, respectively. The increase in gains was significant between diets with Trp:Lys ratios of 0.14 and 0.20 ($P<0.012$ and $P<0.053$ for the respective feeding periods). Even not significant, feed intake increased by 9 and 4.5 % and feed conversion ratio improved by 5.6 and 3 % for the two periods at a Trp:Lys ratio of 0.20, as compared to the ratio of 0.14. These results suggest that pigs from 15 to about 30 kg BW had maximum growth performance at Trp:Lys ratio of 0.20.

Growth performance, nutrient digestibility and intestinal morphology in weanling pigs fed Insoluble Fibre Concentration(IFC) and direct-fed microbials(DFM)

Y.K. Han[1], J.H. Lee[2] and K.M. Park[1], [1]Dept. of Food Science and Biotechnology, Sungkyunkwan University, Seoul Korea, [2]Easybio System, Seoul, Korea*

A study was conducted with 96 weanling pigs(barrows, average BW 6.5 kg) to determined the effect of two levels(0.1%, 0.2%) direct-fed microbilas and 0.5% insoluble fibre concentration in the diet on the growth performance, nutrient digestibility and intestinal morphology. Pigs were allocated in a completely randomized block design with four pigs per pen and 6 pens per treatment. Pigs and feeders were weighed 10-days interval for the 40-d trials to determine ADG, ADFI and feed:gain ratio(F:G). The digestibility coefficients for DM, OM, CP, NDF, ADF, Ca an P were determined by the Celite545 as a marker. At the end of the 40-d experimental period, final BW were determined and then one pig per pen was slaughtered 1 h later. Intestines were obtained immediately after evisceration. Morphometry of intestinal segments, including villus length, villus width and crypt depth were determined on the histological sections. General linear models procedures were used for statistical analysis. Addition of IFC to the diet improved ADG and ADFI during first 20days after weaning($P<0.04$) and overall($P<0.03$). DFM supplementation increased($P<0.01$) DWG during 40days after weaning; however, ADFI and feed:gain ratio did not affected. IFC and DFM supplementation did not affect feed:gain ratio. IFC decreased the digestibility of Ca($P<0.05$) and P(<0.01), whereas DFM 0.2% increased the digestibility of ADF($P<0.01$) and NDF(0.02). IFC and DFM supplementation did not altered intestinal morphology in weanling pigs.

Session 15 Poster 46

Influence of herbal feed additives on intestinal microflora of weanling piglets

S. Galletti, S. Stella and D. Tedesco, Department of Veterinary Sciences and Technologies for Food Safety, Via Celoria 10, 20133 Milano, Italy

Herbs and herbal mixtures are receiving increasing attention as alternatives to additives like antibiotics, that will be banned in the EU in 2006.

In this trial we examined the different effects of four dietary herbal additives on intestinal microflora of weanling pigs. Animals were divided into five groups of 28 piglets each and treated from 21 to 41 days of age as follows: CN (negative control, no additives); AB (positive control, 2 g/kg feed of apramycin and 1 g/kg feed of colystin); LY (1,6 g/kg of *Lycium barbarum*); PE (2 g/kg of *Paeonia lactiflora*); OL (2 g/kg of *Olea europea*); PO (2 g/kg of *Portulaca oleracea*). Faecal samples were collected on 29d and 41d and total bacterial count, *Escherichia coli*, *Enterococcus* spp., total coliforms, anaerobic bacteria, and *Lactobacillus* spp. were cultured in selective media. Results showed a lower value for total bacterial count ($P<0.1$), *E. coli* ($P<0.1$) and total anaerobic count ($P<0.01$) in PE group, with respect to CN group. In PO group *Enterococcus* spp. ($P<0.05$) and total anaerobic count ($P<0.01$) were lower with respect to CN group. No significant effects on intestinal microbiota was observed in animals treated with LY and OL.

These results suggest that some the herbal additives tested could contribute to control intestinal microbiota.

Thanks for supports to Indena S.p.A.

Session 15 Poster 47

Effect of protein level variation of feed ration on slaughtering results of broiler chicken

Minodora Tudorache, I. Custura, Elena Popescu-Miclosanu, Georgeta Dinita, University of Agronomical Sciences and Veterinary Medicine Bucharest, Faculty of Animal Sciences, 59 Marasti Blvd., 1 district, Romania

In an experiment aimed to study influence of protein level on production and slaughtering results were used five flocks of broiler chicken (0 to 6 weeks of age) feed with feeds having protein level on the three phases (starter, grower and finisher) as following 100% for flock 1 as control, 110% and 120% for flocks 2 and 3 respectively, 90% and 80% for flocks 4 and 5 respectively compared to control. Body weigh at slaughtering (42 days) was +5.7% and +6.8% compared to control for flocks 2 and 3 respectively and -5.2% and -13.2% compared to control for flocks 4 and 5 respectively. Breast meat percentage from total carcass was +3.0% and +7.1% compared to control to flocks 2 and 3 respectively and -2.5% and -6.7% compared to control flocks 4 and 5. Abdominal fat percentage was -11.2% and -17.8% compared to control to flocks 2 and 3 respectively and +25.9% and 39.6% compared to control to flocks 4 and 5 respectively.

The effect of different dietary unsaturated to saturated fatty acids ratios on the performance and serum lipids in broiler chickens under feed restriction

B. Navidshad[1], M. Shivazad[1], A. Zareh Shahne[1] and G. Rahimi[2], Karaj, Animal science Department of Tehran University, 31587-11167, Iran, [2]Agriculture University of mazandaran, Sari, Iran

360 male broiler chickens were randomly assigned to a completely randomized design with 4 treatments and 3 replicates. Tallow and sunflower oil were used to formulate four isocaloric and isonitrogenous diets (ME=3200 Kcal/Kg and CP=21% and 19% for grower and finisher periods respectively) with different unsaturated to saturated fatty acids ratios (U/S) (2, 3.5, 5 and 6.5). At 28 and 42 d of age, performance traits, abdominal fat pad percent and serum glucose, cholesterol and triglyceride concentrations were measured. At 28 d, U/S ratio did not have a significant effect on broiler performance, but at 42 d, reducing dietary U/S ratio up to 2 significantly increased feed intake and feed conversion ratio. Serum glucose, cholesterol and triglyceride concentrations were did not affect by dietary treatments. Diet with 5 U/S ratios significantly reduced abdominal fat pad.

Use of long chain calcium salt of fatty acid plant in Holstein calves rations on performance and plasma concentrations of thyroid hormones

Y.J. Ahangari[1] and Y. Roozbahan[2, 1]Islamic Azad University of Ghaemshahr, Iran, [2]University of Tarbiat Modarres, Tehran, Iran

High environmental temperature, above 28°C, coupled with high humidity can cause heat stress in cattle, which can lead to a reduced feed intake, performances and thyroids hormones. Addition of Long Chain Calcium Salt of Fatty Acid Plant (LCCSFAP) in calves rations can improve dry matter intake, daily body weight gain, feed conversion ratio and plasma concentrations of T3 and T4 hormones. 12 Holstein male calves with an average body weight of 200 + 25 Kg were selected for fattening during summer with average temperature of 28.25°C and humidity of 75.50%. Calves were fed with three rations containing 0, 3 and 5% of LCCSFAP, a net energy of 1.76 Mcal/KgDM, for 84 days. Calves body weight measurements and blood samples collection were carried out in every 21 days during experiment. The results showed that rations containing 3 to 5% of LCCSFA has increased daily body weight gains and feed conversion ratios ($p<0.01$). The use of LCCSFAP at 5% showed an increase of T3, a decrease of T4 in blood plasma, an improvement of daily body weight gain and feed conversion ratios significantly ($p<0.01$). Therefore using 5% LCCSFAP in ration of Holstein male calves for fattening is recommended.

Session 15 Poster 50

The effect of different levels of calcium and phosphorous on broiler performance

F. Kheiri[1], H.R. Rahmani[2], [1]Iran Islamic Azad University of Shahrekord, [2]Isfahan University of Technology

In order to study the different levels of calcium & phosphorous on bacterial population of small intestine and broiler performance, during 44 days (of age 12 to 56) with 4 different levels of calcium & phosphorous [to NRC (A) 10%, (B) 20%, (C) 30% (D) less than NRC] with 3 replicates, each replicate consisted 30 broilers. At the end of each week the level of blood calcium, the calcium & phosphorous & ash of tibia, membrane protein of duodenum & jejunum and feed conversion ratio was measured. The different levels of calcium and phosphorous of the diet had no significant effect on the blood calcium & phosphorous., and the phosphorous and ash of tibia, Group D and had lesser ($p < 0.05$) level of calcium of tibia. There was not any different between feed conversion ratio of all groups, in week 3, in group D, the least feed conversion, was observed ($p <0.05$). In week 4, group C had the least levels of feed conversion ($p < 0.05$). In week 5 & 6, statistically there were no significant difference between the groups in feed conversion. In week I & 3, group A had the lowest & in week 2, group B had the highest levels of membrane protein of duodenum.

Session 15 Poster 51

Effect of *Vicia villosa* Roth inclusion on the performance of rabbits

D. Camacho-Morfin, L. Morfin-Loyden, D. G. López Rodriguez, and J. I. Pérez Dosta. Department of Animal Sci. Faculty of Superior Studies Cuautitlán. UNAM. Campo 4. km 2.5 Carretera Cuautitlán-Teoloyucan, Edo. de Méx. México*

The effect of *Vicia villosa* Roth (Vv) on dairy dry matter intake (DMI) and dairy body gain in rabbits was investigated. The test was conducted on twenty seven male of average age of 30 days and weight of 870 g New Zealand x Californian x Chinchilla growing rabbits randomly allocated on nine cages. The cages were allocated in a randomized blocks design on three experimental diets. The diets consisted on a pellet concentrated only (control diet) (D1); 40 % of the control diet plus fresh fodder of Vv *ad libitum* (D2); *ad libitum* both control diet and Vv (D3). The animals were weighted every seven days and every day was register dry matter feed intake. Average of dairy body gain (ADG) and feed conversion were calculated. Differences of total DMI were not significant between diets (116, 118 and 127g). The Vv DMI were 0, 75 and 37 g, respectively. Feed conversion were 4.5, 5.3 and 4.5, D2 was significantly higher ($p<0.001$). ADG were 30.76, 22.12 and 28.25, ADG decreased with the major intake of Vv ($p<0.001$). The results indicate that Vv could replace 30 % of concentrated in growing rabbit ration.

Session 15 — Poster 52

The health status and growth performance of growing rabbits receiving either one or two diets during fattening period

Z. Volek[1], V. Skrivanova[1], M. Marounek[1,2], [1]Research Institute of Animal Production, Prague 10, Czech Republic, [2]Institute of Animal Physiology and Genetics, Prague 4, Czech Republic*

A total of 240 rabbits, weaned at 35 days of age, were used. Two diets (diet A and diet B) were formulated according to recent recommendation for growing rabbits feeding. The diets were similar in the level of crude protein, fibre fractions and fat but differed in the level of starch and digestible energy (144 and 10.0 vs. 119 g/kg and 9.6 MJ/kg in the diet A and B, respectively). The 1st group of rabbits received the diet A from weaning to slaughter at 77 days of age. The 2nd group of rabbits received diet B from weaning to 49 days of age, and then were fed with diet A till slaughter. A higher feed intake in rabbits of the 2nd group before the change of diet was observed but no major differences were observed to the end of the experiment. For the whole fattening period, the weight gain was significantly lower in rabbits of the 2nd group than in rabbits of the 1st group. No significant differences were observed in the mortality rate of rabbits. However, a significantly higher morbidity, mainly caused by diarrhoea, was observed in rabbits fed two diets than in rabbits of the 1st group (22.5 vs. 14.2%, respectively). The highest increase of ill rabbits of the 2nd group was recorded before the change of diet. Then, the morbidity was similar among groups. It can be concluded that during fattening period is possible to use one diet only, without a negative effect on the health status and growth performance of growing rabbits. (Project MZE 0002701403).

Session 15 — Poster 53

Statistical prediction of rumen volume in Friesian bulls using four different markers and live body weight

S.M. Salem, Anim. Prod. Dept., Fac.of Agric., Cairo University. Giza, Egypt

Twenty four Holstien bulls with a mean live body weight of 450±32 Kg, belonging to Sakha experimentral farms were used in the present study under the same dietary regime. The rumen volume and flow rate were estimated from the rate of decline in concenteration of a marker in the rumen fluid following a single ruminal injection. The results of several determination methods of rumen volume using different materiales (lithium sulphate, polyethelienglycol, Cr-EDTA and Cr_2O_3) as markers gave similar values with differences ranging from 6 to 18%. Rumen volumes as determined by different methods are significantly ($p<0.01$) correlated with live body wieght. The correlation coefficient values ranged from 0.60 to 0.79. The direct determination of rumen volum were done by substituting rumen content by water (physical volume) gave rumen volume values ranged from 30 to 36% higher than that estimated by using marker methodes. The flow rate of rumen contentes was affected significantly ($p<0.05$) according markers.The values of flow rate ranged from 3.5% to 4.9% from the total volume per houre. This values equipoise disharge rate ranged from .84 to 1.25 times per day. Using Polyethyleneglycole gave the nearst estemated value to the true rumen volume (Physical volume) r =0.92 comparing with the rest used markers.

Session 15 Poster 54

Estimation of the energy value of high lactating ewes' milk at early lactation

Ch. Milis[1], D. Liamadis[2]. [1]Ministry of Agriculture. [2]Aristotle University of Thessaloniki, Greece

This study was conducted in order to develop equations for the prediction of ewes' milk energy content via milk composition. Twelve multiparous chios breed ewes were examined in a 8-wk study that started at parturition. Ewes were fed a ration containing alfalfa hay (300 g/kg) and a typical concentrate mixture (700 g/kg). Ewes were milked weekly at two consecutive milkings (afternoon and morning), by hand, after removal of lambs in separate pens 12 hours before the first milking. Milk samples were analysed immediately for milk fat, protein, lactose and solids non-fat (SNF) content using a milkoscan 4000 instrument. The same milk samples (96 samples) were lyophilised and then milk's gross energy was measured using an adiabatic bomb calorimeter. Milk's energy and composition correlation coefficients were fitted by the method of least square analysis of variance using general linear model with the GLM SAS procedure. The following equation is proposed when fat content (F; g/kg) is the only variable: EV_L (kcal/kg) = 9.3 F + 496 (r^2 = 0.95). When SNF (g/kg) is an additional variable the following formula is proposed: EV_L (kcal/kg) = 9.3 F + 8.65 SNF - 477 (r^2 = 0.97). Fat corrected milk (FCM) yield can be calculated by the following equation: Y (FCM 6%; kg/d) = L (kg/d) [0.472 + 0.0088 F (g/kg)].

Session 15 Poster 55

Effect of *Physalis alkekengi* fruit feeding on the fertility and reproduction of ewes

M. Yousef Elahi andE. Baghaei, Islamic Azad University of Azadshahr, Iran

The aim of this study was to determine whether *Physalis alkekengi* fruit feeding can affect the fertility and reproduction of ewes. Ten ewes were randomly divided into two groups. The animals in the first and second group were given one and two grams of *Physalis alkikengi* fruit, respectively. Blood samples were collected in both groups for 39 days. The estrogen (E_2) and progesterone (P_4) hormones concentration in blood sampels was determined by Radioimmunoassay. The analysis of results revealed that mean plasma E2 concentration of blood in both groups during and after feeding periods showed significant increased compared to the before feeding period ($p< 0.01$).
In both groups, the mean plasma P_4 concentration during feeding period increased compared to period before feeding. Although, in the period after feeding, it decreased compared to period during feeding ($p<0.01$). Also, no significant difference was observed between the two groups in terms of above experiments.
The results indicated that the *Physalis alkikengy* fruit feeding caused disorders in balance of mean plasma E_2 and P_4 concentration and this maybe the result of existing estrogen compounds in the *Physalis alkikengy* fruit that causes abortion and infertility in sheep.

Effects of concentrate level and starch degradability on milk yield and fatty acid (FA) composition in goats receiving a diet supplemented in sunflower oil

L. Bernard, J. Rouel, A. Ferlay and Y. Chilliard, Herbivore Research Unit, INRA-Theix, 63 122 St-Genès-Champanelle, France*

Fourteen mid-lactating goats were used indoor in a 3 X 3 Latin Square design with 3 diets supplemented with 130 g/d of sunflower oil (SO; *i.e.* 5.8% of diet DM), differing in the level of concentrate: 0.8 kg/d (F-SO) vs. 1.4 kg/d and, for the latter, by the ruminal degradability of starch (DS): corn grain (SlowDS-SO) or flattened wheat (RapidDS-SO) at 1 kg/d, for 3 weeks. High concentrate diets increased ($P < 0.01$) total dry matter intake (+205 g/d), increased milk (+525 g/d), protein (+19 g/d) and lactose (+27 g/d) yields, had no effect on fat yield and decreased milk fat content (-3.8 g/kg). Diets SDS-SO and RDS-SO compared to F-SO increased ($P < 0.01$) 8:0-14:0 and *trans*10-18:1, decreased *cis*9-18:1, *trans*6+7+8, *trans*9 and *trans*11-18:1 isomers, *cis*9,*trans*11-CLA and the sum of *trans*-FA (18:1 and 18:2) percentages. RDS compared to SDS-SO increased ($P < 0.01$) milk lactose content, increased 6:0-16:0, *trans*10-18:1 and 18:2n-6, decreased 18:0, *trans*11- and *trans*12-18:1 isomers, *cis*9-18:1, *cis*9,*trans*11-CLA and the sum of C18-FA percentages. *In conclusion*, increasing concentrate increases the *trans*10 at the expense of the *trans*11 ruminal biohydrogenation pathway and the level of medium- at the expense of long-chain FA. These effects are more pronounced with RDS-SO than with SDS-SO diet *(work funded by EU BIOCLA Project QLK1-2002-02362).*

Effects of concentrate level and starch degradability on expression of mammary lipogenic genes in goats receiving a diet supplemented in sunflower oil

L. Bernard, C. Leroux, Y. Faulconnier, D. Durand and Y. Chilliard, Herbivore Research Unit, INRA-Theix, 63 122 St-Genès-Champanelle, France*

Fourteen mid-lactating goats were fed, in a 3 X 3 Latin Square design, 3 diets supplemented with 130 g/d of sunflower oil (SO; *i.e.* 5.8% of diet DM) differing in concentrate level: 0.8 kg/d (F-SO) *vs.* 1.4 kg/d and, for the latter, ruminal degradability of starch (DS): corn grain (SlowDS-SO) or flattened wheat (RapidDS-SO) at 1 kg/d. From mammary tissue taken by biopsy (end of 3-week periods) and at slaughter (end of experiment) were determined the mRNA levels for lipoprotein lipase (LPL), acetyl-CoA carboxylase (ACC), fatty acid synthase (FAS) and stearoyl-CoA desaturase (SCD) by real-time RT-PCR, and the corresponding enzyme activities. Dietary treatments had no significant effect on mRNA abundance (n=14) and enzyme activities (n=5, n=4 for SDS-SO) of LPL, ACC, FAS and SCD. However, ACC activity increased by 21% and 27%, and milk secretion of C8-C16 fatty acids (FA) by 15 and 30%, respectively for SDS-SO and RDS-SO compared to F-SO. SCD activity slightly decreased (-22%) for RDS-SO compared to other diets, although there was no effect on desaturation ratios for 14:0, 16:0 and 18:0. *In conclusion*, gene expression responses are not always related to the corresponding milk FA responses, due to individual variability and/or other factors (*i.e.* substrate availability). Increasing *trans*10-18:1 secretion was not related to lipogenic activities changes *(work funded by EU BIOCLA Project QLK1-2002-02362).*

Session 15 Poster 58

Chemical characteristics of Kuruma prawn (*Marsupenaeus japonicus*) from two different farming systems

M. Ragni, L. Di Turi, L. Melodia, A. Caputi Jambrenghi, F. Giannico, A. Vicenti, G. Vonghia*
Department of Animal Production, University of Bari, via Amendola, 165/A, Bari, Italy

The work was aimed to determine the influence of the rearing system and the diet on the shrimp meat chemical composition. *Marsupenaeus japonicus*, carnivorous, is cultured in Italy, thanks both to its good adaptability and to its good growth rate. The prawns, belonging to the same genotype, were from semi-intensive (supplemented with an artificial diet - *Ca*) or extensive rearing system (*Le*). They arrived still live. The females were separated from the males. The pH values were measured immediately (pH_1) in correspondence of the hemolymph and the abdomen, then the prawns were refrigerated at 4°C for 24 h. The pH values were detected again (pH_2), for both sexes and rearing systems. The prawns were dissected and muscle tails were analysed. The pH_1 values were significantly lower in *Ca* meat than *Le* one, both in the abdomen and in the hemolymph, the same result was obtained just for pH_2 in the meat. Regarding the chemical composition, the meat from *Ca* prawns showed a lower fat content ($P<0.01$) and a higher undetermined one ($P<0.01$) than *Le*. No difference was noticed between the sexes, since the age of sexual maturity had not been reached yet.

Session 15 Poster 59

Investigating the compensatory growth in finishing lambs

T. Ghoorchi and H. Safarzadeh Torghabeh, Iran, Gorgan, Faculty of Animal Science, Gorgan University of Agricultural Sciences and Natural Resources

An experiment was conducted to investigate the compensatory growth, comparision of carcass characteristics and feed intake in Atabay lambs, using 21 male lambs. 5-6 month lambs with the average of **35±5** kg were devided to finishing treatmants (first treatment), 30 (2nd treatment) and 60 day restriction groups (3rd treatment). At the end of the experiment 4 lambs from each treatment were sloughtered for carcass analysis. It was revealed that there was a significant different in daily feed intake between the control and restriction treatments. The differences in characteristics of skin, head and leg, neck, flank+brisket, shoulder, chest and ribs, feet, back, hot carcass, cold carcass, bone carcass, heart, kidney, lung, liver and fat-taile weights between the treatments was significant ($p< 0.05$). Feed conversion coefficient restricted treatments after the period of restriction compaired with control treatment was decreased . Therefore, the 30 day restriction treatment was prefered for optimum feeding, saving in feed consumption and carcass with less fatness.

Session 15 Poster 60

Study of fattening potential of lambs fed by grazing barley forage

T. Ghoorchi, Z. Karimi and A. Zeinali, Iran, Gorgan, Faculty of Animal Science, Gorgan University of Agricultural Sciences and Natural Resources

In order to study of fattening potential of a lamb by pasturing barley forage,21 Atabay(dalagh)ram lambs ,selected with initial body weight 22.5±0.4 were used in 90- days feeding experiment. The ram lambs were fed diets include: (1)fattening diet (2) 1 month grazing + 2 month fattening deit (3) 2 month grazing +1 month fattening diet. A completely randomized design with 3 treatments,7 replications was used. The results showed that average daily gain of the lambs fed with diets,treatment1,2 and 3 were 184.56,197.11 and 191.88,respectively.There was no significant difference between treatment(P>0.05). At the end of this trial to determine carcase characteristic the lambs were slaughtered. There were no significant differences between warm carcase weights 19.99,20.75 and 19.88 kg and dressing percentage that were 50.36,51.82 and 49.53 on different treatment ,respectively.

Session 15 Poster 61

Milk composition in llamas (*Lama glama*) and the effect of lactational stage

A. Riek and M. Gerken, Institute of Animal Breeding and Genetics, Albrecht-Thaer-Weg 3, University of Göttingen, D-37075 Göttingen, Germany*

Data on llama milk composition are very limited and the determination of llama milk composition during lactation is incomplete. In the present study, milk samples were taken weekly from six llamas during 27 weeks after parturition under controlled stable conditions. Mean values for the contents of the major milk components across the lactation period were 4.77 % fat, 4.16 % protein, 6.01 % lactose and 506 ppm urea. Protein and lactose contents were significantly affected by the stage of lactation (P<0.0001). Fat showed a slight increase during the lactation period (P=0.078), while lactose slightly decreased. Protein decreased in the first four weeks postbirth and then slightly increased towards the end of the lactation. Urea content increased significantly with the stage of lactation (P=0.0001). There was a significant increase in fat:protein ratio (P=0.013) as protein was substituted by fat. The mean gross energy content of the milk was 3.90 kJ/g and increased slightly (P=0.087), while the *ph* decreased significantly during the lactation period (P<0.0001). None of the analysed milk components were affected by the lactation number of the animal, except for urea (P=0.018).

Session 15 Poster 62

Effect of subterranean clover feeding on sheep milk yield and quality

L. Orrù, A. Scossa, B. Moioli, F. Spirito, G. Catillo, C. Tripaldi, V. Pace, Istituto Sperimentale per la Zootecnia, Monterotondo (RM), Italy*

The effects of isoflavonic phytoestrogens on milk production are still unknown.
Milk yield and quality were recorded in 10 first lambing Comisana ewes fed from the time of their weaning on subterranean clover hay (0.2% phytoestrogens on DM). Another group of 10 ewes of the same age and breed, fed on italian ryegrass hay, was also milk recorded as control. Both diets were supplemented with corn grains and sunflower meal to have equal energy and protein content. Milking started immediately after the removal of the lamb, on the 30th day after lambing. Milk yield was recorded weekly for three months; fat and protein analyses were performed at every test. LS means differences for all parameters between the groups were calculated with a linear model including the continuous effect of the days from lambing and the fixed effect of the group. The differences in the daily milk yield between the groups (564.59 g vs. 560.39 g) were not significant; however, milk fat percentage of the clover group (6.09%) was significantly higher than the control (5.75%); milk protein percentage, analysed for half of the samples, was higher in the clover group (5.64 vs. 5.57), although not significantly. These results support the hypothesis that the phytoestrogens could play a role on lipid and carbohydrate metabolism and on blood glucose level.

Session 15 Poster 63

Nutritional value and effect of rapeseed meal on sanjabi lamb performance

M. Souri, M. Moeini and M. Kheirabadi, Department of animal science, Faculty of Agriculture, Razi University, Kermanshah, Iran

Rapeseed meal (RPM) can be used as a high protein supplement for animal nutrition (Hill , 1991, Tripathi *et al.*, 2001). Fourteen Sanjabi lambs (90 days age, 26.0 + - 2.0 kg) randomly divided into two groups , feeding RPM at the rate of % 0 and % 23, DM of dietary which replaced with Soya been meal for 86 days . Dry matter intake, digestibility, average daily gain, nitrogen retention, meat characteristics were measured. Blood samples taken at the end of experiment , monitoring plasma constituent's changes. Data statistically were analyzed using randomized completed design and Toky range tests.
Rapeseed meal contains approximately 92% DM, 34% CP, 6.7%Ash, 11% E.E, 17%CF and 23.3%NFE. The results indicated that the coefficient digestibility of CP, DM, CF, EE and NFE were 84%, 65.3%, 38%, 91% and 61% respectively. Dry matter intake, the average daily gain, nitrogen retention, and meat characteristics of lambs in 23% group were not significantly difference compare to other group. The values of plasma parameters were higher in lambs in 23% group but were not significantly difference. It can be concluded that RPM can be used up to 23% DM of dietary of fattening lambs.
Keywords: rapeseed meal, meat characteristics, digestibility, nitrogen retention

Session 15 Poster 64

The effects of dietary inclusion of organic selenium (Sel-Plex) on ewe performance and milk selenium level in sheep

F. Crosby[1], M. Fooley[1], S. Andrieu[2], [1]University College Dublin, Lyons Research Farm, Newcastle, Co. Dublin, Ireland,[2]Alltech Ireland, Sarney Summerhill Road,Dunboyne,Co. Meath, Ireland

Fifty ewes were used in an experiment to determine the effect of the inclusion of Sel-Plex in the concentrate during the final seven weeks of pregnancy. Ewes were randomly allocated to one of the two treatments: T1=0.875 mg/kg sodium selenite in concentrate; T2= 0.875 mg/kg Sel-Plex. Ten ewes from each treatment were further kept indoors and offered the same experimental concentrates than prior lambing. Milk selenium content was assessed at d10 after lambing. Ewes liveweight was recorded at beginning of the trial and 24 hours post-partum. Lamb liveweight was assessed from birth to weaning.
Ewe liveweight loss from day 98 of gestation to 24 h post-partum was higher for T1 group (-5.5 vs. -3.4 kg; P = 0.07).
Sel-Plex significantly increased the level of selenium in the milk at d10 (60.1 and 98.0 µg/l for T1 and T2 respectively, P = 0.001).
There was no treatment effect on mean lamb birth weight (P>0.05) with birth weights of 4.9 kg (T1) and 5.0 kg (T2) recorded. Lambs from the Sel-Plex group had a higher ADG in the first 35 days post-partum (292 vs. 326 g/day; P<0.05). Although overall lamb ADG from birth to weaning tended to be higher in T2 (231 vs. 249 g/day), the difference observed was not significant (P = 0.08).

Session 15 Poster 65

Effects of Silymarin administration to middle-lactating dairy goats

D. Tedesco, J. Turini and S. Galletti, Department of Veterinary Sciences and Technologies for Food Safety, Via Celoria 10, 20133 Milano, Italy

The standardized extract from *Silybum marianum*, silymarin, is used for the treatment of liver diseases. The silymarin administration in the peripartum period to cows and goats resulted in a higher milk peak, a prevention of hepatic fat infiltration and a better general health condition. This study aimed to determine the effects of Silyvet® in dairy goats in the middle lactation period on milk production.
Thirty dairy goats were divided into two groups according to parity (>2), kidding date, BCS and milk production. Treated goats received 10 mL/d of Silyvet® (Indena S.p.A., Milano, Italy) as a water suspension as oral drenches. Treatment was administered for 15 day in the third month of lactation. Individual milk production was recorded on 0, 4, 10, 15d in the treatment period, and on 21 and 51d. Milk samples were collected on 0 and 15d of the treatment period and analyzed for protein, fat, lactose, urea and SCC. Silyvet treatment significantly increased milk production (P<0.05). Milk yield tended to be higher in treated group also after the end of the Silyvet® administration (day 21 and 51). The milk quality parameters were not affected by treatment. These results show that Silyvet® administration can improve milk production not only in early lactation period, as reported in our previous results, but also in different stages of lactation.

Session 15 Poster 66

Cactus pear (*Opuntia ficus-indica*) as a complement to urea-treated straw in dry season feeding systems of ruminants

Firew Tegegne, K.J.Peters, C.Kijora; Institute of Animal Sciences, Humboldt-University of Berlin, Germany*

Cactus pear is of increasing important as a multipurpose plant under arid conditions. This project identified it's role as forage. A three-months experiment was conducted to assess the effects of cactus pear(C), urea treatment of straw(5% urea) or wheat bran(WB) on intake, digestibility and growth.

The experiment was laid out in a randomised complete block design with eight sheep/treatment. Diets consisted of untreated straw (S)[T_1], S+C[T_2], S+C+WB[T_3], urea-treated straw(UTS)[T_4], UTS+C[T_5] and UTS+C+WB[T_6]. The rate of supplements (C and/or WB) was 40%. Diets were offered, with the aim of having at least 15% refusals daily, in individual troughs twice daily.

Urea-treatment improved crude protein content of straw from 2.68 to 8.69% and apparent dry matter digestibility(DMD) from 55%(T_1) to 65%(T_4). Highly significant differences ($p < 0.001$) were observed for total dry matter intake(DMI), DMD and live weight change. DMI was highest in T_5 and T_6 (90 and 84 g/kg$W^{0.75}$, respectively) and lowest in T_1(55 g/kg$W^{0.75}$) while DMD was highest in T_2 and T_5(65%) and lowest in T_1(55%). Sheep on T_6 had the highest live weight gain(75.5 g/day) followed by sheep on T_3 and T_5(41.5 and 38.0 g/day, respectively). In conclusion, cactus pear significantly increased untreated straw intake. It could substitute wheat bran, provided that straw is urea-treated. Diet six(T_6) seems to be a promising feed package for dry season feeding systems.

Session 15 Poster 67

The effect of lasalocid and monensin on rumen metabolites of sheep

H. Abdoli, A. Taghizadeh, A. Tahmasbi , Dep of Animal Science, Tabriz University, Tabriz, Iran

Twelve sheep (34.6±4.31 kg) are randomly received one of the 3 diet in completely randomized design. The diets content A (control), B (25 g kg^{-1} monensin) and C (25 g kg^{-1} lasalocid) containing predicted DE (3.35 Mcal kg^{-1}DM) and CP (160 g kg^{-1}DM). The period of present study was 60 days. The rumen fluid of each treatment was 60 days. The rumen fluid of each treatment was obtained by stomach tube. The effect of these Ionophores were determined on rumen pH, ammonia-N and total fatty acids (VFA). The pH for diets of A, B and C was 5.9, 6.6 and 6.3, respectively. The ammonia-N (mg/l) was 146.2, 106.2 and 100, respectively. The total VFA (mmol/l) was 124.5, 125.5 and 121, respectively. The results showed that monensin and lasalocid had not significance effect on VFA, while ammonia-N and pH in sheep were fed with diet containing lasalocid and monensin were lower and higher ($P<0.05$) than compared to sheep were fed diet without Ionophore, respectively. These results showed diet containing ionophores with effect on ruminal fermentation process prevented from protein degradation and decreased ammonia -N.

Session 15 Poster 68

Effect of biotin supplementation on milk performance in dairy cows

D.L. De Brabander and V. Wouters, Agricultural Research Centre, Department Animal Nutrition and Husbandry, Scheldeweg 68, 9090 Melle, Belgium*

The effect of biotin supplementation on milk performance was investigated in a field trial carried out in seven high yielding Holstein herds with 425 cows producing on average approximately 9000 kg milk in 305 days. The test was done in 4 periods of 4 weeks each. Biotin was supplemented (20 $mg.cow^{-1}.day^{-1}$) in the second and fourth period. To minimize the disturbing effect of lactation stage, results were analysed as two separate series, i.e. period 1, 2 and 3, and period 2, 3 and 4. Only cows with at least 14 DIM and 15 kg daily milk yield were taken into account. Milk performance records from the official milk control carried out at the end of each period were used as experimental data. Basal diets were composed of maize silage, grass silage and pressed sugar beet pulp and remained unchanged during the trial.

For the control and biotin supplemented diet, milk yield amounted to 29.3 and 30.1 kg, milk fat content to 4.12 and 4.07 %, and milk protein content to 3.47 and 3.50 %, respectively. Milk yield effect was positive in each herd ranging from 0.2 to 1.2 kg, and amounted to 0.4 kg for cows with a daily milk yield between 15 and 20 kg and to 1.0 kg for higher producing cows. The observed tendency for a lower milk fat content and a higher milk protein content was almost consistent for the seven herds.

Session 15 Poster 69

Fibrolytic enzymes in dairy calf starter

G.R. Ghorbani, A. Jafari, and A. Nikkhah, College of Agriculture, Isfahan Uinversity of Technology, Isfahan, Iran 84156

Eighteen 1-d old dairy calves (9 male and 9 female, 47.9 ± 3.1 kg BW; mean ± SD) were used to evaluate the effects of adding fibrolytic enzymes to the calf starter on growth rate, feed intake, and nutrient digestibility. Treatments included: 1) control (no enzyme), 2) enzyme A (*Promote*, 0.6 ml/kg starter), and 3) Enzyme B (*Bvoizyme MT 4001,* 1.9 ml/kg starter). Liquid enzyme supplements were diluted in distilled water (1:10 ratio of enzyme to water) before spreading onto the starter at 1 h prefeeding every morning. Calves were monitored for average daily gain (ADG), dry matter intake (DMI), and nutrient digestibility at the ages of 28, 54, and 84 d, and weaned at 54 d. Data were analyzed as a MIXED model for repeated measurements. Adding fibrolytic enzymes to the calf starter did not ($P > 0.05$) affect ADG, DMI, feed efficiency, and total tract digestibility of nutrients across 3 measurement days. Digestibility of acid detergent fiber (ADF) at 28 d was greater ($P < 0.01$) for calves fed the starter treated with enzyme A as compared to calves on enzyme B and the control group (58.7 vs. 52.7 and 48.4%). Tendencies were noticed for greater digestibility of OM (85.7 vs. 82.4%, $P = 0.06$) and DM (85.1 vs. 81.9%, $P = 0.09$) in calves on control diet than in calves enzyme supplemented diets.

Influence of dry period shortening on colostrum quality and Holstein calves average daily gain

J. Amini, H.R. Rahmani, G.R. Ghorbani, College of Agriculture, Isfahan University of Technology, Isfahan, Iran 84156

Ninety six Holstein cows and their calves were studied to evaluate the effects of dry period length on colostrum quality and calves' average daily gain (ADG). Treatments were arranged in a 3×2×2 factorial design that included dry period (30, 42 and 56 d), parity (primiparous or multiparous) and BCS (BCS<3.2, BCS≥3.2). Calves were blocked by sex. First milking after parturition was collected and its density was recorded. Calves were fed first colostrum meal in nipple-bottles within 1.2 h of birth. Maternal colostrum was fed in amounts of 1.5 L/meal for first two meals. Thereafter, pooled transitional milk was fed for the next four meals. Calves were weighed at 0 h, 42 d, and 90 d. Fecal and health scores was estimated every day by method of Larson. Data was analyzed with proc mixed procedure of SAS. No significant differences due to dry period length were detected for colostrum density (1.0636 ± 0.0086, 1.0582 ± 0.0161, 1.0673 ± 0.0128 gm/ml for 30, 42 and 56 d treatments, respectively), calves' ADG (0.474 ± 0.094, 0.516 ± 0.060 and 0.479 ± 0.152 kg/d, respectively), health, and incidence of diarrhea. Data indicated that a short dry period can be applied as a management tool with no loss in colostrum quality or calf's ADG.

Influence of Baker's yeast (*Saccharomyces cerevisiae*) on digestion and fermentation patterns in the rumen of sheep and milk response in dairy cattle

A.M. Allam, Department of Animal Production, Faculty of Agriculture, Alexandria University, Alexandria, Egypt

The objectives of two experiments were to evaluate the effect of Baker's yeast (*Saccharomyces cerevisiae*, 29%DM) supplementation on dry matter intake, nutrient digestibility, N-balance and rumen fermentation parameters in sheep and on milk response in a commercial dairy heard. In Exp.1, three fistulated Rahmany rams were distributed in 3x3 Latin square design, animal were assigned to three levels of yeast (0, 2.5and 5.0 g/head/day). In Exp.2, 229 Holstein cows were assigned equally to three levels of yeast (zero, 5 and 10 g/head/day). Both sheep and Holstein cows were fed on one experimental ration (TMR, 76% TDN and 17%CP). Cows were milked three times a day, milk yield was recorded, milk analysis was carried using Milk-Scan-133. Results were analyzed by ANOVA using (GLM) procedure of SAS. Yeast supplementation improved dry matter intake, nutrient digestibility, N-balance and the nutritive value of the experimental ration especially with the level of 2.5 g/head/day with sheep. Rumen fermentation parameters were highly affected by yeast supplementation. Although numerical response in milk yield (kg/day), the feed conversion (kg milk yield/kg dry matter intake), milk total solids %, milk fat % and milk protein % were greater for cows fed yeast than did the control group but the response was not significant. In conclusion, a good performance of dairy cattle would be expected by supplementing 5g Baker's yeast in a high concentration ration based on corn silage.

Session 15 Poster 72

Essential amino acid requirements of Dorper lambs estimated by the whole empty body essential amino acid profile

A.V. Ferreira and A.H. Jurgens, Department of Animal Sciences, Faculty of Agricultural and Forestry Sciences, University of Stellenbosch, Stellenbosch 7600, South Africa*

The essential amino acid (EAA) profile of the duodenal digesta and whole empty body of 7 Dorper ram lambs managed under intensive and extensive feeding conditions, was investigated. Significant differences were found between the EAA composition of the duodenal samples, whether protein quantity was taken into account (g AA/100 g crude protein) or not (expressed as % of lysine). The chemical scores indicated that the two most limiting amino acids in the duodenal digesta for whole empty body growth were arginine and histidine. Excess levels of amino acids tended to be present in all three duodenal digesta samples (isoleucine, leucine, lysine, methionine, phenylalanine, threonine and valine). According to the EAA:Lysine ratios of the whole empty body of Dorpers and beef cattle it is remarkably similar. The whole empty body EAA composition (g AA/100 g crude protein) was as follows: 7.10 arginine; 2.40 histidine; 3.31 isoleucine; 7.22 leucine; 6.61 lysine; 1.62 methionine; 3.91 phenylalanine; 3.77 threonine; 4.85 valine and tryptophan 0.72. This composition can serve as an example of the ideal EAA requirements for whole empty body growth between 30 and 40 kg live weight of Dorper lambs.

Session 15 Poster 73

Effect of graded levels of threonine on gimmizah layer hens performance

M. Khalifah and A. Abdella, Animal Production Research Institute, Agriculture Research Center, Ministry of Agriculture, Cairo, Egypt*

This experiment was designed to study the effect of increasing dietary threonine levels on the production performance of Gimmizah layer hens. Ninety-six Gimmizah laying hens (40 wks of age) were weighed and randomly housed in individual cages and allotted for six dietary treatment groups of 16 hens each. The experiment continued for 8 weeks. Threonine (Thr) was added by 0.22 g/kg to its content of the basal diet to obtain 6 graded levels of 5.19, 5.41, 5.63, 5.85, 6.07 and 6.29 g Thr /kg diet.

All the experimental levels of Thr had no significant effect on egg production and feed consumption. Egg weight, egg mass and feed conversion had improved significantly with increasing dietary Thr level up to 6.07 g/kg diet and then reversed with increasing Thr level. Birds fed diet containing 6.09 g Thr/kg during the same mentioned period (40-48 wks of age) recorded the highest egg weight (61.52g); egg mass (37.75 g/hen/day) and the best feed conversion (3.52). The results indicated that Thr concentration equal to 802 mg/h/day was adequate for improving layer performance. According to the input-output analysis, increasing dietary Thr in layer diet increased the economical efficiency (EE) and relative economic efficiency (REE). Moreover, the best records of EE and REE were observed for group of layers fed diet containing 6.07 g/kg Thr.

Session 15 Poster 74

Effects of pre-partum body condition score on milk yield of Holstein dairy cows

A. Pezeshki, G.R. Ghorbani and H.R. Rahmani, College of Agriculture, Isfahan University of Technology, Isfahan, Iran 84156

There are different findings about the effects of body condition score (BCS) on milk yield. The objective of current study was evaluating the effects of BCS on milk yield and composition of dairy cattle. One-hundred twenty two Holstein dairy cows were assigned BCS on 5 point scale in approximately 60 and 30 day before expected calving date. Cows were grouped for analyses into two groups including BCS≥3.2 and BCS≤3.0 (n = 65, n = 57, respectively). The mean pre-calving BCS for low and high BCS group was 2.85 and 3.50. Milk production and composition data were collected for the first 8wk of lactation and were analyzed using the MIXED procedure of SAS. No significant differences due to BCS were detected for milk yield at first 8 wk of lactation (36.19 vs. 37.38). Low BCS cows have higher milk fat than high BCS cows (3.69 vs. 3.31; $p<0.05$). Similarly, milk lactose increased significantly for low BCS cows (5.41 vs. 5.22; $p<0.01$). There was no significant difference between low and high BCS cows for milk protein (3.12 vs. 3.12) and milk SCS (5.69 vs. 5.74). It seems that although different BCS have no significant effect on milk yield at subsequent lactation but high BCS cows intend to produce more milk than low BCS cows.

Session 15 Poster 75

Effect of physical form of the starter on performance of Holstein calves

M. Bagheri, and, G.R. Ghorbani, Deaprtment of Animal Science, Isfahan University of Technology, Iran

Sixteen Holstein calves were paired by birth date (mean age 20 ± 8 d), body weight (mean initial weight 45.5 ± 5.9 kg), and sex (8 male and 8 female) and randomly assigned to either a ground or pelleted starter to study the effect of physical form of diet on average daily gain , feed intake, and weight of weaning. Calves were housed in individual boxes for 5 wk, and weaned at 60 d of age if they reached to a minimum weight of 75 kg, otherwise they continued the trail in group pens separated by treatment. Five l of whole milk were offered 2 times daily for 4 wk and the followed by 4 l of whole milk. Daily feed intake and weekly body weight were measured. Intake of starter was higher for pellet-fed calves. Mean daily starter intake was 573.0 g for calves fed pelleted starter, which was 26% higher than that of ground-fed starter calves (452.7 g, $p>0.05$). Average daily gain was 690 g for pellet-fed vs. 660 g for ground-fed calves ($p>0.05$). However, feed conversion ratio was significantly better ($p<0.05$) at the end of wk 5 for calves on ground vs. pelleted starter (0.65 vs. 0.51). These results indicate that pelleted starter had no significant influence on calf's performance.

Session 15 — Poster 76

The effect of direct-fed fibrolytic enzyme on mid-lactating dairy cow's performance and digestibility

M. Alikhani, M. Shahzeidi, G. R. Ghorbani, and H. Rahmany, Department of Animal Science, Isfahan University of technology

In a duplicated 3 × 3 Latin-square experiment six multiparous lactating Holstein Cow (123 ± 20 DIM) were used to determine the effect of direct-fed fibrolytic enzyme on their performance and feed digestibility. Treatments were 1) control (basal diet with forage to concentrate ratio of 40:60, 2) enzyme (Pro-Mote®; Biovance Technologies Inc., Omaha, NE) containing diet sprayed on concentrate portion of ration (0.9 ml/kg DM), and 3) Enzyme containing diet sprayed on forage portion of ration. Milk and 4% FCM yield were higher in cows used treatment diet 2 than other treatments. The percentages of Milk composition (TS, Fat, CP, and Lac), protein and lactose yield were not affected by any treatments groups, although cows used treatments 2 and 3 tended to have a lower CP percent and fat percent production. Fat yield was higher in cows used treatment diet 2 than other treatments. Digestibility of DM, OM, NDF and ADF and feed and NDF intake were greater for cows used treatments 2 and 3 than treatment 1. Digestibility of CP was not affected by any treatments groups. The result of this experiment indicate that, supplementing mid lactating dairy cows with fibrolytic enzyme has the potential to enhance milk yield and nutrient digestibility and feed intake, without changing milk composition.

Session 15 — Poster 77

The effects of reducing fish meal of diets supplemented with DL-methionine and L-lysine hydrochloride on female broiler performance

N. Eila, B. Hemati and R. Jalilian, Department of Animal Science, Agricultural College, Islamic Azad University of Karaj, Azadi Blvd., Karaj city, post code 3187644511 Tehran province, Iran*

The object of this study was evaluation of reducing fish meal in broiler diets supplemented with DL-methionine and L-lysine hydrochloride. In a Completely Randomized Design: 400 one-day chickens were distributed randomly to 4 treatments x 4 replications of 25 female broiler chicks. The experiment carried out for 49 days. Fish meal level in experimental diets was 0, 1, 3 and 5 percent. Average weight gain, feed consumption, feed conversion ratio, protein efficiency ratio, nutritional cost per kg of weight gain and carcase were measured and at the end of experiment carcass traits; dressing percentage, breast percentage and relative weight of abdominal fat, pancreas and gizzard were measured in one bird of each 25 chicks. Duncan's analysis was used for comparing between treatments. Average weight gain, feed conversion ratio, feed consumption, protein efficiency ratio and carcase traits didn't show any significant difference by reducing fish meal ($P>0.05$).But nutritional cost per kg of weight gain and carcase yield were significantly lower in diets without fish meal comparing with diets with 5 percent fish meal ($P\leq0.05$). Therefore it is possible to reduce fish meal in corn based diets without any adverse effect on performance by adding DL-methionine and L-lysine hydrochloride.

Session 15 Poster 78

Degradability of corn silage in ruminal ambient with different additives[1]

P.A. Katsuki[2], E.S. Pereira[3], B.M.O. Ramos[2], E.L.A Ribeiro[4], M.A. Rocha[4], A.P.Pinto[2], R.Salmazo[2], T.R. Casimiro[2], T.C. Alves[2], M.N. Bonin[2] and I.Y.Mizubuti[4], [1]Partially funded by CNPq; [2]Animal Science Graduation Programme student; [3]Animal Science Department. UNIOESTE. Paraná, Brazil. [4]Animal Science Graduation Programme Professor. Animal Sciences Department, Londrina State University, PO Box, 6001, CEP 86051-990 - Londrina, PR, Brazil*

Dry matter (DM), organic matter (OM), crude protein (CP), neutral detergent fiber (NDF) and acid detergent fiber (ADF) degradabilities of corn silage were evaluated in a Latin square 4x4 assay (four bulls and four incubation periods) in ruminal ambient adapted or not with different feed additives. The treatments were: corn silage in ruminal ambient without inoculation of additive (CCS); corn silage in ruminal ambient inoculated with dehydrated and liofilized ruminal and intestinal bacteria (SLB); corn silage in ruminal ambient inoculated with cellulolytic enzymes (SCE); and corn silage in ruminal ambient inoculated with sodic monensin (SSM). SLB and SCE did not affect the potentially degradable fraction (b) of corn silage nutrients. The sodic monensin reduced the fraction b of DM and OM, as well as the corn silage potential degradability. Among all the additives studied, the sodic monensin provided the largest indigestible fraction of NDF and ADF, reducing the disappearance of these fractions after 48 hours of intra-ruminal incubation. It can be concluded that the different feed additives did not improved the DM, OM, CP, NDF, and ADF effective degradabilities of corn silage.

Session 15 Poster 79

Effect of different feed additives in sugar-cane diets on the productive performance of Limousin x Nelore crossbred heifers[1]

P.A. Katsuki[2], E.S.Pereira[3], B.M.O.Ramos[2], E.L.A.Ribeiro[4], F.B.Moreira[4], M.A.Rocha[4], A.P. Pinto[2], R. Salmazo[2], T.R.Casimiro[2], T.C.Alves[2], I.Y. Mizubuti[4], [1]Partially funded by CNPq; [2]Animal Science Graduation Programme student; [3]Animal Science Department. UNIOESTE. Paraná, Brazil. [4]Animal Science Graduation Programme Professor. Animal Sciences Department, Londrina State University, PO Box, 6001, CEP 86051-990 - Londrina, PR, Brazil*

The objective of this study was to evaluate the effects of feed additives supplied in diets based on sugar-cane. Dry matter intake (DMI), daily average weight gain (DAWG) and feed conversion (FC) were evaluated. Twenty-eight Limousin x Nelore crossbred heifers were distributed in a completely randomized experiment with four treatments and seven repetitions. The diet was based on sugar-cane and concentrate. The treatments contained basic diet and feed additives, defined as: diet with sodic monensin (DSM); diet with dehydrated and liofilized ruminal and intestinal bacteria (DLB); diet with alive yeasts (DAV); and control diet (CLD). The experiment was carried out in 60 days with two evaluation periods of 30 consecutive days. The adaptation period was 20 days prior to the beginning. No differences were observed in DMI, DAWG and FC, showing average values of 5.814 kg/day, 0.920 kg, and 6.719 DM/kg AWG, respectively.

Session 16 Theatre 1

Performance and feeding behaviour of dairy cows in an automatic milking systems with controlled cow traffic

M. Melin, K. Svennersten-Sjaunja and H. Wiktorsson, Swedish University of Agricultural Sciences, Department of Animal Nutrition and Management, Kungsängen Research Centre, 753 23 Uppsala, Sweden*

Automatic milking systems allow for an increased milking frequency and thereby an expected increase in milk yield. Because of the expected milk yield increase, milking in automatic milking systems may lead to a decrease in the energy balance of the early lactating cow. In the present study, the cow traffic in an automatic milking system was controlled with gates at the entrances to the feeding area. Two groups of dairy cows were subjected to two different degrees of controlled cow traffic during early and mid-lactation. By setting different times in the control gates the two groups of cows were separated into one group with a high milking frequency routine (HR) and one group of a low milking frequency routine (LR). The HR cows were milked on average 3.2 times daily, and the LR cows were milked on average 2.1 times daily. The HR group yielded 13 % more milk than the LR cows did, with no difference in milk composition. Despite an apparent change in feeding patterns, the HR group did not increase their dry matter intake, which caused them to come into a somewhat more negative energy balance than the LR cows did. However, the HR cows did not show any signs on severe body tissue mobilization.

Session 16 Theatre 2

Effect of milking frequency on lactation persistency in an automated milking system

K. Svennersten-Sjaunja and G. Pettersson, Kungsängen's Research Centre, Swedish University of Agricultural Sciences,(SLU) 753 23 Uppsala, Sweden*

The lactation persistency is partly controlled by reduced apoptosis (cell death) in early lactation which can be influenced by increased milking frequency before peak lactation. In systems provided with automated milking (AM) where the cows have the possibility to "voluntary" choose milking frequency it is likely that cows with more frequent milking during the first ten weeks of lactation also have a more optimal lactation persistency. Effect of milking frequency on lactation was studied in the AM-system at Kungsängens's research farm, Swedish University of Agricultural Sciences. Lactation curves were studied for 90 SRB cows (Swedish Red and White Breed) with 125 lactations. The cows were divided in two groups based upon average milking frequency the first ten weeks, 2.5 and 3.1 milkings respectively. The criteria for cows included in the calculations were that they should be lactating at least 35 weeks and being pregnant. Results showed that cows with the highest milking frequency had a significant higher lactation production. The effect was most pronounced in primiparous cows who increased lactation yield by 9.4 %, while the cows in second and third lactation increased their production with 6.8 %. The reason why the response was higher on first calvers could be an effect of the udder anatomy, where young cows have less cisternal fraction compared to older cows.

Session 16 Theatre 3

Effect of mastitis on culling in Swedish dairy cattle

M. del P. Schneider[1], E. Strandberg[1] and A. Roth[2], [1]Department of Animal Breeding and Genetics, Swedish University of Agricultural Sciences, P.O. Box 7023, SE-75007 Uppsala, Sweden, [2]Swedish Dairy Association, Box 1146, SE-63180 Eskilstuna, Sweden*

The effect of mastitis on culling in Swedish dairy cattle was analyzed with Survival Analysis. The data included information on 990776 cows calving from 1988 to 1996. Four breeds, Swedish Red and White (SRB), Swedish Friesian (SLB), Swedish Polled Breed, Jersey, and crossbreds (SRB x SLB), were included. Length of productive life was defined as days from first calving to culling (uncensored) or end of data collection (censored). The model (Weibull Proportional Hazard) included the fixed effects of mastitis by parity and lactation stage, peak yield deviation within herd-year, age at first calving, year by season, region, breed, herd production level, and the random effect of herd. The effect of mastitis on culling was modeled as a time-dependent covariate. Lactation was divided in five stages in which mastitis and culling might occur. The risk of culling increased throughout the lactation after the stage in which the disease had occurred, the risk being highest at the end of the lactation. When the disease occurred at the first stage, the risk of being culled was high at that stage, relative to a healthy cow. If mastitis occurred at later stages, culling was delayed until the end of lactation. Similar results were observed for all parities.

Session 16 Theatre 4

Dairy cow health and the effects of genetic merit for milk production, management and interactions between these: udder health parameters

W. Ouweltjes, B. Beerda, J.J. Windig, M.P.L. Calus and R.F. Veerkamp, Animal Sciences Group of Wageningen University and Research centre, The Netherlands*

Milk production per cow has increased significantly as a result of breeding, feeding and other management factors. High producing cows are often compared with athletes that require special care. Therefore concerns about consequences of high production for animal health are increasing. This study aims to address such concerns and deals with health risks for low and high producing dairy cows. In a 2x2x2 factorial design, HF Heifers (n=100) of high or low genetic merit for milk yield that were milked 2 or 3 times a day and fed a Mixed Ration with high or low energy content were compared during the first 14 weeks of lactation. Milk composition and cell counts were determined weekly, quarter milk bacteriology at 3 and teat condition at 4 time points during the experiment. The experimental factors, especially ration composition, resulted in substantial differences in milk production between treatment groups. Preliminary results indicate that ration composition did not affect cell counts or teat condition. Teat condition was impaired but cell counts were lower with increased milking frequency. Cell counts were higher for cows with high genetic merit, teat condition was not related to genetic merit for milk production. Effects of genetic merit on cell counts were highest for 2 times a day milking.

Session 16 Theatre 5

Osteochondrosis in Beef Sires in Sweden

Y. Persson and S. Ekman, Swedish University of Agricultural Sciences (SLU), Uppsala, Sweden*

Swedish insurance companies report that a great number of their insured beef sires are culled because of lameness. To our knowledge, little is known about hind limb problems in beef cattle, but there are indications that osteochondrosis is a common cause of hind limb lameness in Swedish beef bulls used for natural service. The aim of the present study was therefore to examine hind limbs of Swedish beef sires post mortem, regarding the presence of osteochondrosis.
Right and left hind limb bones from 42 beef sires were examined post mortem to identify lesions in the stifle and tarsal joints. The bulls were slaughtered during or after the breeding season due to lameness. The bulls were of five different breeds, Charolais (n=16), Simmental (n=12), Aberdeen Angus (n=9), Hereford (n=4) and Limousine (n=1) and the mean age was 2.5 years (range 1-7 years). Thirty-eight of the bulls (90 %) had lesions in at least one joint. Thirty-four bulls (81 %) had lesions in the stifle and 14 bulls (33 %) had lesions in the tarsus. In a majority of the bulls (67 %), the lesions were bilateral. The most common location of the joint lesions was the lateral ridge of the femur trochlea. Three bulls had no joint lesions. In conclusion, the present results indicate that lesions compatible with osteochondrosis are common post mortem findings in beef sires with lameness.

Session 16 Theatre 6

On the development of asymmetry between lateral and medial rear claws in dairy cows

E. Telezhenko[1], C. Bergsten[1,2], M. Magnusson[3], M. Ventorp[3], J. Hultgren[1], C. Nilsson[3]. [1]Swedish University of Agricultural Sciences (SLU), Skara, Sweden, [2]Swedish Dairy Association, Stockholm, Sweden, [3]SLU, Alnarp, Sweden*

The rate of growth and wear of lateral and medial rear claws were assessed in 120 dairy cows kept in five different flooring systems, but otherwise under identical managemental conditions. The flooring systems were mastic asphalt, rubber mats (KURA-P(tm)) (with and without feed-stalls with rubber mats), and slatted concrete floor. Dorsal wall (toe) and abaxial wall growth and wear were measured over a 4-month period after claw trimming by placing a mark on the dorsal and abaxial wall in both lateral and medial claws of the left rear leg and monitoring its displacement relative the periople and bearing surface.
There were no differences in the rate of toe-growth between lateral and medial claws on either of the floorings except for asphalt without feed-stalls, where toe-growth of medial claws exceeded that of the lateral claws ($P<0.05$), and the growth and wear of lateral abaxial wall exceeded that of the medial claws ($P<0.05$). Abaxial wall growth was also larger in lateral than in medial claws when using feed stalls. In all systems, toe-wear was significantly greater in the medial claws than in the lateral claws, which generally explains longer lateral toes. In contrast, higher net growth of abaxial walls in lateral claws on asphalt was explained by a significantly higher growth rate.

Session 16 Theatre 7

Evaluation of AFB_1/AFM_1 carry-over in lactating goats exposed to different levels of AFB_1 contamination

B. Ronchi, P. Danieli, A. Vitali, A. Sabatini, U. Bernabucci and A. Nardone, Dipartimento di Produzioni Animali, Università della Tuscia, via De Lellis, 01100, Viterbo, Italy

A study was performed to assess AFB_1/AFM_1 carry-over in milk of goats exposed to diet with different level of contamination. A 4x4 Latin-square design was used for comparative carry-over trials. Four lactating Saanen goats were housed in metabolic cages. The experimental period lasted 48 d. Each trial lasted 12 d: 7 d treatment sub-period in which AFB_1 was administrated via naturally contaminated corn at four levels ranging from 0 to 57.4 ± 5.1 µg/kg, and 5 d post-treatment period during which no contaminated feed was given. During trials, samples of milk and urine, and feeds and faeces were collected daily and analysed for AFM_1 (milk and urine) and AFB_1 (feeds and faeces) by IAC-HPLC method. Performance of method was routinely checked. AFM_1 concentration in milk showed was strongly affected ($P<0.01$) by AFB_1 treatment. In contrast subject or trial did not significantly affected AFM_1. Determination of aflatoxins (AF) in feeds, faeces, milk and urine allowed to get interesting results on AF balance. The carry-over of AFM_1, expressed as ng AFM_1 excreted / µg AFB_1 ingested, was quite high ranging between treatments from 0.029 to 0.031. Results on AFM_1 carry-over in goats are novel, and show that carry-over level for goat is higher than that reported for sheep but comparable to levels stated for dairy cows.

Session 16 Theatre 8

The ACTH challenge test to evaluate the individual welfare condition

G. Bertoni, E. Trevisi, R. Lombardelli, L. Calamari - Istituto di Zootecnica, Facoltà di Agraria, Università Cattolica del Sacro Cuore, 29100, Piacenza, Italy*

The ACTH challenge test, despite some different opinions, could be a tool to diagnose stress conditions in the animals. A reason of the disagreements could be the procedure, therefore we have studied the effects of ACTH dosage and of the sampling schedule on the cortisol response and interpretation. Two groups of 4 lactating cows, well trained to blood sampling, were alternatively challenged with a high dose (100 IU) or with low doses (2, 4 and 8 IU) of $ACTH_{1-24}$ (Synacthen). Only the very low dose (2 UI) was also repeated 3 times at 30 min. intervals. Blood samples were taken in the following 60 to 240 min. after i.v. injection.

The highest or the lowest doses induce very similar increase rate and maximum levels of cortisol, if the peak is considered; otherwise higher doses are responsible of more prolonged high values. Furthermore, each cow has a different response that seems unaffected by the basal cortisol level and by the effect of bleeding stress; in fact, repeated low doses of ACTH have only prolonged the cortisol response. It can be therefore concluded that very low doses of ACTH, such as 2-4 IU, can be utilized for ACTH challenge with a bleeding schedule within 45-60 min.

Unfortunately, our results do not allow the explanation of different ACTH sensitivity.

Effect of transport for up to 24 hours followed by twenty-four hours recovery on liveweight, physiological and haematological responses of bulls

B. Earley, D.J. Prendiville and E.G. O'Riordan, Teagasc, Grange Research Centre, Dunsany, Co. Meath, Ireland.*

The objective of the study was to investigate the effect of transport on liveweight, physiological and haematological responses of bulls after road transport of 0, 6, 9, 12, 18 and 24h. Eighty-four continental x bulls (mean weight (s.d.) 367 (35) kg) were randomly assigned to one of six journey (J) times of 0, 6, 9, 12, 18 and 24h transport at a stocking density of 1.02m^2/bull. Blood samples were collected by jugular venipuncture before, immediately after and at 1, 2, 4, 6, 8, 12 and 24h and bulls were weighed before, immediately after, and at 4, 12 and 24h. Bulls travelling for 6, 9, 12, 18 and 24h lost 4.7, 4.5, 5.7 ($P \leq 0.05$), 6.6 ($P \leq 0.05$) and 7.5 ($P \leq 0.05$) percentage liveweight compared with baseline. During the 24h recovery period liveweight was regained to pre-transport levels. Lymphocyte percentages were lower ($P \leq 0.001$) and neutrophil percentages were higher ($P \leq 0.001$) in all T animals. Blood protein and creatine kinase concentrations were higher ($P \leq 0.001$) in the bulls following transport for 18 and 24h and returned to baseline within 24h. In conclusion, liveweight, physiological and haematological responses of bulls returned to pre-transport levels within 24h having had access to feed and water. Transport of bulls from 6 - 24 hours did not impact negatively on animal welfare.

The effect of herbal medicine (RHAM)® on animal dermatic and claw infection treatment

A.H. Ahadi[1], M.R. Sanjabi[1], M.M. Moeini[2] and A. Ghahramani[1], [1]Iranian research organization for science and technology (IROST), Tehran, Iran, [2]Department of Animal science, Razi University, Kerman shah, Iran

There are many studies on the effect of herbal medicine on infection diseases (Morteza et al., 2003; Dixit, 1998). Iran has been endowed with a rich flora variety of medicinal plants and herbs. In this study a mixed herbal plants from some known herbal plants prepared to provide scientific evidence to indicate its wound healing properties. RHAM ® is an indigenous preparation, which its clinical effects on wound healing have been studied. In first phase of the experiment the best formula were selected after standard tests in Lab on different bacteria culture such as *Staphylococcus*, *Bacillus Cereus*, *Streptococcus* and *Saccharamyces cerevisiae*. In second phase a preliminary clinical trial study on twelve infected rabbits were planed. They subjected to infection of subcutaneous wounds for comparison of different gel bases. Then RHAM ® has been used to treat 54 infected dairy cows in 4 herds within 6 month,s period in suburb of Tehran in the final trial.

The results showed that RHAM® are effective on treatment of wound healing, dramatic and claw infections on both rabbits and infected cows and had a potential therapeutic effect on infections. The effect of RHAM ® on healing of skin, claw and wound injuries was noticeable, quick and positive.

Session 16 Poster 11

Free cortisol in milk as an indicator of stress in dairy cows held in barns with automated milking and control gates

M. Melin, K. Svennersten-Sjaunja, H. Wiktorsson, Swedish University of Agricultural Sciences, Uppsala, Sweden*

Monitoring adrenal cortex activity by measuring corticosteroids in milk has been a useful indictor of potential stressors in dairy cows. The motivation to be milked is usually a weak incentive for the dairy cow to visit the milking unit. In order to obtain enough number of visits to the milking unit, control gates are often placed at the entrances to feeding areas. In the present study, it was hypothesized that cows were frustrated from being prevented to enter the feeding area through the control gates (i.e. gates were closed). To test this hypothesis, nine cows were subjected to three control gate routines in a carry-over design: A) gate always closed, B) gate open four hours after each milking event, then closed, C) gate always open. The behaviour of the cows was estimated from measures automatically obtained by the AM-system. A sample was collected at every milking for analysis of free cortisol in milk. Cortisol concentration was significantly different between gate routines in the early morning milk; mean values were 0.5 ng/ml, 1.5 ng/ml and 2.3 ng/ml for gate routine A, B and C, respectively. This was in contradiction to our hypothesis, and may be explained in differences between early morning feeding activity between groups. The analysis of feeding activity and other behaviour is in progress.

Session 16 Poster 12

Effect of milking frequency on milk quality

K. Svennersten-Sjaunja[1], L. Wiking [2], A. Edvardsson[1], A-K Båvius[1] and I.Andersson [3],*
[1]Kungsängens Research Centre, Swedish University of Agricultural Sciences, Uppsala Sweden, [2]Danish Institute of Agricultural Sciences, DK-8830 Tjele, Denmark, [3]Swedish Dairy Association, Scheelev. 18, Lund, Sweden

It has been reported that increased milking frequency results in higher milk yield but decreased or unchanged content of fat and protein, and sometimes an increased content of free fatty acids (FFA). The reason why the milk composition change is not fully evaluated. Both biological factors and technical factors may contribute. In order to increase the understanding of the problem a half udder milking experiment was conducted. The hypothesis was that if a difference in composition was detected it probably would be due to local mammary factors and not systemic factors since the whole udder has the same nutritional and endocrine environment. The experiment included two periods, in period one both udder halves were milked twice daily, in period two one udder half was milked twice daily and the opposite half was milked four times daily. Milk yield was registered and milk samples were analysed for content of fat, protein, casein, whey, lactose and FFA, fatty acid composition, fat globule size, the content of the enzyme γ-glutamyl transpeptidase, and the ions Na and K. The calculation of the results will be done by comparing the difference between the udder halves in period one and two. The calculations are in progress.

Session 16 Poster 13

The welfare of weanling bulls transported from Ireland to Italy

B. Earley, D.J. Prendiville and E.G. O'Riordan, Teagasc, Grange Research Centre, Dunsany, Co. Meath, Ireland*

Twenty-six weanling continental x bulls (414 ± 56kg) were transported (T) from Ireland to Italy on a roll-on roll-off ferry at a stocking density of 1.2m^2/bull, by road for 3-h to a French lairage, unloaded and rested for 24-h, and transported by road on an 18-h journey to Italy. Prior to transport, T bulls were blood sampled (day (d) 0) to provide baseline physiological levels, at arrival in France (d2), and at 12 and 24 hours after arrival, on arrival at the farm in Italy (d 5) and on d7, 9, 11 and 40. Twenty-two weanling continental x bulls (416 ± 60kg) remained in Ireland as controls (C) and were blood sampled at times corresponding to the T bulls. T and C bulls were weighed on d 0, 3, 5, 11 and 40 of the study. Bulls transported to France lost 7.0 % of their bodyweight at d3. T bulls had lower ($P \leq 0.001$) bodyweight on d3, 5 and 11 while C bulls had lower ($P \leq 0.001$) bodyweight on d5 and 11. T bulls spent 63.5% of time lying down during the 24-h sea crossing from Ireland to France. Neutrophil % and creatine kinase concentrations were higher ($P \leq 0.001$) while interferon-γ and lymphocyte % were lower ($P \leq 0.001$) in all bulls at d3. Transport had no adverse effect on the welfare of weanling bulls transported from Ireland to Italy.

Session 16 Poster 14

Cattle metabolic diseases and changes in central nervous system

M. Pilmane[1], I. Zitare[2], A. Jemeljanovs[2], I.H. Konosonoka[2], [1]Institute of Anatomy and Anthropology, Riga Stradins University, Riga, Latvia [1]Research Institute of Biotechnology and Veterinary Medicine "Sigra", Latvia University of Agriculture, Sigulda, Latvia*

The propose of the research was detection of apoptotic cells and correlation of these data with neurodegenerative regions in brain of animals with ketosis. Brain pieces from 8 cow with laboratory affirmed ketosis were examined after animals' compulsory slaughtering. The histochemical and immunohistochemical investigations methods were used. Statistical correlations were investigated between numbers of cells per visual field by use of *Leica DC 3OOF* digital camera, visualisation programme *Image Pro Plus* and programme SPSS'. Student t test was used to evaluate statistically significant differences between grey and white matter in brain cortex. Our results indicated presence of apoptotic cells in both - grey (13.5±4.22) and white matter of animals (11.38±3.34), that showed weak and statistically unsignificant correlation, p amyloid peptide was seen in pyramidal neurons of grey matter, but mainly - in glial cells of white matter. Also deposition of this peptide in the wall of blood vessels in *pia matter* was detected.

In conclusion, main part of cattle with ketosis possess apoptosis and deposition of p amyloid peptide in white, but also grey matter of brain, that might relate to the secondary unspecific changes of the same disease and/or also part of pathogenetical mechanisms in ketosis.

Investigations on microbiological spectrum in subclinical and clinical mastitis of dairy cows in Latvia

I.H. Konosonoka[1], A. Jemeljanovs[1] and I.* Ciprovica[2], *[1]Latvia University of Agriculture, Research Institute of Biotechnology and Veterinary Medicine "Sigra", Sigulda, Latvia, [2]Latvia University of Agriculture, Faculty of Food Technology, Jelgava, Latvia*

Subclinical and clinical udder secretion samples were examined to detected mastitis pathogens. For bacteriological examination, milk was inoculated on different complex and selective media. Microorganisms were further identified using kits of BBL Crystal Identification System. In total, 398 samples were investigated. The acquired data were analysed using analysis of variance.
Microorganisms from the genera *Micrococcus, Staphylococcus, Streptococcus, Corynrbacterium, Enterobacter, Klebsiella, Salmonella, Escherichia,* and *Pseudomonas* were isolated from secretion samples. Microorganisms from the genus *Staphylococcus (S.)* were isolated in 74.1 % of the subclinical and 48.2 % of the clinical cases of mastitis. *S. aureus* and coagulase negative staphylococci were the most frequently isolated species both in subclinical and clinical cases of mastitis.
The average numbers of *S. aureus* were 11 200 and 7 900 cfu ml^{-1} in subclinical and clinical mastitis secretions samples, respectively. Acquired results showed that 82.4 % in subclinical and 75.0 % in clinical mastitis secretion samples numbers of *S. aureus* exceeded 2 000 cfu ml^{-1}. The average somatic cell count (SCC) in subclinical secretion samples with the number of *S. aureus* below 500 cfu ml^{-1} was significantly different ($p < 0.05$) compared to samples with the number of *S. aureus* over 500 cfu ml^{-1}, 700 10^{-3} and 1 620 10^{-3}, respectively

---

Comparative research on the impact of the maintenance system upon milk cows' performance and health

Livia Vidu, Alina Udroiu, I. Raducuta, I. Calin, V. Bacila, University of Agronomical Sciences and Veterinary Medicine, Bucharest, Str. Marasti 59, Romania*

Breeding milk cows includes three maintenance systems: bound, free, and mixed. The choice of a certain system should be made only according to the concrete availability and conditions of the farm on the one hand, and to the (dis)advantages of each maintenance system on the other hand. The maintenance system must provide normal works and technological chain resulting in high milk productions and good health. This study was carried out in 6 Romanian farms, on milk cows belonging to the following breeds: black Baltata, Holstein-Friesian type, and yellow Baltata, Simmental type. In these farms, milk cows are both bound and free. The paper aims at identifying the quantitative and qualitative parameters of the milk production, the share of metabolic, reproduction, and limb diseases, the number of somatic cells in milk. Irrespective of the breed, the largest number of udder diseases is present in the case of bound cows (23.7%, compared to 14.2% in free cows). The largest number of metabolic diseases is found in the bound Holtein - Friesian breed (21.5%). Concerning the hygienic quality of milk, the largest number of somatic cells is present in bound cows.

Session 16 Poster 17

Cubicles height over the floors in passages: implications for hygiene

Anders H. Herlin[1], Madeleine Magnusson[1], Michael Ventorp[1] and Susanna Lorentzon[2], [1]Swedish University of Agricultural Sciences (SLU), P.0. Box 59, SE-230 53 Alnarp, Sweden, [2]Åsmark Norrgård 1, SE-56392 Gränna, Sweden

The hygiene in a cowshed is extremely important for the cows' well being, health and milk quality. Flooring systems used in the passages and the cubicle curb height over the floor was investigated. There were two compartments with mastic asphalt and scrapes with cubicles situated at different curb heights (9 cm vs. 17 cm) and one compartment with concrete slatted floor without scrapes (curb height 9 cm). Ash content of the bedding on area exposed to the udder indicated presence of manure. The hygiene was further compared with bacterial analyses using coliform bacteria and spores from *Bacillus cereus*. The milk quality was assessed by somatic cell counts of the milk analysed every 14 days. All results indicate that the hygiene is best in the compartment with mastic asphalt and scrapes with high cubicle curb height. The amount of manure in the cubicles was lower compared to the other compartments and the number of coliform bacteria was lower. The animals were cleaner in this compartment and the somatic cell count tended to be lower. The compartment with mastic asphalt and low cubicle curb height seemed to be the one with the most poorly hygiene in most hygiene aspects.

Session 16 Poster 18

The effect of short dry periods on health disorders in dairy cattle

A. Pezeshki, H.R. Rahmani, G.R. Ghorbani, M. Alikhani and M. Mohammadalipour, Dept. of Animal Sciences, College of Agriculture, Isfahan University of Technology (IUT), Isfahan, Iran*

The objective of current study was to evaluate the effects of varying dry periods on health of dairy cows. One-hundred eight Holstein cows were arranged to one of three treatments: Traditional (T) dry period (56 day), Moderate (M) dry period (42 day) and Short (S) dry period (35 day). Dystocia was numerically greater in S (11.4%; 4/35) than T (8.57%; 3/35) with M (10.52%; 4/38) being intermediate. Retained placenta was higher in T (17.4%; 6/35) than S (14.28%; 5/35) or M (10.52%; 4/38), but it wasnt significant. There was no difference between S and T groups on incidence of displaced abomasum and milk fever (2.8% vs. 2.8%; 1/35). Clinical mastitis at calving time was greater in S groups (8.57%; 3/35) than M (2.6%; 1/38) or T (2.8%; 1/35), numerically. High-producing cows (milk yield ≥20 kg at drying-off) with 35 days dry tended to have a higher SCS than high-producing cows with 56 days dry (5.84 vs. 5.35; $p = 0.11$). Cows with a higher SCC also had higher days to first service compared to cows with a lower SCC numerically, but it wasn't significant (47.4 vs. 43.6, respectively). It is concluded that with shortening the dry period there is a high risk for dystocia, mastitis and may be lower reproductive efficiency.

Session 16 Poster 19

Longitudinal slope of the cubicle for the dairy cow

Anders H. Herlin[1], Madeleine Magnusson[1] and Cecilia Hagberg[2], [1]Swedish University of Agricultural Sciences (SLU), P.O. Box 59, SE-230 53 Alnarp, Sweden, [2]Länstyrelsen Gotland län, SE-621 85 Visby, Sweden

Improving hygiene is a major concern in intensive dairy production. By altering the longitudinal slopes of the cubicle floors the drainage could be improved and the bedding turnover increased. If the cows position is further back in the cubicles due to the increased slope the effect could be cleaner lying surfaces. This could result in cleaner cows, cleaner udders and a minimization of the risks of mastitis. If the increased slope imply a cow position more diagonal in the cubicles, the effects could be reduced. The increased slope could also alter the animals lying behaviour. From a welfare point of view this may not be acceptable. The aim of this experiment was to examine how an increased longitudinal slope in cubicles, from about 2 % to about 7 %, influenced the hygiene of the cubicles, the diagonal lying positions of the cows and on the lying behaviour. The results showed a better hygiene on the cubicle surface, cows were positioned further to the back of the cubicle and that diagonal positioning of the body depended on the height at the withers of the cows. Total resting time did not differ between the treatments. The 7% slope was a bit too much as the cows seem to slide backwards in the cubicles.

Session 16 Poster 20

Analysis of the factors affecting somatic cell count in milk

H. Kiiman, T. Kaart, M. Henno, O. Saveli. Estonian Agricultural University, Institute of Veterinary Medicine and Animal Science, 1 Kreutzwaldi, Tartu, 51014 Estonia

Increasing awareness of public health and food safety issues have led to a greater interest in milk quality in recent years. The key milk quality element being regulated is somatic cell count (SCC). High SCC levels are not only known to pose a directly public health, but also reflect mammary infection and overall quality of management. Moreover, lower SCC levels have been shown to be related to higher milk yield and better dairy product quality. Milking procedures very a great deal from one farm to another. This is the factor that has the greatest impact on udder health. Data were collected from six dairy cattle farms between 2000-2004. The data on 305-day milk yield, fat and protein content, fat and protein yield and SCC were collected in first, second and third lactation. Cows' sire, breed, calving month, lactation, and milking operator were fixed in a database. Cows were milked using a standard milking routine, including pre-stripping and cleaning with individual cotton towels and teats were post-dipped.

On the basis of data analysis, it appeared that major factors influencing the milk SCC were lactation ($P<0.01$), milking operator, and bull ($P<0.05$). Breed had also a small effect, but it was not statistically significant. Calving month was not a significant factor affecting the milk SCC ($P>0.05$).

Session 16 Poster 21

Characteristics for mastitis incidence in dairy herds in the Czech Republic

M. Stípková, M. Wolfová and J. Wolf, Research Institute of Animal Production, P.O.Box 1, CZ 10401 Praha-Uhríneves, Czech Republic*

Data on clinical mastitis (CM) incidence collected between 1996 and 2003 on five Holstein dairy farms in the Czech Republic were analyzed. The following average values were calculated for the 1st, 2nd and 3rd and subsequent lactations, respectively: 0.35, 0.45 and 0.57 for lactational incidence of CM, 0.63, 0.94 and 1.22 for the number of CM cases per cow and 0.68, 1.00 and 1.27 for the incidence rate of CM per cow-year at risk. The lactational incidence of CM and the number of CM cases per cow were calculated from data with complete lactations only, whereas the incidence rate of CM per cow-year at risk was calculated from the full data set.The analysis of CM incidence based on daily records showed the highest proportion of infected cows during the first 10 days of lactation. The incidence rate of CM per day (or per year) at risk was shown to be the best indicator for mastitis susceptibility because it accounts for the truncated character of the data and for repeated outbreaks of mastitis within a lactation.

The financial support from the project MZE 0002701402 is acknowledged.

Session 16 Poster 22

Milk let-down parameters' association with udder health problem incidence: a case study

N. Livshin[1], E. Maltz[1], M. Tinsky[2] and E. Aizinbud[1,2], [1]Agriculture Research Organization, Volcani Center, PO Box 6, Bet-Dagan 50250, Israel; [2]S.A.E. Afikim, kibbutz Afikim, Israel

This work objective was to relate milk let-down parameters to udder health problems incidence. The study was performed in a commercial Israeli herd with 440-475 high productive Holstein cows (11000 kg/cow/year) milked trice daily. The following milk flow parameters have been monitored by Afiflo system (S.A.E Afikim): flow rates at 0-15, 15-30, 30-60 and 60-120s after milking start, mean flow rate, peak flow timing and rate, milk harvested in 2 first min, low flow duration. Cows' milk let-down parameters were calculated from data for 15-18 successive milkings every two month, during one year. During this year, udder health problems, including clinical mastitis, blood in milk, teat obstruction, swollen udder and flaky milk, were diagnosed for 216 cows. The main associations between milk let-down parameters and udder health problem incidence were found for initial (0-15 and 0-30 s), peak and mean flow rates. From 17 cows exhibiting very high initial flow (2 kg/min or more in first 15s), 16 had udder problems. Also, cows with very high (more than 3.5 kg/min, 12 cows) and very low mean milking flow (less than 1 kg/min, 11 cows) had highest rate of udder problems (about 70 and 50 percent, respectively). The possible explanations of these findings are discussed taken into account physiological and anatomical factors.

Session 16 Poster 23

Teat and teat-end type in three dairy cattle breeds in the Tropics

Mario Riera, Andrea Cefis, Ottavia Pedron, J.C. Alvarez, Universidad del Zulia, Maracaibo, Venezuela; University of Milan, Italy

Teat and teat-end were evaluated in 403 Holstein, 118 Jersey and 1263 Carora cows from Venezuela. Teat type was classified as funnel, cylindrical and bottle and teat-end as pointed, round, flat, disk, funnel shape and prolapsed. Length and diameter were measured for each teat. Our purposes were 1) to evaluate the factors influencing teat traits 2) to verify the relationships between teat traits and 305-d milk yield. Teat length and diameter were analyzed with a mixed model including the fixed effects of breed, cow age, herd-birth year interaction and the random effect of the cow within the herd. A logistic regression was carried out on teat and teat-end types. The effect of each teat trait on milk yield trais was assessed by a mixed model with the fixed effects of breed, age, herd, calving year and month and the random effect of the cow. Breed showed a significant effect on teat length and diameter(5.5, 5.3 and 5.9 cm for length and 2.2, 2.1 and 2.3 cm of diameter for Holstein, Jersey and Carora, respectively). Teat length and diameter increased with age. Milk yield was significantly and negatively influenced by teat length, the regression coefficient ranging from -35 kg/cm for fore right teat to -62 kg/cm for rear left teat. Rear left teat diameter was significantly and positively (+17,5 kg/mm) associated with milk yield.

Session 16 Poster 24

Comparing two concentrate allowances in an automatic milking system

I. Halachmi, Institute of Agricultural Engineering, Agricultural Research Organization (A.R.O.), The Volcani Center, P. O. Box 6, Bet Dagan 50250, Israel, E-mail: halachmi@volcani.agri.gov.il

This study investigated the potential for applying automatic milking system (AMS) to the management of high-yielding cows fed a total mixed ration (TMR). The null hypothesis was that it is desirable to maintain even in AMS, the TMR feeding management practice recommended for high-yielding cows and therefore it can be attained by 'reducing the concentrate allocation in the robot without reducing the number of milkings'. Two feeding regimes were used: the "candy concept", with only 1.2 kg of feed concentrate - the minimum to attract the cow - provided at each visit to the milking robot; and the provision of a maximum of 7 kg of feed concentrate per day. Approximately 100 cows were subjected to one or other of these two treatments. Although the cows in the first treatment consumed approximately 3.5 kg of concentrate per day and those in the second treatment approximately 5 kg per day, no significant differences were observed in the numbers of voluntary milkings.

Session 16 Poster 25

Analysis of the synchronisation of passages out to the pasture in an automatically milked herd with day and night access to the grazing area

E. Spörndly and M. Bergman, Department of Animal Nutrition and Management, Kungsängen Research Centre, Swedish University of Agricultural Sciences, Uppsala, Sweden*

A study was performed on automatically milked cows during the grazing season to study the behaviour of the cows when leaving the barn to go to the grazing area. The objective was to see to what degree the passage out to the pasture area was synchronised and if there was any relationship between synchronisation and other factors such as age, rank, or grazing time.
Observations covering 132 hours and 378 passages were performed during two periods when cows grazed at different walking distances from the barn, Distant (330 m) and Near (60 m). Each passage to the pasture was observed and registrations of number of cows leaving the barn together, position of each cow (first, last or middle) and the time it took for cows to walk the first 60 m from the barn was registered. Correlations between the registrations and other factors were computed. On the Distant pasture 69 % of the passages observed were passages were cows went to the pasture in the company of at least one other cow. Corresponding value for the Near pasture was 49%. No significant correlations were found between synchronization (measured as average number of accompanying cows) or walking speed and the factors age, rank or grazing time.

Session 16 Poster 26

Relationships between temperature, precipitation, and carcass weights of the red deer (*Cervus elaphus* L.) in North-Eastern Poland

P. Janiszewski, T. Daszkiewicz, University of Warmia and Mazury in Olsztyn, Faculty of Animal Bioengineering, Oczapowskiego 5/140A, 10-719 Olsztyn, Poland*

Observations were performed in a hunting ground located in north-eastern Poland (near the city of Olsztyn - 20°30'E; 53°47'N), during 17 years, from 1985 to 2002. A total of 1320 red deer carcasses, including 293 carcasses of stags, 643 carcasses of does and 384 carcasses of fawns, were examined in the study.
The objective of the present study was to determine the relationships between selected weather factors and carcass weights in particular red deer groups.
Coefficients of correlation between mean annual air temperatures, precipitation total, and carcass weights of stags, does and fawns were calculated.
A correlation was found between mean annual air temperature and carcass weights of stags ($r \geq 0.47$) and fawns ($r \geq 0.55$).
Precipitation total is not significantly correlated with carcass weights of the red deer ($r > 0.05$)

Session 16 Poster 27

Identification of strains of mink Aleutian disease virus in Nova Scotia

A. Farid[1], B.F. Benkel[1], F.S.B. Kibenge[2] and G.G. Finley[3], [1]Department of Plant and Animal Sciences, Nova Scotia Agricultural College, Truro, Nova Scotia, Canada. [2]Department of Pathology and Microbiology, AVC, University of Prince Edward Island, Charlottetown, P.E.I., Canada. [3]Veterinary Services, Nova Scotia Department of Agriculture and Fisheries, P.O. Box 550, Truro, Nova Scotia, Canada

Aleutian disease is a serious health problem for the mink industry in Canada. The 4801 bp Aleutian disease virus (ADV) genome displays a high degree of genetic variability, which has resulted in a proliferation of strains of ADV causing varying degrees of severity of the disease. To identify ADV strains, DNA was extracted from frozen tissues of mink that were positive for ADV antibody. A 631 bp fragment of the highly variable portion of the viral genome that codes for the nonstructural NS1 protein was amplified by the polymerase chain reaction (PCR). PCR products from 22 mink from three neighboring ranches were bi-directionally sequenced. Two ADV strains (types I and II) were identified, which differed from each other at three positions on the PCR product, indicating 99.5% sequence similarity. These strains showed 96% sequence similarity with the highly pathogenic ADV-Utah, moderately pathogenic ADV-SL3, and the non-pathogenic ADV-G strain, but showed lower levels of sequence similarity with the highly pathogenic ADV-United (88.5%) and Danish ADV-K (87.4%). The results suggest that viral strain identification is a powerful tool for epidemiology of the ADV on mink ranches.

Session 16 Poster 28

The influence of outdoor raising of pigs on their growth rate and behaviour

V. Juskiene, R. Juska, Institute of Animal Science of LVA, R. Zebenkos 12, LT-82317, Baisogala, Radviliskis distr., Lithuania

Fifty-two crossbred (Lithuanian White x Large White) pigs from 2.5 month of age were used in an experiment to determine the influence of pig raising outdoors on the productivity performance, health and behaviour of pig breed spread in Lithuania. Two analogous groups of pigs of 26 animals each were formed. The control group of pigs was kept in a pig-house, in 20.0 m^2 pens, while the experimental group was kept outdoors in enclosures of 8.5 areas that were fitted with 7.5 m^2 shelters.

The results from the study indicated that the growth of pigs raised outdoors was higher, i.e. the growth rate of growing pigs was 26.8% (P=0.003) higher and the growth rate of finishing pigs was similar for both groups. The growth rate of pigs raised outdoors was higher during the whole fattening period and their average daily gain was 12.7% (P=0.013) higher than that of pigs raised indoors.

The behaviour studies indicated that pigs raised outdoors used to lie 8.9% longer in time than those raised indoors, while pigs raised indoors used to stand and sit 60.2% longer in time than those raised outdoors. Outdoor raising of pigs was beneficial to their significantly higher (1.5 times - P=0.085) activity. Pigs raised indoors were more aggressive during the whole experiment.

Session 16 Poster 29

Short and long slaughter transports increase mortality rates in pigs

C. Werner, K. Reiners and M. Wicke, Research Centre for Animal Production and Technology, Department of the Faculty of Agricultural Sciences of the Georg-August-University Goettingen, Vechta, Germany*

The losses during transport of slaughter animals which are quite important with respect to animal welfare and economical reasons are influenced by exogenic (e.g. climate, duration) and endogenic factors (e.g. stress susceptibility of the animals). In the present study the development of transport losses in Germany was investigated regarding slaughter pigs. For this purpose the mortality data recorded by a big meat company were statistically analysed by (M)ANOVA with a probability of error of less than 0.05 considered significant. The analysis showed that the percentages of animals dying during or after the transport decreased between 1999 and 2003. This reduction was mainly due to changes in the summer months a season with the highest mortality rates over the investigated years. Concerning the transport duration not only long (ø 8h), but also very short durations (ø 1h) affect the welfare of the animals resulting in increasing mortalities as well as circulation problems, fractures etc.. From the presented results it could be suggested that the improvement of transport conditions by the intensification of the EU transport regulations, the decrease of homozygote MHS-positive pigs and the growing sensibility of the consumers positively influenced this development. However, in some points like minimal and maximal transport duration and transport conditions (e.g. ventilation) further changes and regulations are necessary.

Session 16 Poster 30

Dust spatial distribution and seasonal concentration of windowless broiler building

**H.C. Choi[1], K.Y. Yeon[1], J.I. Song[1], H.S. Kang[1], D.J. Kwon[1], Y.H. Yoo[1], C.B. Yang[1], S.S. Cheon[2] and Y.K. Kim[3], [1]Environment Division, National Livestock Research Institute, RDA, 564, Omokchun-Dong, Suwon, 441-350, Korea, [2]Dyne Engineering, 297-6, Ohjung-Dong, Daejeon, 306-820, Korea, [3]Choongnam National University, Agricultural College, Department of Dairy Science, 220, Goong-Dong, Daejeon, 305-764, Korea*

This study investigated the spatial distribution and seasonal concentration of dust originating from tunnel-ventilated windowless broiler building measuring 12-m wide, 46-m long, with a side wall height of 3 m and a capacity of 12,800 birds or a stock density of 23.2 birds/m^2. Dust concentrations in terms of total suspended particles (TSP), and particulate matter of sizes 10 µm (PM10), 2.5 µm (PM2.5), and 1 µm (PM1) were measured at 30-minute intervals using GRIMM Aerosol Monitor (GRIMM AEROSOL, Germany). The spatial distribution of dust in the broiler house showed a lower dust concentration in the inlet than in the outlet of the ventilation tunnel, and a decreasing dust concentration as dust size decreased, as follows: 317.9 µg/m^3 TSP; 74.7 µg/m^3 PM10; 9.7 µg/m^3 PM2.5; and 6.2 µg/m^3 PM1 in the inlet; and 2,678.5 µg/m^3 TSP; 555.5 µg/m^3 PM10; 33.3 µg/m^3 PM2.5; and 10.2 µg/m^3 PM1 in the outlet. On the basis of broiler age, the average dust concentration in TSP did not show any particular pattern during summer, as follows: 1,229; 904.5; 558.8; and 1,053 µg/m^3 on the broilers' first to fourth week of age, respectively. But during winter, the average dust concentration showed an increasing pattern, as follows: 465.4; 1,401; 4,497; 5,097; and 6,873µg/m^3 on the broilers' first to fifth week of age, respectively.

Session 16 — Poster 31

Age-related differences of *Ascaridia galli* egg output and worm burden in chickens following a single dose infection

M. Gauly[2], T. Homann[1] and G. Erhardt[1], [1]Institute of Animal Breeding and Genetics, University of Giessen, Ludwigstrasse 21B, 35390 Giessen, [2]University of Goettingen, Albrecht Thaer Weg 3, 37075 Goettingen, Germany

Four groups of Lohmann-LSL chickens (n = 20) were artificially infected with 250 embryonated *Ascaridia galli* eggs at the age of 6, 12, 18 or 24 weeks. Ten birds were kept as uninfected controls. Six and ten weeks after infection (*p.i.*), individual faecal egg counts (FEC) were performed. The birds were slaughtered after the second sampling and their gastrointestinal tracts were examined for the presence of adult *A. galli*.
The FEC increased from the first to the second sampling significantly in all the infected groups. The highest increase was shown in the group infected at 12 weeks of age, whereas the increase in the other groups was relatively moderate. However, total worm burden and mean FEC at the second sampling were highest ($p < 0.01$) in those birds infected at an age of 12 or 18 weeks.
Age does not seem to play a major role in resistance to *A. galli* infections in layers, whereas a bird's hormonal and immune status, related to laying activity, seems to have a significant negative impact on resistance.

Session 16 — Poster 32

Comparison of two force molting methods on performance of laying hens in second phase of egg production

A. Hassanabadi[1] and H. Kermanshahi[2], [1]University of Zanja, Iran, [2]Uuniversity of Mashhad, Iran*

Effects of force molting methods were studied on performance of laying hens in second phase of egg production. Two treatments involved 20 g/kg zinc oxide to diet for 10 days and feed deprivation for 5 days. 6168 seventy-nine weeks old Leghorn hens, distributed in treatments. There was four replicate in each treatment, 771 hen per each replicate and four birds in each layer cage. The treatments were started at the same time at 80 weeks of hen's age. Two weeks before and thirteen weeks after treatments the performance characters including egg production, feed efficiency, broken eggs, shell quality and feed intake was recorded. Data was analyzed using t test. Egg production in zinc oxide and feed withdrawal treatment was ceased 8 and 7 days after treatment implementing, respectively. Egg production in second phase reached to 5 percent, 35 and 25 days from treatment implementing in zinc oxide and feed withdrawal, respectively. Hen day production didn't show significant differences between the treatments in second phase. However, it was insignificantly higher in feed withdrawal treatment. The treatments had no significant effect on egg weight and eggshell diameter but the broken eggs was significantly higher in feed withdrawal treatment. Feed intake hadn't significant difference between the treatments. Mortality was significantly higher in zinc oxide treatment up to 87 weeks of age.

Session 16 Poster 33

The functional condition of the stomach in goats infected with the alimentary tract nematodes

E. Birgele, D. Keidane, A. Mugurevics, J. Jegorova, Preclinical Institute, FVM, LUA. K.Helmana street 8, Jelgava, Latvia LV - 3002

The task of this investigation was to study if the abomasum pH dynamics change in goats artificially infected with trichosrtrongyloids, and if it does change, then what the changes are. Five 2 - 3 months old kids and five 1 - 2 years old goats were used for the investigation. Before the infestation all the animals were operated fistulae in the abomasum and rumen. The artificial infestation was started on the 10th day after operation. Each of the goats received 5000 Ostertagia circumcincta larvae. In the third and fourth week after the artificial infestation eggs typical of the parasite were found in faeces of all the goats. MacMaster technique was applied for counting parasite eggs. In the group of adult animals the number of eggs varied within the range from 114 to 250 eggs per 1 g of faeces, but in 2 - 3 months old kids the intensity of the infection was higher - 282 to 512 eggs per 1 g of faeces. It was stated that: 1. the intra-abomasal pH medium in kids had become less acid than it was before the infestation of those animals; 2. in adult goats the abomasum pH changes were less expressed than those in kids; 3. the composition of amino acids, fat percentage and urea in the milk obtained from the infected goats were changed little.

Session 16 Poster 34

The functional condition of stomach and some indices of meat quality in bulls in their ontogenesis

A. Ilgaza, E. Birgele, A. Mugurevics, D. Keidane, J. Jegorova, Preclinical Institute, FVM, LUA. K.Helmana street 8, Jelgava, Latvia LV - 3002

The aim of the study is to investigate whether the adaptation processes of the gastro-hepatic system in herbivorous animals in their ontogenesis affects the quality of the meat obtained. There are data from five bull at the age of five months and when they were two years old before slaughtering. The pH-dynamics were measured before feeding by two-electrode pH-probe through the abomasum or rumen chronic fistula. The effect of neural and humoral regulation mechanism on the hydrochloric acid secretion in the abomasum was also studied. After slaughter of the bulls meat was sampled from *m.longissimus dorsi*. It was stated: 1. In both animals at the age of five months and those of two years old reaction in the abomasum two hours before the morning feeding did not differ significantly 2. In five months calves cimetidine caused greater blocking effect of the hydrochloric acid secretion than atropine sulphate, but in two years old bulls the blocking of *n. vagus* secretory effect with atropine sulphate and the blocking of H2 receptors of the fundal glandular epithelium cells caused insignificant intra-abomasal pH changes. 3. The results of meat chemical analyses of two years old bulls, which were fed similarly, differed little. In the meat of non-castrated bulls the cholesterol level was a little lower.

Session 17 Theatre 1

Are time-budgets of dairy cows affected by genetic improvement of milk yield?

P. Løvendahl and L. Munksgaard, Danish Institute of Agricultural Sciences, DK 8830 Tjele, Denmark*

Time is a resource that the cow can spend on feed intake, walking between different locations, waiting and resting. Our hypothesis is that components of dairy cows time budget have genetic variation and those parts closely connected with production traits may be affected by selection for improved yield. We studied time-budgets of 243 first lactation Holstein cows twice in early lactation (mean 86, range 50 - 123 DIM). Estimates of minutes spent eating, lying, standing or walking in alleys were obtained from scan sampling of the activities at 10 min intervals during 24 hours. Daily milk yield (ECM) was recorded at 3 week intervals. Data were analysed by AI-REML using two-trait repeatability models including relationship matrix. The heritability of daily ECM was $h^2 = 0.12 \pm 0.14$, and eating time had similar heritability (0.13 ± 0.13), but lying time had very low heritability (0.02 ± 0.15). Correlations between traits were calculated as individual animal correlations (r_i) based on permanent animal and additive genetic covariance components. Eating time was positively correlated with yield ($r_i = 0.23$), and lying time negatively correlated with yield ($r_i = -0.26$). Although eating time was negatively correlated with lying time ($r_i = -0.38$), the magnitude of this correlation decreased at higher yield. As restrictions on lying time are known to induce stress responses, further selection for higher yield may increase the liability to metabolic disease caused by deficits in time budgets of dairy cows.

Session 17 Theatre 2

Estimates of genetic parameters for milkability from automatic milking

S. Gäde, E. Stamer, W. Junge, E. Kalm*
Institute of Animal Breeding and Husbandry, Christian-Albrechts-University, Hermann-Rodewald-Straße 6, D-24118 Kiel, Germany

Genetic parameters were estimated by restricted maximum likelihood with a multi-trait animal model for three milkability traits with serial data from an automatic milking system from a research farm (401 dairy cows) collected between September 2000 and June 2003 (320834 milkings). Furthermore daily values for milk yield and milkability were calculated from all single milkings and subsequently estimation of genetic parameters was carried out based on daily values once again. The estimated heritability coefficients (based on daily values) are h^2=0.55, h^2=0.55 and h^2=0.39 for average milk flow, maximum milk flow and milking time. The heritabilities are on a high level and thus breeding for good milkability is feasible. The genetic correlations between the three milkability traits are near unity with $r_g = 0.98$ between average and maximum milk flow, $r_g = -0.89$ between average milk flow and milking time and $r_g = -0.86$ between maximum milk flow and milking time. So recording of only one of the traits in performance tests may be sufficient. The genetic correlation between milk yield and average milk flow, maximum milk flow and milking time are $r_g = 0.51$, $r_g = 0.44$ and $r_g = -0.23$.
In future serial data about milkability, already existing in many farms with automatic milk yield recording, should be used to a greater extent for selection towards a good milkability.

Session 17 Theatre 3

Genetic parameters for conception rate and days open in Holsteins

S. Tsuruta[1], I. Misztal[1] and T. J. Lawlor[2], [1]University of Georgia, Athens, 30602, USA, [2]Holstein Association USA Inc., Brattleboro, VT 05301, USA*

The goal of this study was to estimate genetic parameters for conception rate and days open in Holsteins. Data collected in NY from 2001 to 2003 were provided by the DRMS, Raleigh, NC. After editing, it included 89,271 first-parity service records for 43,344 cows. The first model was for conception rate at first service and for days open; effects included herd-year, age class, month of calving, AI status (natural service or AI) and days to first service after calving (for conception rate only), milk yield as a covariate, and additive genetic. Heritability estimates for conception rate at first service and days open were 1.4% and 2.5%, respectively. Genetic correlation between these traits was -0.92. The second model applied to conception rate with repeated records. It included the same fixed effects as the first model plus linear random regression as a function of days to service for the animal additive and permanent environmental effects. Heritability and repeatability estimates were 1.0-1.5% and 3.6-4.2%; they were approximately constant from 50 d to 250 d postpartum. The genetic correlations between the conception at 50 and (150, 250) d were 0.86 and 0.45, respectively. Days open is a good measure of fertility in NY. Conception rate may be a different trait at different days postpartum.

Session 17 Theatre 4

When to farrow? Genetic correlation between gestation length and piglet survival

L. Rydhmer and N. Lundeheim, Swedish University of Agricultural Sciences, Dept. of Animal Breeding and Genetics, Box 7023, S - 75007 Uppsala, Sweden

Piglet survival decreases with increasing litter size. Cross fostering complicates the registration of piglet survival in the field, but gestation length is already included in all data bases. Our hypothesis is that the genetic correlation between gestation length and piglet survival is positive. We studied field data from primiparous Swedish Yorkshire sows, 12000 farrowings. Gestation length was analysed together with no. stillborn and total born piglets. The random effect of herd-year and the fixed effects of farrowing month, litter breed (purebred or crossbred) and mating (natural or AI) were included, together with the random effect of the animal (the sow). Gestation length had a mean of 115.5 d, std 1.6 d. There was no difference in gestation length between purebred and crossbred litters. The heritability for gestation length was estimated at 0.32. There was no significant genetic correlation between gestation length and no. stillborn, r_g = -0.01. The genetic correlation between gestation
length and total born was low, r_g = -0.06. We conclude that gestation length is a heritable trait, but it is not correlated to piglet survival *at birth*. The next steps are to analyse piglet survival after birth, to analyse all parities and to include a genetic effect of the litter in the model.

Session 17 Theatre 5

A method to define sustainable breeding goals for livestock breeding programmes

H.M. Nielsen[1] and L.G. Christensen[2], [1]Department of Animal and Aquacultural Sciences, Norwegian University of Life Sciences P.O. Box 5003, 1432 Ås, Norway, [2]Department of Large Animal Sciences, The Royal Veterinary and Agricultural University, Grønnegårdsvej 8, 1870 Frederiksberg C, Denmark*

The objective was to present a method for the derivation of non-market values for functional traits in livestock breeding goals. A non-market value represents the value of improved animal welfare or societal influences for animal production. Sustainable breeding goals can be defined by adding non-market values to the market economic values in the breeding goal. The consequence of adding a non-market value to a functional trait is less selection response in production traits. Deterministic simulations and selection index theory were used to derive non-market values based on the loss in selection response in production traits, farmers and/or breeding companies are willing to loose to improve functional traits. The method was demonstrated using a breeding goal for dairy cattle with four traits (production, mastitis resistance, conception rate and stillbirth). With five percent loss in response in production and independently derived, non-market values for mastitis resistance, conception rate and stillbirth were .3, 1.4, and 2.9 times estimated market economic values. Increase in response for mastitis, conception rate and stillbirth were 0.18, 0.21 and 0.35 genetic standard deviations. With simultaneously derivation of non-market values, the loss in response in production should be distributed among functional traits to yield maximum response in each trait.

Session 17 Theatre 6

Validation of an approximate multitrait model for prediction of breeding values in dairy cattle: a stochastic simulation study

J. Lassen[1,2], M.K. Sørensen[1] and P. Madsen[1], [1]Danish Institute of Agricultural Sciences, Denmark, [2]The Royal Veterinary and Agricultural University, Denmark*

The aim of this study was to develop and validate an approximate multitrait model for prediction of breeding values. A data structure resembling the Danish Holstein population for a 15 years period was simulated. True breeding values and phenotypic records for: milk yield, SCS, mastitis, days open, non return rate, udder depth and dairy character was simulated. Selection was on a total merit index based on milk yield, mastitis and udder depth.

Univariate BLUP was performed to estimate fixed effects and predict random effects. Records were adjusted for all effects except the animal effect from the BLUP solutions. Variance components were estimated on preadjusted data using a model with mean, animal and residual effect. Parameters estimated on preadjusted data were close to the true parameters.

A multivariate BLUP was performed on preadjusted and raw data. From these two analyses the correlation between true and estimated breeding values was highest when looking at solutions from raw data. Computing time was lowest when analyzing preadjusted data.

A multivariate BLUP would be the best solution, but computer power is a limitation when handling the large number of traits in the total merit index today. Therefore doing multivariate BLUP on subsets of the data is of interest, combining the resulting breeding values using approximate methods.

Session 17 Theatre 7

Another useful reparameterisation to obtain samples from conditional inverse Wishart distributions

I.R. Korsgaard[1], A.H. Andersen[2], P. Madsen[1] and J. Ødegård[3], [1]Department of Genetics and Biotechnology, Research Centre Foulum, PO Box 50, DK-8830 Tjele, Denmark, [2]Department of Theoretical Statistics, University of Aarhus, DK-8000 Aarhus C, Denmark, [3]Department of Animal and Aquacultural Sciences, Agricultural University of Norway, PO Box 5003, N-1432 Ås, Norway*

A method to obtain samples from the fully conditional posterior of the residual variance-covariance matrix in Bayesian multivariate models with Gaussian and binary threshold-liability characters is presented. The models considered allow residuals of liabilities to be correlated, and for an arbitrary number of random effects including animal genetic effects. The method follows by invoking a set of restrictions, suggested by González (2003) for the multivariate probit model, to ensure identifiability, combined with well known properties of the inverse Wishart distribution. Furthermore, for any two models equivalent from a non-Bayesian perspective, we give necessary and sufficient conditions on the prior distribution of the parameters ensuring equivalence of the corresponding posterior distributions.

Session 17 Theatre 8

Properties of random regression models using linear splines

I. Misztal, University of Georgia, Athens, GA 30602, USA

The purpose of this study was to determine rules to select knots in random regression models using linear splines (RRMS). Such models are much easier to implement than models with polynomials because of superior numerical properties and simplicity of obtaining parameters. The variance of an effect with linear splines, relative to a straight line, is concave between the knots. The maximum depression is in the middle of knots and equals 0.5(1-r), where r is the correlation between these knots. For RRMLS, the vector of covariables is [..1-t t..], where t=<0,1>. If this vector is modified to [.. $(1-t)^q$ t^q 0], where q=log[2(1+ r)]/2/log(2), the concavity in the middle can be eliminated. Simulated data included an effect with 5 knots spaced equally; the correlation between the extreme knots (r) varied from 0.0 to 0.99. Predictions were obtained for points corresponding to extremes and the middle of the trajectory by models with 5 knots (K5), 2 knots (K2), and 2 knots with covariables modified (K2M). Compared to K5 and depending on r, the predictions by K2 (K2M) were inflated at the extremes by 0-16% (0-3%) and deflated in the middle by 0-39% (0-2%). Accuracies by K2M and K2 were similar and 0-0.04 below those by K5 but less than 0.01 for r ≥ 0.6. In practical application of RRMS, one can select the minimal number of knots with correlations between the adjacent knots ≥ 0.6.

Session 17 Theatre 9

Genetic improvement in broilers using indirect carcass measurements

S. Zerehdaran[1], A.L.J. Vereijken[2], J.A.M. van Arendonk[1], H. Bovenhuis[1] and E.H. van der Waaij[1,3], [1]Animal Breeding and Genetics Group, Wageningen University, PO Box 338, 6700 AH Wageningen, The Netherlands, [2]Nutreco Breeding Research Center, P.O. Box 220, 5830 AE, Boxmeer, The Netherlands, [3]Department of Farm Animal Health, University of Utrecht, Yalelaan 7, 3584 CL Utrecht, The Netherlands*

The objective of the study was to determine the consequences of using different indirect carcass measurements (ICM) on genetic response and rate of inbreeding in broiler breeding programs. In the base scheme, selection candidates were evaluated based on direct carcass measurements on relatives. Genetic response and rate of inbreeding for different schemes were investigated using deterministic simulation. In the base scheme the genetic response for breast muscle percentage (BMP) was 0.3% and rate of inbreeding was 0.96% per generation. A scheme with ICM for male and female selection candidates resulted in a 66.2% increase in the response for BMP compared to the base scheme and the rate of inbreeding was reduced to 0.79% per generation. The improved gain resulted from an increase in the accuracy of selection. The accuracy of ICM had consequences on the response for BMP and the rate of inbreeding. In most cases, an accuracy of 30% was already sufficient for higher BMP and lower rate of inbreeding compared to the base scheme.

Session 17 Theatre 10

Base populations in fish breeding programs: A simulation study

M. Holtsmark[1], J.A. Woolliams[1], A.K. Sonesson[2], G. Klemetsdal[1]. [1]Norwegian University of Life Sciences, Department of Animal and Aquacultural Sciences, Ås, Norway, [2]AKVAFORSK (Insitiute of Aquaculture Research), Ås, Norway*

Fish-breeding programs may have to be based on fish from wild sources, and knowledge of the structure and the genetic properties of these populations are often poor. One alternative is to assume that wild fish are unrelated across and within subpopulations. Alternatively, one can assume that the fish within subpopulations are related with an additive relationship of 2F, where F is the average inbreeding coefficient in the population. F can be derived from the additive genetic variances within and between subpopulations, (1-F)Vg and 2FVg respectively, where Vg is the additive genetic variance in an ancient base population. Given this approach we examined the effect of (1) increasing the number of subpopulations the fish were sampled from (1 to 8), and (2) mating design, when establishing the base (mating within, or both within and between subpopulations), on inbreeding, genetic variance and genetic gain. Stochastic simulations assuming an infinitesimal model were used. Selection was performed over 10 generations, using optimal contributions based on phenotypes. The additional genetic gain diminished with increasing number of subpopulations, and sampling fish from five subpopulations resulted in 99 percentage of the accumulated additive genetic gain obtained with eight subpopulations. Mating randomly across and within subpopulations forming the base was slight better than mating within subpopulations.

Session 17 Poster 11

Evaluation of heterosis, general and special combining ability for some biological characters in six silkworm lines

A. Qotbi[1]., A.R. Seidavi[1]., M.R. Gholami[2], A.R. Bizhannia[3], [1]Islamic Azad University, Branch of Rasht [2]Iran Silkworm Corporation, Rasht. [3]Iran Silkworm Researches Center, Rasht

Maternal effects were studied by double crossing between three Chinese (32, 104, 110) and three Japanese (31, 103, 107) pure lines of silkworm *Bombyx mori* L. All produced pf offspring were reared in same conditions and calculated heterosis, general combining ability and special combining ability for some biological characters. From obtained results, cocoon weight, shell cocoon weight and shell cocoon percentage are affected by genetically groups ($P<0.05$). Average of cocoon weight of three Chinese and three Japanese pure lines were 1.67 and 1.64 gr respectively. Heterosis of cocoon weight was between +32.5 to +45.95 for commercial hybrids 104×103 and 32×31 respectively. Special combining ability for cocoon weight at commercial hybrids of 31×32, 32×31, 103×104, 104×103, 107×110 and 110×107 were -0.247, -0.245, -0.031, -0.074, +0.308 and +0.288 respectively. General combining ability of cocoon weight was between +0.246 to -0.298. Finally pure lines of 32 and110 (Chinese) and 31 and 107(Japanese) had the highest and lowest of general combining ability at each genetical groups respectively.

Session 17 Poster 12

Non-additive breed effects on milk production in cattle

J. Wolf, L. Zavadilová and E. Nemcová, Research Institute of Animal Production, P.O.Box 1, CZ 10401 Praha-Uhríneves, Czech Republic*

Crossbreeding effects on milk production traits of Czech dual-purpose and dairy cattle breeds were estimated. Nearly 370000 cows with known gene proportions from Czech Pied, Ayrshire or Holstein cattle were selected from the national milk recording data base. Single-trait animal models with exact solutions including standard deviations for estimates of fixed effects were calculated for milk, fat and protein yield, fat and protein content. The model of Dickerson including additive, additive maternal, heterotic and recombination effects was used for the part of the animal model describing the crossbreeding effects in all calculations. For milk yield, the additive genetic effect (defined as deviation from Czech Pied cattle) was 850 to 900 kg for Holstein and 240 to 480 kg for Ayrshire. The maternal effects were low and negative. Low significant positive heterotic effects were observed being up to approximately 100 kg for Czech Pied x Holstein. The recombination effects were negative and statistically significant for Czech Pied x Holstein. The results for fat and protein yield were similar to the results for milk yield. For fat and protein content, nearly no statistically significant crossbreeding effects were found.
The financial support from the project MZE 0002701401 is acknowledged.

Session 17 Poster 13

Estimates of genetic parameters of final weight at slaughter, yield grade and marbling scores in beef cattle

A.P. Márquez, A.Correa, J.F. Ponce, L. Avendaño, S.C. Ochoa, Instituto de Ciencias Agrícolas-Universidad Autónoma de Baja California, Avenida Obregón y Julián Carrillo S/N, Mexicali, B.C. México. CP 21100*

Genetic parameters for final weight at slaughter FSW, yield grade YG,and marbling score MS were estimated. Carcasses(n=500) of progeny Charolais C, Brahman B, Brangus Br, Hereford H, Angus A, and Holstein Hf inheritance were evaluated. Measurements on rib eye area REA, fat thickness on the ribeye area FTREA, and percentage of kidney, pelvic, and heart fat KPH% were measured. The analytical model included, fixed main effects: of sire breed, sex, slaughter group, dam breed and interaction sire breed x dam. Random nested components of sires within sire breed, and dams within dam breed and residual. FSW values 530, 520, 518, 476, 475, 470, 456, 448, 436, and 435 kg involved inheritance of CxBr, CxB, HfxHf, BrxBr, CxA, BrxA, HxBr, HxB, BxB, and HxA, respectively. The values in REA 87.93, 80.21, 80.26, 78.38, 72.80, 72.80, 65.66, 65.71, 62.2, and 56.28 cm^2 corresponded to progeny CxB, CxBr, HxA, BrxBr, CxA, BrxA, HxBr, HfxHf, HxB, and BxB, respectively. YG values 1.1, 1.2, 1.2, 1.3, 1.4, 1.4, 1.8, 1.8, 1.9, and 2.1 corresponded to carcasses of progeny CxB, CxBr, HxA, BrxBr, CxA, BrxA, HxBr, HfxHf, HxB, and BxB inheritance, respectively. The heritability values (h^2=.45±.05, h^2=.47±.04, and h^2=.36±.06) corresponded to FWS, YG, and MS, respectively.

Session 17 Poster 14

Genetic parameters for type traits of Brazilian Holstein cattle

C.N. Costa[1], J.A. Cobuci[2], A.F. Freitas[1], N.M. Teixeira[1], R.B. Barra[3] and A.A. Valloto[4] [1]Embrapa Gado de Leite, Juiz de Fora - MG, Brazil, [2]CNPq RD Fellow, [3]Minas Gerais Holstein Association, Juiz de Fora - MG, Brazil, [4]Paraná Holstein Association, Curitiba - PR, Brazil*

The Brazilian Holstein Association provides linear type classification services based on the Canadian System in order to guide breeders' decisions for corrective matings and to improve cows longevity under Brazilian herd environments. Data consisting of 64,075 linear classification records from cows classified between 1994 to 2003 from 751 herds were edited for herd size, contemporary group, classifier, stage of lactation, number of classifications/cow and availability of milk production in current lactation. After editing, records of 22 linear traits including Final Classification of 21,208 cows, progeny of 842 sires were used to estimate variance components using univariate analyses by REML. The model included fixed affects of herd-year, season of calving, classifier, age of calving, stage of lactation at classification, and the random additive genetic animal and residual effects. Genetic variances ranged from 0.09 for udder texture to 0.58 for stature. Heritability estimates were lower for udder texture (0.09), foot angle (0.11) and chest width (0.15) and larger for stature (0.40), teat length (0.38) and pin width (0.32). There is a potential for improvement of type traits of Holstein cattle under Brazilian herd environments.

Session 17 Poster 15

Effect of genetic potential for immunocompetence on vaccination efficiency in broilers

B. Ask[1,2], [1]Utrecht University, Department of Farm Animal Health, P.O. Box 80151, 3508 TD Utrecht, The Netherlands, [2]Wageningen University and Research Centre, Animal Breeding and Genetics Group, P.O. Box 338, 6700 AH Wageningen, The Netherlands*

Efficacy of vaccination in neonate chicks is especially in broilers important because of their short lifespan. The efficacy depends on age and the genetic potential for immunocompetence in interaction with environmental factors, e.g. ongoing or previous infection, which may suppress the immune response. Depending on age, vaccination efficacy is complicated by immaturity of the immune system, which influences the ability to respond to vaccination, and by maternal immunity, which inhibits the stimulation of the immune system.
Vaccination efficacy to provide protection in a broiler flock as a whole additionally depends on how large a proportion is protected, and therefore on the genetic composition regarding immunocompetence and infection status of the flock. An increasing proportion of vaccinated broilers will reduce the probability of the spread of disease.
It is possible that timing and dose of vaccination should be different for broilers with different maternal immunity, genetic potential for immunocompetence and immune response, and for broiler flocks with different genetic composition and infection status. A transition-state model, which predicts the immune system development and immune response in a broiler flock given genetic potential and infection status, was therefore developed to gain insight into the optimisation of vaccination schemes.

Session 17 Poster 16

A PC program to analyse the genetic and environmental trends in the Italian Holstein dairy herds

S. Biffani, F. Canavesi, M. Marusi, Associazione Nazionale Allevatori Frisona Italiana (ANAFI), Via Bergamo 292, 26100 Cremona, Italy*

One of the roles of a National Breeders Association is to provide the farmers with tools and information that might help enhance profitability and genetic progress on their herds. Over the last years the Italian Holstein breeders Association has been developing some PC programs to help breeders choose sires, mate cows and heifers, analyse inbreeding level, evaluate environmental and genetic progress. Among the programs offered by ANAFI winPGA, an herd analysis program, represents a useful tool which can help the farmers in the day-to-day management. The Herd Genetic Profile program (winPGA) reviews the genetic, environmental and phenotypic trends for a group of traits over a period of 5, 10 or 15 years as well as the level of inbreeding. It allows a constant monitoring of the genetic and environmental situation of a particular herd, province, region or for the whole country. Thanks to several graphics and tables, which can be printed for a further analysis, the farmer can have a closer look on what it has happened and what it is going on in his herd, offering insight for future decisions. Special attention has been focused on inbreeding and apart from a 10 year trend with its relative annual averages, the program provides additional information about the current inbreeding herd level.

Session 17 Poster 17

Gene flow in animal genetic resources: a study on status, impacts, trends from exchange of breeding animals

K. Musavaya, A. Valle Zárate, Institute of Animal Production in the Tropics and Subtropics, University of Hohenheim (480a), 70593 Stuttgart, Germany*

Baseline data on exchange of livestock genetic resources among countries and within and among regions over time are compiled to create a sound information basis for political negotiations at national, regional and global level.
Bases are qualitative and quantitative analysis of publications and project reports, national and international statistics and reports of country representatives for animal genetic resources of the Food and Agriculture Organisation for cattle, sheep, goats and pigs. Systematic expert consultations lead to complementary information and crosschecking and validation of data.
Accessible information differs considerably in quantity and quality between species, breeds and regions. Animal movements from the perspectives of selected northern and southern countries for selected breeds are quantified over time. Human migration led to animal transfers in the past with high impact. Today, breeding and trade regulations and organisations enhance or inhibit animal movements. Increasingly veterinary regulations constrain transfers. Commercially driven activities tend to have higher impacts than governmental decisions or isolated development projects.
The impact of animal exchange cannot globally be valued positive or negative. Only within specific case studies conclusions on success and failure can be drawn. Main determinant is breed suitability for prevailing production systems. For different stakeholders and aspects, like impact on biodiversity, environment and food security, conclusions differ.

Session 17 Poster 18

Optimization of the Bavarian PIG Testing and Breeding Scheme

D. Habier[1,2] and L. Dempfle[1], [1]Technical University of Munich-Weihenstephan, Department of Animal Science, Alte Akademie 12, 85354 Freising, [2]Bavarian Institute of Agriculture, Institute of Animal Breeding, Prof.- Dürrwaechter-Platz 1, 85586 Poing-Grub, Germany

Previous deterministic studies, using a multi stage truncation selection approach to determine the genetic progress, revealed, that a half sib design is substantially superior to a progeny testing design. The shortcomings of the theoretical approach used here is, that it assumes an infinite population size, neglects the Bulmer-Effect and gives no exact information of the effective population size with different selection schemes. Therefore a Monte Carlo Simulation was performed to overcome these difficulties and to verify the deterministic results.
The results suggest that the additive-genetic variances and the accuracy of estimated breeding values deviate more with the half sib design than with the progeny testing design. This might be due to the Bulmer-Effect. Within the progeny testing designs one has to distinguish between designs where progeny of test sires enter the breeding population or where the progeny are discarded. The genetic progress per generation is in fact reduced in the former case, but the generation interval is also shortened. Thus the half sib design is superior only in comparison to that progeny testing design, where the progeny of test sires do not enter the breeding population. Moreover the half sib design implies a much higher rate of inbreeding. Consequently the results suggest that in the long run there is no advantage from a half sib design.

Session 17 Poster 19

Genetic variability of growth traits in performance testing of bulls

V. Bogdanovic, R. Djedovic, P. Perisic, Institute of Animal Sciences, Faculty of Agriculture, University of Belgrade, Nemanjina 6, 11080 Zemun - Belgrade, Serbia*

Performance test of young bulls represents the first step in the process of genetic improvement of cattle. This procedure is important especially for beef and dual-purpose breeds and for those traits that are measurable in the live animals. In the most case those traits are suitable for direct selection and also they have influence on the level of production later in the life of animals. One of the most important groups of traits in performance testing of bulls is growth traits. In order to estimate heritability and genetic variability of growth traits, data on 371 Simmental performance tested bulls was used. Analysed traits were birth weight, body weight at different ages, average and relative daily gains. Since the structure of data disables implementation of Animal Model procedure, additive genetic variances and heritability of traits were obtained using restricted maximum likelihood (REML) methodology applied to sire model. Heritability estimates for birth weight, test-on (120 days of age) and test-off (365 days of ages) body weight were 0.23, 0.25 and 0.30, respectively. For pre-test average daily gain, average daily gain in test and lifetime average daily gain heritability were 0.27, 0.39 and 0.29, respectively, while heritability for pre-test relative daily gain, relative daily gain in test and lifetime relative daily gain were 0.29, 0.22, and 0.26, respectively.

Session 17 Poster 20

Genetic analysis of the fertility in Hanoverian Warmblood horses

H. Hamann[1], H. Sieme[2], and O. Distl[1], [1]Institute for Animal Breeding and Genetics, University of Veterinary Medicine Hannover, Foundation, Germany, [2]National State Stud of Lower Saxony Celle, Germany*

In a retrospective field study data of artificial insemination in horses were collected over six breeding seasons. The analyses included 13,192 Hanoverian Warmblood mares and 125 stallions, which belonged to the National Stud of Lower Saxony and were only used for artificial insemination. In total 57,950 oestrus cycles with 128,538 inseminations were recorded. The trait in the analyses was the pregnancy rate per oestrus (PRO). The objective was the estimation of genetic parameters for the influence of the paternal, maternal and the direct genetic component of PRO. Beside these genetic components the model regarded the permanent environmental effects of the stallions and mares. The estimated direct genetic variance component explained about 4% of the total variance. The influence of paternal and maternal genetic components were found to be smaller and did not exceed a value of 0.2%. A reduced model including only the paternal or maternal component gave heritability estimates of h^2 = 1.1% for the paternal component and of h^2 = 1.5% for the maternal component, respectively. Genetic correlations between the paternal, maternal and the direct effects were close to zero. Due to the intensive use of AI-stallions in the Hanoverian breeding region and the large number of progeny available for these stallions a breeding value evaluation seems feasible.

Session 17 Poster 21

Genetic relationship between different measures of feed efficiency and its component traits in Wagyu (Japanese Black) bulls

Md. Azharul Hoque, Tetsue Kunieda and Takuro Oikawa, Graduate School of Natural Science and Technology, Okayama University, Okayama 700-8530, Japan*

Genetic parameters of daily gain (DG), metabolic weight (MWT), body weight at finish (BWF), feed intake (FI), feed conversion ratio (FCR), and residual feed intake (RFI) as phenotypic RFI (RFI_{phe}) and genetic RFI (RFI_{gen}) were estimated with REML for the test periods of 112 days in 740 Wagyu bulls. The mean for RFI_{phe} was close to zero and RFI_{gen} was negative. RFI_{phe} and RFI_{gen} were regarded as the same traits and the genetic and phenotypic correlations of FCR with RFI (RFI_{phe} and RFI_{gen}) were high. The genetic correlations among DG, MWT and BWF were unity, while the phenotypic correlations between them were moderate to high. RFI_{phe} has weak genetic relations with MWT and BWF, whereas RFI_{gen} has no genetic relations with them. The genetic correlations between RFI and FI were high, while between FCR and FI was moderate and the heritabilities for RFI were higher than that of FCR, indicating that selection against RFI would lead to a large reduction in FI with little change in growth traits, and selection against FCR would lead to a small reduction in FI with small increase in growth traits. These results suggests that RFI could be an alternative trait for FCR as selection criterion, however, expected correlated response is suggested to be different between them.

Session 17 Poster 22

Genetic and environmental relationships between milk yields at different parts of lactation in Iranian Holsteins

H. Farhangfar and H. Naeemipour, Animal Science Department, Agriculture Faculty, Birjand University, Birjand, P.O. Box 97175-331, Iran*

A total of 17 946 complete first lactation records was utilised to estimate genetic and environmental correlations between total milk yields at different parts of lactation in 17 946 Iranian Holsteins first calved from 1986 to 2001 and distributed at 287 dairy herds of Iran. Three parts of the lactation (days in milk) were defined as M90, M180 and M270. A multivariate animal model considering fixed effects of herd-year-season of calving, age at first calving and lactation length, and also additive genetic random effect, was used to estimate genetic parameters through AIREML algorithm applied in ASREML package. The results obtained in the present research showed that the genetic correlations between M90 and M180, M90 and M270, and M180 and M270 were 0.97, 0.93 and 0.99 respectively. The corresponding environmental correlations were found to be 0.88, 0.79 and 0.95 respectively. The results also revealed that M90, M180 and M270 had the heritability estimates of 0.16, 0.24 and 0.28 respectively which regard to high genetic correlation between M180 and M270, it can be concluded that M180 may be used instead of M270 in genetic evaluation of Iranian Holsteins for lactation milk yield as reduced time of cows maintaining in the herds as well as early proof of youn sires in the progeny test programmes are needed.

Session 17 Poster 23

Non-genetic influence on test day milk and fat yields data of Moroccan Holstein-Friesian cows

A. Tijani, Ecole Nationale D'agriculture de Meknes, B.P. S/40,50001, Meknes, Morocco

A total of 116,634 monthly test-day records of milk and fat yields from 11,278 lactations of 7725 Holstein-Friesian cows from 192 herds in 17 areas of Morocco involved in the official milk recording program were used to investigate environmental effects on test-day milk and fat yields. Mean and standard deviation for milk and fat yields were 17.6 kg and 6.5 kg, 636.8 g and 222.8g, respectively. Mixed linear model was used in order to explain total variation. The effects of the area of production, herd, age at calving, year and season of calving, stage of lactation were very highly significant ($p<0.0001$) for test day milk and fat yields. Day in milk (DIM) was the most contributing effect to the variation followed by region of production and age at calving. Milk and fat yields were high at the beginning in the lactation and decreased since the 54 day in milk. Differences between the beginning and the end of lactation for milk and fat yields were 10.1 kg and 123.8 g, respectively. Test day milk and fat yields were high in spring and low in Summer months.

Session 17 Poster 24

Genetic parameters of fur coat and reproduction traits for Polish arctic foxes

H. Wierzbicki[1] and W. Jagusiak[2], [1]Agricultural University, Wroclaw, Poland, [2]Agricultural University, Krakow, Poland

Genetic parameters for fur coat and reproduction traits in Polish arctic foxes were estimated using records of 5540 individuals. An animal model with DFREML procedure was used. Random effects were additive genetic, common litter environment and residual. Estimates of direct heritability and portion of litter variation ranged from 0.11 and 0.08 for skin length to 0.28 and 0.14 for hair length, respectively. Estimates of direct genetic correlations ranged from 0.95 between skin length and animal size to -0.31 between colour purity and general appearance, while estimates of phenotypic correlation ranged from 0.64 between skin length and animal size to -0.26 between colour purity and hair length. Estimates of correlation between common litter environment effects ranged from 0.94 between skin length and animal size to -0.37 between skin length and hair length. Estimates of heritability for reproduction traits ranged from 0.08 for pregnancy length to 0.17 for litter size at birth. High favourable genetic correlations were found between litter size at birth and litter size at weaning (0.95), and between litter size at birth and number of dead pups (0.78). High unfavourable genetic correlation was estimated between pregnancy length and pup weight at weaning (-0.97). Phenotypic correlation estimates ranged from 0.92 between litter size at birth and litter size at weaning to -0.39 between litter size at birth and pup weight at weaning.

Session 17 Poster 25

Effect of mating ratio on response for a selection index

J.L. Campo, S.G. Dávila and I. Peña, Instituto Nacional de Investigación Agraria, Departamento de Mejora Genética Animal, Apartado 8111, 28080 Madrid, Spain*

Three lines (I, IH5, and IH10) were selected to improve simultaneously the weights at 21 and 28 days in *Tribolium*, using three alternative mating ratios (1:1, 1:5, and 1:10). The selection criteria were conventional indexes based on individual performances of both weights (I line), or on individual and sibs performances (IH5 and IH10 lines). The sibs performance was a mixture of full sibs and half sibs. The selected proportion was 25% in females, whereas in males it was 25%, 5%, and 2.5% (I, IH5, and IH10 lines). Genetic change in each generation was calculated from the breeding values of individual animals, which were estimated using an animal model incorporating the best linear unbiased prediction. There were significant differences among lines, selection response being smaller in the IH5 line (1.31 ± 0.16). Responses in the I and IH10 lines were similar (2.77 ± 0.15 and 2.81 ± 0.10). The expected responses to selection predicted similar results in the IH10 and IH5 lines followed by the I line. Results indicate that the advantages of using mating ratios of 1:5 or 1:10 for balancing selection response and inbreeding, and sibs information for increasing the accuracy of selection, were not apparent, the individual index line with a mating ratio of 1:1 being similar or better than the family indexes.

Session 17 Poster 26

An investigation on the erythrocyte potassium polymorphism and relation between several Mohair characteristics in Angora Goat *(Capra hircus)*

M.I. Soysal[1], E.K. Gürcan[1], E. Özkan[1], M. Aytaç[2] and S Özkan[3], [1]Trakya University , Tekirdag Agricultural Faculty , Department of Animal Science, 59030 Turkiye, [2]Lalahan Livestock Central Research Institute, Lalahan, Ankara, Turkiye, [3]Güdül Ankara, Turkiye*

The main purpose of this research is to study the blood sodium and potassium levels of Angora goat population in terms of genetically structure raised at Boyali Village of district of Güdül of province of Ankara and Araç district of Kastamonu province.The Angora Goats in this region were famous for their hair. By taking 10 cc blood from the neck vein (W. Jungularis) of 73 Angora goats the sodium (Na), potassium (K) levels types were determined.The blood potassium concentration less than 20 Meq/l called lower potassium concentration (LK) and over the same level called as higher potassium concentration (HK).

The gene frequency for (LK) and (HK) were estimated. At the result of the tests for conformity with the genetically balance the conformity with Hardy-Weinberg rule were observed .The relationship between several fibre characteristics such as tenacity (gr/des), work to rupt (gr/cm), force to rupture (g), length (cm), slenderness (micron), elongation (percentage of fibre), body weight (kg) and (K) types were also analysed.

Session 17 Poster 27

Genetic parameters for carcass traits of field progeny and their relations with feed efficiency traits of their sire population for Japanese Black cattle

Md Azharul Hoque[1], Takuro Oikawa[1] and Keiji Hiramoto[2], [1]Faculty of Agriculture, Okayama University, Okayama 700-8530, Japan, [2]Okayama Prefecture Animal Industry Center, Okayama, Japan*

Genetic parameters for carcass traits of 1,774 progeny (1,281 steers and 493 heifers), and its genetic relations with feed conversion ratio (FCR) and residual feed intake (RFI) of their sire population (740 bulls) were estimated with REML. Progeny traits were carcass weight (CWT), rib eye area, subcutaneous fat (SFT), yield estimate (YEM), marbling score (MSR), meat quality grade, meat color (MCL) and meat texture (MTX). The estimated heritability for CWT was high (0.70) and heritabilities for all other traits were moderate (ranging from 0.33 to 0.47), except for MCL and MTX (low heritable). YEM genetically highly correlated with MSR (0.60), suggesting that simultaneous improvement of high yield carcass and beef marbling is possible. Genetic correlations of RFI of sires with CWT and MSR of their progeny were favorably negative and between RFI and SFT was positive indicating that selection against RFI of sires may have contributed to produce heavier carcass and increase in marbling without increasing unexpected subcutaneous fat of their progeny. RFI correlated favorably stronger with CWT and fatness than those of FCR with them. This experiment provides evidence that selection against RFI would be better than selection against FCR in sire population for getting higher correlated responses in carcass traits of progeny.

Session 17 Poster 28

Genetic parameters of body weights and carcass traits in two quail strains

N. Vali, M.A. Edriss, H.R. Rahmani and A. Samie, Dept. of Animal Sciences, Isfahan University of Technology (IUT), Isfahan, Iran*

In order to estimate genetic parameters for body weights and carcass yields of two strains of quails; 32 pairs of Japanese quails (coturnix Japanese) and 26 pairs of range quails (coturnix ypsilophorus) were randomly selected from the base populations of parents. Produced progenies (650 full and half-sibs) were used to estimate genetic parameters of body weights and carcass characteristic traits. Data were analyzed using SAS (1997) and DFREML (Meyer, 1998). Body weights of two strains at 35, 42, 49, days of ages were significantly different, while body weights at 63 days of age were not significant ($P>0.05$). Carcass weight, carcass percent, breast weight and thigh weight were significantly affected by strain source of variations ($P<0.01$). Estimated heritability for different traits were from 0.030 ± 0.090 for percent of breast weight of coturnix Japanese to 0.787 ± 0.406 for thigh weight of coturnix ypsilophorus. Genetic correlations among body weights at 35, 42, 49 and 63 days of age and also among carcass traits (carcass weight, breast weight, percent breast and thigh weight) were all positive and high while genetic correlations for breast percent and carcass percent were low. Genetic correlation of thigh percent with the other considered traits were negative except body weight at 42 days of age which tended to be low.

Session 17 Poster 29

Weaning performance of Hungarian Grey Calves

Barnabás Nagy[1], Zoltán Lengyel[1], Imre Bodó[2], István Gera[2], Márton Török[1], Ferenc Szabó[1],
[1]*University of Veszprém Georgikon Faculty of Agriculture Hungary,* [2]*Association of Hungarian Grey Cattle breederseHungary*

Weaning performance of 2857 purebred calves (660 male and 2197 female) born from 1498 cows mated with 78 sire were analised in seven farms. Genetic- and environmental variance and heritability, breeding value of weaning weight (WW), preweaning daily gain (PDG) and 205-day weight (CWW) were calculated. Farm, year of birth, season of birth, sex, number of calving as fixed, while sire as a random effect was treated. Data were analyzed with SPSS 9.0 and Harvey's (1990) Least Square Maximum Likelihood Computer Program. The environmental factors examined had an effect on all traits. The overall mean value and standard error (SE) of WW, PDG and CWW were 208±3.31 kg, 887±15.66 g/day and 199±14.774 kg, respectively. The heritability of the investigated traits was 0.24, 0.25 and 0.25. The results of the examination show that the 205-day weight was increased to seventh calving. Additive and multiplicative correcting factors were calculated. The phenotypic correlations between WW and the age of calves at weaning was r_p=0.47. The male calves were hevier than females, the difference was 22 kg (10,5%).

Session 17 Poster 30

Polymorphism detection in bovine Stearoyl-CoA desaturase (SCD) *locus* by means of microarray analysis

G. Conte[1], B. Castiglioni[2], M. Mele[1], A. Serra[1], S. Chessa[3], G. Pagnacco[3] and P. Secchiari[1],
[1]*DAGA, Sezione Scienze Zootecniche, Università di Pisa, via del Borghetto 80, 56124 Pisa, Italy*
[2]*IBBA-CNR, via Bassini 15, 20133 Milano, Italy,* [3]*VSA, Università di Milano, via Trentacoste 2, 20100 Milano, Italy*

Variability in the SCD genotype is one of the major sources of genetic variation in milk fatty acid composition. Two types of this gene have been observed, in which three SNPs in the ORF region were predicted to cause an amino acid replacement from valine (V) to alanine (A). We describe an application of the microarray technology for the polymorphism identification at the SCD gene. This method, based on the discriminative properties of DNA ligation detection reaction (LDR), combines enzymatic processing with "tag sequences" hybridisation, in Universal array. Genomic DNA was extracted from 11 Somba (African breed) cows milk samples and the 5th exon of the SCD gene was amplified by PCR using locus-specific primers. The resulting PCR products are subjected to LDR reaction and then hybridised onto Universal Array. We found 5 samples homozygous for the allele V and 6 heterozygous (AV). No samples were detected homozygous for the allele A. Genotype was assigned by comparing the microarray results with those obtained both by SSCP and direct sequencing. This work showed that such microarray assay is an efficient method to the SNPs identification at the SCD *locus.*

250.

Value of traits in beef cattle breeding

J. Pribyl[1], J. Pribylova[1], L. Stadník[2], P. Safus[1], M. Stipkova[1], Z. Vesela[1], M. Wofova[1], [1]Res.Inst.Anim.Prod., P.O.Box 1,Uhríneves 10401 Czech Rep., [2]Czech Agricultural Univ. 16521 Prague, Czech Republic*

Selection index was constructed for bulls of beef breeds. Breeding aim is composed of direct and maternal effects for calving ease, calf losses, growth, carcass value, fertility and longevity of cows (20 items in total). Breeding values routinely calculated in the Czech Republic are a source of information: direct and maternal effects for calving ease and growth (10 values), daily gain of bulls in performance-test stations (1 value) and body conformation (10 values). Economic values are used with discount of 10% and without discount. Calf losses, fertility and longevity of cows, for which suitable information from performance testing are missing, were not included in total genotype for the index. Several indexes were constructed according source of information; only for the most important traits, or for all traits with known breeding values. Discount did not have a significant influence on selection indexes. Selection according to indexes is almost exclusively expressed by genetic gain of direct effects. Maternal effects have low significance in the breeding aim. Among all traits, direct effects for weight gain until weaning and after weaning have the highest significance in the selection aim, accounting jointly for 95-97% of total effect. As a source of information direct and maternal effects of weaning weight accounts for 90-97% and 3-7%.

New microsatellites assignment using a hamster-sheep cell hybrid panel

N. Tabet-Aoul[1], R. Ait-Yahia[2], A. Derrar[2], N. Boushaba[2], N. Saidi-Mehtar[1], [1]Département de Biotechnologie, Université d'Oran, Es-Senia, Algeria, [2]Laboratoire de Génétique Moléculaire et Cellulaire, USTO, Oran, Algeria

A panel of 24 hamster-sheep somatic cell hybrids had been previously obtained by Saïdi-Mehtar et al., 1981 and it was investigated for sheep gene mapping. Characterization of this panel was performed using PCR method with 229 markers (genes or microsatellites) selected from sheep, cattle and goat maps. 146 chromosomal fragments of different lengths were defined.

In the present study, 21 additional microsatellites were analyzed by PCR in this hybrid panel. Synteny was identified according statistical rules from Chevalet and Corpet, 1986.

The results allowed new localization of these microsatellites on 9 different sheep chromosomes: OAR2, 3, 5, 6, 11, 21, 24, 26 and X.

This study contribuated to hybrid panel characterization by detecting new breakpoints on the following ovine chromosomes: OAR2, 5, 11, 21, 24 and X, defining new chromosomal fragments. Thus, the total number of chromosomes fragments was increased to 152.

Session 17 Poster 33

General combining ability estimates in silkworm inbreed lines (*Bombyx mori l.*)

Georgeta Dinita[1], Carmen Antonescu[2], A. Marmandiu[1], Minodora Tudorache[1], [1]University of Agronomical Sciences and Veterinary Medicine Bucharest, Faculty of Animal Sciences, 59 Marasti Blvd., 1 district, Romania, [2]Institute of Research and Developement for Apiculture, Bucharest 41, Ficus Blvd., 1 district, Romania

The general combining ability of the 30 silkworm inbreed lines in I_4 has been evaluated. For that purpose the inbreed lines were crossed with Tester and 60 silkworm hybrids resulted. According to the results obtained in hybridization process, 8 inbreed lines were selected, with the following performances: pupation percentage 79.76-84.10 %, cocoon yield/10.000 larvae 15.0-18.3 kg, raw cocoon weight 1.87-2.38 g, shell weight 0.414-0.550 g, silk content 21.46-24.16 %, filament length 1132-1218 m, filament size 2.36-3.10 denyer, raw silk percentage 16.40-18.10 %.

Session 17 Poster 34

Algebraical and geometrical interpretation of restricted best linear unbiased prediction of breeding values

M. Satoh and M. Takeya, National Institute of Agrobiological Sciences, Tsukuba-shi 305-8602, Japan*

The restricted best linear unbiased prediction (restricted BLUP) procedure estimates what are defined as restricted breeding values (RBV). The aim of the present study is to represent the concept of RBV algebraically and geometrically. The results showed that RBV based on a restricted BLUP procedure imposing the same restrictions on all animals should be the same as those imposing restrictions on only some animals. The set of RBV was represented as the $(q - r)$-dimensional restricted hyperplane with a specific RBV represented as the point of intersection between the restricted hyperplane and the r-dimensional vector space that parallels $\mathbf{G}_0\mathbf{C}_0$ and passes through the breeding value (BV), where q and r are the numbers of traits and restrictions, respectively, $\mathbf{G}_0$ is an additive genetic variance-covariance matrix for the q traits and $\mathbf{C}_0$ is a q x r restriction matrix with full column rank. The relationships between BV and RBV for two and three traits are illustrated with several examples. The relationship between RBV and BV was found to be applicable to four traits or more.

Session 17 Poster 35

Direct genetic, maternal genetic and common environmental effects on Landrace and Duroc piglet growth

K. Suzuki[1], H. Kadowaki[2] T. Shibata[2], H. Uchida[3] and A. Nishida[1], [1]Graduate School of Agricultural Science, Tohoku University, Sendai, Japan, [2]Miyagi Prefecture Animal Industry Experiment Station, Iwadeyama, Miyagi, Japan, [3]Miyagi Agricultural College, Sendai, Japan*

This study investigates direct (h^2) genetic effect, maternal (m^2) genetic effect, and litter environmental (c^2) effect on early growth for Landrace (L) and Duroc (D) breed piglets. Body weights at birth, 1, 2, 3, 4, 5, and 8 weeks of age were available for 2998 Landrace piglets and 2606 Duroc piglets derived from selection experiments during seven generations. Daily weight gain data during 0-3, 3-5 and 5-8 weeks were used. Piglet body weights after three weeks of age and daily gains over 3-5 weeks of age, and during 5-8 weeks of age were significantly greater for Duroc than for Landrace. Landrace piglet body weights did not change with selection generation except at 5 weeks of age, but those of Duroc increased with selection generation after 4 weeks of age. The h^2 in Landrace changed from 0.02 to 0.10; Duroc h^2 changed from 0.10 to 0.40. The Landrace m^2 decreased from 0.20 to 0.05; that for Duroc changed from 0.2 and 0.35. The c^2 in Landrace changed from 0.2 to 0.35, whereas that for Duroc changed from 0.10 to 0.25. After weaning, h^2 and m^2 were reversed ($h^2 > m^2$) in both breeds. Negative genetic correlation between direct and maternal effects increased with age in Duroc, but the direction changed in Landrace from negative to positive with age after three weeks.

Session 17 Poster 36

Simulation of multiple trait data for testing breeding value estimation programs

M. Wensch-Dorendorf[1], H.H. Swalve[1] and J. Wensch[2], [1]Institute of Animal Breeding and Husbandry, University of Halle, Halle/Saale, Germany, [2]Institute of Mathematics, University of Potsdam, Potsdam, Germany*

Computer programs to be used for routine genetic evaluations need to be thoroughly tested and checked before use. Testing of such programs may require simulated data with known phenotypes and breeding values. Using a method under which the breeding values for base animals are generated and the breeding values of the offspring are derived from their respective parents such that the solutions to the mixed model equations yield the exact values of the simulated ones, we extended this method from the univariate to the multivariate case. For the multivariate case, covariance matrices are needed such that the data generated satisfy the Mixed Model Equations. The covariance matrices can be chosen arbitrarily but include variances that can be fixed to desired values. In order to guarantee the covariance matrices to be positive definite, we apply similarity transformations with random Householder reflections combined with a final diagonal scaling to a suitable initial matrix.

Session 17 Poster 37

Estimation of the milk urea course during lactation

E. Nemcová[1], M. Stípková[1], F. Jílek[2] and M. Krejcová[1], [1]Research Institute of Animal Production, P.O.Box1, 104 01 Praha 10, Czech Republic [2]Czech University of Agriculture Prague, Czech Republic*

This work was part of a project focused on the description of the relationship between milk urea and reproduction parameters. The aim of this study was to model the course of the urea concentration in the milk during the lactation. Approximately 19,000 test-day records from official milk performance recording from 5 selected Holsteins herds in Czech Republic were available. Several test-day models differing in the definition of the factors and factor levels were compared. All effects were considered to be fixed and each parity was treated in a separate analysis. The model including fixed effects of herd-test-day, season of calving, age at first calving (first lactation only), length of previous calving interval (2nd and subsequent lactations only) and the animal effect modeled by a 4th order Legendre polynomial was chosen to be the best. For all lactations, the milk urea concentration showed the lowest values in the beginning and at the end of lactation, an increase of these values during the first 3 months of lactation and a slow decrease in next months were observed. The smallest level of milk urea concentration was found in the first lactation and the highest in the second one. This analysis confirms our previous findings from animal model about the statistical significance of the animal effect, days in milk and HTD.

Session 17 Poster 38

Estimates of genetic parameters for reproduction and production traits of purebred Berkshire in Japan

M. Tomiyama[1], T. Oikawa[1], T. Sano[2], T. Arakane[2] and H. Mori[2], [1]Faculty of Agriculture, Okayama University, Okayama, Japan, [2]Okayama Prefectural Center for Animal Husbandry and Research, Okayama, Japan.*

Pork produced from three-way cross is common for daily meat market and annual production of purebred Berkshire is about 180,000 (2.2% of total population) in Japan. The aim of this study is to initiate genetic improvement of this Berkshire was population by estimated parameters. Records on 2956 (1,537 males, 1,419 females) pigs from Okayama prefecture were used in this study. Traits studied included reproduction traits of boars and sows, and body weight, daily gain, meat production and number of teats (TEAT) of piglets. Genetic parameters were estimated by VCE-5. Heritabilities of back fat thickness (BFT), loin eye area (LEA) and TEAT were 0.46, 0.70 and 0.55, respectively in piglet. With bi-variate model heritabilities of direct genetic effect for birth weight, weaning weight and body weight at 60 days of age (W60) were 0.09, 0.11 and 0.33, respectively. Heritabilities of maternal genetic effect for the above traits were 0.18, 0.12 and 0.03, respectively. Thus influence of maternal effect was larger than direct effect before weaning. Estimated genetic correlation between W60 and final weights at average age of 200 days and between BFT and LEA were 0.70 and -0.93, respectively. Moreover, heritabilities for reproduction traits ranged from 0.13 to 0.25 for sows and from 0.15 to 0.35 for boars.

Non-genetic factors effect on body weights and grease fleece weight in Loribakhtiary sheep flock

E. Asadi-khoshoei[1], S.R. Miraei-Ashtiani[2], [1]Department of Animal Science, University of Shahrekord, Iran, [2]Department of Animal Science, University of Tehran, Iran

Data of 3218 lamb records obtained over a 9-year preiod (1989-1998) from the Loribakhtiary sheep flock at the Loribakhtiary Experimental Station of Iran, were used to investigate factors influencing birth weight, weaning weight and body weights at 6, 9, 12, 18 months as well as fleece weight at 16 months of ages. Year of birth, sex, birth/weaning status of lamb were significant ($p<0.01$) source of variations for fleece weight and body weights at all ages. Weights at birth, weaning, 6, 9 and 12 months was significantly ($P<0.01$) influenced by age of dam. Age of lamb was significant ($P<0.01$) on all body weights and fleece except weight at 18 months of age. Significant ($P<0.01$) two-way interactions were found between year and sex on weaning weight, 6, 9 as well as fleece weight and also type of birth and sex and type of birth and age of dam($P<0.01$) and year of birth and type of birth ($P<0.05$) on birth weight. Least square means(kg) for body weight were 4.60± 0.02, 26.01± 0.20, 37.8± 0.27, 51.11± 0.47, 57.23± 0.50 and 77.77± 0.57 respectively at birth, weaning, 6, 9, 12 and 18 months of age, and 2.1± 0.35 for fleece weight. This results was compared to reported in the scientific sources.

Keywords: Fat-tailed sheep, non-genetic effects, body weight

Genetic parameters of calving interval in Japanese Holstein

O. Sasaki[1], M. Aihara[2], K. Hagiya[3], K. Ishii[1] and Y. Nagamine[1], [1]National Institute of Livestock and Grassland Science, Tsukuba, Japan, [2]Livestock Improvement Association of Japan, INC., Tokyo, Japan, [3]National Livestock Breeding Center, Nishigo, Japan

The control of calving intervals is important to manage dairy farming. The objectives of this study were to estimate the influence of management on the calving interval and genetic correlations between calving interval and milk production traits. Data of milk yield, fat yield and calving interval were obtained from Livestock Improvement Association of Japan. They included 2,994,826 records on 1,353,914 Holstein cows calving between 1991 and 2000. Multi-trait animal model was applied to analyze all traits and genetic parameters were estimated using the statistical software MTC. The herd was divided randomly into ten groups and the parameters were estimated within group. The mean values of milk yield, fat yield, and calving interval were 8,321±1,728kg, 319±68kg, and 415±80 days, respectively. The heritabilities of milk yield, fat yield, and calving interval were 0.278±0.007, 0.256±0.007, and 0.063±0.004, respectively. Genetic correlation, milk yield and fat yield 0.358±0.015, milk yield and calving interval 0.437±0.031, and fat yield and calving interval 0.431±0.026, were estimated. The genetic ability of calving interval increased with increment of milk yield and fat yield. The reliability on the estimator of genetic ability of calving interval was increased by using multi-trait model with milk yield and fat yield.

Session 17 — Poster 41

Estimation of genetic parameters for early growth traits in the Fat-tailed Lori-Bakhtiary lambs

E. Asadi-Khoshoei, Department of Animal Science, Shahrekord university, shahrekord, Iran

This study was conducted on 2522 records of Lori-Bakhtiary lambs during of nine years (1989-1997). Genetic parameters were estimated for birth weight (BW), weaning weight (WW) and pre weaning average daily gain (ADG) using derivative-free restricted maximum likelihood (DFREML) procedures with animal models. Six different models were fitted with or without maternal (genetic or permanent environment) effects. Estimates of direct heritabilities were 0.23, 0.10 and 0.08 for BW, WW and ADG, respectively. Estimates of maternal heritability for BW was 0.24 when only maternal genetic effect fitted in the model, but decreased to 0.16 when maternal permanent environmental effect was fitted. The maternal heritability for WW and ADG were 0.04 and 0.09, respectively. The maternal permanent environmental variance as a proportion of phenotypic variance were 0.07, 0.10 and 0.09 for BW, WW and ADG, respectively. The estimation of correlation between direct and maternal genetic were 0.25, 0.20 and 0.01 for BW, WW and ADG, respectively. This results indicated that the maternal component should be accounted to the selection of Lori-Bakhtiary lambs (males and females).

Session 17 — Poster 42

Prediction of the genetic progress applying embryo transfer into the breeding programs in cattle

A. Marmandiu[1], Mihaela Nedelcu[2], Monica Pîrvu[1], Georgeta Dinita[2], I. Raducuta[2] University of Agronomical Sciences and Veterinary Medicine Bucharest, [1]Faculty of Veterinary Medicine, [2]Faculty of Animal Sciences, 59 Marasti Blvd., 1 district, Romania

Using modelling into an active population of 100000 cows, the genetic progress for milk quantity applying MOET on the way of bull dams, maintaining bulls selection on progenies and applying collateral selection (full sibs; full sibs and half sibs) was estimated. The annual genetic progress presented a maximum of ΔG=2.7479%, using $MOET_{PROGENIES}$ as comparison to ΔG=2.4560% for $MOET_{FULL\ SIBS}$, and ΔG=2.6154% using $MOET_{FULL+HALF\ SIBS}$. Although the generation intervall decreased by about 3.6 years on the way of bull sires, MOET and bulls selection on collateral diminished the accuracy in bull sires selection, fact that determined a lower value of the annual genetic progress. The result recommends application of MOET in bull dams and maintenance of bulls selection on progenies. MOET programs constitute a viable alternative to the classical programs, the gain in the annual genetic progress varying from ΔG=+16.95% (progenies selection) to ΔG=+4.52% (full sibs selection), respectively ΔG=+11.30% (full sibs and half sibs selection).

Session 17 — Poster 43

Genetics of size traits, fur quality traits and litter size in blue fox

J. Peura, I. Strandén and E.A. Mäntysaari, MTT Agrifood Research Finland, Animal Production Research, 31600 Jokioinen, Finland*

The goal in Finnish blue fox breeding is to increase litter size and pelt size and improve fur quality. To set proper breeding goals, heritabilities and genetic correlations for fertility, size traits and fur quality traits were estimated. One fertility trait (litter size), four pelt character traits (pelt size, pelt colour darkness, pelt colour clarity and pelt quality) and six grading traits (animal size, grading colour darkness, grading colour clarity, under fur density, guard hair coverage and grading quality) were analysed. The data included 54 680 animals born between years 1987-2002 from 7 farms. The heritabilities were high for pelt colour darkness and grading colour darkness, moderate for pelt size and low for other traits. In general, heritabilities were higher for pelt character group than in grading group. Moderate antagonistic genetic correlation was found between litter size and pelt size. Genetic correlations between litter size and fur quality traits were low. Genetic correlations within pelt character group were low (~0.10) and within grading group mainly moderate or high (~0,44). High genetic correlations between pelt darkness and grading darkness, between pelt quality and grading density, between pelt size and animal size, between pelt quality and grading quality and between pelt colour darkness and grading guard hair coverage suggest, that selection of pelt character traits via grading traits is relatively effective.

Session 17 — Poster 44

Effect of small-herd clustering on the genetic connectedness of the Portuguese Holstein cattle population

J.Vasconcelos[1,2] and J.Carvalheira[1,2], [1]Centro Investigação em Biodiversidade e Recursos Genéticos (CIBIO), Universidade do Porto, Rua Padre Armando Quintas, 4485-661 Vairão, Portugal, [2]ICBAS-Universidade do Porto, Portugal*

Contemporary groups in test-day models may have unacceptable small number of observations (e.g., less then 4) depending on the herd size. Clustering these herds based on, for example, the simultaneous similarity between phenotypic mean and STD may prevent the loss of this information. However, these techniques may induce a false increase in genetic connectedness (GC), inflating the accuracy of EBV. The aim of this study was to evaluate possible effects of clustering herds on the GC of the Portuguese Holstein cattle. Test-day data from 1997 to 2003 (194,980 animals) was used to compute the degree of GC by 2 methods: number of direct genetic links between management groups (GLT) and genetic drift variance (GDV). Both methods were applied to 3 datasets: D1-containing all herds, D2-same as D1, but small-herds were clustered according to the criteria above and D3-same as D1, but small-herds were removed from the analysis. Results from GLT for all datasets indicate an increase in GC from 1997 to 2001, levelling since then. Correlations for GLT between all datasets were high (between 0.9 and 1), suggesting that clustering had a negligible effect on GC. The GDV method gave similar results as GLT, reinforcing the previous conclusion. Further studies are necessary to validate the use of clustering techniques on genetic evaluations.

Session 17 — Poster 45

The interdependence of some morphological traits in Frasinet carp breed

C. Nicolae, L.D. Urdes and I. Cringanu, University of Agricultural Sciences and Veterinary Medicine, 59, Marasti Blvd., sector 1, Bucharest, Romania*

Genetic progress per generation depends on heritability, number of considered traits and also the correlations among them, especially the genetic ones. Concerning fish breeding programmes, one of the major problems is the selection for the body shape and weight, which affect the meat production. We have studied how body weight is correlated with the main traits, which determine body shape in Frasinet carp breed (body length, maximum body height and maximum height/length ratio). The experiment was carried out for three years. From the genetic point of view, at 137 days age, body weight was positively correlated with body length and with maximum body height but negatively with the maximum body height/length ratio, showing the existence of some antagonistic relations between the two traits. Over the next two years of life, we found the same tendencies as at the first age. The obtained results have consequences for the selection response expected when selecting on body weight.

Session 17 — Poster 46

Impact of pedigree errors on the genetic gain in a dairy cattle population

K. Sanders, J. Bennewitz and E. Kalm, Institute of Animal Breeding and Husbandry, University of Kiel, D-24098 Kiel, Germany*

Wrong (WSI) and missing sire information (MSI) are well known pedigree error problems in the dairy breeding value estimation. In the present study a marker assisted estimation of WSI in the German Angeln dairy cattle population was implemented. The red Angeln breed is a small dairy cattle breed located in the North of Germany. Five paternal half-sib families with a total of 805 daughters were selected. The family size ranged between 123 and 199 daughters. Blood samples of the daughters were collected on different farms and semen samples of the sires were taken. Sixteen high polymorphic microsatellite markers were used to determine the genotypes of the daughters and their putative sires. The estimated proportion of WSI in the Angeln population was 7% and the proportion of MSI was around 10%. The effect of WSI and MSI on the reliability of estimated breeding values and on the genetic gain was investigated using deterministic calculations. Different values for heritability, the number of progenies, WSI, and MSI were taken. A higher number of progenies and a higher value of heritability reduced the influence of WSI and MSI. The effect of WSI on the reliability was higher than of MSI. It was shown that WSI and MSI accumulated their effect on the genetic gain.

Session 17 Poster 47

Modelling post-weaning growth of the Avileña Negra-Ibérica beef cattle breed under commercial using random regression

C. Díaz, C., A. Moreno, A., M. Serrano., M.J. Carabaño*
INIA, Dpto de Mejora Genética Animal, Ctra de la Coruña km 7,5. 28003, Madrid, Spain

The goal of this study was to define a model for the genetic evaluation of animals of the Avileña Negra Ibérica beef cattle of the performance of animals during fattening. Data consisted of 22,186 post-weaning weights recorded on 5113 males calves collected under commercial conditions. Due to the commercial conditions, age of entry and length of fattening period are highly variable. Regressions on age at recording using Legendre Polynomials were used to describe changes of weight along time in terms of systematic environmental factors and random animal and permanent environmental components. Regression on days in fattening at date of recording was also included Statistical analysis was carried out in a Bayesian framework. Several models were compared. Models differ in the contemporary group definition, in the way conditions previous to fattening (herd of origin and weight and age at entry classes) were considered, and in the degree of the polynomial (2 or 3). Models were compared using Bayes Factors (BF) and cross-validation predictive densities (PD). Both criteria determined that the third degree polynomial was the best. However, BF and PD identified different models as the best ones when comparing models within the same degree of polynomial. Models also showed major differences in terms of estimates of genetic parameters.

Session 17 Poster 48

Responses to seven generations of selection for ovulation rate or prenatal survival in Large White pigs

A. Rosendo[1], J. Gogué[2], T. Druet[1], J.P Bidanel[1], INRA, [1]Station de Génétique Quantitative et Appliquée, 78352 Jouy-en-Josas Cedex, [2]Domaine de Galle, 18520 Avord, France

Effects of selection for ovulation rate (OR) or prenatal survival (PS) were examined using data from three pigs lines derived from the same base Large White population. Two lines were selected for seven generations on either high OR at puberty (OR line) or high PS corrected for ovulation rate in the first two parities (PS line). The third line was an unselected control (C) line. Genetic parameters for OR at puberty (ORP) and at mating (ORM), PS either corrected (CPS) or not for ORM (PS) and total number of piglets born per litter (TNB) were estimated using REML methodology. Responses to selection were estimated by computing differences between OR or PS and C lines at each generation using both least-squares and mixed model methodology. Average genetic trends were computed by regressing line differences on generation number. Heritabilities estimates were 0.36±0.03, 0.31±0.03, 0.14±0.02, 0.14±0.03 and 0.15±0.03 for ORP, ORM, PS, CPS and TNB respectively. ORP and ORM were strongly correlated (0.83) and had moderately negative genetic correlations (-0.25 and -0.45, respectively) with PS and TNB (0.41 and 0.42, respectively). Average genetic trends in OR and PS lines were, respectively, 0.42±0.079 and 0.15±0.15 for ORP; 0.46±0.02 and 0.04±0.09 for ORM. Responses to selection were similar for both ovaries. No significant difference was found for PS, CPS and TNB in the two lines.

Session 17 Poster 49

New nucleotide sequence polymorphism within 5'noncoding region of the bovine receptor estrogen α (ER α) gene

Tomasz Szreder, Lech Zwierzchowski, Institute of Genetics and Animal Breeding, Polish Academy of Sciences, Jastrzebiec, 05-552 Wólka Kosowska, Poland

Estrogens play an important role in female reproduction, in cellular metabolism, and in the development, growth and differentiation. Estrogen receptors and their genes are considered candidates for the markers of production and functional traits in farm animals, including cattle. Known are two isoforms of the estrogen receptor - α and β. Each of them is coded for by separate genes, localised on different chromosomes; in the human genome on the chromosomes 6 and 14, respectively. In the present study, basing the sequences of the human, ovine, and porcine ER genes, available in the GenBank database, we designed sets of PCR primers and used them to amplify the bovine Er α gene 5'-region. Seven overlapping fragments of 5' region of the bovine Erα gene were amplified and then sequenced. Altogether, these fragments were composed in the 2853-bp sequence which was deposited in the GenBank database under accession no AY340597. The sequenced fragment included the noncoding exons A, B, C, their putative promoters, and a part of the coding exon 1. Using this sequence we identified, for the first time, a polymorphism within 5' region of the bovine Erα gene - A/G transition, which could be recognized with RFLP *BglI*, lying upstream to the exon C. In the promoter preceding the exon B we identified another polymorphism - an A/G transition at position -967, recognised by *SnaBI*.

Session 17 Poster 50

Effect of breeding value of bulls and performance traits of their daughters

B. Sitkowska, S. Mroczkowski, University of Technology and Agriculture, Genetics and Animal Breeding, Zootechnical Department, Mazowiecka 28, 85-084 Bydgoszcz, Poland

The research was carried out based on breeding documentation about 102 bulls and their 1562 daughters. The research covered 305-day lactations and lifetime milk yield of daughters. The numerical data were verified statistically with variance analysis of the GLM procedure, incorporating the effect of basic factors such as: herd, successive lactation, year of calving, share of HF in the cow genotype, month of calving and breeding values of bulls.There were also estimated correlation between breeding values of bulls and dairy milk performance traits of their daughters.

Dependences among breeding values of bulls and performance traits of their daughters were qualified and was mostly highly significant. A discrepancy was notice between the estimated breeding value of bulls and the real milk yield of their daughters in the high milk heard. This calls for a modification of the currently available methods of the breeding value evaluation and selection of sires for cattle improvement.

Session 17 Poster 51

Influence of paternal country origin on chosen performance traits of their daughters

B. Sitkowska, S. Mroczkowski, University of Technology and Agriculture, Genetics and Animal Breeding, Zootechnical Department, Mazowiecka 28, 85-084 Bydgoszcz, Poland

The aim of the studies was to analyze the effectiveness of utilization of bulls, whose semen was used in the OHZ Osieciny in 1990-2002. The evaluation bulls utilization involved the milk yield and lifetime milk yield, reproductive performance and functional traits of female offspring. The 1132 cows were milk-utilised over 1990-2002. The numerical data were verified statistically with variance analysis of the GLM procedure, incorporating the effect of basics factors such a: herd, successive lactation, year of calving, share of hf in the cow genotype, month of calving and paternal country origin.
The effect of the paternal origin on almost all the traits of their female offspring was highly significant. The best origin in milk traits was observed in the daughters of Canadian and Dutch bulls.

Session 17 Poster 52

Carcass analysis in Japanese quail lines divergently selected for shape of growth curve

L. Hyánková, J. Lastovková and Z. Szebestová, Research Institute of Animal Production, Prátelství 815, P.O.Box 1, 104 01 Prague-Uhríneves, Czech Republic*

Carcass composition changes following 22 generations of divergent selection for the shape of the growth curve were investigated. A total of 800 males from the HG (high relative gain from 11 to 28 d of age) and LG (low relative gain) lines were subjected to two dietary treatments differing in protein level of starter diet (standard or suboptimal). All males were weekly weighed from hatching until 70 d of age. At the same ages, 8 or more males per line/dietary treatment were sacrificed and their defeathered and eviscerated carcasses were analysed for dry matter, protein and total lipid contents. The carcass composition changed significantly with age. Both percentage of dry matter and total lipids increased with age during the whole experimental period. The percentage of protein significantly accelerated, however, only to age at the inflection point of the growth curve. As expected, the inter-line differences in carcass composition were confirmed for the early growth . The LG males exhibiting a higher growth rate than the HG males immediately post-hatch were characterized by a significantly higher percentage of dry matter and lipids during the acceleration phase of the growth curve on both dietary regimes. However, different levels of protein in diets significantly influenced the carcass composition of both lines during the early growth period.

Session 17 Poster 53

Evaluation of imported gene, direct and maternal heterosis, and estimation of genetic parameter and in the Iranian crossbred dairy cattle population

A. Ghorbani[1], S. Miraii Ashtiani[2] and M. Moradi Shahrebabak[2], [1]Genetic and breeding animal, Science and research unit of Eslamic Azad university, [2]Genetic and breeding animal, university of Tehran, Iran

The performance of crossbred dairy cattle in Iran was considered across 1991-2003 years. The crossbred animals were results of Holsteinx Indigenous (HxI) and Brown Swiss x Indigenous (BxI) crosses and their backcross progeny. Milk, fat yield, fat percent and milk day traits were considered in this research. The average performance for mentioned traits were 2688, 123.53Kg, 3.89 percent and 260.1days in (HxI) and 2493,118.5 Kg, 3.38 percent and 252.56 days in (BxI) respectively. Variance component were estimated using animal model (single trait) and Derivative-Free Restricted Maximum Likelihood method for different traits. The estimation of heritability for milk yield, fat yield, fat percentage and milk days were 0.274, 0.153, 0.32 and 0.13 in (HxI) population and (BxI) population were 0.2007, 0.2089, 0.38 and0.1089 respectively. Breed, individual and maternal heterosis effect were calculated using Least Squares method in two populations. These effects for (HxI) were 10.14, -1.7 and 3.34 kg for milk yield , 0.0061,0.00017 and 0.008 kg for fat yield, -0.0015,-0.00015 and 0.00017 percent for fat percentage and -0.241,0.0579 and -0.135 for milk days respectively and for (BxI) were 10.38, 3.16 and -2.19 kg for milk yield 0.12, 0.0342 and 0.029 kg for fat yield, 0.00457, 0.00131 and -0.00132 percent for fat percentage and -0.54, -0.174 and 0.277 days for milk days.

Session 17 Poster 54

Estimation of genetic parameters of growth traits in Golpayegan's calves

M.F. Harighi, Azad Islamic University of Kermanshah, Iran

In this study, the traits of birth weight (BW), 3- month (3MW) 6MW, 12MW, 15MW, 18MW of Golpayegan's Calves (it is an Iranian native cow) were used to estimate effects of environmental factors and heritability, phenotypic, genetic and environmental correlations among these traits. The effect of, cattle year of birth, season of birth number of calving, the sex of calf, as a fixed effect and the effect of sir within herd and error as a random effect were considered.

Heritability and standard error estimates for aforementioned weight were 0.29 ± 0.08, 0.499 ± 0.10, 0.57 ± 1.34, 0.78 ± 2.7, 0.78 ± 2.65, 0.62 ± 1.63. respectively. However, whole phenotypic, genetic and environmental correlation were positive among weight's, BW, 3MW, 12MW, 15MW, 18MW. Due to obtained results of this research the ideal age of selection for growth traits of Golpayegan's Calves is at sixth month.

Selection Indices and sub indices for improvement milk composition traits in Friesian cattle in Egypt

A.M. Hussein[1], M.N. El-Arian[2], A.S. Khattab[3], E.A.Omer[1], and F.H.Farrag[3], [1]Animal Production Research Institute, Ministry of Agric. Egypt. [2]Department of Animal production, Faculty of Agriculture, Mansoura University, Egypt. [3]Department of Animal production, Faculty of Agriculture, Tanta University, Egypt*

A total of 623 normal first lactation records of Friesian cows in Egypt sired by 60 bulls, were used to estimate genetic and phenotypic parameters of milk yield (MY), fat yield (FY) and protein yield (PY).In addition, four selection indices were constructed using all combinations of two or three traits studied. two sets of relative economic values (actual economic values (set_1) and one standard deviation (set_2)) were used. The animal model used in the analysis, included, the fixed effects of year and month of calving, lactation length as a covariate and random effects of individual and errors. Means of MY, FY and PY were 2570, 99 and 77 kg, respectively. Heritability estimates were 0.39, 0.12 and 0.30 for MY, FY and PY, respectively. All genetic and phenotypic correlations among all traits ranged from (0.41 to 0.49). The expected genetic change per generation ranged for set_1 from 294 to 323 kg for MY, 2.48 to 5.13 kg for FY and from 6.69 to 8.25 kg for PY. While for set_2 ranged from 294 to 322 kg , 2.08 to 5.13 and 6.74 to 8.60, respectively. The two different sets of economic weight are succeeded in constructing different selection indices.

The genetic and environmental effects on milk yield and fat percentage in Isfahan dairy farms

Mehdi Babaei, Department of animal science ,Shahrekord Azad university, Shahrekord, Iran

First lactation milk yield and fat percentage of 3145 Holstein dairy cows in Isfahan from 9 herds between 1990 and 1997 were used to study genetic and environmental effects on milk yield and fat percentage with an sire model. Sires were included 69 Iranian and 124 exotic sires .Dams were 2556 cows .Genetic factors were sire groups (exotic or Iranian) and sire within groups .Non genetic factors were herd, year, season and regression of age at first calving (AFC).Means ± SD were 6878 ± 1956,5522 ± 1578,5262 ±1278 and 2.69 ± 0.44 for total milk(TM) ,fat corrected milk 4% (FCM 4%),fat corrected milk 4% and 305 day(FCM 4%&305)and fat percentage(FP) respectively .AFC was 26.65 month. Exotic group daughters were Produced more TM, FCM 4%,FCM 4% &305,($P<0.05$). Group daughters hadn't significant difference for FP. Herd effects difference were significant ($p<0.05$). AFC at 28 month produced most amounts of TM,FCM 4%, FCM 4%& 305 and FP. Heritabilities ± SE were 0.19 ± 0.04,0.08 ± 0.03 and 0.45 ± 0.06 for TM, FCM 4%, FCM 4% &305 and FP. Breeding value for all traits, sires and dams were estimated with BLUP- sire model. The best sires were 800139, 600538, 802462, 500001, for TM, FCM 4%, FCM 4% &305 and FP respectively.

Session 18 — Theatre 1

Effect of sex and slaughter weight on pig performance and carcass quality

J. Mullane[123], P.G. Lawlor[1], P.B. Lynch[1], J.P. Kerry[2] and P. Allen[3], [1]Teagasc, Moorepark, Fermoy, Co. Cork, Ireland, [2]Department of Food Technology, University College, Cork, Ireland, [3] National Food Centre, Ashtown, Dublin 15, Ireland*

The aim of this trial was to examine the effect of sex and slaughter weight on performance and carcass quality in pigs of a lean genotype. Single sex pairs (n = 45) of pigs (L x LW) were used in a 3 (sex; boar (B), gilt (G), castrate (C)) x 3 (Liveweight at slaughter; 80, 100 and 120kg) factorial design. The experimental period was from weaning (mean =26 days and 8.6kg) to slaughter. All pigs were fed the same sequence of dry pelleted diets. Daily gain, feed conversion ratio (FCR) and carcass lean meat content (by Hennessy Grading Probe) were 737, 753 and 710g (s.e. 9.6; P<0.01); 2.30, 2.45 and 2.47 (s.e. 0.04; P<0.05) and 563, 544 and 567g/kg (s.e. 4, P<0.01) for B, C and G respectively. Daily gain, FCR and carcass lean meat content were 717, 735 and 748g (s.e. 10, P = 0.11); 2.15, 2.43 and 2.64 (s.e. 0.04; P<0.01) and 568, 557 and 549g/kg (s.e. 4; P<0.01) for slaughter weights of 80,100 and 120kg respectively. Boars grew faster than gilts, more efficiently than gilts and castrates and had a greater lean content than castrates. Daily gain increased, FCR deteriorated and lean meat content reduced as slaughter weight increased.

Session 18 — Theatre 2

Estimation of whole body lipid mass in finishing pigs

M. Kloareg, J. Noblet, J. Mourot, J. Van Milgen, UMR SENAH, INRA, 35590 Saint-Gilles, France*

Most pig nutritional growth models are based on the deposition of protein (P) and lipid (L) at the whole body level, which can be determined by chemical analysis. However, this method is expensive, time consuming and the carcass is lost. Alternatively, P and L may be estimated using simple indicators that should be precise and easily accessible. Although empty body weight (EBW) is a good indicator for P, L is more difficult to estimate. A study was carried out to evaluate the relationship between simple carcass measurements and L. Measurements included backfat thickness *in vivo* and at slaughter in the hot and cold carcass and the weight of carcass, organs and primal cuts. A total of 30 pigs (females and castrates, and two genotypes) were slaughtered at body weights typically used in Europe (90 to 150 kg). All backfat thickness measurements were highly correlated (R^2 ranged from 0.86 to 0.90) and appeared good indicators for total backfat mass (B). Nevertheless, B (in combination with EBW) was the best indicator for L (L = 0.0567xEBW + 1.500xB, R^2=0.86). The second best indicator was backfat thickness measured in the hot carcass between 3rd and 4th last lumbar vertebra at 8 cm off the mid line (F34LV; L = (0.0814 + 0.0072xF34LV)xEBW, R^2=0.80). Combined with EBW, backfat mass or thickness can be used to estimate L under practical conditions.

Preliminary study of the effect of the IGF-II genotype on meat quality in pigs

K. Van den Maagdenberg[1], A. Stinckens[2], E. Claeys[1], N. Buys[2], S. De Smet[1], [1]Laboratory of Animal Nutrition and Animal Product Quality, Department of Animal Production, Ghent University, Proefhoevestraat 10, 9090 Melle, Belgium, [2]Centre for Animal Genetics and Selection, Department of Animal Production, K.U.Leuven, Kasteelpark 30, 3001 Heverlee, Belgium*

Recently a new QTN, located in the regulatory sequence of the paternally imprinted IGF-II gene was discovered in the pig. Effects of the IGF-II polymorphism on muscle growth, fat deposition and carcass lean meat percentage have been reported. The aim of this study was to investigate the effect of the IGF-II paternal allele (Qpat/qpat) on meat quality parameters, like pH 40 minutes and 24 hours pm, water holding capacity (drip loss and filter paper method), colour (CIELAB values) and tenderness (Warner-Bratzler shear force), and on bioelectrical impedance as a measure of carcass lean content . These parameters were measured in three muscles from 29 boars of 26 weeks of age (Rattlerow Seghers). The number of Qpat/qpat animals were respectively 15/14, 12/11, 13/10 for the Longissimus, Semimembranosus and Triceps brachii. Data were analysed with a univariate general linear model with IGF-II genotype, muscle, slaughterhouse and sire as fixed factors and live weight as covariate. There were significant effects of muscle and sire. The electrical volume was significantly higher and the L* value was significantly lower in Qpat animals. There were no significant differences for the other meat quality traits.

Genetic parameters for meat quality and production traits in Finnish Landrace and Large White pigs

M.-L. Sevón-Aimonen and A. Mäki-Tanila MTT Agrifood Research Finland, Animal Production Research, FIN-31600 Jokioinen, Finland

Loin meat quality has been breeding goal in Finnish pig breeding programme since 1983 but quality of ham has not been routinely recorded. Preliminary study of heritability for ham quality was made at 1999, where ultimate pH and colour of three different ham muscles were measured from totally1000 station test pig. Due to results of that study, the pH and colour in light part of *semimembranosus* decided to be include into breeding programme at year 2000. The main purpose of the current study was to reliably estimate (co)variance components for meat quality and other traits from a large data set. Data contained 7685 records from Finnish Landrace and 7467 records from Finnish Large White pigs. Breeds were analysed separately using multitrait animal model. Heritability estimates for pH varied from 0.15 to 0.25, for lightness from 0.10 to 0.34, for redness from 0.26 to 0.53, for average daily gain from 0.28 to 0.30 and for meat percentage from 0.51 to 0.56. Genetic correlations between average daily gain and meat quality traits were in most cases low (under 0.15).Genetic correlations between meat quality traits (especially ham quality) and meat percentage were high (between 0.2 and 0.6) and unfavourable. Due to high unfavourable correlation between meat quality and economically important meat percentage, the quality must be included to breeding goal to prevent the deterioration of quality.

Session 18 Theatre 5

Influence of genetic markers on the drip loss development of case-ready pig meat

G. Otto[1], R. Roehe[1], G.S. Plastow[2], P.W. Knap[2], H. Looft[2], E. Kalm[1], [1]Institute of Animal Breeding and Husbandry, University of Kiel, D-24098 Kiel, Germany, [2]PIC International Group, Ratsteich 31, D-24837 Schleswig, Germany*

Phenotypic data on pig meat quality traits (including drip loss development of case-ready meat measured daily from first to seventh day post-mortem) and molecular information on 12 genetic markers including *RYR1* ("halothane") was obtained for 1155 market pigs slaughtered under commercial abattoir conditions. For each pig a case-ready meat drip loss development curve was fitted using linear-quadratic regression. An association study between curve parameters and genetic markers revealed lower intercept, linear and quadratic terms for "stress resistant" pigs at the *RYR1* locus and also at a second marker locus. This indicates a lower starting point at one day of measurement and a slower and more linear increase over time. Furthermore, stress resistant pigs showed superior performance in almost all meat quality traits, except ultimate pH value, initial conductivity and redness of the meat. Animals homozygous for the favourable genotype at the second locus showed less drip loss regardless of the measurement (Bag-method, EZ-DripLoss, case-ready meat) as well as higher ultimate pH and darker meat. In total seven of the markers showed significant effects on the target traits. These results show the potential of marker-assisted selection for development of reduced drip loss, a trait of increasing importance to retailers of fresh meat.

Session 18 Theatre 6

Genetic parameters of direct and ratio traits of Hungarian pig populations

I. Nagy[1], L. Csató[1], J. Farkas[1], K. Tisza[1], L. Radnóczi[2], [1]University of Kaposvár, Kaposvár, Hungary, [2]National Institute for Agricultural Quality Control, Budapest, Hungary

Genetic parameters of several growth and carcass traits were estimated for the Hungarian Large White and Hungarian Landrace pig breeds and for their cross (F_1). The objective of the analysis was to compare the direct (days of station test, consumed feed, valuable cuts and age) and ratio/composite (meat quality score, net daily gain, feed conversion, proportion of valuable cuts, carcass fat content, lean meat percentage and average daily gain) traits, which were collected in the course of station and field tests. The analysis was based on the national database (1997-2003) using univariate and bivariate animal models. Estimated heritabilities for station test traits ranged between 0.34-0.58 (except for meat quality score, where the heritability was low) and exceeded that of the field test traits (0.18-0.23). Relative importance of random litter effects was low for the station test traits (0.05-0.29) but moderate for the field test traits (0.22-0.48). The unfavourable genetic correlation between lean meat percentage and meat quality score is worth mentioning. In both performance tests the direct and ratio test counterparts showed similar heritabilities and their genetic correlation were close to unity (0.74-0.95). Based on these results selection on either the direct or on the ratio traits would possibly result similar selection response.

Session 18 Theatre 7

Does the Swedish animal welfare legalisation influence the working-hours spent in pig production?

B. Mattsson[1], Z. Susic[2] and N. Lundeheim[2], [1]Praktiskt inriktade grisförsök (Pig), S-532 89 Skara, Sweden, [2]Swedish University of Agricultural Sciences (SLU), Uppsala, Sweden

The aim of this study was to estimate the time spent in Swedish pig production and on which tasks this time is spent. The study is based on information from 39 Swedish herds. Average herd size was 210 (range 96 to 400) sows and 4400 (range 1200 to 22700) fatteners produced per year. On average,15 h/sow/year (range 8 to 28) was spent in piglet production. Of this time, 41 % was spent during farrowing to weaning, 20 % on mating and tending of dry sows, 19 % on the piglets from weaning to 12 weeks of age, 5 % on gilts and 15 % on miscellaneous tasks. Daily pen cleaning and providing of straw accounted together for 34 % of the total time spent in piglet production. During the fattening period, on average 10 minutes/pig (range 9 to 12.5) was spent in herds with specialised fattening pig production and 14 minutes/pig (range 6.5 to 34.5) in farrow-to-finish herds. Daily pen cleaning and providing of straw accounted for 37 % of the total time. It was concluded that the Swedish animal welfare legalisation and its influence on stable design and management routines has increased the need of working-hours in pig production, but also increased the welfare and comfort of the pigs.

Session 18 Theatre 8

Sow behaviour and litter size in first parity sows kept outdoors

A. Wallenbeck[1], L. Rydhmer[1] and K. Thodberg[2], [1]Swedish University of Agricultural Sciences, Dept. of Animal Breeding and Genetics, BOX 7023, S-75007 Uppsala, Sweden [2]Danish Inst. of Agricultural Sciences, Dept. of Animal Health and Welfare, P.O. Box 50, DK-8830 Tjele, Denmark*

This study is based on 40 first parity YxL sows kept outdoors. Sow behaviour was recorded on videotapes 4 days postpartum and direct-monitored 4 and 6 weeks postpartum. Statistical analyses were done with SAS-GLM. Correlations were found between nursing frequency at 4 days, 4 and 6 weeks (4d-4w: 0.4**; 4d-6w: 0.3*; 4w-6w: 0.6***) and also between duration of nursing at 4 and 6 weeks (0.7***). Proportion of time the sow was lying at 4 and 6 weeks was correlated (0.4*). Proportion of time spent outside the hut, unavailable for the piglets, and proportion of time laying down, available for the piglets, had no significant correlation to nursing frequency at day 4. Litter size day 4 affected nursing frequency at both 4 and 6 weeks, indicating that Large litters (>10) are nursed less frequently than Average (9-10) and Small (<9) litters (4w*: L20 ^{b}, A27 ^{a}, S28^{a}; 6w**: L16^{b}, A24 ^{a}, S22^{a}). Sows with large or average litters spent more time outside the hut than sows with small litters (4d*: L9^{a}, A10 ^{a}, S6^{b} % of time). Sows with large litters laid down less than sows with small litters (6w*: L35^{a}, A45 ^{ab}, S50^{b} % of observations).

Session 18 Theatre 9

Behaviour, health and performance of piglets exposed to atmospheric ammonia

E. von Borell[1], A. Özpinar[2], K.M. Eslinger[2], A.L. Schnitz[2], Y. Zhao[2] and F.M. Mitloehner[2], [1]Institute of Animal Breeding and Husbandry, Martin-Luther-University, Halle, Germany, [2]Department of Animal Science, University of California, Davis, USA*

Chronic effects of 35 and 50 ppm of atmospheric ammonia (NH_3) on the welfare of piglets were studied. Four groups of 24 piglets in two environmental chambers were exposed for 20 d to 50 vs. 0 ppm (control) and 35 vs. 0 ppm of NH_3, respectively. Performance (BW, DMI, ADG, and F:G) was measured on d -1, 8 and 20 at which blood samples were obtained for analysis of haematological (blood cell differentials), metabolic parameters (BUN, glucose, lactate, ammonia), cortisol, and haptoglobin. Behaviours (body posture, feeding, and aggression) were recorded on d 3 and 19. Total WBC, lymphocytes, and monocytes were increased ($P<0.05$) in pigs exposed to 35 ppm vs. control animals. Other haematological and metabolic parameters were similar between treatments. Haptoglobin was increased ($P<0.05$) in 50 ppm ammonia exposed pigs compared to the control at d 8 and 20. Pigs that were exposed to both levels of NH_3 showed increased ($P<0.05$) cortisol concentrations on d 20 and spent more time in upright postures than the control ($P<0.001$ and $P<0.05$). Pigs exposed to 50 ppm of ammonia vs. the control tended to decrease DMI ($P=0.096$) and feeding behaviour ($P=0.059$). In summary, exposure to 35 and 50 ppm of NH_3 elicits systemic responses to injury and inflammation like haptoglobin, total white blood cells, lymphocytes, and monocytes. Atmospheric NH_3 also increased cortisol concentrations and tended to decrease feeding behaviour resulting in a trend for lower dry matter intake.

Session 18 Theatre 10

Ranking of discrete choice

Ulrich Halekoh[1], E. Jørgensen[1], M. B. Jensen[2], L.J. Pedersen[2] and M. Studnitz[2], [1]Deparment of Genetics and Biotechnology, [2]Department of Animal Health, Welfare and Nutrition, Danish Institute of Agricultural Sciences, Tjele, Denmark*

In an experiment to identify adequate rooting materials for pigs, animals were confronted with three different materials positioned in the arms of a maze and the choice or reluctance of a choice was recorded. The experiment was repeated four times for each animal, and different combinations of materials and the orientation of the arms were used. This experiment illustrates a common problem in applied ethology to derive a ranking of discrete choices. Because of the simultaneous presentation of more than two options a simple pairwise comparison of the observed choices is not sufficient to derive a ranking. Additionally, an analysis has to consider the correlation introduced by repeated observations on the same animal. We present a Bayesian analysis of the experiment based on a generalization of the well know logistic regression model for binary data. This model allows incorporating the repeated structure of the observations and provides an estimate of the ranking together with an evaluation of its uncertainty. For the choice among soil-like materials the pigs choose compost with an average probability of 62%, peat with 34%, wood-shavings and reluctance of choice each with 2%. The ranking where the first two materials were preferred had a posterior probability of more then 95%. The computations were performed with generally available software.

Session 18 Theatre 11

Influence of emergency vaccination on the course of classical swine fever epidemics

I. Witte[1], J. Teuffert[2], G. Rave[3], J. Krieter[1]*

[1]Institute of Animal Breeding and Husbandry, Christian-Albrechts-University, D-24098 Kiel, [2]Federal Research Center for Virus Diseases of Animals, D-16868 Wusterhausen, [3]Institute of variation statistics, Christian-Albrechts-University, D-24098 Kiel, Germany

A spatial and temporal simulation model, using Monte-Carlo methods, was developed to analyse the course of classical swine fever epidemic with special emphasis to emergency vaccination. In the present study a certain region with 2986 pig farms and a high farm density was assumed and simulations varying time until vaccination starts (0, 24, 47, 69 days) were compared to simulations without vaccination.

The epidemic duration (days) is not affected (-1, -2, -16, -17). In contrast the number of diagnosed farms is different for late vaccination starts (-1, -2, -8, -9) as well as the number of farms with movement restrictions (-19,- 80, -387, -550) and culled farms (-2, -7, -57, -70). Taking the number of vaccinated farms into account as culled farms, the results differ for early and late vaccination (29, 18, -28, -42). For the farm density assumed the size of the vaccination circle had no influence on the course of the epidemic.

Assuming a high farm density emergency vaccination can decrease the number of infected animals but has no effect on the epidemic duration.

Session 18 Poster 12

Performance traits of improved Lithuanian White pigs

R. Klimas, A. Klimiene, Siauliai University, P. Visinskio 25, 76351 Siauliai, Lithuania

Having a purpose to preserve the base of Lithuanian White pigs in conditions of up-to-date market, till 2005 it is necessary to increase their muscularity no less than by 3-5%. This became especially relevant after introducing EUROP standard for pigs being slaughtered. For more rapid improvement of genetic potential of the population of purebred Lithuanian Whites (LW) the most purposeful is to use the boars of English Large White (ELW) breed. Whereas namely ELW had the biggest influence on nurture of LW, therefore additional infusion of blood of the above-mentioned breed may be considered as pure breeding. According to prepared programme this work has been started in 2002 in nine breeding centres of purebred LW pigs.

It was indicated that ELW had no significant influence on the litter size and milk yield of LW pigs, but improved their fattening performance, and especially carcass traits. According to the data of control fattening and slaughtering, lean meat content of improved first generation (LW x ELW- F_1, n=227) progeny, loin lean area and ham weight were, respectively, by 3.7%, 4.4 cm^2 and 0.4 kg higher (P<0.001) and the backfat thickness at the last rib was by 5.8 mm thinner (P<0.001) than that of old type LWxLW (n=172). Improved LW, having 50% of ELW blood, by their leanness are becoming comparable to Yorkshires and Large Whites, bred in the country.

Session 18 — Poster 13

The sequence of myostatin in double-muscled pigs

A. Stinckens[1], J. Bijttebier[1], T. Luyten[1], K. Van den Maagdenberg[2], N. Harmegnies[3], S. De Smet[2], M. Georges[3], N. Buys[1], [1]Centre for Animal Genetics and Selection, Department of Animal Production, K.U.Leuven, Kasteelpark 30, 3001 Heverlee, Belgium, [2]Laboratory of Animal Nutrition and Animal Product Quality, Department of Animal Production, Ghent University, Proefhoevestraat 10, 9090 Melle, Belgium, [3]Centre for Molecular Genetics, Faculty of Veterinary Medecine, University of Liège, Rue de Colonster 20, 4000 Liège, Belgium*

Myostatin, a TGF-ß family member, is a negative regulator of muscle growth related with the double-muscling phenotype in cattle. The similarity of muscular phenotypes between the double-muscled cattle and Piétrain suggested that myostatin is also a candidate gene for muscular hypertrophy in pigs. In this study, we used PCR and sequencing to determine the entire myostatin DNA sequence of 22 extreme-muscled Piétrain pigs. ChromasPro was used to compare this sequence with the myostatin sequence of 6 normal-muscled control animals and previously published sequences (EMBL accession number AY527152, AJ133580, AF019623, AJ237662, AJ237920). Several polymorphisms were identified in the promoter-region and both introns. Two polymorphisms found in the promoter-region were present in all 22 extreme Piétrain but lack in the control animals. Whether these polymorphisms have any influence on the muscular hypertrophy found in Piétrain pigs requires further study.

Session 18 — Poster 14

Implementation of a selection and mating strategy to optimize genetic gain and rate of inbreeding in the Swiss pig breeding program

H. Luther and A. Hofer, SUISAG, CH-6204 Sempach, Switzerland

In small nucleus populations inbreeding may increase fast and could affect the long-term genetic gain. An optimal selection and mating strategy should account for the breeding values of sires and dams, their relationship to the nucleus population and the breeding values of the potential offspring. SUISAG implements a new procedure based on the optimal contribution theory to select and to mate sires and dams of the nucleus herds within the Swiss pig breeding program.

Nucleus sows in heat within the next month and all available boars within each nucleus line are pre-selected according to minimal requirements on breeding values. The selection of boars and sows to be mated is performed by the GENCONT software, that maximizes the genetic gain (total breeding value: TBV) while setting a restriction on the average relationship of the expected population. All possible combinations of matings are evaluated. Matings with favourable estimates in different breeding values (e.g. production, reproduction, exterior, TBV) and a low coefficient of inbreeding of the offspring get higher scores, favouring corrective matings. A linear programming technique is used to find the optimal mating scheme, that maximizes the average score of the matings.

At first, we will use this strategy for elite matings to select new AI-boars from the male offspring. We will extend the number of planned matings to produce all female replacements in nucleus herds.

Session 18 Poster 15

Results concerning the use of the compactness index in boars selection

V. Bacila [1], M. Vladu[2], I. Calin[1], Livia Vidu[1], P. Tapaloaga[1], R.A. Popa[1], [1]University of Agronomic Sciences and Veterinary Medicine Bucharest, Animal Science Faculty, Technology Department, 59 Marasti boulevard, sect. 1 , 011464, Romania; [2]University of Craiova, Agricultural Science Faculty, Phytotechny - Animal Science Department, 15 Libertatii street, 200583, Romania*

The compactness index, is calculate with the formula: $C_I = \dfrac{W}{\pi \times \frac{a}{2} \times \frac{b}{2} \times l}$

In which: C_I = compactness index; W = corporal weight; a = thorax depth; b= thorax width; l = body length.

This index was calculated for 138 boars: 62 boars from 2 maternal breeds (Large White and Landrace) and 76 boars from 2 paternal breeds (L.S.345 Peris and L.S.P. 2000). The analyze of the compactness indexes obtained for these 4 breeds bring to the conclusion that between breeds exists differences corresponding to the hypothesis that the compactness index (the body density) for maternal breeds has less values than the compactness index calculate for paternal breeds. This suggested compactness index can be used for the selection of boars from maternal breeds for reproduction traits, (will select boars whereat this index value is most little). For paternal breeds is preferred and kept to reproduction boars whereat the value of the compactness index is eldest.

Session 18 Poster 16

The growth rate of piglets from primiparous and multiparous litters with equalized pre- and postnatal effect of maternal environment conditions related to litter size

R. Czarnecki[1], M. Rózycki[2], M. Kamyczek[2], A. Pietruszka[1], B. Delikator[1], [1]Department of Pig Breeding, Agricultural University in Szczecin, Poland, [2]Institute of Animal Husbandry in Kraków, Poland

The study was aimed at determining accurately the growth rate of piglets reared by primiparae or multiparae considering equalized "maternal effect" related to litter size both before and after the birth.

The study included 178 litters of 990 synthetic line sows which delivered 13 to 14 piglets per litter (63 primiparous litters and 115 multiparous litters). On the first day after the birth, 76 litters were standardized to 8 piglets and 102 litters to 12 piglets per litter. The body weight of piglets from primiparae or multiparae in litters numbering 8 piglets amounted respectively to 5.45 and 5.93 kg on day 21, 7.37 and 7.83 kg on day 28, 18.51 and 19.40 kg on day 63, whereas in those numbering 12 piglets it was, correspondingly, 4.85 and 5.34 kg on day 21, 6.48 and 7.08 kg on day 28, and 16.18 and 17.07 kg on day 63.

The obtained results show clearly that piglets from multiparous litters grow faster when compared to those from primiparous litters. These results also show clearly that piglets reared in smaller litters grow faster than those reared in larger litters.

The study was financed by Committee for Scientific Research (grant No 6 P06E 035 20).

Session 18 — Poster 17

Pig osteochondrosis in Lithuania and its relationship with fattening traits and meatiness

A. Klimiene, R. Klimas, Siauliai University, P. Visinskio 25, 76351 Siauliai, Lithuania

The monitoring of pig osteochondrosis (OC) in Lithuania was started in 2001. 1596 pigs (791 gilts and 805 castrates) of various breeds and with an average weight of 95 - 100 kg were tested. Housing and feeding conditions were the same for all pigs. The pigs were selected at a control fattening station, slaughtered, and OC was measured according to the methods applied in Sweden by the cut surface of distal femur and humerus. The severity of this disease was scored in elbow and knee joints on a 0-5 point scale. The occurrence of OC among all the tested pigs of various genotypes made up 47.4 %. The study indicated that castrated males were more inclined to have the leg weakness syndrome than gilts (50.4 % vs. 44.4 %). The investigation data indicated that OC should be controlled in the course of selection of pigs raised at the breeding centers of Lithuania. When analyzing influence of OC on the fattening performance of Lithuanian White, Swedish Yorkshire and German Landrace pigs, significant differences were not found (P>0.1-0.5). However, dependence of this defect on muscularity and other carcass traits of pigs were indicated. Lean meat percentage, loin lean area and ham weight of pigs, having OC lesions of leg joints were higher than that of pigs not having this defect, when backfat thickness was lower.

Session 18 — Poster 18

The comparison of prediction abilities of pig carcass dissection methods

J. Pulkrábek, J. Pavlík, L. Valis and M. Vítek, Research Institute of Animal Production, Prague - Uhríneves, 104 01 Czech Republic

Results of full and simplified dissections of pig carcasses were analysed. The dissections are conducted on a sample of carcasses representing the pig production in a certain geographic area and time period. Earlier only full dissections based on cutting all the carcass parts except for head and feet were used. Muscle is defined as striated muscles that can be separated by knife from other tissues. This approach is, however, extremely time and labour consuming. Therefore, a simplified dissection method was developed. It is based on dissections of leg, shoulder, tenderloin, loin and belly with bones. According to this method, connective tissue is also included into the muscle weight. The coefficient 1.3 is used to recalculate the value for the whole carcass. Totally 20 carcasses from pig final hybrids fattened under common conditions were included into the analysis. Both full and simplified dissections were conducted on each carcass. The muscle proportions determined by full and simplified dissections were 56.85 and 56.58 %, respectively. The difference between the two values was rather low. High reliabilities of the results were also confirmed by the high correlation between the two results (r±sr = 0.97±0.053). The obtained results suggest that the simplified dissection method is utilisable in the process of developing regression equations.

The study was supported by the Institutional project MZE 0002701403.

Session 18 Poster 19

MRI as a reference technique to assess carcass composition in pig performance testing

U. Baulain[1], E. Tholen[2], R. Hoereth[3] and M. Wiese[2], [1]Institute for Animal Breeding Mariensee, Federal Agricultural Research Centre (FAL), D-31535 Neustadt, Ger¬many, [2]Institute of Animal Breeding Science, University of Bonn, D-53115 Bonn, Germany, [3]Institute of Meat Production and Market Research Kulmbach, Federal Research Centre of Nutrition and Food, D-95326 Kulmbach, Germany*

A total of 202 pigs were investigated to evaluate the suitability of Magnetic Resonance Imaging (MRI) as a reference technique to determine body composition in performance testing. Carcass sides of sire line Piétrain and dam lines German Landrace and Large White, as well as carcasses of two hybrid lines usual in the market were scanned by MRI and manually dissected afterwards. Mean percentage of lean (dissection) was 51.1% for the dam lines, 58.0% for hybrids and 64.8% for the sire line. Correlations between MRI muscle volume and muscle weight determined by dissection were r > 0.98 for dam lines and hybrid pigs. A slightly lower correlation (r = 0.94) was found for the extremely lean sire line. To estimate muscle weight from MRI data multiple regression equations were established. The estimation error was less than 1.9% for all breeds. The highest estimation accuracy was found for the group of the dam lines (1.3%). For percentage of lean corresponding errors of estimation were calculated.
In conclusion, MRI is suitable as a reference technique in performance testing. It can be applied instead of full dissection, when i. e. new measuring techniques and/or measuring positions have to be evaluated for its benefit in performance testing.

Session 18 Poster 20

Effect of vitamin E administration on the qualitative characteristics of the meat of Nero Siciliano pig

B. Chiofalo, D. Piccolo, L. Liotta, A. Zumbo, V. Chiofalo, Dept. MO.BI.FI.PA., University of Messina, Italy.*

Eighteen Nero Siciliano pigs, living in plein-air system in the Nebrodi area, were divided into two groups (B.W.=35±2kg, age=180 days) called CTR and Vit.E, to evaluate the effect of vitamin E on the quality of meat. Animals fed on spontaneous fruits of the undergrowth integrated (3% of B.W.) with concentrate. Every 15 days, the pigs of the Vit.E group were treated with an intramuscular administration of d,l-α tocopheryl acetate (200 U.I./head). After the slaughter (age=250 days) the physical, chemical and acidic characteristics of the *Longissimus dorsi* muscle were determined.
Colour (L*, a*, b*, Chroma and Hue) and oxidative stability (TBARs) measured at 2, 5 and 8 days *post mortem* were slightly better in Vit.E group, while Warner-Bratzler shear force values (CTR=8.50kgf/cm^2, Vit.E=8.37kgf/cm^2) were comparable. The chemical composition was not influenced by the treatment. Among the fatty acid classes, SFAs (CTR=40.08%, Vit.E=42.21%) and PUFAs (CTR=8.45%, Vit.E=9.40%) were similar, MUFAs showed a significant difference (P<0.01) between the CTR (51.47%) and Vit.E group (48.39%); the UFA/SFA ratio was slightly lower in the Vit.E (1.38) than that of the CTR group (1.50). Atherogenic index (0.50 *vs* 0.54) and thrombogenic index (1.26 *vs* 1.39), respectively in the CTR and Vit.E group, were comparable. Results suggest a further optimisation of dosage and time of vitamin E administration, considering the interest of farmers and technicians for this production.

Session 18 Poster 21

Ability of fresh thigh evaluation to predict cured ham quality

G.L. Restelli[1], A. Stella[2], G. Pagnacco[1], [1]Dipartimento di Scienze e Tecnologie Veterinarie per la Sicurezza Alimentare (VSA), Via Grasselli 7, 20137 Milano, Italy, [2]FPTP - CERSA, Palazzo L.I.T.A., Via F.lli Cervi 93, 20090 Segrate (MI), Italy*

Dry-cured ham is the most important Pdo's product for pig industry in Italy. The quality of the fresh thigh is the main factor in affecting the defects in the finished product.
The aim of this work was to investigate the ability of evaluation of fresh thigh to predict quality of the cured ham. 5098 fresh thighs were evaluated after trimming to identify defects like marbling, excessive fat, an uncovered crown, excessive thinness, off-colouring, irregular shape, and haematomas. At the end of curing period the hams were evaluated for the same defects. The presence of a given defect was indicated with a code = 1, and absence with 0. Logistic regression was applied to compare the evaluation of fresh thighs with the evaluation of dry cured ham.
The analyses revealed that excessive fat in the thighs was a significant factor for marbling and excessive fat in hams (odds ratio, 15.95 and 2.85; p = 0.0001), the same result was observed for marbling in the fresh thighs and in the finished hams (odds ratio, 5.77; p = 0.0001). Finally colour defects in the thighs seems was associated with abnormal ham colour (odds ratio, 1.63; p = 0.049).

Session 18 Poster 22

Influence of lairage time and genotype on levels of stress in pigs

K. Salajpal[1], M. Dikic[1], D. Karolyi[1], Z. Sinjeri[2], B. Liker[1], I. Juric[1], [1]Faculty of Agriculture University of Zagreb, Croatia, [2]Central Medical Laboratory,Koprivnica, Croatia

Pigs from crosses of a (♀Duroc x ♂Swedish Landrace) x ♂Pietrain, (genotype A, n=24) and (♀Swedish Landrace x ♂Large White) x ♂Pietrain, (genotype B, n=26) were used to investigate the effects of different lairage time (2 h and 24 h) on levels of stress and meat quality traits. Blood samples were collected before loading and at exanguination to measure cortisol, lactate, glucose, electrolytes (sodium, potassium and chlorides), lactate dehydrogenase (LDH), creatine kinase (CK), alanine aminotransferase (ALT) and aspartate aminotransferase (AST). On the *Longissimus dorsi* muscle the initial and ultimate pH values (pH_i and pH_u) as well as water holding capacity (WHC) were measured. Long lairage pigs showed lower level of pH_i, glucose, lactate, LDH and higher CK, AST and ALT activity (P<0.01). No effect of the lairage time on cortisol, electrolytes and other meat quality parameters was observed. Pigs from genotype B showed higher level of K, Na, glucose and lactate (P<0.01), lower pH_i (P<0.01) value and higher drip loss (P<0.05) then pigs in the genotype A. These results suggest that lairage period had a smaller effect on meat quality traits compared with the genotype but a long lairage period without food intake may produce hipoglycemia and muscle damage signs.

The effect of outdoor raising on meat quality of pigs

R. Juska, V. Juskiene, Institute of Animal Science of LVA, R. Zebenkos 12, LT-82317, Baisogala, Radviliskis distr., Lithuania

A study involving crossbred (Lithuanian White x Large White) pigs was carried out to determine the influence of pig raising outdoors on the quality of carcass and meat. Two groups of pigs were used in the experiment: control pigs were raised in a pig-house according to the routine on-farm technology, and experimental pigs were raised in outdoor enclosures of 8.5 areas. Eight analogous animals (4 gilts and 4 castrates) were taken from each group for control slaughtering.
Samples of *M. longissimus dorsi* and backfat were taken for the analysis of the physicochemical indicators of meat. Chemical composition of meat was analysed by standard methods.
The results from the study indicated that the carcass weight of experimental pigs was 7.3% higher than that of control pigs. Backfat thickness at all measured points was also 4.6-7.3% lower for pigs raised outdoors, but the differences were statistically insignificant. The loin lean area was 10.9% ($P>0.233$) higher for pigs raised outdoors.
The analysis of the chemical composition of *M. longissimus dorsi* indicated that there were no significant differences between the groups for the contents of dry matter, fat, ash, trytophan and oxyprolin, but the protein content was 0.76% ($P=0.088$) higher.
There were no significant differences between the groups for the pH-value of meat, colour intensity, cooking losses, water binding capacity and backfat hydrolysis.

Fatty acid composition in different tissues of Mangalitsa crossbreds

J. Seenger, Cs. Ábrahám, M. Mézes, H. Fébel, E. Szücs, Szent István University, Gödöllö, [2]Research Institute for Animal Breeding and Nutrition, Herceghalom, Hungary*

Fatty acid and composition in different tissues of two *Mangalitsa* crossbreds was compared within a series of experiment. Genotypes used were as follows: *(A) Dumecoα x Mangalitsa* (n=5) and *(B) Dumecoα x (Duroc x Mangalitsa)* (n=5). The animals were raised under similar housing and feeding conditions. Samples were taken from two layers of backfat, leaf-fat as well as loin, and analysed for saturated fatty acids *(SAFA)*, monounsaturated fatty acids *(MUFA)* and polyunsaturated fatty acids *(PUFA)*. The intramuscular fat in loin and leaf-fat of *crossbred A* contained more *SAFA* than that of *genotype B* ($P<0.05$). However, the outer layer of backfat of *genotype A* showed higher *MUFA* values ($p<0.05$). On the contrary, leaf-fat samples taken from *crossbreds B* contained more *MUFA* as compared to *genotype A* ($P<0.01$). The extent of variation among fatty acids was the highest in the leaf-fat, i. e. the largest differences among different fatty acids as well as *SAFAs*, *PUFAs* and *MUFAs* were shown in leaf-fat samples. Differences were also established in all of the three fatty acid categories for the external layer of the backfat, as well ($P<0.05$; $P<0,01$). In the intramuscular fat of loin, and the internal layer of the backfat also high variations were found for several *SAFAs* ($P<0.05$; $P<0.01$), but differences were shown in the case of the remaining fatty acid categories.

Session 18 — Poster 25

The effect of driving pigs to stunning prior to slaughter on their stress status and meat quality

Cs. Ábrahám[1], M. Weber[1], K. Balogh[1], J. Seenger[1], M. Mézes[1], H. Febel[2] and E. Szücs[1] [1]Szent István University, Gödöllö, Hungary, [2]Research Institute for Animal Breeding and Nutrition, Herceghalom, Hungary*

Pigs of identical genotype at the HAL loci, i.e. NN were raised under same environmental conditions. Forty animals in all were transported from the pig farm to abattoir and slaughtered under commercial conditions. The aim was to establish effects of driving animals [with (Group A) or without electric goad (Group B)] to stunning.. In blood samples taken 1 hour prior to stunning and during exsanguination, physiological parameters were determined as follows: cortisol, glucose, lactic acid. In addition, NEFA, MDA as well as indicators of the antioxidant defence system (ascorbic acid, GSH, GSHPx) were also analysed. In LD, following meat quality parameters (pH_{45}, pH_u, colour) were measured. Findings reveal significant differences between treatments only for cortisol (Group A: 122.75±48.11; Group B: 162.39±41.18; P<0.01), and NEFA (Group A: 0.52±0.22; Group B: 0.30±0.10; P<0.001). It was stated that the use of electric goad in driving pigs to stunning cannot be abandoned under commercial conditions. Electric goad helps in moving up animals to stunning faster, which results in lower stress level. For meat quality no significant differences were found. Thus, further evidence has been established that the detrimental effect upon meat quality in MHR-negative pigs is not as high as expected.

Session 18 — Poster 26

Effect of sex and slaughter weight on carcass measurements of pigs

J. Mullane[123], P.G. Lawlor[1], P.B. Lynch[1], J.P. Kerry[2] and P. Allen[3], [1]Teagasc, Moorepark, Fermoy, Co. Cork, Ireland, [2]Department of Food Technology, University College, Cork, Ireland, [3] Teagasc, Ashtown, Dublin 15, Ireland*

The aim of this trial was to examine the effect of sex and slaughter weight on carcass traits. Single sex pairs (n = 45) of pigs (L x LW) were used in a 3 (sex; boar (B), gilt (G), castrate (C)) x 3 (LW at slaughter; 80, 100 and 120kg) factorial design. Carcass length, leg length and ham circumference were 845, 832 and 836mm (s.e. 3.3; P<0.05), 388, 383 and 383mm (s.e. 2.5; P>0.05) and 712, 720 and 719mm (s.e. 3.4; P>0.05) for B, C and G respectively. Weight of hind leg, shoulder, loin and belly were 9.30, 9.73 and 9.59kg (s.e. 0.153; P>0.05); 5.48, 5.56 and 5.40kg (s.e. 0.096; P>0.05); 6.39, 7.17 and 6.59kg (s.e. 0.159; P<0.01) and 3.46, 3.67 and 3.47kg (s.e. 0.060, P<0.05) for B, C and G respectively. Carcass length, leg length and ham circumference were 793, 837 and 884mm (s.e. 3.3; P<0.01); 364, 385 and 405mm (s.e. 2.5; P<0.01) and 672, 720 and 760mm (s.e. 3.4; P<0.01) for slaughter weights of 80,100 and 120kg respectively. Carcases were longer for B than G or C and the four primal cuts were heavier for C than B. All measurements including the weight of the 4 primal cuts increased sequentially as slaughter weight increased.

Genetic correlation of intramuscular fat content with performance traits and litter size in Duroc pigs

F.X. Solanes[1,2], J. Reixach[2], M. Tor[1], J. Tibau[3] and J. Estany[1], [1]Departament de Producció Animal, Universitat de Lleida, Rovira Roure 191, 25198 Lleida, Spain; [2]Selección Batallé, S.A., 17421 Riudarenes, Spain, and [3]Centre de Control Porcí, IRTA, Veïnat de Sies s/n, 17121 Monells, Girona, Spain

Data on a commercial Duroc pig line were used to estimate the genetic correlation between intramuscular fat content (IMF, n=1,313) with on-farm performance test weight (WT, n=19862), ultrasonic backfat thickness (BF, n=19,099), weight of hams (WH, n=1,305) and number of piglets born alive at first parity (BA, n=5,114). IMF was measured in the glutaeus medius muscle of females and estimated with the Near Infrared Transmittance technique. Restricted maximum likelihood estimates of genetic correlations were obtained using a multivariate animal model, in which WT and BF were adjusted for age at test (180d) and WH and IMF for carcass weight. Estimates of heritability for IMF, WT, BF, WH and BA were 0.56, 0.18, 0.35, 0.41 and 0.06, respectively. Estimates of the genetic correlation between IMF and WT, BF, WH, and BA were 0.34, 0.64, -0.34 and 0.11, respectively. The predictive ability of the model for IMF was assessed using an independent set of 212 castrated pigs for the cured ham market (220 d of age), in which IMF was determined in the glutaeus medius (IMFGM) and longissimus dorsi (IMFLD) muscles. The regression coefficient of IMFGM on the average predicted additive values of the parents for IMF was 2.05±0.62, which indicate that observed values were consistent with predicted genetic values. No significant response was observed for IMFLD (0.65±0.67). The results suggest that selection for IMF can be effective, although an uneven response across muscles may occur.

Session 18 Poster 28

Correlation coefficients between productive traits of Polish Large White, Duroc and Pietrain gilts

B. Orzechowska, M. Tyra, Institute of Animal Production. 31-083 Balice. Poland

Relationships between meat quality and quantity characteristics were analysed based on data on detailed dissection of half-carcasses from 400 Polish Large White (PLW), Duroc and Pietrain fatteners, which were tested in slaughter testing stations in Melno and Pawlowice during 2000-2003. Indicators of meat quality (pH, water holding capacity, colour) were measured on the *longissimus dorsi* muscle. For each breed, correlation coefficients were determined between traits of meat quality, and muscling and fatness of carcass and carcass cuts. The r values obtained indicate that meat quality traits bear little relationship with traits of fatness of carcass and carcass cuts. Higher relationships were found with traits of carcass meatiness. In PLW pigs, pH of meat measured 24 h postmortem (pH_{24}) showed a relationship with meat for most of the carcass cuts. In Duroc pigs, there were significant relationships between pH of meat measured 45 min postmortem (pH_{45}) and weight of meat in ham, and between pH_{24} and weight of meat in neck and half-carcass. In the Pietrain breed, a high correlation was found between pH_{45} and weight of meat in ham, loin and primal cuts, and between pH_{24} of meat and weight of meat of ham and belly. Significant correlations were also shown between water holding capacity and meat of ham and primal cuts, and between colour lightness (L) and meat of belly.

Session 18 Poster 29

Relationships between ultrasound measurements of the *m. longissimus dorsi* in pigs using Piglog 105 and Aloka SDD 500 devices

M. Tyra, B. Orzechowska, Institute of Animal Production. 31-083 Balice. Poland

This study was conducted using 386 gilts (Polish Large White, Polish Landrace, Pietrain and Line 990) originated from Pig Testing Stations and tested in 2001-2004. Ultrasonic measurements of the *m. longissimus dorsi* (LD) and backfat thickness were made at 100 kg of body weight. Measurements were taken at the boundary of thoracic and lumbar vertebrae (behind the last rib) on the right side using ALOKA SDD 500 with a 3.5 MHz linear array transducer type UST-5011U. Concurrently, backfat thickness at P2 and P4 points and height of LD muscle at P4M point were measured with PIGLOG 105 ultrasound device.

Mean measurements of backfat thickness by ultrasound (U2 and U4) and *in vivo* (P2 and P4) were similar except measurement U2 which was higher than P2 for all the breeds tested. Slight differences were observed for the height of LD muscle. Higher values were observed for Aloka measurements (6.89 cm on average) compared to Piglog live testing (6.31 cm).

Correlations were also estimated between the two measurement techniques. For backfat thickness measurements with PIGLOG 105 and ALOKA SSD 500, the correlations ranged from r = .403 for P2 and U2 to r = .612 for P4 and U4 measurements. The correlation between the height of LD muscle between P4M and U2 was r = .603. These correlations were lower than the respective correlations between such measurements and actual measurements taken after dissection of the half-carcasses (r = .511 to r = .817).

This work was carried out as part of grant no. 3 PO6Z 054 23

Session 18 Poster 30

Cd level in organs and tissues of pigs of early ripe meat breed (SM-1)

S.A. Patrashkov, V.L. Petukhov and O.S. Korotkevich, Research Institute of Veterinary Genetics and Animal Breeding of Novosibirsk State Agrarian University, 160, Dobrolubov Str., Novosibirsk 630039, Russia

There are several types of pigs of Early Ripe Meat breed (SM-1) in Russia. One of them is szonal Siberian type of SM-1, well-adapted to severe environmental conditions. Cd is a very toxic heavy metal (HM) that can affect important systems of organism such as nervous, cardiovascular, haemopoietic, respiration, reproduction and others.The objects of our research were 32 Early Ripe Meat pigs at the age of 6 month from the farm located near Novosibirsk city. Cd level in muscle, liver, kidney, lung, heart, spleen, bristle and hoof was determined by the atomic absorption spectrometry (AAS) using AAS-3 analyzer. It was revealed that Cd was accumulated in kidney to the highest degree (0.37 ± 0.01 mg/kg). The content of Cd in muscle and liver was 1.9 and 2.1 times less ($P < 0.001$), respectively. The ranking of the organs and tissues regarding the Cd concentration was as follows: kidney > bristle > hoof > spleen > heart = lung > muscle > liver. Positive Cd - Pb correlations in kidney and spleen (r = 0.537 and 0.738, respectively) were detected.

Session 18 Poster 31

The influence of heavy metals on some biochemical indices in pigs

O.S. Korotkevich, S.A. Patrashkov, V.L. Petukhov and E.V. Gart, Research Institute of Veterinary Genetics and Animal Breeding of Novosibirsk State Agrarian University, 160, Dobrolubov Str., Novosibirsk 630039, Russia

Heavy metals (HM) significantly influence many processes of animal organisms. The content of any HM in animal blood, urine, hair, bones, teeth, milk, etc. shows the load to an organism on the whole. Metal content in different organs and tissues may correlate to their level in the environment and animal diseases. The aim of our investigation was to find the correlation between HM levels in pigs bristle and main biochemical and hematological indices in blood. The objects of our research were 32 Early Ripe Meat pigs aged of 6 months from the farm next to Novosibirsk city. The samples of blood were investigated to establish some biochemical and hematological indices. The bristle samples were studied by the stripping voltammetric analysis (SVA) method with TA-2 analyzer to determine Zn, Cu, Pb and Cd concentrations.

The high positive connection between Pb in bristle and Hb in blood ($r = 0.61$) was detected. The indices of Pb in bristle and Ca in blood as well as Cd in bristle and bilirubin in blood had positive correlation ($r = 0.78$ and 0.73, respectively). Zn in bristle correlated negatively with alkaline and acid phosphatase ($r = -0.41$ and -0.38, respectively).

Thus, in unfavorable ecological regions animal hair can be used as a marker of the biochemical status of animals.

Session 18 Poster 32

Digestibility and net energy value of wheat bran and sunflower meal: pregnant sows versus fattening pigs

M.J. Van Oeckel, J. Vanacker, N. Warnants, M. De Paepe, and D.L. De Brabander. Agricultural Research Centre, Department Animal Nutrition and Husbandry, Scheldeweg 68, 9090 Melle, Belgium*

The nutrient digestibility, digestible energy (DE) and net energy (NE) of wheat bran (WB) and sunflower meal (SFM) for pregnant sows and fattening pigs was studied by means of in vivo trials. Within each animal category, two cross-over design trials were carried out with a control diet and a diet composed of 75% control diet + 25% WB or 25% SFM. In each trial 6 pregnant hybrid sows (on average 2.8th parity, 208 kg for WB and 4.3th parity, 240 kg for SFM) and 6 Piétrain x hybrid pigs (on average 50 kg) were involved. Sows were fed at requirement and pigs at 3 times maintenance level. The nutrient digestibility coefficients are not significantly different between sows and pigs for WB and SFM. The crude fibre digestibility tends to be higher for sows compared to pigs ($P = 0.051$) for WB. The DE and NE of WB is 7 and 6% higher (not significant) for sows than for pigs, with respectively a DE-value of 10.6 and 9.9 MJ/kg and a NE-value of 6.9 and 6.5 MJ/kg. A similar DE and NE of SFM for sows and pigs is found, with respectively a DE-value of 9.7 and 9.6 MJ/kg and a NE-value of 5.6 and 5.6 MJ/kg.

Session 18 Poster 33

Energy intake limiting effects of wheat bran and sunflower meal in gestation diets of sows

M.J. Van Oeckel, J. Vanacker, N. Warnants, M. De Paepe, and D.L. De Brabander, Agricultural Research Centre, Department Animal Nutrition and Husbandry, Scheldeweg 68, 9090 Melle, Belgium*

The effect of two levels of wheat bran (WB: 17 sows) and sunflower meal (SFM: 15 sows) on the ad libitum feed intake of pregnant hybrid sows was studied in two separate trials. A control diet, a medium fibre diet with 25% WB or 17% SFM and a high fibre diet with 50% WB or 34% SFM were compared in a Latin square design. All diets had a protein level between 14.5 and 15%. The net energy level of the diets decreased with increasing levels of WB (8.6, 8.0 and 7.4 MJ/kg feed) and SFM (8.6, 7.9 and 7.3 MJ/kg feed). The adaptation and feed intake registration period lasted 14 and 7 days. The average energy intake for the control, medium and high fibre diet was, respectively, for WB 2.2 (s.d. 0.7), 1.9 (s.d. 0.5) and 1.7 (s.d. 0.5) and for SFM 2.0 (s.d. 0.5), 1.8 (s.d. 0.5) and 1.5 (s.d. 0.5) times the daily energy requirement (CVB, 2004). Seventy-six% and 60% of the sows on the medium WB and SFM diets and 71% and 60% of the sows on the high WB and SFM diets ate more than 150% of the energy requirement. In conclusion, WB and SFM are not suited as energy intake restrictors for voluntary fed pregnant sows.

Session 18 Poster 34

Comparison between wet/dry feeders and liquid feeding in a trough for growing-finishing pigs

J.A.M Botermans[1], L. Meijer[2], M.A. Andersson[1] and D. Rantzer[1], [1]Swedish University of Agricultural Sciences, Alnarp, Sweden, [2]Wageningen Agricultural University, Wageningen, The Netherlands*

It was found previously that feeding growing-finishing pigs with feed dispensers in comparison to trough feeding led to a higher average daily feed intake (ADFI), higher average daily gain (ADG) and lower feed conversion ratio (FCR). The present study was done to determine the effects of using feed mixed with water under the above conditions. Eight pens (10 pigs per pen) were fed with a wet/dry feeder (5 pigs/feeding place), while the other 8 pens were fed with fresh liquid feed (2 feedings/day, 39 % dry matter) using a trough. During the first 18 days after introduction, the pigs were permitted to adapt to the new feeding system. Thereafter until 70 kg, the pigs were fed ad lib. During the ad lib period, pigs fed with the wet/dry feeders showed an 18 % higher ADFI (2640 vs. 2230 g) and 17 % higher DWG (966 vs. 823 g). During the subsequent restricted feeding period, from 70 kg until slaughter, no differences were observed. In all, from 18 days after introduction until slaughter, pigs fed with wet/dry feeders showed a 6 % higher ADFI (2690 vs. 2530) and 7 % higher ADG (899 vs. 840). No statistical differences in FCR and carcass meat percentage were detected.

Session 18 Poster 35

Effect of diet supplementation with grass-meal on performance and carcass composition of Duroc and Landrace cross bred pigs

P.G. Lawlor[1], P.B. Lynch[1], J. Kerry[2], and S. Hogan[2], [1]Teagasc, Moorepark, Fermoy, Co. Cork, Ireland, [2]Dept. of Food Technology, University College, Cork, Ireland*

The objectives here were to assess (1) the effect of diet dilution with grass-meal and (2) the effect of supplemental antioxidant on pig performance. Pigs (n = 1080) produced from Duroc (D) or Landrace (D) boars mated to crossbred sows were weaned (mean =26 days and 8.2kg), into same-breed, single-sex groups of 15 pigs (n = 72). At 21 days post -weaning, pigs were allocated at random to the following treatments: (1) High density diets (HD) to slaughter, (2) Diets with grass-meal (GM) to slaughter, (3) Diets with GM to 50kg followed by HD. (4) Diets with GM to 80kg followed by HD. (5) Diets with GM to 80kg followed by a vitamin E enriched (200 mg a-tocopherol/kg) diet with 50g/kg of 00-rapeseed oil, (6) As E with tea extract instead of a-tocopherol. Pigs were slaughtered at c. 105kg liveweight. Pigs from L boars grew marginally faster (707 v. 694g/d; NS) and more efficiently (2.55 v. 2.65; $P<0.01$) from weaning to slaughter and had leaner carcasses than pigs from Duroc sires (596 v. 592 g/kg; $P<0.01$). Feeding GM depressed pig performance (growth rate and FCR). Supplementation of the diet in the final stages of finishing with vitamin E or tea extract had little effect on pig performance.

Session 18 Poster 36

An in vivo model development to know feed ingredient preferences in weanling pigs

D. Solà-Oriol[1], E. Roura[2] and D. Torrallardona[1], [1]IRTA-Centre de Mas Bové, Reus, Spain, [2]Lucta SA, Barcelona, Spain*

The palatability of feed ingredients may be a tool to improve voluntary feed intake in weanling pigs. Four trials were conducted to develop a model to quantify feed preference. The trials consisted of a double choice test between the diet under evaluation and a reference diet based on rice (50%) and a soy protein product low in ANF (15%). In each trial, newly weaned (28d) and post-weaned (48d) piglets were used (12 pens per age). For the first three trials three cereals were tested (sorghum, maize and rye) at inclusion levels of 25, 50 and 100%. Similarly, in the fourth trial lupine (*Lupinus angustifolius*) was tested at inclusion levels of 7.5, 15 and 100%. To prepare the test diets, the testing cereals and lupine replaced rice and soy protein from the basal diet, respectively. It was observed that piglets preferred rice to the other cereals tested and the low ANF-soy protein product to lupine. It was also observed that the use of 28d-old weaned pigs was associated with a large variability and the use of pure ingredients with inconsistent observations. Therefore for better consistency and lower variability of the results in in vivo preference tests, 48d-old over newly weaned pigs and mixed rations over pure ingredients are recommended.

Session 18 Poster 37

Reduction of the crude protein content in diets for growing-finishing pigs

D. Torrallardona[1], M. Cirera[2], D. Melchior[2] and L. LeBellego[2] [1]IRTA, Centre de Mas Bové, Reus, Spain, [2]Indukern SA, Barcelona, Spain, [2]Ajinomoto-Eurolysine, Paris, France

Reducing the protein content of the diets for growing-finishing pigs contributes to reduce nitrogen excretion. However for this to be effective this reduction should not affect the performance of the animals. A trial was conducted in order to study the effect of reducing the protein content with or without correcting the corresponding Thr:Lys and SAA:Lys ratios on the performance of growing-finishing pigs. Ninety-six Landrace pigs, (48 males and 48 females) of about 21kg BW were offered four experimental treatments consisting of: high-protein diet (HP); low-protein diet (LP) without added Thr or Met; LP diet with added Thr (LP-Thr) and LP diet with added Thr and Met (LP-Thr-Met). It was observed that reducing the protein content of diets for G/F pigs resulted in a Thr and Met deficiency during the grower phase and in a Thr deficiency during the finisher phase. Over the whole trial, the pigs on LP had the poorest ($P<0.05$) growth and feed to gain ratio, and no statistically significant differences were observed between treatments HP, LP-Thr and LP-Thr-Met. It is concluded that the protein content in grower and finisher diets can be reduced from 19.4 to 15.4% and from 17.8 to 14.5% respectively, without significantly affecting performance if the right Thr and Met ideal ratios are maintained with the addition of free amino acids.

Session 18 Poster 38

Effects of amino acid type on odour from pig manure

P.D. Le, A.J.A. Aarnink, A.W. Jongbloed, C.M.C. van der Peet-Schwering, M.W.A. Verstegen and N.W.M. Ogink, Wageningen-UR, Wageningen, The Netherlands.

The objectives of this study were to determine effects of amino acid type on odour concentration, intensity and hedonic tone, ammonia emission, and manure characteristics from pig manure. An experiment was conducted with growing pigs in a randomised complete block design with three treatments in six blocks. Treatment groups were 1) three times requirement methionine and cysteine supplementation, 2) two times requirement tryptophan, tyrosine, and phenylalanine supplementation 3) without supplementation of these amino acids. Pigs (n=18), initial body weight of 41.2 ± 3.4 kg, were kept individually in partly slatted floor pens. Daily feed allowance was adjusted to 2.8x maintenance. Feed was mixed with water, 1/ 2.5 (w/w). Faeces and urine of each pig was accumulated in a manure pit under the slatted floor. The experiment lasted for seven weeks. At the end of the experiment one odour, one ammonia, and one manure sample was collected from each manure pit. Data was evaluated using analysis of variance. No significant differences were observed in daily feed intake, daily gain and feed conversion ratio among treatments. Supplementing S-containing amino acids in surplus of pigs' requirement significantly increased odour emission, odour intensity and odour hedonic tone ($P <0.05$). Remarkably supplementing trytophan, tyrosine, and phenylalanine in surplus of pigs'requirement did not increase odour emission, odour intensity and odour hedonic tone. The analyses of manure characteristics and feed composition are still going on. Data on manure characteristics and feed composition may explain reasons behind this observation.

Session 18 Poster 39

Water consumption of pigs at weaning: effect of mixing water with pelleted feed and of using a water nipple or a water bowl

D. Rantzer, M. Andersson, L. Stålhandske, J. Svendsen, Swedish University of Agricultural Sciences (SLU), Alnarp, Sweden*

Among the causes of post-weaning problems in pigs may be a lack of water. The feed and water consumption of two pigs per pen were studied during three days after weaning. Half of the pens studied had water added to the pelleted dry feed (1.8:1) at feeding during the day (WF=wet feed, DF=dry feed). During the night, all pigs had access to dry feed. Feed consumption during day and night time were registered daily. Within each treatment, the pigs in half of the pens had free access to water from a nipple (WN) and the other half from a water bowl (WB). These were connected to one bucket each, and every morning the remaining water was weighed and new water given and weighed. Underneath the water nipples and bowls were boxes (40*30*14 cm) with perforated tops placed to collect waste water.

The WF-pigs had a significantly higher daily feed consumption ($p=0.0018$) than the DF-pigs and a tendency of higher daily gain ($p=0.0981$). The greatest difference in feed consumption was during day time. Water intake was not significantly different. No significant difference was found between WN- and WB-pigs regarding daily gain or feed or water consumption. There was significantly more waste water for WN-pigs compared to WB-pigs.

Session 18 Poster 40

Relation between activity of enzymes in boar sperm plasma and breeding value of the boar

J. Owsianny, B. Matysiak, A. Konik and A. Sosnowska. Department of Pig Breeding, University of Agriculture in Szczecin, ul. Dr Judyma 10, 71-460 Szczecin, Poland

The research objective has been to determine a relation between activity of AspAT and GGTP in sperm plasma and sexual activity and sperm characteristics of boars. The tested group included 95 boars (14 of Pietrain breed, 35 of Duroc breed and 46 of the 990 Polish synthetic line). Tests of libido and sperm started on the 180th day. Three ejaculates collected at 7-days' intervals were tested. The collected ejaculates were subjected to quantitative and qualitative analyses. Activity of tested enzymes in all three ejaculates of every boar was determined with the kinetic method. A statistically significant relation of AspAT activity ($r= -0.261^{**}$) and GGTP activity ($r= -0.239^{**}$) and the ejaculation time was found. AspAT was significantly correlated ($r= -0.228^{*}$ and $r= -0.224^{*}$) and GGTP was very significantly correlated ($r= -0.357^{**}$ and $r= -0.374^{**}$) with the general volume of sperm and the volume of sperm fraction. It has been found that activity of tested enzymes was more correlated with quantitative characteristics of sperm (volume, concentration of spermatozoa, general number of spermatozoa in ejaculate) than with qualitative characteristics (percentage of major and minor changes, percentage of changes in acrosome). The achieved results cannot be a base for an univocal conclusion that measurements of enzymes activity are useful while determining future breeding usefulness of boars.

Session 18 Poster 41

The reproductive value of 990 line young boars after primiparae or multiparae sows considering equalization of "maternal effect" related to litter size both before and after the birth

R.Czarnecki[1], M. Rózycki[2], M. Kawecka[1], A. Pietruszka[1], B. Delikator[1], M. Kamyczek[2], [1]Department of Pig Breeding, Agricultural University in Szczecin, Poland, [2]Institute of Animal Husbandry in Kraków, Poland

The study was aimed at determining the effect of age of the boar's dam (primipara or multipara) on his reproductive value with equalized effect of pre- and postnatal maternal environment conditions related to original litter size.

The study included 135 young boars (39 after primiparae and 96 after multiparae). All boars were delivered in litters numbering 13 to 14 piglets per litter. All litters were standardized to 12 piglets per litter on the first day after the birth. The assessment of reproductive value of these boars was initiated when they turned half a year and was made on three ejaculates. Mean values for the analysed traits of reproductive value in boars after primiparae or multiparae were respectively: volume of both testes on day 180 - 275.25 vs. 237.93 cm^3, ejaculation volume after filtration - 109.86 vs. 106.96 cm^3, spermatozoa concentration in cm^3 $x10^6$ - 208.93 vs. 167.99, total spermatozoa number in ejaculate $x10^9$ - 22.71 vs. 18.12, spermatozoa percentage with normal acrosome - 85.00 vs. 81.11.

The study was financed by Committee for Scientific Research (grant No 6 P06E 035 20).

Session 18 Poster 42

Workers influence on reproductive performance of gestating sows

J. Morales, L.M. Ramírez, S. Ayllón, M. Aparicio and C. Piñeiro, PigCHAMP Pro Europa, S.A., Segovia, Spain*

Swine producers need to computerize records to track production traits and costs quickly and efficiently, diagnose problems and make informed decisions, accordingly Management decisions taken at the farm level may affect sow productivity and the future of the herd's overall performance. In this study, we used data provided by database PigCHAMP® of a 2000 sow farm to assess differences in fertility depending on the worker due to the influence of the insemination. Seven workers were evaluated and only sows with a weaning to service interval of less than 7 d were included. Fertility (range 87.8-80.0%; P=0.04) and regular returns (range 10.4-3.9%; P=0.08) were affected by the worker, whereas irregular returns and abortions were not. This observation was expected because regular returns are mostly consequence of problems in oestrus detection, and time and quality in the insemination procedure. Assuming a performance of 2.35 parturitions/sow/year and regular returns length of 21 d, the differences due to the influence of the worker reaches a total of 6,500 (3.25 per sow) non productive days (NPD) in this farm. Depending on the value of a NPD, the minimum cost per year is of 9,750 (4.9 per sow). We conclude that the control and assessment of the capability of the workers for the mating process is a key factor to improve performance and decrease cost in the gestation unit.

Session 18 Poster 43

Optimal standardised ileal digestible (SID) lysine level in hybrid meat pigs (70-110 kg)

*N. Warnants, M.J. Van Oeckel, M. De Paepe and D.L. De Brabander**
Ministry of the Flemish Community, Department Animal Nutrition and Husbandry, Scheldeweg 68, 9090 Melle, Belgium

A growth trial was performed with Pietrain x Hybrid pigs between 70 and 110 kg live weight. Six diets with SID lysine levels from 0.47 to 1.04% were fed to 55 barrows and 55 gilts per treatment. Barrows and gilts were housed separately in pens of 5. Diets were cereal-soybean meal-based and balanced in threonine, methionine + cystine and tryptophan relative to lysine, according to the ideal protein. The diets were isoenergetic, with a net energy of 9.2 MJ/kg feed. The SID amino acid content of the diets was verified in an ileal digestion trial with 4 pigs. For barrows, there is only a significant lysine effect from 70 to 80 kg, above 80 kg the lowest SID lysine level of 0.47% fulfils the requirement. The average daily gain (ADG) and feed conversion ratio (FCR) data, from 70 to 80 kg for barrows and from 70 to 110 kg for gilts, as a function of %SID lysine are described by quadratic curves. The optimum SID lysine levels in the above intervals, derived from the ADG and FCR curves, are respectively 0.77 and 0.82% for barrows and 0.83 and 0.94% for gilts. The observed requirements are, compared to other recommendations, elevated for gilts and rather low for barrows.

Session 18 Poster 44

Feeding activity and other activities during the grazing journey of the Iberian pig in the montanera fattening period

V. Rodríguez-Estévez, A. Martínez, A.G. Gomez and C. Mata, Departamento de Producción Animal, Facultad de Veterinaria, Campus de Rabanales, Univesity of Cordoba, 14004 Córdoba, Spain

A group of Iberian pigs have been studied in the montanera fattening period (between 100 to 160 Kg. and since 1th november to 15th january) at the dehesa agrosystem (clear forest of *Quercus ilex*) in order to study their feeding activity and grazing behaviour. From this group a sample of 5 randomly choosen and different pigs have been followed during 7 whole grazing journeys ((10 ininterrupted hours, since 8:30 to 18:30) every 10 days. The results show the following means: 7 hours and 6 minutes of grazing activity, 97,2 visited holm oaks, 1468,5 consumed acorns, 1447,9 grass bites, 67,7 rooting times, 9,1 water consumptions, 1,3 mood baths, 5,8 scratches, 5 urinations and 10 defecations. Statistical analysis of data shows a great individual variations. These could explain final fattening performance differences and quality when this is measured by the fatty acid profile analysis.

Session 19 Theatre 1

Ideal protein to improve protein utilization of sows during gestation and lactation

S.W. Kim[1], F. Ji[1], G. Wu[2], and R.D. Mateo[1], [1]Texas Tech University, USA, [2]Texas A&M University, USA*

Lactating sows (40) and pregnant gilts (20) were used to evaluate if ideal amino acid (AA) formulation improves protein utilization of sows during gestation and lactation. Ideal dietary AA profiles for lactating sows and pregnant gilts were obtained from our previous studies. Lactating sows (or pregnant gilts) were allotted to two dietary treatments: CON-Lac (or CON-Prg) for corn-soybean meal formulation or IP-Lac (or IP-Prg) for ideal AA formulation. AA ratios were 100:67.9:86.4:97.6 (lys:thr:val:arg) for CON-Lac (or 100:76.5:100.0:127.0 for CON-Prg) and 100:63.6:78.4:53.0 for IP-Lac (or 100:71.8:69.0:94.4 for IP-Prg). Sow and litter weight measurements as well as blood sampling were done at d 0, 7, 14, and 21 of lactation for lactating sows. For pregnant gilts, body weight measurements and blood sampling were done at d 30, 60, 90, and 109 of gestation and d 5 and 17 of subsequent lactation. Blood urea nitrogen (BUN) contents were analyzed as an indicator of protein utilization. Sows in IP-Lac had lower ($P < 0.05$) BUN contents than sows in CON-Lac during the entire lactation period. Gilts in IP-Prg had lower ($P < 0.05$) BUN contents than gilts in CON-Prg on d 90 and 109 of gestation. These results indicate that the ideal amino acid formulation can improve protein utilization of sows during lactation and pregnancy.

Session 19 Theatre 2

Effect of organic selenium on tissue Se content in growing-finishing pigs

R.D. Mateo[1], J.E. Spallholz[1], F. Ji [1], R.. Elder[2], I.K. Yoon[3], and S.W. Kim[1], [1]Texas Tech University, [2]Seaboard Farms, [3]Diamond V Mills Inc., USA*

This study was done to determine the efficacy of organic Se (OSe; Seleno*Source*™ AF, Diamond V Mills, Inc., Cedar Rapid, IA) on loin and liver accumulation relative to inorganic Se (ISe; sodium selenite). A total of 180 pigs at 34.4±0.06kg were allotted to five dietary treatments (6 pens/treatment and 6 pigs/pen) with different Se supplementation levels: 0, 0.135, 0.243 and 0.361ppm of supplemental Se from OSe, and 0.288ppm supplemental Se from ISe. Experimental diets were changed at 66.1±0.53kg and 99.0±0.93kg. At 129.9±1.35kg, pigs were transported to a local abattoir (Seaboard Farms Inc., Guymon, OK) to obtain loin and liver samples for Se analysis. Se contents were 0.323, 0.367, 0.388, 0.431, and 0.400ppm in the liver and 0.131, 0.195, 0.208, 0.335, and 0.200ppm in the loin from five treatments listed above, respectively. Contents of Se in liver ($y=0.324+0.292x$, $P<0.001$, $R^2=0.33$) and loin ($y=0.122+0.511x$, $P<0.001$, $R^2=0.56$) increased linearly as dietary OSe level increased. Liver Se content was similar for both OSe and ISe. The effect of dietary ISe supplementation at 0.288ppm Se on loin Se accumulation is estimated to be equivalent to the dietary supplementation of OSe at 0.153ppm Se based on the equation obtained above. The results of this study indicate that OSe was 188% more efficient in accumulating Se in loin compared to Ise.

Preference of piglets for mash or granulated cereal diets

D. Solà-Oriol[1], E. Roura[2] and D. Torrallardona[1], [1]IRTA-Centre de Mas Bové, Reus, Spain, [2]Lucta SA, Barcelona, Spain*

Pelleting modifies the characteristics of cereal feed ingredients and it may also change their palatability. A trial was conducted to study how granulation may affect the palatability of barley or oats at different levels of inclusion compared to rice based (reference) diet in double-choice feeding tests. A total of 36 four-piglet pens were used and offered simultaneous access to two diets; the reference and the test diet containing increasing amounts (0, 20, 40 and 60 %) of barley or oats in replacement of rice from the reference diet in three consecutive 4d periods. All the diets were tested in mash and pelleted form. The preference for each treatment was calculated as the percentage contribution of the test diet to total feed intake. It was observed that piglets preferred pellets to mash ($P<0.05$) and barley resulted in better preferences than oats ($P<0.01$). An interaction between technological process and ingredient was reported ($P<0.01$); so that the improvement in palatability due to granulation only occurred for barley-based diets. Furthermore there was an interaction ($P<0.01$) between ingredient and its inclusion level so that barley improved palatability of the reference diet at 20% but reduced it at 40% and 60%, whereas oats reduced the palatability at all levels. It is concluded that pelleting improves palatability in barley but not in oats based diets for pigs.

Time-length of postweaning starvation period affects adaptation of pancreatic secretion in piglets

A. Huguet, G. Savary, E. Bobillier, Y. Lebreton and I. Le Huerou-Luron, UMR INRA-Agrocampus SENAH, 35590 Saint-Gilles, France*

At weaning abrupt dietary changes result in a typical low feed intake period, the length of which is highly variable between individual. These changes require digestive function adjustment for nutrient digestion. However less is known on effects of the length of the starvation period on the adaptibilty of pancreatic secretions. Piglets weaned at 33 d of age were either fed a daily increasing amount of feed from d1 to d5 (high level; H group) or were highly restrained for the first 3 postweaning d and then given increasing amount of feed (low level; L group). In L group feed intake at d5 was equivalent to intake of H piglets at d3. Pre-and postprandialy (30 min periods) pancreatic juice flow was measured and partially collected for enzyme analysis. Effects of time and feed intake were tested by variance analysis using GLM procedure of SAS. Pre-and postprandial protein and trypsin flows increased with increasing amounts of feed intake. At d5, postprandial values were 17- and 412-fold higher ($P<0.05$) than preprandialy in H group whereas they were only 7- and 3- fold higher in L group. Compared to H piglets at d3, postprandial trypsin flow and concentration were 40 times lower ($P<0.06$) in L piglets at d5, i.e. with similar feed intake. Pancreatic secretions adjusted to feed intake but a long starvation period immediately after weaning interfered with enzymatic adaptation.

Session 19 Theatre 5

Increased litter size as an effect of fat and polyunsaturated fat in sow diets

I.J. Wigren, M. Neil and J.E. Lindberg, Swedish University of Agricultural Sciences (SLU), Uppsala, Sweden*

There are a limited number of studies available concerning the effects on sow and piglet performance as regards dietary fat level in relation to fatty acid composition. The objective of this study was to investigate the influence of dietary fat level and fatty acid composition on sow and litter performance. In this study 146 Swedish Yorkshire sows in parity 2-8 were used. The experimental period reached from weaning of the sows' previous litter until weaning of the experimental litter. Sows were fed one of three experimental diets: a conventional low fat (3%) sow diet (LF), a high fat (6%) diet with saturated to monounsaturated fat (HFS) or a high fat (6%) diet with mono-/polyunsaturated fat (HFO). Composition of all diets included oats. The HFO diet also included a high-fat (10% in DM) oats variety. Treatment significantly affected litter size ($p<0.05$) and number of piglets born alive ($p<0.01$) where HFO fed sows gave birth to the largest litters and highest number of liveborn piglets. As could be expected, piglets in these litters were lighter at birth ($p<0.05$). If the differences in litter size and live weight remain up to and after weaning, and to what extent, will be further analysed in this study. Neonatal survival of piglets will be particularly considered.

Session 19 Theatre 6

Effect of dietary fat of different sources on growth and slaughter performance of growing pigs and fatty acid pattern of back fat and intramuscular fat

H. Boehme[1], R.. Kratz[2], E. Schulz[1],G. Flachowsky[1] and P. Glodek[3], [1]Federal Research Centre of Agriculture (FAL), Institute of Animal Nutrition, Bundesallee 50, D-38116 Braunschweig, [2]Adisso GmbH, Frankfurt; [3]University of Goettingen, Institute of Animal Breeding and Genetics, Germany

To study the transfer of dietary fatty acids into body fat a feeding trial was conducted with 96 pigs, which differed in their genetic construction (progeny of sire lines Du, Ha x Du, Pi x Ha). Dietary treatments were supplements of 2.5% tallow, olive oil, soyabean oil or linseed oil (increasing percentage of poly-unsaturated fatty acids) to a cereal/soyabean meal diet.

The 2-factorial analysis of performance data demonstrated that the treatment "fat source" had no significant effect on liveweight gain and energy conversion ratio. These parameters were significantly affected only by the treatment 'genetical construction'. Fat source and genetic construction also did not influence lean meat percentage as well as protein and fat content of the carcass. However, the fatty acid composition of backfat and intramuscular fat of the *M.longissimus dorsi* was significantly affected by PUFA-intake. For the concentration of PUFA in backfat (y) and PUFA consumption (x) a close relationship was analysed ($y = 3.73x + 0.910$; $R^2 = 0.85$) and also for the PUFA content in the *M. longissimus dorsi* ($y= 3.165x + 0.605$; $R^2 = 0.91$).The increased PUFA content was found to have a negative effect on sensoric parameters of meat and a decreased oxygen stability of back fat.

Effect of feed processing on size of (washed) faeces particles from pigs measured by image analysis

P. Nørgaard[1], L. F. Kornfelt[1], C. F. Hansen[2] and T. Thymann[1], [1]Department of Basic Animal and Veterinary Sciences, KVL, Grønnegårdsvej 2, DK-1870 Frederiksberg, [2]The National Committee for Pig Production, Axeltorv 3, DK-1609 Copenhagen

Our aim was to study the effect of dietary physical characteristics on size of washed faecal particles from finishing pigs (98±6 kg). The pigs were fed either a finely ground pelleted diet (FP) or a coarsely ground meal diet (CM) comparable in all ingredients. The diets were based on barley, wheat and soybean meal as the main ingredients and were either ground through a 2 mm screen before pelleting or ground to pass a 4.5 mm screen. Faecal samples were collected from 7 FP and 8 CM fed pigs and washed in nylon bags before drying and sieving into 3 fractions: >1 mm, 0.5-1 mm and bottom bowl. Samples of particles from the sieving fractions were scanned and the length and width of individual particles were identified using image analysis software. The overall mean, mode and median were estimated from a composite function. The dietary physical characteristics significantly affected the proportion of faecal particles retained in the bottom bowl and >1mm sieving fractions ($P = 0.001$). Particle dimensions in each sieving fraction ($P = 0.05$) were altered as well as the overall means (P = 0.01) 2.7±0.2 vs. 0.77±0.03 (length) and 1.45±0.22 vs. 0.47±0.04 (width) for CM and FP treatment, respectively.

Investigations on the tryptophan requirements of weaned piglets

F.X. Roth[1], T. Ettle[1], C. Relandeau[2], L. Le Bellego[2] and J. Bartelt[3], [1]Technical University of Munich, Department of Animal Science, Division of Animal Nutrition and Production Physiology, Hochfeldweg 6, 85350 Freising-Weihenstephan, Germany, [2]Ajinomoto Eurolysine, 153 rue de Courcelles, 75817 Paris Cedex 17, France, [3]Lohmann Animal Health, P.O.B. 446, 27472 Cuxhaven, Germany*

The present study investigated the dietary tryptophan requirement in 48 ad libitum fed weaned crossbred piglets (7.2 ±1.0 kg), subdivided into 5 experimental groups. Dietary tryptophan concentrations for groups 1 to 5 were 1.75; 1.96; 2.18; 2.34 and 2.59 g/kg for days 1 to 25 and 1.68; 1.88; 2.08; 2.29 and 2.61 g/kg during days 26 to 47, respectively. Daily feed intake, average daily gain and feed to gain ratio responded in a linear manner to the increasing tryptophan supply during the first 25 experimental days. Therefore, the dietary tryptophan requirements for this period were estimated to be at or above the highest dietary tryptophan concentration on test (2.59 g/kg; Trp:Lys ratio of 23%). As in the first experimental period, performance in the second half of the experiment was as well significantly increased by the increasing dietary tryptophan supply. In the second part of the experiment, dietary tryptophan requirement to maximize feed intake, growth and feed to gain ratio in parallel was estimated from nonlinear regression analysis to be 2.49 g/kg diet (Trp:Lys ratio of 23%). Data of the present experiment indicate higher dietary tryptophan requirements compared to current recommendations.

Session 19 — Theatre 9

Effects of breeder age and egg storage on hatchability, time of hatch and supply organ weights in quails

I. Yildirim and A.Aygun, Selcuk Universitesi, Ziraat Fakultesi, Zootekni ABD, 42031, Kampus, Konya, Turkey

Two thousand four hundred quail (*Coturnix coturnix japonica*) eggs were used to determine the effects of breeder age and storage time on egg traits, hatching traits and some selected organ weights in newly hatched chicks. Eggs from two commercial flocks (Young; 10 wk-old-age and Old-40 wk of age) of the same strain (*Coturnix coturnix japonica*) under the same management and nutritional regimen were incubated. Eggs were stored for 4 or 14 d prior to setting in incubator. Percent wet albumen, yolk and shell weights, conductance (G) were recorded immediately after the collections of eggs. All eggs were set an experimental setter and incubated in uniform conditions for around 19 d (448 h). Fresh egg weight, chick weight and percent yolk weight were significantly higher in eggs obtained from breeders at 40 wk of age, whereas percent albumen weight was significantly higher in eggs obtained from breeders at 10 wk of age. There was a significant storage treatment x hen age interaction for water loss in chicks, early death, late death, internal pipping, external pipping and hatchability of fertile eggs. Percent of hatched chicks were influenced the storage treatment x hen age interaction from 405 h to 441 of incubation. The jejunum length from 40-wk-old breeders was higher when eggs stored for 4 d rather than 14 d storage. At hatch, percentage liver weight from 10 wk old was higher when eggs stored for 4 d rather than 14 d storage. It was concluded that the effects of extended storage of fertile eggs on hatching traits were different for eggs laid by young and old breeders.

Session 19 — Theatre 10

Growth performance, meat yield and abdominal fat of broilers subjected to early feed restriction

M.A. Mohamed[1], Amal E. El-Sherbiny[2], Akila S. Hamza[3] and T.M. El-Afifi[3], [1]Department of Animal Production, National Research Center, Dokki 12622, Egypt, [2]Animal Production Department, Fac. of Agric. Cairo Univ, [3]Central Lab. For Food and Feed (CLFF), Agric. Res. Center, Giza, Egypt

A number of 420 day old male Ross chicks raised on a commercial starter diet from 0 to 6 days of age were distributed among three groups. The control group had access to feed. The other groups were subjected to feed restriction using a maintenance diet (1.5 x $BW^{0.67}$ Kcal ME/day, 40 Kcal/bird/day) or diluted diet (50% sawdust) from 7 to 12 days of age. Different refeeding regimens were applied. At market age, birds subjected to early feed restriction failed to compensate their growth and gave significant less body weight compared to the control birds. Significant effect ($P<0.05$) of feed restriction method was observed on abdominal fat (% of body weight). Birds fed the diluted diet during feed restriction period gave higher ($P<0.05$) abdominal fat compared to birds fed the maintenance diet. No significant effects were detected on meat yield (breast and dark meat) and weights of giblets due to either feed restriction or refeeding methods. It could be concluded that broilers severely restricted in feed early in their life were unable to catch-up in body weight at market age.

Session 19 Theatre 11

Effect of nonphytate-P level and phytase supplementation on performance and bone measurements of broiler chicks

Amal E. El-Sherbiny[1], H.M.A Hassan[2], Y.A.F.Hammuda[2] and M.A Mohamed[2], [1] Animal Production Department, Fac. of Agric. Cairo Univ [2]Department of Animal Production, National Research Center, Dokki 12622, Egypt

A basal diet containing 0.14% nonphytate-P (NPP) was supplemented with three levels of NPP to reach the chick requirements (0.45%). Phytase enzyme (500 U/Kg) was added to such diets. One day old Ross broilers were fed these diets from 1 to 35 days of age.
Weight gain and feed intake increased with increasing level of dietary NPP. Also feed conversion ratio significantly ($P<0.05$) improved. Addition of 500 U phytase/ Kg improved body weight, feed intake and feed conversion ratio equal to that occurred when the dietary NPP was increased by 0.1%.
Length of tibia and weights of tibia and toe significantly ($P<0.001$) increased with increasing dietary NPP level. The addition of phytase enzyme increased ($P<0.01$) length of tibia and improved weights of tibia and toes ($P>0.05$) and ash percentages ($P<0.001$) in tibia and toe.
It could concluded that the addition of 500 U phytase/Kg of broiler chicks fed corn-soybean meal diets containing different levels of NPP improved growth performance and bone measurements.

Session 20 Theatre 1

Dairy cow health and the effects of genetic merit for milk production, management and interactions between these: blood metabolites and enzymes

B. Beerda*, W. Ouweltjes, J.J. Windig, M.P.L. Calus and R.F. Veerkamp, Animal Sciences Group of Wageningen University and Research centre, The Netherlands

Outbreaks of, sometimes inexplicable, diseases trigger societal concerns about the health and welfare of high producing dairy cows. This study investigates health risks in high producing dairy cows and addresses different aspects of high yield, namely genetic merit, management and interactions between these. In a 2x2x2 factorial design, we compared HF Heifers (n=99) of high (HGM) or low genetic merit (LGM) for milk yield that were milked 2 or 3 times a day and fed a Mixed Ration with high or low energy content. Blood samples were taken from prepartum till 12 weeks postpartum, and analysed for glucose, non-esterified free fatty acids (NEFA), ß-hydroxybutyrate, albumin, calcium and enzymes. Energy balances were poor around week 2 and compromised by low energy ration and 3 times milking per day. NEFA levels were especially high during week 2 in HGM cows fed low energy ration. In general, HGM animals had relatively high levels of NEFA and aspartate aminotransferase, but low albumin. Calcium levels were most reduced in HGM cows, milked twice a day, that were fed low energy ration instead of high energy ration. Results show that interactions between genetic merit for high milk production and management affect the relationship between dairy cow production and health.

Methodology of breeding value estimation for functional longevity in Czech Republic

E. Páchová[1,2], L. Zavadilová[1], J. Sölkner[3], [1]Research Institute of Animal Production in Prague, [2]Czech University of Agriculture in Prague, [3]University of Natural Resources and Applied Life Sciences Vienna

The aim of this study was to finish a methodology of breeding value estimation for functional longevity in Czech Republic. An analysis of productive life in Holstein cattle was performed using programs package Survival Kit, based on Weibull. Data included 230 028 cows. LPL was observed between years of first calving 1990 and 2004. Records on cows that were still alive in the end of study were treated as censored (19.4 %). The model included four time-dependent effects (interaction of lactation number and stage of lactation, interaction of year and breed, interaction herd-year-season and class of milk production within herd and year), time-independent fixed effect of age at first calving and random effect of sire. The highest risk of being culled was found for the cows on the first lactation in the first (0 - 60 days) and last part of lactation (240 and more days) and for the cows on further lactations on last part of lactation. The hazard decreased with parity. The lowest risk of being culled have cows in the best class for milk production. Cows which are younger at first calving had lower risk of being culled. Breeding values of sires were expressed through hazard of daughters. They varied between 0,4 and 2,5.

Relationship of metabolic traits and breeding values for milk production traits

G. Freyer[1], R. Staufenbiel[2], E. Fischer[3] and L. Panicke[1], [1]FBN 18196 Dummerstorf, [2]Freie Universität 14163 Berlin, [3]Universität 18059 Rostock, Germany*

Breeders are interested in early information on evaluating growing dairy bulls. The contemplation of genetic and physiological aspects seems to be important. Metabolic reaction in young sires contains information on genetic potential of daughters performance. Insulin has a central function in regulation of energy metabolism. Methodically, the insulin function can be measured using the glucose tolerance test (GTT). Individual physiological reaction to glucose challenge in young dairy bulls, being probed at one year of age, was tested by means of the GTT. The heritability coefficients of glucose half life time and glucose area for original and logarithmic values range from 0.33 to 0.59. Calculating breeding values (CBV) from 83 bulls with complete observations on these traits and estimated breeding values, based on linear regression. Several combinations, involving metabolic parameters and ancestral predicted breeding values (PBV) were considered. Comparing correlation coefficients (r) between CBV and EBV of 15 combinations enabled us to find the most promising combination for predicting breeding values. Protein is being used as equivalent of metabolism evaluation. A medium correlation of about r= 0.38 is reached between glucose and protein yield. Involving PBV from the pedigree leads to low increase of r=0.40. An additional recommendation for the evaluation of breeding bulls before starting the test could be given, if the results will be confirmed.

Session 20 Theatre 4

Genetic effects of embryo transfer in dairy cows

F. Bosselmann, S. König and H. Simianer, Institute of Animal Breeding and Genetics, Georg-August-University, Göttingen, Germany*

Different genetic and environmental traits have an impact on a successful flushing of a donor and finally on developing a viable calf with MOET. The objective of this study was to estimate the environmental and genetic effects for the number of flushed, the number of suitable and the number of fertilised embryos. For this analysis 4695 flushings were available, done during five years in four breeding organisations. The fixed effects of the breeding organisation and hormone type were highly significant compared to the season. However more embryos were developed in the cold season. Additional information about physiological stress like time of lactation and milk rating, the feed composite as measured by the fat/protein percentage and the milkurea and the udder fitness in form of the somatic cellcount were entered into the models. Data was Cox-Box transformed to achieve normal distribution. In a synergistic model, the genetic influence of the donor, the sire and the recipient were estimated. There is a negative correlation between the number of suitable embryos and the milk rating as a factor of physiological stress. The estimated heritability of the maternal part for the number of flushed embryos is 0,22 and 0,12 for the suitable embryos. Heritability for the fertilised embryos is 0,05. The phenotypic correlation between flushed and suitable embryos is 0,67 and the genetic correlation between these two traits is 0,83.

Session 20 Theatre 5

Duration of mobilization period in dairy cows

D. Bossen and M.R. Weisbjerg, Danish Institute of Agricultural Sciences, Foulum, Denmark*

Duration of the mobilization period after parturition for 239 lactations was analysed using Proc Lifetest in SAS for influence of breed (Danish Red, Danish Holstein, Danish Jersey), parity (P1: first lactation, P2: second and later lactations) and energy concentration in the ration. Beside 3 kg of concentrate fed separately, cows were fed ad lib. with a mixed ration with either medium (T1: 26 %) or high (T2: 45%) concentrate proportion, resulting in energy concentrations in total rations of 7.4 and 7.7 MJ NE_L/kg dry matter, respectively. Data originates from an ongoing production trial with live weight registration at each milking. Duration of mobilization period expressed as days after parturition to minimum live weight continued longer for P2 compared to P1, and for T1 compared to T2. There was no influence of breed. 25, 50 and 75% of the cows in P1T1 obtained their weight minimum after 35, 56 and 80 days respectively. Similarly quartiles were 25, 45 and 64 days for P1T2; 55, 70 and 93 for P2T1 and 34, 54 and 89 days for P2T2, respectively. Total average live weight loss corresponding to P1T1, P1T2, P2T1 and P2T2 were 69, 42, 53 and 41 kg, respectively. This analysis demonstrates that duration of the mobilization period as well as weight loss in kg is highly variable between individual cows, and is highly affected by the ration energy concentration.

Session 20 Theatre 6

Fatty acid composition of intramuscular and subcutaneous fat from Limousin and Charolais heifers supplemented with extruded linseed

L. Barton, V. Teslík, V. Kudrna, M. Krejcová, R. Zahrádková, D. Bures, Research Institute of Animal Production, Prátelství 815, 104 00 Prague - Uhríneves, Czech Republic*

Fatty acid composition of the longissimus thoracis and subcutaneous fat from totally 46 Limousin (LI) and Charolais (CH) heifers was determined. The animals were fed two isocaloric and isonitrogenous diets differing in content and source of dietary fat (EXP and CON). Both mixed diets consisted of maize silage, hay, straw and concentrates. The EXP diet was, in addition, supplemented with extruded linseed (high in C18:3n-3). CH heifers had higher proportions of C16:0 and C18:0 ($P<0.05$ and $P<0.001$, respectively) while LI heifers had higher proportions of C18:1c9, C18:2n-6 and CLA ($P<0.001$) in the longissimus thoracis. These differences resulted in significantly higher MUFA:SFA, PUFA:SFA and n-6/n-3 PUFA ratios in LI ($P<0.001$). Including linseed in the EXP diet increased C18:3n-3, CLA, EPA, DPA and DHA ($P<0.05$) and resulted in a lower n-6/n-3 PUFA ratio ($P<0.001$) in the longissimus thoracis. Subcutaneous fat generally contained more MUFA and less SFA and PUFA. The study was supported by the project MZE0002701403.

Session 20 Theatre 7

Bovine intramuscular adipose tissue has a higher potential for fatty acid synthesis from glucose than subcutaneous adipose tissue

J.F. Hocquette[1], C. Jurie[1], M. Bonnet[1], D.W. Pethick[2], [1]INRA, Herbivore Research Unit, Theix, France, [2]CRC, Murdoch University, Perth, 6150, Western Australia

Intramuscular fat (IMF) deposition (marbling) influences many quality attributes of beef meat. In ruminants, acetate (which is converted into acetyl-CoA) is the major precursor of fatty acids (FA) synthesised *de novo*. It has been suggested, however, that glucose is an important precursor of acetyl-CoA, and hence FA, within intramuscular adipocytes. To test this hypothesis, enzyme activity or gene expression of rate-limiting steps in glucose metabolism were assessed in intramuscular (IMAT) and subcutaneous (SCAT) adipose tissues. Eight Angus and 12 Limousin steers were slaughtered at 23-28 months of age after a long finishing period with a cereal-rich diet to increase IMF deposition. IMAT was dissected from *Longissimus* muscle and was not contaminated by fibres based on specific enzyme activities. On average, GLUT4 mRNA level (a transporter regulating glucose uptake by cells), phosphofructokinase (PFK) and ATP-citrate lyase (ATP-CL) activities (two enzymes of the pathway converting glucose into acetyl-CoA) were higher in IMAT than in SCAT (at least x1.4, x25, x5.0 for GLUT 4, PFK and ATP-CL, respectively, $P<0.05$). In conclusion, compared to SCAT, IMAT has a higher potential to take up glucose from blood and to convert it into acetyl-CoA, which supports earlier work showing that a combination of glucose and lactate were more effective substrates than acetate for *de novo* lipogenesis in IMAT.

Session 20 Theatre 8

Lean meat estimate on half carcass crossbreed bulls

M. Iacurto and D. Settineri, CRA, Animal Science Research Institute, via Salaria, 31, 00016 Monterotondo (Rome), Italy

A good prediction of carcass lean meat content (LMC) is both of experimental and commercial interest. The objective of this study was to found the simplest equation that allowed a good estimate of lean meat utilising the dissection data of the numerous trials performed in our Institute: 261first generation crossbreed (F1, meat breed x Friesian) and 146 second generation (F2, meat breed x F1) bulls. The weights of the total anatomical regions and their dissection components (muscles, bones, external and internal fat, other tissues) were considered. The data were processed developing a linear regression that take into account the relative importance of all the single utilised factors (forward option, SAS Inst., 8). The best estimate of LMC (407 observations R^2=0.985, RMSE=2.737) was based on 4 weights: neck meat and bone and two proximal pelvic limb muscles (*semimembranosus, gluteobiceps*); but these values supposed a complete and sometime experimental dissection; whereas a first estimate, satisfying economic and commercial purposes, that considered as variables only neck and proximal pelvic limb weights outlined 4 subjects (F2 group) which discarded heavily in respect to the other data. The final equation (403 observations; R^2 = 0.9979, RMSE = 3.23431) was: LMC_{kg} = 2.45201 ($P \leq 0.0828$) +2.73507*$neck_{kg}$ ($P \leq 0.0001$) +1.33176* proximal pelvic limb $_{kg}$ ($P \leq 0.0001$). The data mean values were: LMC_{kg} = 116.16±19.40; $neck_{kg}$ =20.14±4.20; proximal pelvic $limb_{kg}$ = 44.01±6.27; the estimated lean meat was: LMC_{kg} = 116.16±19.02 with a standard deviation between real and estimated values of ±3.85.

Session 20 Theatre 9

Body condition scoring in double-muscled Belgian Blue beef cows

L.O. Fiems, W. Van Caelenbergh, S. De Campeneere and D.L. De Brabander, CLO-Department Animal Nutrition and Husbandry, Scheldeweg 68, 9090 Melle, Belgium*

Body condition scoring (BCS) is a non-destructive technique to assess reserve tissue in cattle. We investigated the value of a 6-point scale ranging between emaciated and very fat. BCS was assessed in 24 non-lactating and non-pregnant cows, sacrificed with diverging parity and condition. It was correlated with empty body composition (g/kg): water, fat, energy, ash ($P<0.001$; $r>|0.8|$) and protein ($P<0.05$; r=-0.540). Variations of BCS and BW (body weight) during winter and subsequent summer periods were measured in 126 breeding cows, yielding 546 observations. Cows were subjected to different energy levels during the winter period: 100, 90, 80 or 70% of their requirements. A close correlation between BCS and BW was found ($P<0.001$; r=0.734), ranging from 0.737 to 0.746 for primiparous and multiparous cows ($P<0.001$). Primiparous cows had a lower BCS than cows with 2 or more calvings: 1.46, 1.71 and 1.84, respectively ($P<0.001$). BCS at parturition exerted no clear effect on subsequent calving interval (r=0.063; P=0.378). Within the same cow herd we checked the effect of BCS on calf survival within 4 months after calving. Dams whose calves were stillborn or died, exhibited a lower BCS than dams of surviving calves (1.45 vs. 1.75; P=0.044). Results suggest that BCS is a useful technique to estimate body composition in vivo in Belgian Blue double-muscled cows, allowing to make management decisions.

Session 20 Poster 10

A long-period study for the improvement of quantitative traits in a Hungarian Holstein-Friesian herd

I. Györkös[1], E. Báder[2], K. Kovács[1], J. Völgyi Csík[1], E. Szücs[3], T. Petró[1], A. Kovács[2], [1]Research Institute for Animal Breeding and Nutrition, Gesztenyés str. 1., 2053 Herceghalom, Hungary, [2]University of West Hungary, Vár 2., 9200 Mosonmagyaróvár, Hungary, [3]Szent István University, Páter K. str. 1., 2100 Gödöllö, Hungary*

15 basic body conformation traits, 6 artificially formed conformation category points and 6 milk production traits of 1118 Holstein-Friesian cows were scrutinized during a 3-years period study. Correlations were looked for between the registered conformation and milk production traits. Statistical analysis of database was carried out by analysis of variance and regression analysis with the help of SPPS 10.0. for Windows software. On the basis of description data of the population, an improving tendency could be found in conformation and production traits in the subsequent years. Analyses proved that there was a significant difference between the year-groups in case of all conformation and production traits. The ways of getting out and culling reasons were stated. The most significant culling reasons include infertility, mastitis, foot and leg disorders, ketosis and low milk production. Significant differences were found in milk production traits between groups created on the basis of final conformation scores. A loose, positive linear, but significant correlation could be detected between milk production and conformation. The results confirm the importance of the parallel improvement of functional and milk production traits.

Session 20 Poster 11

Growth and mature weight of female beef cattle of different breeds

Sz. Bene, M. Török, F. Szabó, University of Veszprém Georgikon Faculty of Agriculture Hungary

Growth and mature weight of female beef cattle of different breeds such as Hungarian Simmental, Hereford, Angus, Lincoln Red, Shaver and Blonde d'Aquitaine were evaluated and compared. The studied animals were born between 1983 and 2003 kept in the same environmental condition on peat bog soil pasture of the experimental farm of the Georgikon Faculty of Agriculture at Keszthely. During the experiment weight data of 203 female animals were taken permanently. Heifers were weighed in each second, while cows in each sixth months. Weight data were evaluated with analysis of variance moreover growth equations were determined. As a result it was found that the age data of heifers when reached the breeding maturity that is the 2/3 of the mature weight were: 20-, 19-, 20-, 18-, 22-, 20 month, respectively. Lincoln Red seemed to be the earliest while Shaver the latest maturity. The mature weight of the cows of studied breeds were: 634-, 515-, 602-, 627-, 672-, 721 kg, respectively. When Hungarian Simmental considered as 100% the relative mature weight of Hereford 81.2%, Angus 95.0%, Lincoln Red 98.9%, Shaver 106.0%, Blonde d'Aquitaine 113.7%. Age data of cows of different breeds when reached mature weight were: 5,50-; 4,85-; 5,00-; 4,05-; 4,70-; 3,80 year, respectively.

Preliminary data on body measurements and temperament of Aubrac heifers in Hungary

A. Szentléleki[1], J. Tözsér[1], Z. Domokos[1], R. Zándoki[1], C. Bottura[2], A. Massimiliano[2], [1]Ábrahám Cs. [1]Szent István University, Faculty of Agricultural and Environmental Sciences, H-2103 Gödöllö, Páter K. 1., [2]La Garonnaise Ltd, H-3773 Sajólászlófalva*

Authors' aim was to examine relationships between body measurements of Aubrac heifers, and as a new breed in Hungary, to evaluate its temperament. Experiment was carried out in a Hungarian farm in 2004 with 54 Aubrac heifers (age: 1.78±0.10 year, live weight: 415.5±44.31 kg). Body measurements (height at withers= 118.5±3.62 cm; hip height= 126.8±3.31 cm; chest circumference= 177.2±7.36 cm; slanting body length= 154±7,82 cm) were taken by measuring tape. Temperament (1.65±0.80 scores) was assessed in a 1-5 scale (1:calm, 5:nervous) while spending 30 seconds on weighing scale. All body measurements, age and live weight showed normal distribution. Significant positive correlation were observed between live weight and body measurements ($r=0.58$-0.85, $P<0.05$), and between all body measurements, in every relations ($r=0.5$-0.79, $P<0.05$). As a result of the regression analysis, height at withers was affected mostly ($R^2= 61\%$,$P<0.001$) by hip height, and slanting body length by live weight ($R^2= 54\%$,$P<0.001$). Among the parameters investigated, temperament showed correlation only with age ($r_{rank}=-0.27$,$P<0.05$). Examining these traits of Aubrac breed in Hungary for the first time, it was concluded that heifers showed calm behaviour. Correlation coefficients showed similar tendency to those obtained in other beef breeds.

Ethnic study of Minhota cattle breed: biometric and liveweight analysis

J.P. Araújo[1], H. Machado[2], J. Pires[1], J. Cantalapiedra[3], A. Iglesias[3], F. Petim-Batista[4], J. Colaço[4] and L. Sánchez[3], [1]Escola Superior Agrária Ponte de Lima, IP Viana Castelo, Portugal, [2]Associação Criadores Bovinos de Raça Minhota, Portugal, [3]Universidad Santiago de Compostela, Lugo, España, [4]Universidade Trás-Montes e Alto Douro, Vila Real, Portugal*

Data on 503 Minhota cows were analyzed, considering live weight, twelve zoometric measurements and eleven indices (2 ethnological and 9 functional). Two principal components analyses were computed: 1) weight and zoometric measurements, and 2) shape indices.

The measurements means of cows with 7.4±3.59 years were 560.4±91.78 kg (weight), 165.6±9.59 cm (body length), 189.9±11.07 cm (thorax perimeter) and 130.5±4.60 (front height).

Relatively to weight and zoometric measurements, the first principal component (CP1) accounted for 54.5% of the variation in the 13 variables, and provided a linear function of body size. The second principal component (CP2) accounted for an additional 10.9% of the generalized variance and was mainly related to height and chest width.

Relatively to shape indices, the CP1 accumulated 24.6% of the variance, being the most correlated variables the corporal, thoracic, shank circumference\thorax perimeter and shank circumference\chest width. The CP2 represented 21.4%, mainly obtained from the relative depth of chest, front height\body length, previous and posterior indices.

This analysis allowed us to evaluate the breed beef condition, due to the good relationship between the bones and muscle masses of the back, rump and leg. The proportion between thoracic capacity and diameter of the tubular bones reflects its aptitude for grazing.

Session 20 Poster 14

Genetic improvement of cattle population of Yayladag district of Hatay province by crossing of South Yellow x Brown Swiss breed and first results obtained

Ö.Sekerden, Y.Z.Güzey, Mustafa Kemal University, Faculty of Agriculture, Dept. of Anim. Sci., Antakya, Turkey*

In order to obtaining of cattle genotype (South Yellow x Brown Swiss) that it would have 2500-3000 kg lactation yield, good fattening and draft power performance by using Brown Swiss breed at the limited conditions of Yayladag District of Hatay Province were aimed in the Project which have been started at April of 2001 by financial support of Prime Ministry State Planning Organization of Turkish Republic

The material of the research was formed by cattle population at 2 seperate villages of Yayladag and two Brown Swiss bulls which had been given to the two villages .

Growth performance have been determined on F_1 animals. Fattening and milk production characteristics also will be determined on on the the F_1 and G_1 material of the project. Taking into consideration of the data, breeding animals will be selected. F_1 animals that have good performances will be kept, others will be inseminate with Brown Swiss bulls again for G_1 generation.

It is though that the Project will be completed in 2007. Until now (28 December, 2004) as a total of 100 F_1 calves were born in the two willages which were included in the Project. Most of females from F_1 generation are pregnant and most of them will be calve in 2005 year.

Session 20 Poster 15

Progeny performance of bulls differing in genetic index

M.G. Keane[1] and M.G. Diskin[2], [1]Teagasc, Grange Research Centre, Dunsany, Co. Meath, Ireland, [2]Teagasc, Athenry Research Centre, Athenry, Co. Galway, Ireland

Genetic indices for growth and carcass traits are available for Irish beef bulls. The objective was to compare feed intake, growth and carcass traits of progeny from low and high growth genetic index bulls. A total of 70 progeny, predominantly out of Holstein-Friesian cows, and 7 (3 low and 4 high index) Limousin bulls were reared together. Mean expected progeny differences for the low and high index bulls were 10 and 29 kg carcass weight. After slaughter the 6th to 10th ribs joint was separated into fat, muscle and bone. Growth rate was numerically higher for the high index progeny and kill-out proportion was significantly higher. This resulted in a significantly greater carcass weight. Carcass grades were not affected by genetic index but carcass measurements scaled for carcass weight indicated more compact carcasses for the high index progeny. Neither hind quarter proportion, ribs joint composition nor feed intake were affected by genetic index. It is concluded that two thirds of the extra carcass weight of the high index progeny was due to higher growth rate and one third was due to a higher kill-out proportion. While the mean difference between the high and low genetic index groups was in the expected range, the relationship between growth index and actual growth rate was poor.

Session 20 Poster 16

Comparison of productivity in mixed herds of local and commercial dairy cattle breeds

R. Rizzi, C. Maltecca, F. Pizzi, A. Bagnato, G. Gandini, Department of Veterinary Science and Technologies for Food Safety, University of Milan, Italy*

In many cases local breeds decline in number because of lack of economic profitability compared to other breeds. Performance evaluations of local breeds may enlighten possible strong points and provide elements for conservation decisions. The Reggiana, a local endangered Italian cattle, represents a typical process of displacement of a local breed by the more productive commercial Holstein. In the 1940' the Reggiana cattle counted more than 40,000 cows, then progressively declined to 500 cows in the early eighties. Since the 1990' the breed is progressively recovering. Profitability of Reggiana (RE) and Italian Holstein (IH) cows farmed in mixed herds were compared. Both milk traits and functional traits, including fertility (RE vs. IH: 387 vs. 417 d. of calving interval), milkability (RE vs. IH: 39 vs. 55 seconds/kg milk yield), longevity (RE vs. IH: 59 vs. 44 months of productive life) and diseases resistance were considered to estimate costs and revenues. Two possible strategies aimed to reduce the observed lower productivity of Reggiana cows were analysed: the development of a brand of Parmigiano Reggiano cheese linked to the local breed and the genetic improvement of some functional and productive traits. Finally the current and future significance of EC economic incentives to reduce the profitability gap between the two breeds are discussed.

Session 20 Poster 17

Functional polymorphism within the bovine growth hormone receptor (GHR) gene

L. Zwierzchowski[1], A. Maj[1] and A. Gajewska[2], [1]Institute of Genetics and Animal Breeding, Polish Academy of Sciences, Jastrzebiec, 05-552 Wólka Kosowska, Poland; [2]Institute of Animal Physiology and Nutrition, Polish Academy of Sciences, 05-110 Jablonna, Poland

The genes encoding GH and its receptor are promising candidate markers for selection purposes in cattle breeding. In this work, the effect of T/A transition in exon 8 the bovine GHR gene was studied on the receptor function and on meat production traits of the Black-and-White (Friesian) bulls. This mutation results in F/Y amino-acid substitution within the transmembrane domain of GH receptor and was reported by Blott et al (2003) as a strong QTL for fat content in cows milk. No marked effect was found, except for body weight at slaughter (TT bulls appeared heavier by 11.58 kg on average than those of AT genotype; $P \leq 0.05$), and weight of bones in the carcass (10.20 vs. 9,99 kg). Studied was also the relationship between the ligand-binding activity of growth hormone receptor and the A/T polymorphism in the bovine GHR gene. Receptor binding capacity (B_{max}) and dissociation constant (K_d) for GH-R were determined by the Scatchard method using ^{125}I-labelled GH as a ligand. Significant differences were found in the liver GH-R B_{max} between TT and TA genotypes (B_{max} = 16.613 and 10.993 fmol/mg tissue for TT and AT genotype, respectively; $P \leq 0.009$). No significant differences were found in the K_d values between genotypes.

Session 20 Poster 18

The effect of different calf rearing systems on calf's growth, feed consumption and behaviour and the cow's milk production

S. Fröberg[1], L. Lidfors[2], I. Olsson[1], A. Herrloff[1], K. Svennersten-Sjaunja [1], [1]SLU, Department of Animal Nutrition and Management, SE- 753 23 Uppsala, [2]SLU, Department of Animal Environment and Health, P.O. Box 234, 532 23 Skara*

The effect of restricted of restricted suckling and automatic milk-feeding in group-housed calves (A and B-calves) and milk feeding in teat-buckets in calves kept in single pens (C-calves), on weight gain and behaviour of the calves, and milk production in the cows was investigated in a pilot study. 23 Swedish Holstein cow-calf pair were followed from parturition. A-cows were milked and suckled twice daily and B+C-cows milked three times daily. Milk production, milk constituents, intake and composition of suckled milk were recorded. Amount and milk constituents for the entire lactation were available through the official milk recording scheme. Calves were weaned at weeks of age. General behaviour of A and B-calves were recorded two consecutive days every second week. A-cows tended to produce less energy corrected milk (ECM, kg) than B+C-cows but showed slightly improved udder health. Despite a higher daily intake of metabolisable energy from milk and concentrate until 9 weeks of age for the A-calves compared to the B- and C-calves, their weight gain was less. Contrary, the postweaning growth until 24 weeks of the A-calves was higher. B-calves tended to drink water more frequently and showed more explorative behaviour.

Session 20 Poster 19

Evaluation of fat depth of rump (P8) measured by real-time ultrasound machine in polled and horned Charolais young bulls

J. Tözsér[1], Z. Domoko[1], G. Holló[2], I. Holló[2], A. Szentléleki[1], R. Zándoki[1], M. Bujdosó[3], M.L. Wolcott[4], [1]Szent István University, Faculty of Agricultural and Environmental Sciences, H-2103 Gödöllö, Páter K. 1., [2]University of Kaposvár, Faculty of Animal Science, Kaposvár, [3]Charolais Ltd. Lajosmizse, [4]University of New England, Armidale,Australia*

Authors' aim was to test Falco100 (Pie Medical, 3.5 MHz head) ultrasound machine for measuring the P8 (fat depth of rump) in horned (n=13) and polled (n=23) Charolais young bulls, and to compare the scrotum circumference (SC) of the two groups. Research was carried out on a Hungarian farm at the end of self performance test, in 2004. Bulls were kept on deep litter, in small groups, fed on corn silage, hay and concentrate. The depth of rump subcutaneous fat (P8) was measured and calculated by Falco100 real time ultrasound machine. Results of LDS test showed that no significant effect of polledness either on P8 (horned=0.46 cm, polled=0.47 cm) or SC (horned=36.5 cm, polled=35.0 cm). P8 showed significant correlation ($r=0.42$, $P<0.05$) with live weight only in polled bulls. SC correlated positively with age (polled:$r=0.48$, $P<0.05$) and live weight (polled:$r=0.51$, $P<0.05$, horned:$r=0.51$, $P<0.05$). According to their results, ultrasonic measurements of P8 can be involved into the selection system of beef sire candidates, and gives additional information about fattening intensity of animals.

Weaning effect on the Galician veal quality acceptable for the protected geographical indication "Ternera Gallega"

T. Moreno, L. Monserrat, J.A. Carballo, L. Sánchez, N. Pérez, Mabegondo Research Centre, 15080 A Coruña; Spain*

The objective was to differentiate beef of "Normal" from "Suprema" class calves in the Protected Geographical Indication Galician Veal "Ternera Gallega", depending on whether they were weaned two months before or at slaughter time, respectively.
Fiftyeight "Rubia Gallega" breed calves were reared with their mothers on pasture and fed with complementary concentrate during one month before being finished indoors during two more months previous to the slaughter time at eight months of age.
Animals were assigned to a 2 (class) x 2 (sex) factorial experiment to analyse the chemical composition (humidity, ashes, fat and protein) and the fatty acid profile of the intramuscular fat on the *Longissimus thoracis* muscle.
An ANOVA (SAS) and discriminating analysis (Statsoft) were performed on the data. "Suprema" class of males relative to females showed higher humidity and ashes (76.52 vs 75.28 %, $P<0.001$ and 1.21 vs 1.17 %, $P<0.01$, respectively) and lower fat (0.86 vs 1.66 %, $P<0.01$). Significant differences between classes for the fatty acid profile were found. "Suprema" class showed lower PUFA (5.27 vs 6.15 %, $P<0.1$) and SFA (44.28 vs 48.69 %, $P<0.01$) and higher MUFA (50.45 vs 45.16 %, $P<0.01$) mainly oleic acid (1021.98 vs 794.78 mg/100 g muscle, $P<0.05$) than "Normal" class. The discriminating analysis correctly separated between the two classes and sex in 93.33% of the calves.

Physical, compositional and organoleptic properties of beef from Charolais and Limousin heifers fed different diets

D. Bures, L. Barton, V. Teslík, M. Krejcová, R. Zahrádková, Research Institute of Animal Production, Prátelství 815, 104 00 Prague - Uhríneves, Czech Republic*

The effect of breed and diet on physical, compositional and organoleptic properties of beef from *musculus longissimus lumborum et thoracis* were investigated in totally 46 purebred Charolais (CH) and Limousin (LI) heifers. The animals were fed two isocaloric and isonitrogenous mixed diets (EXP and CON) based on maize silage, hay, straw and concentrates. EXP diet was, in addition, supplemented with extruded linseed. The heifers were slaughtered at the target live weight of 490 kg and at the average age of 540 days. Meat color was significantly affected by breed when the meat from CH heifers was lighter ($P<0.001$) and less reddish ($P<0.001$). Dry matter and protein content was higher for LI ($P<0.05$ and $P<0.001$, respectively). Fat and cholesterol contents were similar for both breeds and dietary treatments. Eating quality was assessed by a trained panel using a 7-point scale. There were no differences between breeds with the exception of better scores for texture of CH samples ($P<0.05$). Odor, flavor and texture were rated lower ($P<0.001$) for samples from heifers fed EXP diet supplemented with linseed. The study was supported by the project MZE 0002701403.

Session 20 Poster 22

Carcass and meat quality of "Vitellone Bianco dell'Appennino Centrale" (IGP)

G. Preziuso, C. Russo, M. D'Agata, P. Verità, Department of Animal Production, University of Pisa, 56124 Pisa, Italy*

The European Economic Community has assigned the IGP label "Vitellone Bianco dell'Appennino Centrale" on meat produced from Chianina, Marchigiana and Romagnola cattle that were entered on the Genealogical Register and raised according to strict regulations. The aim of this study was to evaluate various carcass traits as well as the quality of meat from 40 Chianina bullocks that had been raised according to regulations regarding the "Vitellone Bianco dell'Appennino Centrale", slaughtered at the average age of 19 months and marketed with the IGP label. Twenty-four hours after slaughter, the carcasses were evaluated for conformation and state of fattening and a duoble steak was taken from each hindquarters, vacuum packaged and stored at 4°C. After a 14-day ageing period the *longissimus dorsi* muscle was isolated and analyzed to determine several quality traits: pH, meat colour, water holding capacity (drip loss and cooking loss), tenderness of cooked meat and chemical composition. The carcasses, weighing an average of 458 kg, showed good conformation (R) and adequate state of fattening (2.98), in line with the guidelines established by the regulations; even meat quality traits were in the ranges reported by regulations. The results confirmed the valuable characteristics of carcass and meat from Chianina bullocks, as well as the validity of the rearing methods.

Session 20 Poster 23

Effects of storage on colour and water holding capacity of meat

C. Russo, G. Preziuso, M. D'Agata, P. Verità, Department of Animal Production, University of Pisa, 56124 Pisa, Italy*

Meat colour and water holding capacity are very important to the consumer, both at time of purchase and at the moment of consumption, which generally occurs after a period (usually short) of domestic storage. To evaluate the aptitude for conservation of Chianina beef, 54o meat samples taken from the *caput longum triceps brachii* (180 samples), *longissimus dorsi* (180 samples) and *semitendinosus* (180 samples) were used and subjected to the following analyses: colour measured both before (Time 0h) and after 48 h of storage under standard conditions at 4°C (Time 48h) and water holding capacity evaluated as drip loss. The analyses of the results allowed us to identify some differences between the three muscles analysed: meat derived from *semitendinosus* always appeared to be lighter and paler and had significantly less drip loss. Storage did not seem to have any negative effects on the colour; the significant increase in values for b* and H* showed that the beef tended to become slightly paler, an attractive for the Italian consumer. Drip loss, indicating the amount of liquid lost by the meat during refrigeration, was generally very small. These results showed the good aptitude of Chianina meat for domestic storage, since colour and drip loss are closely related to the appearance of the meat.

Session 20 Poster 24

Castration effect on fatty acid profile of intramuscular fat in Charolais cattle

G. Holló[1], R. Zándoki[2], G. Pohn[1], E. Varga-Visi[1], J. Tözsér[2], Z. Domokos[3] and J. Csapó[1], [1]University of Kaposvár, Faculty of Animal Science, Kaposvár, H-7400 Guba Sándor str. 40, [2]St. István University, Faculty of Agricultural and Enviromental Sciences, Gödöllö, H-2103 Páter K. str. 1., [3]Charolais Ltd. Lajosmizse, Hungary*

During the last decades the consumer requirements for beef have been basically changed. Consumers prefer the meat of young animals, which are poor in fat and choose the adequately marbling meat because of its tenderness and eating quality. Within the frame of this study our aim was to examine the fatty acid composition of longissimus intramuscular fat of charolais bulls (n=6) vs. steers (n=5). The slaughter weight of animals was 651.63±60.32 kg. The dressing percentage came out at about 60.86±3.86 %, which is characteristic for the breed, however between the genders there were significant differences. We found out from our results, that the fat content of the right half carcass was significantly higher in case of steers. Considering eating quality the crude fat content of longissimus in steers can be seen more prosperous. The effect of gender on SFA and the n-3 fatty acid content were significant. In general steers produced more beneficial beef due to the higher quantities of n-3 fatty acids and PUFA.

Session 20 Poster 25

Glucose-6-Phosphate Dehydrogenase and leptin are linked to marbling differences between Limousin and Angus or Japanese Black x Angus steers

M. Bonnet[1], Y. Faulconnier[1], C. Leroux[1], C. Jurie[1], I. Cassar-Malek[1], D. Bauchart[1], P. Boulesteix[2], D.W. Pethick[3], J.F. Hocquette[1], Y. Chilliard [1], [1]INRA, Herbivore Research Unit, Theix, France, [2]UPRA France Limousin Sélection, Lanaud, [3]Murdoch University, Perth, 6150, Australia*

The aim of this work was to understand the metabolic basis for the variability of intramuscular adiposity (marbling) in cattle. We studied pathways related to lipid accumulation in two oxydo-glycolytic (*Longissimus thoracis*: LT, *Rectus abdominis*: RA) and one glycolytic (*Semitendinosus*: ST) muscles, from 23-28 month-old steers of three genotypes (n=8-12), two with a high (Angus and Japanese Black x Angus) and one with a low (Limousin) marbling. Compared to Angus or Angus cross, Limousin steers were characterised by i) a 4- (RA and ST) to 10-fold (LT) lower intramuscular triacylglycerol (TG) content, ii) a 2- to 4.6-fold lower glucose-6-phosphate dehydrogenase activity (G6PDH, involved in *de novo* lipogenesis) in the three muscles, iii) a 6-fold lower leptin mRNA level in LT (the sole muscle assayed). Activities or mRNA levels of other enzymes involved in *de novo* lipogenesis (acetyl-CoA carboxylase, fatty acid synthase, malic enzyme), fatty acid esterification (glycerol-3-phosphate dehydrogenase) and circulating TG uptake (lipoprotein lipase) differed between genotypes depending on muscle types ($P < 0.05$), but these measures were not related to the variability in marbling. In conclusion, it is hypothesised that leptin and G6PDH play a significant role in the expression of intramuscular fat in beef cattle.

Session 20 Poster 26

The influence of linseed supplementation on carcass quality traits of Hungarian Simmental and Holstein-Friesian young bulls

G. Holló[1], J. Seregi[1], J. Tözsér[2], L. Nagy[3], Cs. Ábrahám[2], B. Húth and I. Holló, [1]University of Kaposvár, Faculty of Animal Science, Kaposvar, H-7400, [2]St Istvan University, Faculty of Agricultural and Enviromental Sciences, Gödöllö, H-2103, [3]Institute for Small Animal Research, Gödöllö, H-2100, Hungary*

In this trial the effect of concentrate with and without linseed supplementation on carcass quality traits was compared using Hungarian Simmental (n=10) and Holstein-Friesian (n=12) growing finishing bulls. The animals were assigned to a 2x2 factorial experimental design in four groups. Bulls were kept on deep litter, fed on corn silage, hay and concentrate with or without linseed supplementation. The supplementation was fed in the last two months of the growing-finishing period. Average final live and slaughter weights were actually identical in all groups (550.36±44.75 kg, 510.91±41.11 kg). Statistical differences were recorded among groups for hot carcass weight. The effect of breed and breed x diet interaction were significant on pH_1 and pH_u in all three measured muscles (longissimus, semitendinosus, psoas major) as well as on meat colour values of longissimus (b* value), semitendinosus (L*, a*, b*) and psoas major (L*, b*) measured by Minolta Chromameter. The content of lean meat was significantly influenced by breed and breed x diet interaction (P<0.001), fat and bone in the right half carcasses only by breed, but feeding. In conclusion, the effect of linseed supplementation demonstrated mainly in the meat quality traits (pH, colour).

Session 20 Poster 27

Comparison of concentrate feeding strategies for growing bulls

K. Manni[1,2], M. Rinne[1] and P. Huhtanen[1], [1]MTT Agrifood Research Finland, 31600 Jokioinen, Finland, [2]Present: Häme Polytechnic Degree Programme in Agricultural and Rural Industries, 31310 Mustiala, Finland*

The aim of this study was to examine the effects of periodic allowance and level of concentrate feeding on performance and meat quality of Finnish Ayrshire bulls. In a 2x3 factorial arrangement, level of concentrate feeding (50 or 100 g/$LW^{0.6}$) and periodic concentrate allocation (steady, or lower or higher during part of the growing period) was studied. The concentrate used was ground barley and the animals had ad libitum access to grass silage. The average initial and final live weights of the bulls were 94 kg and 554 kg, respectively. When the concentrate level increased from 25 % to 40 % of total dry matter (DM) intake, the growth rate increased from 1090 to 1187 g/d. Increased concentrate intake decreased silage intake but increased total DM intake from 6.52 to 6.72 kg/d. The substitution rate was 0.86 kg silage DM per concentrate DM and the growth response per additional 1 kg concentrate DM was 73 g/d. Increased concentrate intake resulted in a more efficient feed conversion. Periodic concentrate allocation decreased carcass fatness but did not decrease growth rate or carcass quality significantly. Increased concentrate level increased DM and fat contents of Musculus longissimus, and periodic concentrate allowance reduced tenderness of it. There were no significant interactions between the level of concentrate and periodic concentrate allocation.

Session 20 Poster 28

Production and economic comparison of milking F_1 Holstein x Gir cows with and without the stimulus of the calf

F.S. Junqueira, F.E. Madalena and G.L. Reis, Federal University of Minas Gerais, School of Veterinary Science, Cx.P. 567, 30123-970, Belo Horizonte, MG, Brazil*

F1 Holstein x Gir cows were allocated to two treatments as they calved: C- (53) were milked without their calves (bucket-fed four kg milk/d) and C+ (52) were milked with the stimulus of the calf for approximately 60 days and without the calf thereafter. Analyses of (co)variance were performed by least squares or maximum likelihood (repeated observations) and discrete variables were analysed by chi-square. The milked out milk means in C+ and C- were, respectively, 2383 ± 176 and 2184 ± 176 kg and the milk consumed by the calves 268 ± 75 and 195 ± 88 kg. The concentrations of protein, fat and solids in the milked out milk were not significantly different between treatments (P>0.05). C+ cows had longer lactations than C- cows (251 ± 12 and 216 ± 12 days). No significant differences were found between treatments in somatic cell count, calf mortality and morbidity, two-mo calf weight, cow weight and body condition score, calving to first oestrus and calving to conception intervals. The additional time needed to milk with the calf in C+ was similar to the time spent on artificial feeding in C- (4.0 and 4.3 min/calf/day). Saleable milk was 410 kg higher in C+ than in C-, resulting in a US$ 33.5/lactation/cow superiority in gross margin.

Session 20 Poster 29

Influence of λ-casein and β-lactoglobulin genotypes on the milk coagulation properties

I. Kübarsepp, M. Henno, H. Viinalass, D. Sabre, and O. Saveli, Estonian Agricultural University, Institute of Animal Science, Kreutzwaldi 1, 51014 Tartu, Estonia*

Purpose of this study was to find connections between milk renneting properties of dairy breeds in Estonia and genetic variants of λ-casein and β-lactoglobulin. Milk (n=2234) and blood (n=87) samples were taken from Polula Research Farm. All measured coagulation parameters were significantly better for the λ-casein BB and worse for the λ-casein AA genotype. λ-Cn BB exhibited also the lowest percentage of noncoagulated milk samples and samples that did not reach K_{20} 30 min after enzyme addition. β-Lg genotypes had no significant effect on milk coagulation parameters, only milk coagulation time was the shortest and percentages of noncoagulated milk samples and samples with poor coagulation properties were lower for the β-Lg AA genotype. Coagulation properties of milk were higher for Estonian Native breed. No noncoagulated milk samples were observed in this group. Estonian Red breed (EPK) has second-best coagulation properties of milk. Percentage of noncoagulated milk samples in group of EPK (3.6%) was lower than in groups of Estonian Holstein and Red-and-White Holstein (percentage of noncoagulated milk samples 5.0% and 7.7% respectively). Better milk coagulation properties among native breeds are explicable with higher frequency of λ-Cn B allele. Milk from cows having λ-Cn AB and BB genotype were throughout the lactation more suitable for cheese making than milks on the average.

Session 20 Poster 30

Differences in milk urea content in dependency on selected non-nutritive factors

F. Jílek[2], M. Stípková[1], M. Fiedlerova[1], D. Rehák[1], J. Volek[1] and E. Nemcová[1], [1]Research Institute of Animal Production, P.O.Box 1, CZ 104 01 Praha 10 - Uhríneves, Czech Republic [2]Czech University of Agriculture Prague, Czech Republic*

The aim of this work was to detect differences in milk urea content and milk and protein yield in dependency on different levels of selected non-nutritive factors. Altogether, ca 17,800 monthly measurements from the official milk performance recording stemming from 1,603 cows of selected Holstein herds in Czech Republic were analyzed. The traits considered were milk yield, protein yield, protein content and milk urea content. The data were evaluated using multiple-way analysis of variance. The mean values of milk urea content and milk production parameters differed significantly between herds and years. Significant differences were found between parities and stages of lactation as well. The results of this work were used as a basis for additional analyses of the relationship between milk urea content and cows' reproduction performance.
The financial support from the project NAZV 1G46086 is acknowledged.

Session 20 Poster 31

The causes and extent of variability in lactation curve parameters in British commercial dairy herds

B. Albarrán-Portillo and G. E. Pollott, Department of Agricultural Sciences, Imperial College London, Wye campus, Ashford, Kent, TN25 5AH, United Kingdom*

Recent interest in biological models of lactation as a means to simplify the complex test-day genetic evaluation methods and as an aid to modeling dairy cow management strategies have highlighted the lack of information on the causes and level of variability in lactation curve parameters in commercial dairy herds. This study reports an analysis of 818,902 lactations from 209,422 cows kept in 488 large British Holstein-Friesian herds. Test-day records from 4,385,805 cow events were used to construct the datasets. A recently proposed biological model of lactation was fitted to these records and a range of lactation curve parameters and calculated values were obtained for 447,676 lactations. These records were analysed by mixed-model REML methodology to investigate the factors which affected each variable and the extent of within cow, herd, age group and year variability. Herd, year cow, and lactation number significantly affected the majority of lactation curve parameters. Between-herd variability was the major source of differences in most variables, reflecting the wide range of management practices and climatic conditions across Britain. Within-cow variability was the least-important source of variation but differences between cows in most parameters indicated considerable scope for selection on lactation shape. These preliminary analyses indicate that lactation curve parameters may be linked to critical health and welfare aspects of dairy cow production.

Session 20 Poster 32

Comparative study of body condition scores and blood parameters in dairy cows

E. Báder[1], Z. Gergácz[1], E. Brydl[2], I. Györkös[3], A. Kovács[1], [1]University of West Hungary, Vár 2., 9200 Mosonmagyaróvár, Hungary, [2]Szent István University, Faculty of Veterinarian Science, 1078 Budapest, István u. 2., Hungary, [3]Research Institute for Animal Breeding, Gesztenyés str. 1., 2053 Herceghalom, Hungary

A study was conducted to investigate the relationship between body condition score (BSC) and blood profile in dairy cows at the end of dry period and postpartum. Blood samples were taken from a total of 1922 clinically healthy dairy cows on 49 dairy-farms, in 2003.
Blood profile of cows with thin condition (BCS 1-2.8) indicated negative energy balance, increased lipid mobilisation, decreased glucose level, and metabolically overloaded liver function. Furthermore, the extremely high blood urea concentration of thin cows can further damage their liver function. (FFA=0,18mmol/l, acetoacetate=0.08mmol/l, AST=93U/l, urea=6.1mmol/l, glucose=2.5mmol/l). In the normal group (BCS 3.0-3.8) cows had high AST and blood urea with low glucose level. Liver dysfunction was also present in cows with normal BCS. (FFA=0,16mmol/l, acetoacetate=0,07mmol/l, AST=90U/l, urea=5.5mmol/l, glucose=2.9mmol/l). In the obese group (BCS 4-5), the major blood parameters were in the range of normal physiological values. More than 70% of obese cows were in the preparturition period. Therefore, the over condition at the beginning of lactation may partly explain the metabolic disorders found in the thin and normal group of cows. Achieving an optimal condition according to the lactation state can decrease the incidence of certain metabolic problems in dairy cows.

Session 20 Poster 33

Effect of sensory stimulation, feeding or brushing, during milking on milk yield and hormone release in dairy cows

E. Wredle[1], M. S. Herskin[2], L. Munksgaard[2], R. M Bruckmaier[3], K. Uvnäs-Moberg[1] and K. Svennersten-Sjaunja[1], Swed. Univ. Agic. Sci (SLU), Uppsala, Sweden, Research Centre Foulum, Tjele, Denmark, Tech. Univ. Munich, Freising Weihenstphan, Germany*

The objective of the present study was to examine the effect of sensory stimulation (feeding and abdominal brushing) during milking on milk yield and release of oxytocin and cortisol. Twelve dairy cows, Swedish Red and White, 9-24 week into lactation and with a daily milk yield of 26-34 kg were kept in tie-stalls. The experiment was a 2 x 2 factorial design with different sensory stimulation during milking, 1) tactile stimulation by a hard brush at the abdomen; 2) provision of 2.1 kg of concentrate. All cows received all treatments in a balanced order. The treatments began at the start of pre-stimulation of the udder and continued until the milking cluster was taken off. The experiment was divided into four periods lasting three days each. During the morning milking day three, following registrations were done; blood sampling for analyse of plasma concentration of oxytocin and cortisol, milk yield, milk flow, milking time and milk composition.
Preliminary data suggests that feeding increased milk yield compared with no feeding (18.5 ±0.7 kg vs. 16.4 ±0.7 kg; $P < 0.01$), while the yield was not effected by brushing (17.3 ±0.7 kg vs. 17.5 ±0.7 kg). More results will be presented during the conference.

Session 20 Poster 34

Time of first insemination in Hungarian dairy herds

E. Báder[1], Z. Gergácz[1], I Györkös[2], P. Báder[1], A. Kovács [1], N. Boros, [1]University of West Hungary, Vár 2., 9200 Mosonmagyaróvár, Hungary, [2]Research Institute for Animal Breeding and Nutrition, Gesztenyés str. 1., 2053 Herceghalom, Hungary*

Reproductive performances of 22004 cows kept on 6 dairy farms between 1990 and 2004 were analysed. Cows conceived for first time were inseminated between 63 and 108 days after calving. Time of first insemination of cows conceived for the second or third insemination was between 54 and 91days. The average of first insemination date was between 55 and 101 days after calving. The number of services per conception was between 1.82 and 2.70 and the average number of days open was between 122 and 155 days. 20-68% of cows that conceived for first service were inseminated less than 60 days after calving. Only 15-27% of cows were inseminated in optimal time (60-70 days after calving). 10-40% of cows were first inseminated more than 100 days after calving. 24-76% of cows conceived for second or more inseminations were first inseminated within 60 days after calving and 15-27% of them were first inseminated at ideal time. 4-31% of cows were first inseminated more than 100 days after calving. It can be established that ideal time of first insemination after calving could not be determined. It is influenced by the biological status of cows. The main task is to decrease the date of first insemination after calving.

Session 20 Poster 35

Growth rate of scrotal circumference in males of the Galician Blond cattle breed

J.M. Ferreiro[1], J.M.G. Pires[2], J.P.P. Araújo[2], J.Cantalapiedra[1], L. Sánchez[1], A. Iglesias[1], [1]Dpto. Anatomía y Producción Animal. Facultad de Veterinaria. 27002 Lugo, España, [2]Escola Superior Ágraria de Ponte de Lima. Inst. Politécnico de Viana do Castelo, Ponte de Lima, Portugal*

The objective of this work was to estimate the scrotal circumference (SCA) in males of Galician Blonde cattle breed, from 8 to 16 month of age (A) and body average weight (W) between 385 and 718 Kg respectively. Using data from 36 males property of A.C.R.U.G.A, (Nucleus of Control of Performance in the Improvement Program for the Galician Blond beef cattle) they were measured monthly during the years 2003 and 2004, and considering the multivariate normal distribution were generated 10000 records by simulation. Using the original and the simulated data set, estimates of SCA in function of A and W, confidence intervals of 95% of probability and instantaneous growing rate were obtained by cubic polynomial equations. The diameter of SCA ranged from 30.3 ± 2.9 (month 8) with 385,0 kg of body weight, to 38.1 ± 4.7 and 718.8 kg (month 16), with the estimates obtained from the simulated data set more realistic. The availability of this information about the growth rate of the circumference scrotal will be of immediate utility for the technical personnel and producers of cattle.

Session 21 Theatre 1

Selection for lean weight based on CT and ultrasound in a meat line of sheep

T. Kvame and O. Vangen, Department of Animal and Aquacultural Sciences, Norwegian University of Lifessciences, P.O. Box 5003, Arboretveien 6, N-1432 Ås-NLH

Genetic parameters for carcass traits in lamb estimated by CT (n=234) and ultrasound (n=1743), and the response to selection for ultrasound muscle depth (UMD) and proportion of LEAN were estimated. The data included animals from a meat line (ML) and a conventional breed line (CL) of Norwegian White Sheep (NWS). The ML was in-crossed with Texel from 1998, and selected for UMD from 1993 until 2001, and for LEAN weight from 2001 to 2004. Average age at ultrasound and CT recordings was 128 and 123 days, respectively. An average of 23 CT images were taken per animal. Traits analysed were UMD and ultrasound fat depth (UFD), carcass LEAN, FAT and BONE, subcutaneous fat (SFAT), intermuscular fat (IeFAT) and noncarcass fat. Esitmations were made by univariate and bivariate mixed-animal model by methods of AIREML, including all animals with records and their relatives. The statistical models included fixed effects, weight or age of lamb at weaning (covariate), and random direct genetic effect.
Heritability estimates found for weight of LEAN, FAT and BONE was 0.57, 0.29 and 0.51 from the model corrected for weight. All estimates of heritabilities were lower for traits adjusted for age. High genetic correlation was found between LEAN and UMD (0.70) and between FAT and UFD (0.82). The genetic trend regressed on year of birth was significant greater in ML for the traits UMD and LEAN.

Session 21 Theatre 2

Changes in, and relationships among, lamb growth and carcass traits measured by computer tomography (CT)

N.R. Lambe, E.A. Navajas, K.A. McLean, L. Bünger and G. Simm, SAC, Sustainable Livestock Systems Group, West Mains Road, Edinburgh, EH9 3JG, Scotland, United Kingdom*

Male and female Texel (TX, n=130) and Scottish Blackface (SBF, n=103) lambs were CT scanned four times between 8 weeks of age and slaughter (live-weights ~13-47kg). CT traits studied included: total weights and proportions of fat, muscle and bone (predicted from CT reference scans in the leg, loin and chest); tissue distribution (areas of fat and muscle in each reference scan as a proportion of total area in the three scans); muscle shape in the hind-leg (depth:width ratio); eye muscle area; tissue densities. After adjusting traits for fixed effects and sire (random), relationships among CT traits and between CT traits and live-weight were investigated within breed, to assess changes with growth. Increased live-weight (per 10kg when numbers given) was associated with (i) increases in: total fat proportion (TX=28%, SBF=25%), fat:muscle ratio, muscle:bone ratio, proportion of fat in the chest region (TX=4%, SBF=8%) and bone density (TX=5%, SBF=7%), in both breeds, as well as proportion of muscle in the loin (3%), and leg-muscle shape (4%) in TX lambs; (ii) decreases in: proportion of fat in the leg in both breeds (TX=-3%, SBF=-9%), and leg-muscle shape (-6%) and fat density (-4%) in SBF lambs. Similar trends between CT traits and total fat proportion suggest that associations hold for increasing maturity, as well as size.

Session 21 Theatre 3

Using X-ray computed tomography to predict intramuscular fat content in sheep

J.M. Macfarlane[1], M.J. Young[1,2], R.M. Lewis[3], G.C. Emmans[1] and G. Simm[1], [1]Scottish Agricultural College, West Mains Road, Edinburgh, United Kingdom, [2]Sheep Improvement Ltd, PO Box 66, Lincoln University, Canterbury, New Zealand, [3]Department of Animal and Poultry Sciences (0306), Virginia Tech, Blacksburg, Virginia, USA*

Intramuscular fat content is an important factor in producing tender, juicy meat. Sheep breeding programmes are using increasingly accurate tools such as X-ray computed tomography (CT) to select for carcass lean and fat contents in the live animal. Such tools may also be useful to predict other carcass attributes such as intramuscular fat content. Terminal sire lambs of three breeds (Suffolk n=100, Texel n=40, Charollais n=20) were CT scanned, one fifth at each of 14, 18 and 22 weeks and two fifths at 26 weeks of age. CT scanning produced cross-sectional scans at 7 anatomical landmarks [ischium, femur, hip, 5th and 2nd lumbar vertebrae (LV), 6th and 8th thoracic vertebrae (TV)] from which fat, lean and bone areas and densities were determined. After CT scanning, lambs were slaughtered and the *M. longissimus dorsi* dissected out and chemically analysed for fat %. All-subsets regression of intramuscular fat % on CT tissue areas and densities and pre-scan live weight showed that intramuscular fat % can be predicted with moderate accuracy by only two variables from a single anatomical position, 2nd LV fat area and 2nd LV muscle density (R^2=0.566). Including more predictors or a breed effect did not improve prediction accuracy.

Session 21 Theatre 4

Genetic parameters for birth weight, growth and litter size for Danish Texel and Shropshire

J. Maxa[1,2], E. Norberg[1], P. Berg[1] and J. Pedersen[3], [1]Danish Institute of Agricultural Sciences, Department of Genetics and Biotechnology, P.O. Box 50, 8830 Tjele, Denmark, [2]Czech University of Agriculture, Department of Cattle Breeding and Dairying, Prague, Kamycka, 165 21, Czech Republic, [3]Danish Cattle Federation, Udkærsvej 15, Skejby, 8200 Aarhus N, Denmark.*

Texel and Shropshire are the most common sheep breeds in Denmark. In this study, heritabilities and (co)variance components of birth weight (BW), average daily gain from birth until two months (ADG2) and litter size (LS) were estimated. Data from 1990 to 2004 were extracted from the sheep recording database at the Danish Agricultural Advisory Centre. Average values for BW, ADG2, LS were 4.55 kg, 318 g, 1.38 for Texel and 4.20 kg, 281 g and 1.36 for Shropshire, respectively. Direct and maternal heritability for Texel were both 0.19 for BW and 0.14 and 0.10 for ADG2, respectively. For Shropshire the direct and maternal heritability were both 0.18 for BW and 0.21 and 0.14 for ADG2. The heritability for LS were 0.06 for both breeds. The direct genetic correlations between BW and ADG2 were 0.27 for Texel and 0.46 for Shropshire. Genetic correlations between LS and direct and maternal effects of BW were 0.04 and 0.16 for Texel and 0.14 and 0.34 for Shropshire. Genetic correlations between LS and direct and maternal effects of ADG2 were positive, ranging from 0.07 to 0.26 for Texel and Shropshire.

Session 21 Theatre 5

Effect of birth type on milk production traits in East Friesian ewes

B. Fuerst-Waltl, R. Baumung and J. Sölkner, University of Natural Resources and Applied Life Sciences Vienna, Department of Sustainable Agricultural Systems, Division Livestock Sciences, Gregor Mendel-Str. 33, 1180 Vienna, Austria.

A total of 5592 240d-lactation records of 2131 East Friesian ewes collected between 1990 and 2003 were used to analyse the effect of birth type and other environmental and genetic effects on milk production (milk, fat, and protein yield; fat and protein percentage). In a separate analysis for first and higher lactations, effects accounted for were herd*year*season, lactation number, birth type at lambing, birth type of the ewe, lambing interval, stillbirth, and age at lambing. Birth type at lambing significantly affected all production traits in higher lactations, and milk yield, fat yield and fat percentage in first lactations. Yield traits and protein percentage increased with higher number of lambs whereas fat percentage slightly decreased. The birth type of the ewe showed a less significant effect on milk production traits. Ewes born as singles had significantly higher protein percentages in higher lactations and were in tendency superior to ewes born as multiples for other production traits. Estimated heritabilities for yield traits ranged from 0.25 to 0.37 and from 0.26 to 0.27 in first and higher lactations, respectively. Heritabilities for protein and fat percentage were 0.63 and 0.42 in first, and 0.53 and 0.45 in higher lactations.

Session 21 Theatre 6

Year-round lamb production systems on pasture: theory and practice

P.C.H. Morel, S.T. Morris, P.R. Kenyon, G. de Nicolo and D.M. West. College of Sciences, Massey University, Palmerston North, New Zealand*

Traditional lamb production systems in New Zealand are driven by the seasonal pattern of pasture growth with ewes lambing only once a year. An alternative is the implementation of year-round lamb production systems such as the Star system, where an individual ewes lamb five times over a three year period. In addition the STAR flock is sub-divided into three groups resulting with 5 lambing periods within a given year. This paper describes some initial modelling undertaken to assess the economic viability of the Star systems in New Zealand. Preliminary results of a field trial conducted at Massey University with 480 mixed aged Romney and East Friesian ewes are also presented. The model shows that, at the same lambing percentage (160%) and with a 10% premium for out of season lambs the Star system earned an extra 56% in profit. Over the first 5 lambing periods (one calendar year) in the field trial, ewes in the Star system had a 10% lower pregnancy rate, weaned 47 % more lambs and generated an extra 10 % income from lamb sales compared to the traditional once a year lambing flock. However, due to higher inputs cost (hormonal treatment and labour) the Star system is uneconomic without significant premiums for the product, even with the more productive East Friesian ewes.

Session 21 Theatre 7

Evaluation of in-farm *versus* weather station data as heat stress indicator in Mediterranean dairy sheep

R. Finocchiaro[1], A. Di Grigoli[1], J.B.C.H.M. van Kaam[2], A. Bonanno[1], B. Portolano[1]. [1]Dipartimento S.En.Fi.Mi.Zo. - Sez. Produzioni Animali, Viale delle Scienze, 90128 Palermo, Università degli Studi di Palermo, Italy, [2]Istituto Zooprofilattico Sperimentale della Sicilia "A. Mirri", Via Gino Marinuzzi 3, 90129 Palermo, Italy.*

Heat stress is a limiting factor impairing growth, milk production and reproduction. In-farm (IF) *versus* weather station (WS) data were evaluated for use as heat stress indicator in dairy sheep. Maximum temperature (T) and relative humidity (RH) were monitored 24h before milk recording in November 2002-July 2003 by one weather station and with thermo-hygrographs in three farms. Data contained 1.056 test-day records belonging to 309 Valle del Belice ewes. The correlation of WS-T was 0,83 with IF-T. The correlation of WS-RH was 0,78 with IF-RH. The correlation of milk production with WS-T was -0,47, with WS-RH 0,43, with IF-T -0,56 and with IF-RH 0,28. The correlation of fat+protein production was -0,57 with WS-T, 0,52 with WS-RH, -0,60 with IF-T and 0,38 with IF-RH. A GLM model including the fixed effects flock, DIM, and temperature-humidity index (THI) resulted in a decrease of -32,9g milk and -4,8g fat+protein per unit increase of THI for IF data *versus* -30,0g and -4,7g respectively using WS data. Models were always slightly better when using WS than IF information based on the R^2 and root MSE. These results show that WS data can replace the IF collection.

Session 21 Theatre 8

Flight test as detector of transport stress in sheep

S. Diverio[1], A. Barone[1], G. Tami[1] and N. Falocci[2], [1] Dpt. Scienze Biopatologiche, Igiene delle Produzioni Animali e Alimentari, Perugia University, Via San Costanzo 4, 06126 Perugia, Italy, [2] Dpt. Scienze Statistiche, Faculty of Economy, Via Pascoli, 06100, Perugia, Italy*

This study is part of a multidisciplinary project aimed to assess transport stress in sheep, by evaluating several physiological, immunological and behavioural parameters. Aim was to assess the effectiveness of a Flight Test (FTest) as detector of behavioural changes following transportation. On a group of 10 sheep with lambs at foot, a 2 minute FTest was performed before loading (FTest1) and after unloading (FTest2), which followed a 4.30 hour transport on a truck. Each sheep-lamb pair was placed in a 2x8 m arena where it was twice approached by an unknown operator. First Reaction Time (FRT), Flight Time, (FT), Flight Distance (FD) and a series of behaviours, position and distance from the operator were recorded. Data were analysed by Instantaneous Scan Sampling and statistically analyzed by ANOVA and PCA. While FRT and FT revealed no significant difference before and after transportation, in FTest2 sheep reacted when the operator was closer compared to FD recorded in FTest1 ($P<0.01$). After transportation, locomotor activity significantly increased ($P<0.05$), while sheep reduced their interaction with the lambs ($P<0.05$). PCA confirmed differences between FTests, highlighting a higher behavioural reactivity in FTest2. FTest proved to be a useful mean to assess transport stress response in sheep.

Session 21 Theatre 9

Body posture and orientation in sheep during transportation

S. Diverio[1], A. Barone[1], G. Tami[1] and N. Falocci[2], [1] Dpt. Scienze Biopatologiche e Igiene delle Produzioni Animali e Alimentari, Perugia University, Via San Costanzo 4, 06126 Perugia, Italy, [2] Dpt. Scienze Statistiche, Faculty of Economy, Perugia University, Via Pascoli, 06100, Perugia, Italy*

This study is part of a multidisciplinary project carried out to assess the stress response to transport in sheep, by evaluating several physiological, immunological and behavioural parameters. The aim was to monitor sheep behaviour during a 4.30 hour journey on a truck. The experiment was carried out during July on a group of 20 sheep. Subjects' behaviour and road conditions were simultaneously video-recorded during the whole journey. Videos were analysed by Instantaneous Scan Sampling at 1 minute interval, and body posture, head position, body and head orientation compared to march direction (i.e. towards, against and perpendicular to) were recorded. Transport route was scored according to vehicle status (Stationary *vs* Movement) and road type (3-point rating scale for cornering; 4-point rating scale for slope). Data were analysed by Chi Square and polynomial analysis. Sheep significantly preferred to stand ($P<0.001$) compared to lying, independently from vehicle movement. However, time spent lying significantly increased ($P<0.001$) during stops. Road type seemed not to significantly affect sheep behaviour. Head position, body orientation and movement of the vehicle resulted to be significantly associated ($P<0.001$), with sheep preferring standing with the head down, forward or lateral facing the direction of the travel.

Session 21 Theatre 10

Low-cost feeding strategies for fattening Awassi lambs in Syria

B.W. Hartwell[1], L. Iniguez[1], M. Wurzinger[2] and W. Knaus[2], [1]ICARDA (The International Center for Agricultural Research in Dry Areas), P.O. Box 5466, Aleppo, Syria. [2]University of Natural Recourses and Applied Life Sciences, Division of Livestock Sciences, Vienna, Austria*

Awassi lamb fattening systems in Syria have a comparative advantage in marketing their lambs in local and neighboring countries due to favorable prices. A limiting constraint to these systems is cost of feeding, which have a direct influence on the income margins obtained by farmers. ICARDA investigated low- cost diets to reduce feeding expenses by combining low-cost feed resources available locally as opposed to expensive concentrates used traditionally by farmers. In a 90-day fattening trial, 2 treatments combining concentrates and agricultural byproducts (barley, wheat bran, cotton seed cake, lentil, faba bean or maize mixed with molasses) with urea-treated straw (5% urea solution) were tested against the traditional diet (control). Thirty-six Awassi lambs (mean BW 23.04 ± 3.5 kg, 4 months) were randomly allocated into the three experimental groups and individually fed *ad libitum*. In addition to daily measurements of intake and refusal, a digestibility trial was conducted (10 days). Results showed that using molasses and urea-treated straw reduced the cost of feeding and that the lowest-cost diets can succefully sustain 291 g/d weight gains, as opposed to 274 g/d (control) with less cost. Using maize (14%), molasses (14%) and urea treated straw was found to be the most cost efficient diet.

The effect of supplemental feeding duration on performance of Balouchi ewes

V. Kashki, M.R. Kianzad, M. Raisianzadeh, M. Nowrozi, H. Tavakoli and A. Davtalabzarghi, Agriculture and natural resources research center of Khorasan. P. O. Box: 91735-1148, Mashhad, Iran

The available pastures in Khorasan province are low in energy and protein contents. Therefore supplemental feeding is one of important management factors for sheep production. One hundred fifty Balouchi ewes, aged higher than 3 years were selected for uniformity in BW, age and production from a flock of approximately 3000 Balouchi ewes and studied in incomplete random design. 250 g barley grain was fed once daily for supplementing in periods from pre-breeding to weaning. Treatments included 1) no supplement (control) 2) supplemented from 30 d prior to breeding to weaning 3) supplemented from breeding to weaning 4) supplemented from 30 d prior to lambing to weaning 5) supplemented from lambing to weaning. Ewe body weight changes, grease fleece weight, milk production, lamb birth weight, lamb weaning weight and average daily gain for lambs were measured. Experiment data were analysed using the GLM procedure of SAS (6.12). Treatments had significant effects ($P<0.05$) on ewe weight 30 d after lambing, and control had less weight. All ewes lost weight 30 d after lambing. The ewes weight was affected by treatments at the time of weaning ($P<0.05$). Daily milk production was not affected by treatments. Fleece weight was affected by treatments ($P<0.05$). The lamb birth weight was not affected by treatments. The lamb weaning weight and ADG was significantly different between treatments ($P<0.05$) and receiving feed treatments had lambs with higher weight than control.

Kinetics of responses of goat milk fatty acids to dietary forage:concentrate ratio and/or high doses of sunflower or linseed oil, or extruded mixture of seeds

Y. Chilliard[1], J. Rouel[1], P. Guillouet[2], K. Raynal-Ljutovac[3], L. Leloutre[4] and A. Ferlay[1], [1]URH, INRA-Theix, 63122-France, [2]INRA-Lusignan,[3]ITPLC-Surgères, [4]INZO-France*

Eighty-four multiparous mid-lactation goats received indoor one of 7 diets differing in either forage:concentrate ratio (HF-70:30 or LF-30:70) and/or lipid intakes (0 *vs.* 180 g/d, *i.e.* 7% of diet DM, of either linseed oil, LO, or 18:2-rich sunflower oil, SO, or an extruded 40:30:30-mixture of linseeds:sunflower-seeds:wheat, EX). After 5 weeks of treatment, rumenic/palmitic acid % in milk fatty acids were 0.5/26.5, 2.7/19.4, 2.9/20.9, *0.3/31.3*, 3.5/17.0, *5.1/16.9* and 3.2/16.5, respectively, for LF, LFLO, LFSO, *HF*, HFLO, *HFSO* and HFEX diets (concentrate-lipid interaction, $P<0.05$). LF (*vs.* HF) diet decreased 18:3n-3, and increased 18:0, *cis*9-18:1 and 18:2n-6. LO diets decreased 10:0-14:0 and increased 18:0, 18:3n-3 and *trans*-18:1+*trans*-18:2 isomers. HFSO diet induced the greatest decrease in 10:0-14:0 (-44%) and increased *cis*9-18:1, *trans*10+*trans*11-18:1 and 18:2n-6. HFEX diet maximized 18:0 and 18:3n-3. The sum of non-*trans*11(18:1+18:2) was highest with LO diets, high with HFEX, medium with SO diets, and lowest with basal diets. *LFSO*, LF and HFSO diets maximized the 18:2n-6/18:3n-3 ratio. All the responses were fully expressed 2 weeks after the beginning of treatments, and stable thereafter. *In conclusion*, the studied dietary treatments rapidly, strongly and long-lastingly changed goat milk fatty acid profile. Goat's responses differ markedly from cow's ones to similar diets, particularly by the stability of rumenic acid response (*Work funded by EU BIOCLA Project QLK1-2002-02362*).

Session 21 Poster 13

Research regarding the morpho-productive parameters of F_1 crossbreds resulted from crossing Lacaune breed with Tigaia breed

I.Raducuta[*]*,Livia Vidu, A. Marmandiu, University of Agronomical Sciences and Veterinary Medecine, Bucharest, Str. Marasti 59, Romania*

In Romania the direction in sheep breeding and exploitation is focussed on milk production because it is more profitable than wool or meat. Unfortunately, our breeds are featured through a small milk yield (with limits between 60 and 90 liters, according to the breed performance), which recommends a fast improvement. This work aims at investigating the main morpho-productive parameters of F_1 crossbreds resulted by crossing the Lacaune breed with the Tigaia breed. The obtained data showed that even at their first lactation the F_1 crossbreds' milk yield is higher by 60.3% comparative with the Tigaia breed (138.7 liters vs. 86.5 liters). We can also notice that this crossing variant did not spoil the Tigaia wool quality (which possesses semi-fine and uniform wool) because the Lacaune breed has a similar fibrous composition of wool staples and the features of F_1 crossbreds' wool are situated between the parental breeds. Another advantage of this crossing variant is the decreasing of the unproductive time period of Tigaia by about 10 months by distributing the F_1 crossbred females to an early reproduction at the age of 9-10 months instead of 18, without further negative consequences. These results suggested that, to increase the milk yield of the Tigaia breed by the crossbreeding method, it is recommended to use the Lacaune breed successfully.

Session 21 Poster 14

Heterosis analysis of *Haemonchus contortus* resistance and productions traits in Rhönschaf, Merinoland sheep and crossbred lambs

A. Hielscher[1], M. Gauly[2], H. Brandt[1] and G. Erhardt[1], [1]Institute of Animal Breeding and Genetics, University of Giessen, Ludwigstrasse 21B, 35390 Giessen, [2]University of Goettingen, Albrecht Thaer Weg 3, 37075 Goettingen, Germany

A crossbreeding program was conducted to estimate the resistance against *Haemonchus contortus* in reciprocal crossbred lambs and their pure breed counterparts (Merinoland (Ml), Rhönschaf (Rh)). Faecal egg counts (FEC) and haematocrit values (Hc) of 449 lambs were evaluated four and eight weeks after an artificial infection (*p.i.*). Worm counts were obtained after slaughtering eight weeks *p.i.*. The Rh group consistently showed the highest FEC and the highest Hc values when compared with Ml. Comparing the two crossbreeds RhxMl showed the lowest FEC whereas the reciprocal cross breed showed the highest FEC. Hc ere not significant different. The effect of heterosis for FEC was negative but low after four weeks and almost zero after eight weeks *p.i.* Heterosis of HC was weak favourable four weeks and almost zero at eight weeks *p.i.*. Worm counts showed little or almost none heterosis. For body weight parameters the crosses fell intermediate to the pure breeds and showed heterosis between 3.2 and 4.6 percent.

Heterosis analysis proofs that the measured parameters of the F1 favoured the group RhxMl. Therefore, crossbreeding Rh to Ml sheep may be a suitable way to produce lambs with improved resistance to *H. contortus* infection, but production may be compromised.

Session 21 — Poster 15

Alpine network for sheep and goat promotion for a sustainable territory development

F. Bigaran[1], D. Kompan[2], C. Mendel[3], A. Feldmann[4], F. Ringdorfer[5], G. De Ros[6], S. Venerus[7] and E. Piasentier[8], [1]Autonomous Province of Trento, Italy, [2]University of Ljubljana, Biotechnical Faculty, Slovenia, [3]Bayerische Landesanstalt für Landwirtschaft, Germany, [4]Gesellschaft zur Erhaltung alter und Gefährdeter Haustierrassen in Deutschland, Garmisch-Partenkirchen, Germany, Agricultural Research and Education centre Raumberg-Gumpenstein, Department of Sheep and Goats, Irdning, Austria, [6] Istituto Agrario di San Michele all'Adige (Trento), Italy, [7]Province of Pordenone - Settore Agricoltura Aziende Sperimentali Dimostrative, Italy, [8]Department of Animal Science, University of Udine, Italy*

A network among Alpine shepherds' Associations of Bavaria, Austria, Slovenia and North Italian Regions, regional administrations and research institutions has been recently built in the framework of the Interreg IIIB Alpine Space Programme, by means of the project "Alpinet Gheep" (Contract code I/III/1.2/10). The objectives of the 16 partners involved in the project under the coordination of the Autonomous Province of Trento is the organisation of common, trans-national and coordinated activities of research and development to promote and to preserve sheep and goat sector, its social involvement and its capacity to participate in the sustainable spatial development of the Alpine area. The main activities planned in the next three years are: data collection on spreading and asset of breeding activities, identification and inventory of areas suitable for grazing, evaluation of economic and ecological impact of pastoralism, implementation of a trans-national monitoring system for breeds, evaluation of the traditional sheep and goat products.

Session 21 — Poster 16

Alternative sheep breeding schemes in Norway

L.S. Eikje[1], L.R. Schaeffer[2], T. Ådnøy[1] and G. Klemetsdal[1], [1]Department of Animal and Aquacultural Sciences, Norwegian University of Life Sciences, P.O. Box. 5003, 1432 Ås, Norway, [2]Centre for Genetic Improvement of Livestock, Department of Animal and Poultry Science, University of Guelph, Guelph, ON, Canada N1G 2W1*

The effect of three alternative national sheep breeding schemes was investigated by stochastic simulation. Evaluation was based on genetic response for all 10 traits that make up the aggregate genotype in the present breeding scheme, and inbreeding. The schemes assessed were: (i)Young test rams and elite rams used by natural mating within ram circles. This corresponds to the Norwegian breeding scheme of today, with about 120 000 ewes, and was used in the simulation as a basis for comparison with: (ii)Young test rams used by natural mating within ram circles and elite rams used by AI both within and across circles, and: (iii) Both young test rams and elite rams used by AI. Within scheme (ii) and (iii) additional experimentation was done for: 1) number of ewes mated / inseminated per test ram, 2) number of ewes inseminated per elite ram and 3) percent ewes mated to elite rams versus test rams. Advantageous results were obtained when few ewes were mated / inseminated to each test ram (on average 10 and 100 ewes in scheme (ii) and (iii), respectively), many ewes were inseminated to each elite ram (on average 1100 ewes) and when using a low percentage of elite mating (30 percent). Highest genetic response in the aggregate genotype was found for scheme (ii). The response was found to be about 26 percent higher than in scheme (i). Rate of inbreeding per year was then 0.076 percent giving an effective population size of 658 individuals.

Morphologic characterization and body measurement of Hungarian goats

T. Németh[1], A. Molnár[1], G. Baranyai[2] and S. Kukovics[1], [1]Research Institute for Animal Breeding and Nutrition, Herceghalom, Hungary, [2]Hungarian Goat Breeders' Association, Herceghalom, Hungary*

The objective of this study was to phenotyically (morphologically) characterize and compare the local goat breeds (Hungarian Milking White /HMW/, Hungarian Milking Brown /HMB/, Hungarian Milking Multicolour /HMM/) and the adult offspring of the imported breeds, bred in Hungary, like Alpine and Saanen. This survey is the first step of the comparison of breeds, and can be the base of development of Hungarian goat selection indices. The farms were chosen from the register of the Hungarian Goat Breeders' Association, concerning each breed. In the case of native breeds, most of the farms are keeping all of these breeds. The 1,157 measured female goats, in 24 farms were with age varying between 1 and 8 years. The following measurements were taken: body weight, wither height, body length, thorax depth, thorax width, pelvic width, hip width, head length, ear length and width, and distance between eyes.
In the survey, there were 18% Alpine of the measured goats, 30% Saanen, 13% HMW, 16% HMB and 23% HMM.
The collected data was analysed by SPSS for Windows 10.0 software. According to the results received no significant differences (in $P<0,05$) were observed in each body measures, between the HMW and Saanen, or HMB and Alpine does, even in considering the age and the herd effect. The HMM breed has several kinds of breeds of its background including brown and white goats, but its measurements do not differ in morphology from the other 4 breeds.

Evaluation of Boer goat performances in two climatic environments

E. Láczó, P. Póti and F. Pajor, Department of Cattle and Sheep Breeding, Szent István University, Gödöllö, Hungary*

In the examination, we evaluated 33 and 25 does kidding performance and their kids daily weight gains recorded on 30. and 100. day. The flock in Hungary was grazing grass during the daytime, in the goat house at night, except one month post-partum when goats were kept indoor, supplemented with hay and maize grains (300-400 g). The flock in North Carolina was penned on the millet pasture with hay and a concentrate mix (500 g) supplement.
The Boer goat appeared just a bit more prolific in the North Carolinian than Hungarian environment concerned fecundity (96 % vs. 94 %), percentage multiple birth (83 % vs. 77 %) and litter size (2.17 vs. 2.00). The kids gained daily 156 g and 206 g in the American ranch and 145 g and 201 g in the Hungarian one during the test, respectively. Gross meat production (litter weight weaned per doe) was higher by circa 7 kg in the American than Hungarian ranch (54 kg vs. 47.1 kg).
The better performances showed by Boer goats in the North Carolinian flock were rather due to dietary differences (earlier availability of grazing forage, high protein concentrate-mix in a higher daily portion vs. maize grains) and a greater flock homogeneity than to climatic differences. Nevertheless, the Boer goat is worthy of breeding in Hungary providing correct nutrition conditions.

Session 21 Poster 19

Effect of winter shearing during late pregancy in the Latxa dairy sheep

R. Ruiz, J. Arranz, I. Beltrán de Heredia, A. García, L.M. Oregui. NEIKER, Vitoria-Gasteiz, Spain*

Whereas winter shearing is a relatively common practice in meat production systems, there is an evident lack of data regarding its effect and interest for milk production systems. Recently some local farmers have become interested in this practice for dairy sheep. In November 2003 a two-year project begun to estimate the effect of winter shearing late-pregnant ewes upon food intake and productive parameters in the *Latxa* dairy sheep. In the present paper, the methodology and preliminary data regarding food intake and birth weight of lambs for two productive campaigns (2003-04 and 2004-05) will be discussed. A total number of 56 sheep were used in 2004 and 114 in 2005. Each year animals were allocated into either one of two sets of animals comparable in live weight and genetic value for milk production: then one set was shorn 6 weeks prior to the expected lambing date (SH), and the non-shorn (NSH) was used as control. Each set was divided into 4 subsets according to their live weight. Forage consumption in SH ewes was significantly higher at the end of pregnancy in comparison to NSH ones. A slightly delay in the lambing date of shorn sheep was observed, but no significant differences were found in the birth weight of new-born lambs during the first year. Data related to milk production are not still fully available.

Session 21 Poster 20

Influence of weaning age on lamb growth and animal health in Boer goats

G. Das, Eva Moors and M. Gauly, Institute of Animal Breeding and Genetics, Albrecht Thaer Weg 3, University of Goettingen, 37075 Goettingen, Germany

The aim of the study was to estimate the effects of age at weaning on health and growth rates of Boer goats. Therefore 60 kids from 34 mothers were divided into two groups and weaned at an age of 6 or 12 weeks, respectively.

Vocalisation of lambs and mothers were recorded for four days after weaning. Kids were weighted in weekly intervals until an age of 18 weeks. Does were weighed at week 6, 12 and 18 after birth. At that time udder health and body condition scores (BCS) were recorded. Faecal samples were taken from all lambs at an age of 6, 12 and 18 weeks to determine oocysts/g faeces (OpG) using a modified McMaster method.

OpG increased immediately after weaning in both groups. Daily weight gains were significantly higher in kids weaned at an age of 12 weeks and total vocalisation was lower. However total vocalisation in the mothers was significantly higher in this group. No significant differences in udder health, body weight and body condition score were found.

Changes in somatic cell counts of sheep milk and their effect on rennetability and quality of rennet curdling during lactation

P. Zajicova and J. Kuchtik, Mendel University of Agriculture and Forestry Brno, Department of Animal Breeding, Zemedelska 1, Brno 613 00, Czech republic*

The evaluation of changes in somatic cell counts (SCC) of sheep milk and their effect on renetability (R) and quality of rennet curdling (QRC) during lactation was carried out using samples of milk obtained from 10 ewes (crosses of East Friesian and Improved Valachian breeds). All ewes under study were on the second lactation. Milk was sampled on the average 74th, 102nd, 130th, 160th, 190th and 222nd day of lactation. The average values of SCC, R and QRC varied between 147,2 - 2419,9 thousands/1 ml.; 183 - 282 s and 1,2 - 2,6 during lactation. The average contents of SCC; R and QRC for whole lactation were 656,6 thousands/1 ml; 238 s and 1,7. As far as the values of the SCC and QRC depending on the average day of lactation, these were relatively well balanced from 74th to 190th day of lactation, while in the end of lactation there was registered high significant increase of SCC accompanied by significant deterioration of QRC. Significant positive correlations ($P \leq 0{,}01$) between somatic cell counts and quality of rennet curdling were registered only on the 130th and on 190th day of lactation. On the other hand between SCC and R there were registered any significant correlations.

Breeding for scrapie resistance and controll strategie in Hungary

B. Nagy[1], L. Fésüs[2], L. Sáfár[1], [1]Hungarian Sheepbreeders' Association, Löportár u.16., 1134 Budapest, Hungary, [2]Research Institute for Animal Breeding and Nutrition, Gesztenyés u.1., 2053 Herceghalom, Hungary

There are some researches and studies a long time in Hungary, that to reveal the scrapie resistant of Hungarian flocks (L. Fésüs *et al*, 2004). These assays show that the presence of ARR/ARR prion protein (PrP) gene in the each breeds is sufficient to start a selection for the resistant genotype in flocks. By right of decrees of European Union and the Hungarian Agricultural Ministry, a national scrapie programme was started, in 2004. This programme based on the selection for homozygous ARR animals. Our aim to increase the resistant allele frequency of a flock and a population level, in particular of the rams used for insemination. According to the European Union's and the Hungarian decrees we get identification of all Hungarian bred ewes and rams PrP genotypes in the Research Institute for Animal breeding and Nutrition of Herceghalom. In 2003 and 2004 about 662 blood samples were analyzed by RFLP, and there are samples under analyzing too. We adopt this selection programme in the Breeding Programme of Hungarian Sheepbreeders' Association, although the scrapie is not pre-sent in Hungary (only a single case was reported, 1964).

Our study provides base-line data on genotype frequencies of Hungarian bred stocks for the national scrapie programme to increase the resistant allele, without losing out on some relevant productive characteristics.

Development of a generic database for sheep and goat in Germany

U. Müller[1], R.. Fischer[1], F. Rikabi[2], R.. Walther[1], E. Groeneveld[3] and U. Bergfeld[1], [1]Saxon State Institute for Agriculture, Department of Animal Production, Am Park 3, 04886 Köllitsch, [2]Association of German Sheep Breeding Organizations, Godesberger Allee 142 - 148, 53175 Bonn, [3]Agricultural Research Centre, Institute for Animal Breeding, Hoelty-Strasse 10, 31539 Neustadt, Germany

In accordance with the European Union guideline 2003/100/EG each country has to introduce a breeding programme on scrapie resistance for sheep. Furthermore, an European Union regulation exists concerning the identification of sheep and goats. Given the current herdbook systems used in the Federal Republic of Germany, both guidelines can only partly be implemented. Under co-ordination of the Association of German Sheep Breeding Organizations a generalized data base structure on the basis of APIIS (Adaptable Platform Independent information system) was developed for so far 13 breeding societies. Different animal identification systems among the societies could be integrated and the equally different coding systems be unified. Following Groeneveld (2004) the data from different data collecting systems were loaded into a central data base. On this platform all required tasks concerning animal registration, population analysis, genetic evaluation, and management of scrapie resistance breeding programmes can be solved. The presentation deals with the data base structure, the information flow and the possibilities of use.

Effect of parity and number of kids in yield and milk composition in Sarda goat *G.M. Vacca*, V. Carcangiu, M.C. Mura, M.L. Dettori and P.P. Bini, Dipartimento di Biologia Animale, via Vienna 2, 07100 Sassari, Italy*

Sarda goat is the most numerous Italian breed goat. Very rustic, it is reared on natural pasture and is characterized by good longevity and production of milk with a high fat and protein content. The amount of individual milk produced and prolificity are not high. In order to assess the influence of parity and number of kids born on yield and milk composition, an investigation was carried out on 500 goats, aged between 1 and 14 years, belonging to 20 herds from the Central Sardinia. On milk samples collected in the middle of lactation were determined: fat, protein, lactose and urea content and SCC (infrared method), cryoscopic index (cryoscope), pH value, TMC (flow cytofluorimeter). Data analysis (ANOVA) showed lower milk yield ($P<0.05$) in primiparous goats, and a higher lactose content ($P<0.01$). Fat and proteins did not reveal significant differences between the different parity, while SCC ($P<0.01$) and TMC ($P<0.05$) showed a growing increase with the advance of parities. Goats that gave birth to only one kid produced lower amounts of milk with higher fat and lactose ($P<0.01$) content and lower urea ($P<0.01$) and SCC ($P<0.05$). These results, besides confirming the influence of the kind of birth, single or multiple, on yields, shows how the high longevity of Sarda goats goes with good quanti-qualitative productions.

Session 21 Poster 25

Estimation of lamb carcass composition using real-time ultrasound

F. Pajor, P. Póti, Láczó E. and J. Tözsér, Department of Cattle and Sheep Breeding, Szent István University, Gödöllö, Hungary*

Real-time ultrasound measurements (Falco 100 device, linear probe 3.5 MHz) were used to estimate in vivo carcass composition of Hungarian Merino (n=9) and British Milk-sheep (n=10) ram-lambs. Animals weight were 29.73±1.11 kg and 28.98±0.95 kg at 110-120 days age, respectively. There were little differences between breeds in the ultrasound records for subcutaneous fat thickness (0.33±0.10 cm; 0.36±0.05 cm), longissimus muscle depth (LMD) (2.44±0.12 cm; 2.29±0.13 cm) and area (LMA) (17.65±0.89 cm; 16.63±0.95 cm) taken between the 12th and 13th rib. In our sample, simple correlation of live-weight with ultrasound LMD and LMA, respectively, was 0.52 and 0.44 in Hungarian Merino and it was 0.57 and 0.58 in British Milk-sheep rams. The corresponding correlation of cold carcass weight with ultrasound LMD and LMA, respectively was 0.43 and 0.48 in Hungarian Merino and it was 0.49 and 0.50 in British Milk-sheep rams.
According to these results, ultrasonic measurements can be envolved into the selection system of lamb breeding programs, because these measurements give additional pieces of information about growth intensity, -capacity, and conformation.
Additional experiments are planned to compare live ultrasound and carcass ultrasound records for accuracy.

Session 21 Poster 26

Evaluation of carcass quality of different sheep breeds in performance testing

U. Baulain[1], W. Brade[2], A. Schoen[2] and S. von Korn[3], [1]Institute for Animal Breeding Mariensee, Federal Agricultural Research Centre (FAL), Hoeltystraße 10, D-31535 Neustadt, Germany, [2]Chamber of Agriculture Hannover, Johannssenstr. 10, D-30159 Hannover, Germany, [3]Nuertingen University, Neckarsteige 6-10, D-72622 Nuertingen, Germany*

To improve efficiency of performance testing in German sheep breeding detailled information about carcass quality of different breeds is of particular interest. In this study a total of 100 carcasses of German Blackheaded Mutton, German Mutton Merino, German Bleu du Maine and Leine sheep from stationary progeny testing were investigated by Magnetic Resonance Imaging (MRI) to ascertain morphological differences and conformation of legs with loin as a primal cut. After taking regular measurements as required for the test, legs with loin were separated from the carcass between 5th and 6th rib. A set of parallel transverse MR images (slices) was acquired covering the entire cut. Loin muscle areas were measured in consecutive slices and the volume of the entire muscle was calculated. Significant breed differences were found in muscle scores of loin and leg, of loin muscle area measured between 5th and 6th rib and of leg girth. The middle to large sized Leine sheep was inferior in all these traits. No breed differences could be observed for the weight of legs with loin and for the volume of the loin muscle estimated by means of volumetric MRI.

Session 21 Poster 27

Quality of meat from Pomeranian lambs and crossbreeds by Berrichon du Cher and Charolaise rams, stored under modified atmosphere conditions

H. Brzostowski, Z. Tanski, University of Warmia and Mazury in Olsztyn, Faculty of Animal Bioengineering, Oczapowskiego 5/140A, 10-719 Olsztyn, Poland*

The quality of meat from 100-day-old single ram lambs of the Pomeranian breed and F_1 crossbreeds by rams of meat breeds, Berrichon du Cher and Charolaise, was analyzed in the study.
Samples of the quadriceps muscle of the tight (*quadriceps femoris*) were stored in modified atmosphere (80% N_2 /20% CO_2) for 10, 20 and 30 days. The evaluation of their quality included an analysis of chemical composition, physicochemical and sensory properties, and the fatty acid profile of intramuscular fat.
Storing lamb meat under modified atmosphere conditions (80% N_2 /20% CO_2) resulted in an increase in the concentrations of dry matter, protein, fat and ash, especially after 20 days of storage. As the time of storage was extended, meat color became slightly lighter, acidity increased (lower pH), water-holding capacity decreased, and lower grades were given for some sensory parameters, especially the intensity and desirability of taste. Storing lamb meat in modified atmosphere had a positive effect on changes in the fatty acid profile of intramuscular fat.

Session 21 Poster 28

Ram odour: does it affect Meat quality in Norwegian lambs?

L.O. Eik[1], J.E. Haugen[2], O. Sørheim[2] and T. Ådnøy[1], [1]Department of Animal and Aquacultural Sciences, P.O. Box 5003, N-1432 Ås, Norway, [2]Matforsk AS - Norwegian Food Research Institute, Norway*

In the traditional Norwegian sheep farming system, sheep and lambs are grazed on unimproved mountain pastures during summer and slaughtered in September. In recent years, off-season slaughtering has become more popular. Castration of ram lambs is not permitted in Norway and delayed slaughtering might therefore affect meat quality of lambs. The objective of this study was to compare quality including flavour of meat from off-season slaughtered lambs.
The experiment was undertaken in three different locations and included slaughtering of 270 lambs of the Norwegian Cross Bred Sheep (Norsk Kvitsau) in mid September, October, November, March and April. Slaughtering and grading were undertaken at a commercial slaughterhouse. Thereafter, loin samples of *m. longissimus dorsi* were analysed for sensory traits and meat quality. Significant differences in sensory attributes between ram and female lambs were observed for all months except September, indicating a possible off-flavour quality problem related to sexual maturation of ram lambs. To avoid off-flavour on Norwegian lambs meat it is concluded that male lambs should be slaughtered in September and hence consumer demand for fresh lambs in winter, spring and summer should preferentially be met by marketing ewe lambs.

Session 21 Poster 29

Growth performance of Awassi, Charollais-Awassi and Romanov-Awassi ram and ewe lambs

A.Y. Abdullah[1], R.T. Kridli[1], A.Q. Al-Momani[1] and M. Momani-Shaker[2], [1]Department of Animal production, Faculty of Agriculture, Jordan University of Science and Technology, Irbid 22110, Jordan, [2]Institute of Tropical and Subtropical Agriculture, Czech University of Agriculture, Prague, Czech Republic*

The aim of this experiment was to study growth performance of both ram and ewe lambs of Awassi (A), Charollais-Awassi (CA), and Romanov-Awassi (RA). Twenty, six-month old lambs (10 males, 10 females) of each genotype were used in this study. Body weights, body dimensions, and body condition scores (BCS) were recorded at monthly intervals between 6 and 16 months of age. Birth weights were higher ($P < 0.01$) in CA than A and RA lambs, while weaning weights were higher in A lambs than their CA and RA counterparts. The overall mean square for biological growth parameters including body weight, wither height (H1), rump height (H2), diagonal body length (DBL) and body girth (BG), and hand girth (HG) were significantly different ($P<0.05$) between the two sexes. All variables were higher in ram lambs. Hip height (H3) and BCS were similar between sexes. Genotype significantly affected H1, H2, DBL, HG, while other parameters were not significantly different. The CA lambs were shorter and had longer bodies than their A and RA counterparts. Results of this study indicate that CA lambs had body conformation similar to meat type breeds, while RA lambs had similar conformation to pure Awassi.

Session 21 Poster 30

Experiences on S/EUROP meat qualification system on sheep breeds in Hungary

A. Molnár[1], Gy. Toldi[2], S. Kukovics[1] and T. Németh[1], [1]Research Institute for Animal Breeding and Nutrition, Gesztenyés u. 1. H-2053 Herceghalom, Hungary, [2]University of Kaposvár, Faculty of Animal Sciences P.O. Box 16. H-7400 Kaposvár, Hungary

The aim of the study was to analyse the carcass quality of difference sheep breeds according to the requirements of two - Northern (heavy weight) and Southern (light weight) - S/EUROP methods. The results received using the two S/EUROP methods were compared to the results of previously used Hungarian methods on meat/carcass qualification. This latter method was based on cutting of carcasses and on the appraisement of the rations of valuable and less valuable parts.

The lambs of Hungarian Merino, German Mutton-Merino, British Milk Sheep, Ile de France , Tsigai, Milking Tsigai and Suffolk breeds were represented in the study (20 heads from each breed). Based on the data received within breeds and between the breeds differences were observed concerning the two (S/EUROP and old Hungarian) studied methods. The two kinds of Tsigai sheep showed significantly lower ($P<5\%$) results than that of the others using the both methods. Among the other three breeds the Suffolk was proved to be the poorest one ($P<5\%$), however, the ratio of the valuable meat parts in this breed was the highest using the old method ($P<0.1$). According to the authors' opinion the breed should be considered during classification, along with the costumers' point of view.

Session 21 Poster 31

The effect of mannanoligosaccharide supplementation on the performance of ewes in late pregnancy and on lamb performance

F. Crosby[1], M. Fooley[1], S. Andrieu[2], [1]University College Dublin, Lyons Research Farm, Newcastle, Co. Dublin, Ireland, [2]Alltech Ireland, Sarney Summerhill Road,Dunboyne,Co. Meath, Ireland

Twenty-four individually fed ewes were used in an experiment to determine the effect of the inclusion of MOS (Bio-Mos 1g/ewe/day) in the diet of the final seven weeks of pregnancy. Ewes were allocated at random to two identical rations as follow: T1= control, T2=control plus Bio-Mos.All ewes were hand milked completely at 1, 10 and 18 h post-partum to evaluate colostrum yield. Blood samples were taken from lambs at 24 h for serum IgG determination. Growth performances of lambs were recorded.

Total colostrum yield to 18 h has a trend to be higher in T2 group (respectively 2193 ml vs 1694 ml; P= 0.053) There was no treatment effect for colostrum yield at 1 h or 18 h; however T2 ewes had higher colostrum yield (P<0.05) at 10 h (571 vs. 831 ml).

Lambs from T2 group had higher mean serum IgG levels than T1 lambs (16.9 vs. 13.8 g/l; P<0.05). Similarly, lambs from T2 absorbed a higher percentage of colostral IgG than lambs from T1 (17.4% vs. 14.7%; P<0.05).

Lambs ADG from T2 group tended to have numerically higher ADG in all of the recorded periods, but did not reach significance. ADG from birth to weaning at fifteen weeks of 233 and 262 g/d were recorded for T1 and T2 respectively (P<0.1).

Session 21 Poster 32

Effect of feeding pistachio skins on performance of lactating dairy goats

A.A. Naserian and P. Vahmani, Animal Science Department of Ferdowsi University of Mashhad, Khorasan, P.O.Box:91775- 1163, Iran*

Pistachio skins (seed coats) are a by-product of the pistachio processing in order to produce green kernels. The objective of this study was to determine the effect of pistachio skins (PS) as a Feed Ingredient for lactating dairy goats in the early lactation. Four multiparous lactating Saanen goats (55±7.2 days in milk and BW = 47.12±6 Kg) were used in a replicated 4×4 Latin square design (4 treatments & 4 periods). treatments were 1) 0% PS (Control diet), 2) 7% PS, 3) 14% PS, and 4) 21% PS in dietary dry matter (DM). The PS were substituted for wheat bran in the control diet. Each experimental period was for 21 days (adaptation, 14d; sample collection, 7d). Chemical analysis of PS showed that this by-product contained 21.3% crud protein, 19.6% ether extract, 37.6% NDF, 22.08% ADF, and 4.20% ash. Increasing levels of PS in diets had no significant effect on DMI, milk yield, milk protein, lactose and SNF (P>0.05).Milk fat was increased (P<0.05) from 3.1% to 3.96% as the level of PS in the diet increased from 0 to 14% in dietary DM. Dry matter intake as a percentage BW was increased (P<0.05) for goats fed 21% PS compared to goats fed 7% PS in dietary DM or control diet. The result of this study showed that PS can be a desirable feed ingredient for dairy goats.

The effect of roasted cereals on growth and blood parameters of lambs fattening

Z. Antunovic[1], M. Domacinovic[1], M. Speranda[1], B. Liker[2], D. Sencic[1], Z. Steiner[1],. Faculty of Agriculture, Trg sv. Trojstva 3, 31000 Osijek, Croatia, [2]Faculty of Agriculture, Svetosimunska 25, 10000 Zagreb, Croatia*

The study examined the effect of raw and roasted cereals (corn, oat) in diet on growth performance and blood parameters of lambs fattening. Totally 40 lambs of Merinolandschaf breed, after ablactation at average of 50 days, were tested and equally divided into two groups: an experimental (E) and a control (C) group. The fattening period lasted 57 days. The lambs were weighed at the beginning, on 28th day and the end of the experiment (57th day) when collected individual blood samples from the jugular vein. The lambs from E group grew faster (by 15%), gained higher final body weight (by 6.72%) although feed consumption was lower (by 2.32%) in comparison with the lambs from the C group. The blood serum concentrations of cholesterol-total, cholesterol-HDL and AP activity were lower for lambs E group. There were no differences among groups for the serum minerals content (Ca, P-inorganic, K, Na, Cl, Fe), concentration glucose, urea, creatinine, bilirubin, albumine, total protein and enzymes activity (ALT, AST, CK, GGT, LDH). Concerning the gained production results and a higher price of feed mixtures with roasted cereals in relation to the price of standard mixtures it may be concluded that the use of roasted cereals in feed mixture in lambs fattening is in terms of economics justified.

Influence of the diet on the productive performances of Gentile di Puglia lambs

A. Vicenti, M. Ragni, C. Cocca, L. Di Turi, F. Toteda, G. Vonghia, Department of Animal Production, University of Bari, via Amendola 165/A, Bari, Italy*

The research was aimed to evaluate the productive performances *ante* and *post mortem* of 36 Gentile di Puglia lambs raised from the age of about 50 days to the age of 169 days following three different diets. The first one (group A) was exclusively fed on complete feed *ad libitum*, the second one (group B) was fed on both hay *ad libitum* and 50% group A complete feed, the third one (group C) was fed on pasture with a feed supplementation as the group B. Live traits parameters were assessed in the periods 50-106 days and 107-169 days. The group C lambs showed the greatest live weights and daily average gains, as well as the lowest consumptions and the best food conversion indexes. At slaughtering, the heaviest body weights resulted in the group C animals. After the dissection, the same group showed a good weight of the half carcass and the highest percent incidence of the steaks. In the group A the incidence of the lumbar region ($P < 0.01$) resulted the greatest; moreover the lambs at pasture (group C) presented carcasses with reduced kidney fat.

Session 21 Poster 35

Influence of zinc-methionine supplementation on milk composition, somatic cell counts and udder health in sheep

M.El-Barody[1] and S. Abd El-Razek[2], [1]Anim. Prod. Depart., Fac. of Agric. El-Minia Univ., El-Minia, Egypt, [2]Dairy Sci. Depart. Fac. of Agric. El- Minia Univ., El- Minia, Egypt*

Twenty pregnant Ossimi ewes were assigned randomly into two equal groups. The first group (G1) served as control. The other group(G2) was supplemented daily with 1 g Zinc-Methionine (Zn-Met) in the concentrate mixture for one month prepartum. Zinc- Metionine supplementation of ewes significantly (P<0.01) increased serum immunoglobulin G (Ig) values. Ewes of treated group tended to have higher (P<0.05) percentages of total protein and Casein. Conversely, they tended to have lower (P<0.05) percentages of whey protein, non protein nitrogen and milk fat. However, lactose was not significantly affected by treatment. The Zn-Met supplementation resulted in a significant decrease (P <0.01) on the overall somatic cell counts arithmetic and geometric means by 32 % and 22 %, respectively. The percentages of positive milk samples and infected ewes were significantly lower (P<0.05) in Zn-Met group than in the control group. In conclusion, Zn-Met. Could be a useful tool as immunostimulator and improving the protein utilization in sheep.

Session 21 Poster 36

Goat dairy performances according to dietary forage:concentrate ratio and/or high doses of sunflower or linseed oil, or extruded mixture of seeds

J. Rouel[1], E. Bruneteau[2], P. Guillouet[2], A. Ferlay[1], P. Gaborit[3], L. Leloutre[4] and Y. Chilliard[1], [1]INRA-Theix, 63122-France, [2]INRA-Lusignan,[3]ITPLC-Surgères, [4]INZO-France*

Eighty-four multiparous mid-lactation goats received indoor one of 7 diets differing in either forage:concentrate (HF-70:30 or LF-30:70) and/or lipid intakes (0 *vs.*180 g/d, *i.e.* 7% of diet DM, of either linseed oil, LO, or 18:2-rich sunflower oil, SO, or an extruded 40:30:30-mixture of linseeds:sunflower-seeds:wheat, EX). After 5 weeks, net energy intake was 17-18 *vs.* 22-23 MJ/d in HF *vs.* LF diets. Daily milk(kg)/fat(g) yields were *4.3*/108, 4.0/128, 4.1/138, 3.4/98, 3.3/114, 3.5/129 and 3.5/*139*, respectively, for *LF*, LFLO, LFSO, HF, HFLO, HFSO and *HFEX* diets. Milk fat content (25-40 g/kg from LF to HFEX) increased with lipid diets (+ 6-11 g/kg). Milk lactose content was lower (-2 g/kg) in HF than LF, and increased (+ 1-3 g/kg) with lipid diets (more markedly with HF diets). Milk fat post-milking lipolysis was strongly decreased (- 46-77%) by lipid diets. Almost all changes were fully expressed one week after the beginning of treatments. *In conclusion*, dairy performances rapidly and strongly changed. Energy intake, milk yield and lactose content were strongly increased by LF diet, and milk yield slightly decreased by lipid diets. Milk fat yield was not decreased by LF diet, and was strongly increased by lipid diets, as lactose content was. Goat's responses differ markedly from cow's ones to similar diets, particularly for milk fat yield and lactose content (*Work funded by EU BIOCLA Project QLK1-2002-02362*).

Session 21 Poster 37

Milk yield and it's constituents of local barbary sheep in Libyan Arab Jamahirya

I.A. Azaga and K.M. Marzouk, Dept. of Anim. Prod., Fac. Of Agric., Sebha Univ., Jamahirya, Libiya

Forty local Barbary ewes belong to " Maknousa" Project in Fazan area- South Jamahirya were used to study both milk production and it's composition as well as factors affecting them. The ewes were hand-milked once daily and milk production was estimated at 7-day intervals according to suckling technique. Average milk yield was 39.32 kg during the lactation period of 24 wks. Peak yield occurred at 4 th week. The mean values for Percentages of fat, protein, lactose, ash, dry matter and milk energy were 5.80, 5.20, 5.40, 0.91, 17.33 and 4.19 MJ , respectively. Age of ewe had a significant effect ($P \leq 0.01$ or 0.05) on milk yield and both of %fat and % dry matter. On the other hand effect of sex of lambs had not significant effect on all traits studied. The highest value of correlation coefficient was 0.81 between milk yield and lactation length and the lowest value was -0.82 between milk yield and dry matter. It was concluded that milk yield in Barbary ewes was low, Ewes need to improve genetic source through crossing with dairy breeds as well as improve the management and environmental factors especially nutrition. It's better for more accurate to determine milk yield as twice daily milking rather than once.

Session 22 Theatre 1

Estrus synchronization by $PGF_{2\alpha}$ and $PGF_{2\alpha}$ + PMSG applications on Awassi sheep and their effect on fertility

Ü. Yavuzer[1] and F. Aral[2], [1]Department of Animal Science, Faculty of Agriculture, Harran University, Sanlıurfa, Turkey, [2]Department of Reproduction and Artificial Insemination, Faculty of Veterinary Medicine ,Harran University, Sanlıurfa, Turkey*

This study has been conducted in order to compare pregnancy rates of Awassi sheep on which have been performed estrous and ovulation synchronization during their breeding season by using only prostaglandin F2 alpha ($PGF_{2\alpha}$), and those which have been synchronized during their breeding season by using both prostaglandin F2 alpha and pregnant mare serum gonadotropin (PMSG). Ewes in their mating season (n=60) which formerly gave birth to a single litter, and which are from various age groups formed the material of the study. The sheep in Group 1 (n=20) were injected $PGF_{2\alpha}$ two times, with 12 day intervals. The sheep in Group 2 (n=20) were subject to the same procedure with the exception of being injected 500 I.U. of PMSG, 24 hours prior to the last injection. 20 sheep which were not injected any hormones formed the control group, Group 3. Pregnancy rate was higher in the sheep from Group 2, in comparison with the groups 1 and 3 ($P<0.05$). This study has shown that $PGF_{2\alpha}$ or $PGF_{2\alpha}$ + PMSG can be successfully applied on Awassi breed sheep in their breeding season.

Session 22 Theatre 2

Cryopreservation and insemination of ejaculated and epididymal semen from Dutch rare sheep breeds

H. Woelders[1], C.A.Zuidberg[1], H. Sulkers[1], M.Pieterse[2], K.Peterson[2] and S.J. Hiemstra[1]. [1]Centre for Genetic Resources, the Netherlands (CGN), Wageningen UR, Lelystad. [2]Veterinary Faculty, Utrecht University

In response to the 2001 FMD crisis and the threats of the scrapie sensitivity elimination programme, cryopreservation of semen of five Dutch rare sheep breeds was undertaken in order to be able to re-establish these breeds if necessary. 7000 doses (0.2 x 10^9 sperm/dose) of semen of 55 rams was stored in the gene bank. However, semen collection was not very successful for two breeds, because the rams were not used to human handling. As an alternative, semen was collected from the epididymides of rams after slaughter. In this way, 3660 doses of semen of 34 rams were frozen. Epididymal semen (Epi) from three rams and ejaculated semen (Eja) of five rams of Veluwe Heath Sheep was selected for an insemination experiment. Post-thaw % motile sperm and % live sperm was 42 ±4.5 and 48.8±2.1 for Eja and 60±0 % and 62.3±5.6 % for Epi (mean ±s.d.). Synchronized one year old ewes were inseminated cervically (C), or by laparoscopy (L) with 0.2 and 0.09 x 10^9 sperm, respectively, using either Eja or Epi semen (10 ewes per group). Pregnancy rates detected by ultrasonography at day 36 were 0, 40, 50, and 70% for Eja-C, Eja-L, Epi-C, and Epi L, respectively. Freezing of epididymal ram semen is cost-effective, compared to ejaculated semen and fertilizing ability of epididymal semen seems to be at least as good as that of ejaculated semen.

Session 22 Theatre 3

Divergent selection lines for spontaneous spring ovulatory activity in Mérinos d'Arles sheep: results of the first generations

L. Bodin[1], J. Teyssier[2], B. Malpaux[3], M. Migaud[3], S. Canepa[3], P. Chemineau[3], [1]INRA, SAGA, BP 52627, 31326 Castanet-Tolosan, France, [2]UMR ERRC, INRA-Agro.M, 34060 Montpellier, France, [3]UMR PRC, INRA Nouzilly, 37380 Monnaie, France*

A divergent selection for spontaneous ovulatory activity in spring is processed since 1998 in Merinos d'Arles ewes. The ovulatory activity of ewes (cyclic or non-cyclic) was determined each year in April, before any reproductive event, by the level of progesterone in two blood samples collected 9-10 days apart. At the beginning (generation G0), two groups of animals (about 100 ewes and 5 rams, each) were selected from a large flock for their extreme breeding value estimated over 3-year records. Selection was conducted in each line in separated generations from G0 to G3. From 1998 to 2003, mating were made after hormonal treatment while in 2004, natural mating are performed after induction by a male effect.

From 2000 to 2004, the cycling percentage remained significantly higher in the High line with a difference ranged from 20.8% to 40.7% for the G1, and higher than 20% for the G2 ewes. This confirms the efficiency of a selection for spontaneous spring ovulatory activity in this breed.

In 2004, results of fertility after a natural mating with a ram effect, showed no difference in the total fertility (85.4% and 81%, respectively), but ewes of the High line lambed significantly earlier than the Low line.

Session 22 Theatre 4

Honeybee royal jelly: an alternative source to serum for in vitro maturation of ovine oocyte

A.G. Onal[1], M. Kuran[2], I. Tapki[1], E. Sirin[2] and O. Gorgulu[1], Department of Animal Sciences, Universities of [1]Mustafa Kemal, Hatay, [2]Ondokuz Mayis, Samsun, Turkey*

Serum supplementation to culture media has been shown to alter embryonic development with possible detrimental implications during the culture of embryos and oocytes. Therefore, the aim of this study was to investigate whether honeybee royal jelly can be an alternative protein source to serum when employed in in vitro maturation of sheep oocytes. Ovine oocytes were aspirated from 2 to 6 mm diameter ovarian follicles and a total of 585 oocytes were matured either in 10% (v/v) foetal calf serum (Control; n=106 oocytes) or different levels of royal jelly ((2.5% (w/v) royal jelly (RJ1) or 1.25% (w/v) royal jelly (RJ2)) supplemented TCM-199 in the presence or absence of FSH (5µg/ml) and LH (5µg/ml) under a humidified atmosphere of 5% CO_2 at 38.6 °C for 22 h. Royal jelly supplementation to culture media decreased ($P<0.01$) the ratio of oocytes reaching Metaphase-II stage of nuclear maturation when compared to control group (70.8% in serum-supplemented group, 30% in RJ1 and 40% in RJ2 groups). Addition of FSH and LH to royal jelly supplemented groups significantly increased ($P<0.05$) the number of oocytes reaching Metaphase-II stage of nuclear maturation (40% in RJ1 and 55% in RJ2 groups). These results show that royal jelly supplementation some potential in the nuclear maturation of ovine oocytes in vitro, however culture conditions should be optimized with the supplementation of gonadotrophins.

Session 22 Theatre 5

Puberty occurrence in Awassi (A), F_1 Charollais-Awassi (CA) and F_1 Romanov-Awassi (RA) ram and ewe lambs

R.T. Kridli[1], A.Y. Abdullah[1], A.Q. Al-Moman[1]i and M. Momani-Shaker[2], [1]Department of Animal production, Faculty of Agriculture, Jordan University of Science and Technology, Irbid 22110, Jordan, [2]Institute of Tropical and Subtropical Agriculture, Czech University of Agriculture, Prague, Czech Republic*

The aim of this experiment was to evaluate puberty age of both ram and ewe lambs of Awassi (A), F_1 Charollais-Awassi (CA), and F_1 Romanov-Awassi (RA). Twenty, six-month old lambs (10 males and 10 females) of each genotype were used in this study. Body weights were recorded at monthly intervals between 6 and 12 months of age. Blood samples were collected at weekly intervals from ewe lambs and biweekly from ram lambs to monitor progesterone and testosterone profiles, respectively. Puberty weights in ewe and ram lambs were not significantly different among the three genotypes and ranged around 35 kg and 42 kg, respectively. In ewe lambs, puberty age was significantly higher ($P<0.01$) in A ewe lambs than RA and CA where it was 280±11.5, 232±11 and 255±11.5 days, respectively. Ram lamb puberty age was also significantly different ($P<0.01$) among the three genotypes being higher in A and CA than RA ram lambs where it was 312±14, 284±15 and 236±14 days, respectively. Results of the present study indicate that crossing Awassi with Romanov advances puberty age in ram and ewe lambs and improves semen characteristics in ram lambs.

Session 22 Theatre 6

An investigation on the effects of gossypol in cottonseed meal on scrotal circumference and spermatozoa quality in Atabay rams

F. Ghanbari[1], Y.J. Ahangari[2], T. Ghoorchi[2] and S. Hasani[2], [1]Islamic Azad University of Baft, I.R. Iran,[2]Faculty of Animal Science, Gorgan University of Agricultural Science and Natural Resources, I.R. Iran*

An experiment was conducted to investigate the effects of gossypol in cottonseed meal(CSM) on scrotal circumference(SC) and spermatozoa quality in Atabay rams. Eight Atabay rams of 2 years old with an average body weight of 58±6.09 were used. The isocaleric and isonitrogenous diets were calculated for the control and experimental groups contained %10 soybean meal and %15 CSM(containing 850 ppm free gossypol) respectively. Semen was collected for 12 weeks, and percentages of motility, live and normal spermatozoa were measured. SC was measured during the experiment every week. The data were analyzed with the Nested Design using SAS software. The results showed that the differences of SC and spermatozoa quality between the control and experimental groups were significant($P<0.01$). The percentages of motility, live and normal spermatozoa, and SC were greater in control than in experimental group(%69.57, %77.25, %80.61 and 33.55cm vs %62.48, %74.59, %59.36 and 32.72cm). The results showed that feeding Atabay rams with 15% cottonseed meal containing 850 ppm free gossypol, should be limited for long time use.

Session 22 Theatre 7

Sugarcane bagasse silage treated with different levels of urea for improvement sheep production: *II. Body weight changes and ewes` reproductive performance*

M.A. Kobeisy[1], M. Zenhom[2], I.A. Salem[1] and M. Hayder[2], [1]Anim. Prod. Dept., Fac. of Agric., Assiut Univ., Assiut, Egypt, [2]Mallawi Animal Prod. Station, Anim. Prod. Institute, Agric. Res. Center, Egypt

This study was to investigate the influence of sugarcane bagasse silage treated with different levels of urea on body weight and ewes reproductive performance. Animals (179) were four treatment groups, a control group fed concentrates with wheat straw and three silage fed groups, T0, T1.5 and T3 receiving silage containing 0, 1.5 and 3 % urea, respectively. Data were statistically analyzed using GLM of SAS. During pregnancy, T0, T1.5 and T3 had higher body weight and body weight changes ($P< 0.05$) and feed intake ($P< 0.01$) compared with control group. Number of ewes exhibited estrus or lambed were lower in T1.5 and T3 than control and T0. Number of service per conception was adversely affected by urea treatment. Silage fed groups had higher body weight at lambing and their lambs had higher average and total birth weight and weaning weight than control group. Urea had adverse effect on fertility of treated ewes as compared with fertility before treatment. About 18 % of urea fed-groups had estrus length more than 48 hours. In conclusion, feeding urea may had a negative effect on reproductive performance of ewes.

Session 22 Theatre 8

Reproductive and endocrine characteristics of delayed pubertal ewe lambs after melatonin and L-Tyrosine administration

K.A. El-Battawy, National Research Center, Animal Reproduction and A.I. Dept., Cairo, Egypt

This investigation was carried out to study the impact of melatonin and L-Tyrosine administration on the onset of cyclicity in delayed pubertal ewe lambs.
Fifteen delayed pubertal ewe lambs (age >16 month) were used in this study after being assigned randomly into three groups. First group (melatonin treated group, n=5), each lamb was administered 3 mg melatonin orally at 1600 from 1st July to 15th September while the second group (L-Tyrosine treated group, n=5), each lamb was administered L-Tyrosine at the level of 100 mg/kg b. w. as a single oral dose. A third group (n=4) served as control. Lambs were exposed to mature, fertile rams daily and blood samples were collected twice weekly.
The progesterone concentrations (P_4 evaluations) were significantly higher ($p<0.05$) in treated groups than control one. Ovarian activity, assessed by P_4 evaluations , showed that all animals in the first group came in estrus from them four got pregnant (80%) while three lambs from the second group came in heat from them two became pregnant (40%). On the other hand, non of the control lambs showed estrus
In conclusion, this study confirms that the oral administration of melatonin and L-Tyrosine played essential physiological roles to induce cyclicity in delayed pubertal ewe lambs and improved their reproduction.

Session 22 Poster 9

The ventricullar ependyma fine structure at Gallus domesticus, in Scanning Electron Microscopy

L.D. Urdes and C. Nicolae, University of Agricultural Sciences and Veterinary Medicine, 59, Marasti Blvd., sector 1, Bucharest, Romania*

In all vertebrates, the walls of cerebral ventricles and ependymar channel are lined by the ependymar layer, sustained on a basal layer, which limits it from the blood vessels and sustaining tissue. The fine structure of ventricullar ependyma has been described in numerous speciality studies at mammals. The present study comes to complete the literature data, with suggestive aspects of the two lateral cerebral ventricle in Gallus domesticus species. For this experiment, it has been taken eight different ages of White Leghorn chicks, from one to eight months old. The chicks was bred in farming unities, being submitted to immuno-prophylactic programm. For the Scanning Electron Microscopy, fresh fragments of cerebrum was sectioned in the smallest possible fragments, then introduced at once in prefixation liquid (2,5% glutharaldehide, 0,1 M Caccodilate buffer liquid with 2mM Ca Cl2) for 24 hours. The pieces was investigated by using Electron Microscope type XL-30 ESEM TMP with EDAX analysor, and Pentium IV Personal Computer.

Session 22 Poster 10

Peripheral blood growth hormone level can influence temperament in mithun (*Bos frontalis*)

M. Mondal[1], A. Dhali[1], B. Prakash[1], C.Rajkhowa[1], B.S. Prakash[2], [1]Animal Endocrinology Laboratory, National Research Centre on Mithun (ICAR), Jharnapani, Medziphema, Nagaland-797 106, INDIA,[2] Dairy Cattle Physiology Division, National Dairy Research Institute, Karnal-132 001 (Haryana), India*

The aim of the present study was to find out relationship, if any, between peripheral blood plasma levels and behavioral temperament in mithun (*Bos frontalis*), a semi-wild ruminant. For the purpose, a total of sixty-nine mithuns from four different strains viz., Arunachal, Nagaland, Mizoram and Manipur, were divided into six age groups (Group I - 0-6 months; Group II - >6 -12 months; Group III - >1-2 years; Group IV - >2-2.5 years; Group V - >2.5-3.0 years and Group VI - >3.0 years). Blood samples collected weekly for consecutive six weeks were assayed for plasma GH. The temperament score was awarded on 6-point scale; 6 for very aggressive and 1-point for the docile. It was found that the GH concentrations and temperament scores for each strain between different age groups and different strains within a group differ significantly (P<0.001). The interaction between age and temperament score was also significant (P<0.001). The strain with higher plasma GH levels had greater temperament scores and vice versa. The Manipur and Arunachal strains were found to have the highest and lowest temperament scores in all age groups with the highest and lowest plasma GH levels, respectively. In conclusion, peripheral blood GH levels can influence the temperament in mithuns. The underlying mechanism of GH on alteration of behavior is not clear. The studies on GH action on central nervous system and metabolic activity and energy status at cellular level for four different strains of mithun await a better understanding for role of GH on behavior in this species.

Session 22 Poster 11

A multi-threshold approach to analyze sensory panel data: an example of rabbit meat quality

P. Hernández[1] and L. Varona[2], [1]Departamento de Ciencia Animal. Universidad Politécnica de Valencia. 46071. Valencia. Spain. [2]Area de Producció Animal. Centre UdL-IRTA. 25198. Lleida, Spain

A multi-threshold Bayesian procedure with a probit approach is presented to analyse data from sensory panels. The procedure assumed that the analysed trait follows an underlying gaussian distribution, and each taster applied individual thresholding. The method provides estimates of the effects affecting a the underlying measure, and it also estimates the ability to discriminate of each taster. The procedure avoids the use of any technique of standardization, and the results are expressed in standard deviations of the underlying trait. As an example, the procedure has been applied to previously analysed data from a selection experiment in rabbits. Animals from generation 7th with animals from generation 21st were compared. Embryos belonging to generations 7th were frozen, thawed and implanted in does to be contemporary of animals born in generation 21st. Control and selected group were contemporary. The sensory analysis was carried on samples of the *Longissimus dorsi* muscle. The parameters evaluated were: Intensity of rabbit flavour (IRF), aniseed odour (AO), aniseed flavour (AF), liver flavour (LF), tenderness (T), juiciness (J) and fibrousness (F). There were differences between the selected group and control group for IRF, AO, AF, LF, confirming the previous results. Moreover, substantial differences between the ability of the tasters are obtained.

Effect of Silymarin on plasma and milk redox status in lactating goats

M.S. Spagnuolo[1], P. Abrescia[2], S. Galletti, D. Tedesco, F. Sarubbi[1], L. Ferrara[1], [1]ISPAAM-CNR, via Argine 1085, 8014-Napoli, Italia, [2]Dipartimento delle Scienze Biologiche-Università di Napoli Federico II, via Mezzocannone 8, 80134-Napoli, Italia*

Reactive oxygen species (ROS) are produced by metabolic processes. Their damaging action, if not counteracted or prevented by antioxidant defences, causes tissue oxidative stress. During lactation, the increase in metabolic processes is associated with increased ROS production. Retinol and α-tocopherol, major liposoluble antioxidants, scavenge free radicals, and possess immunomodulatory activities. Milk is the most important source, for calves, of these antioxidants, that save food quality and prevent lipid oxidation in growing tissues.
Silymarin possess antioxidant, and antiinflammatory properties. Ten goats received 1 g/d of silymarin from 7 days before calving until 15 days after calving. Ten goats were not treated. Milk and plasma samples, collected weekly for 21 days, were analysed for retinol and α-tocopherol concentrations. Nitro-tyrosine content in plasma was evaluated, and used as marker of protein oxidation by peroxynitrite. Plasma titres of liposoluble antioxidants did not differ between the two groups. Nitro-tyrosine plasma level was higher in control than in treated goats ($P < 0.005$). Milk levels of retinol and α-tocopherol were found higher in treated goats ($P < 0.01$). These results demonstrate that silymarin administration does not affect plasma levels of liposoluble antioxidants. Moreover silymarin seems accumulate antioxidants in milk, thus improving its quality. Our data also suggest that silymarin might play a role in preventing plasma protein oxidation by peroxynitrite.

Peripheral blood growth hormone level can influence temperament in mithun (*Bos frontalis*)

M. Mondal[1,], A. Dhali[1], B. Prakash[1], C.Rajkhowa[1], B. S. Prakash[2], [1]Animal Endocrinology Laboratory, National Research Centre on Mithun (ICAR), Jharnapani, Medziphema, Nagaland-797 106, India, [2] Dairy Cattle Physiology Division, National Dairy Research Institute, Karnal-132 001 (Haryana), India*

The aim of the present study was to find out relationship, if any, between peripheral blood plasma levels and behavioral temperament in mithun (*Bos frontalis*), a semi-wild ruminant. For the purpose, a total of sixty-nine mithuns from four different strains viz., Arunachal, Nagaland, Mizoram and Manipur, were divided into six age groups (Group I - 0-6 months; Group II - >6 -12 months; Group III - >1-2 years; Group IV - >2-2.5 years; Group V - >2.5-3.0 years and Group VI - >3.0 years). Blood samples collected weekly for consecutive six weeks were assayed for plasma GH. The temperament score was awarded on 6-point scale; 6 for very aggressive and 1-point for the docile. It was found that the GH concentrations and temperament scores for each strain between different age groups and different strains within a group differ significantly ($P<0.001$). The interaction between age and temperament score was also significant ($P<0.001$). The strain with higher plasma GH levels had greater temperament scores and vice versa. The Manipur and Arunachal strains were found to have the highest and lowest temperament scores in all age groups with the highest and lowest plasma GH levels, respectively. In conclusion, peripheral blood GH levels can influence the temperament in mithuns. The underlying mechanism of GH on alteration of behavior is not clear. The studies on GH action on central nervous system and metabolic activity and energy status at cellular level for four different strains of mithun await a better understanding for role of GH on behavior in this species.

Expression of CD147 and monocarboxylate transporters MCT1, MCT2 and MCT4 in porcine small intestine and colon

K. Sepponen[1,], M. Ruusunen[2] and A.R. Pösö[1], [1]Department of Basic Veterinary Sciences, University of Helsinki, Finland, [2]Department of Food Technology, University of Helsinki, Finland*

Considerable amounts of lactate and short-chain fatty acids (SCFAs) are formed in the gastrointestinal tract of pigs. The epithelium of the colon derives 60-70% of its energy from SCFAs, which has also other functions in the intestinal epithelium. In the neutral pH, these monocarboxylic acids are dissociated and require transporter proteins, such as monocarboxylate transporters (MCTs), to facilitate their influx into epithelial cells. We measured with immunoblotting the amounts of MCT1, MCT2, MCT4 and co-protein CD147, present in the small intestine (SI) and the colon (C) of 40 pigs (Finnish Landrace, Finnish Yorkshire and Landrace x Yorkshire) and determined the amounts of these isoforms 1 m and 12 m from pylorus. Differences were calculated either by paired t-test or with one-way analysis of variance with Tukey's post-test. Differences were regarded as significant at $P<0.05$. MCT1 and MCT4 were found in both SI and C, but MCT2 was expressed only in SI. The amount of MCT1 decreased significantly towards caudal end of SI. In both SI and C, Yorkshire pigs had more CD147 than Landrace pigs. In SI, the expression of MCT2, which has high affinity for lactate, may be of physiological importance for the complete absorption of lactate. Further studies are needed to explain the significance of interbreed differences observed.

Interrelation between milk urea concentration and reproductive performances in dairy cows under farm condition

A. Dhali, D.P. Mishra, R.K. Mehla and S.K. Sirohi, National Dairy Research Institute, Karnal, Haryana- 132 001, India*

The study was conducted on 88 crossbred Karan-Fries cows. Milk urea (MU) was analysed from noon milk samples (1200 to 1300 hr) to interrelate with parturition to first service interval, number of insemination per conception, first service conception rate and service period. Milk progesterone and MU concentration were analysed in noon milk samples at day 1, 10, 20, 30 and 60 post insemination to study the relationship between MU concentration and early embryonic mortality. The interval between parturition to first service was found significantly ($p<0.01$) higher (77.2 ± 5.5 days) when MU concentration was ≥ 63.4 mg/ dL. MU concentrations (mg/ dL) were found 42.5 ± 2.5, 47.9 ± 1.5 and 50.9 ± 3.0, respectively in the animals conceived at 1st, 2nd and 3rd insemination. First insemination conception rate (68.7 %) was found significantly ($p<0.01$) higher when MU concentration was ≤ 32.4 mg/ dL. Service period was also increased (125.4 ± 8.8 days) significantly ($p<0.05$) when MU concentration was ≥ 45.1 mg/ dL. Milk progesterone level indicated that the cows, those were detected as not pregnant at day 60 post insemination were actually pregnant till day 30 post insemination and pregnancy was terminated between day 30 to day 60 post insemination. MU level was found significantly ($p<0.01$) higher in the non pregnant cows.

Session 22 Poster 16

The association of first 60-d cumulative milk yield and embryo survival in Iranian Holstein dairy cows

A. Heravi Moussavi[1], M. Jamchi[2], R. Noorbakhsh[3], M. Danesh Mesgaran[1] & M. E. Moussavi[4], [1]Ferdowsi University, Dept of Animal Science, Mashhad 91775-1163, Iran, [2]Kenebist Razavy Farm, Mashhad, Iran, [3]Institute of Standards and Industrial Research, Mashhad, Iran, [4] DamRayaneh Co., Mashhad, Iran*

The study was designed to study embryo survival (ES) trends during the last decade along with the association of the first 60-d cumulative milk yield and ES. According to the calving date, 6500 records from one dairy farm were used. The days in milk (DIM) when cows got pregnant were used for embryo survival data. The whole milk produced in the first 60-d corresponding lactation period in which cow got pregnant was used for analysing the production data. Milk production (1826±43, 1820±40, 1886±38, 1908±36, 1821±37, 2052±35, 1999±36, 2034±37, 2030±37 and 2080±36 kg/first 60-d lactation period, respectively for years from 1994 to 2003) was significantly different among years ($P<0.0001$). The average postpartum interval for pregnancy to occur (50% of all pregnancies) was 88, 98, 104, 106, 104, 97, 103, 120, 125 and 113 DIM, respectively. Open days and the milk yield were correlated significantly ($P<0.001$, $r=0.06$). The results showed a significant reduction in reproductive efficiency along with significant increase in milk production. The results also demonstrated that in addition to milk production might be other factors contributed in reduced reproductive efficiency.

Session 22 Poster 17

The relationship between plasma Leptin and FSH concentrations in sheep with different lambing rate

M. Moeini[1], A. Towhidi[2], H. Solgi[1] and M.R. Sanjabi[2], [1]Razi university, Kermanshah, [2]Tehran university, IROST,Animal science Dep., Iran

Reports have been demonstrated a strong positive correlation between plasma leptin and FSH concentrations. Leptin has been known as a metabolic signal for reproduction (Barb C.R. 1999, Towhidi et *al.*, 2002, Kosior and Bobowiec, 2003). The aim of this study was to determine the relationship between Leptin and FSH concentrations during follicular phase of two breeds of sheep with different lambing rate. Forty 1- year Iranian native sheep (Mehraban and Sanjabi breed) were being used. Animals were cyclic and synchronized with PG $F_{2\alpha}$. Blood samples were collected 24 and 72 hours after second injection of PG from the jugular vein. Animals mated and pregnancy, lambing rates were recorded. Plasma leptin concentrations measured using radioimmunoassay Kit for mutlispecies Leptin (Mediagonst Ltd, Aspenhaustr,/Germany, 2003).
The results indicated a significant correlation between Leptin and FSH level ($R= 0.78$ in Mehraban and $R=0.83$ in Sanjabi ewes), no significant relationship has been observed between FSH/Leptin concentrations with lambing rate but the responses were different between two breeds. Plasma leptin concentrations were significantly higher in all animals at 72^h (Mehraban: 4.74±0.15; Sanjabi: 4.68±0.10 ng/ml Human Equivalent) than at 24^h (Mehraban: 2.635±0.11 and Sanjabi: 2.56±0.04 ng/ml H.E.). FSH plasma was also significantly higher at 72^h in both breeds (Mehraban: 2.752±0.17, Sanjabi:
2.74±0.15 ng/ml) than at 24^h (Mehraban: 1.19±0.05, Sanjabi: 1.19±0.04ng/ml).

Session 22 Poster 18

Embryo survival and its association with first 60-d cumulative milk yield in Holstein dairy cows

A. Heravi Moussavi[1], M. Jamchi[2], R. Noorbakhsh[3], M. Danesh Mesgaran[1] and M.E. Moussavi[4], [1]Ferdowsi University, Dept of Animal Science, Mashhad 91775-1163, Iran, [2]Kenebist Farm, Mashhad, Iran, [3]Institute of Standards and Industrial Research, Mashhad, Iran, [4] DamRayaneh Co., Mashhad, Iran*

The study was designed to study embryo survival (ES) trends during the last decade along with its association with first 60-d cumulative milk yield. According to the calving date, 6500 records from one dairy farm were used. The days in milk (DIM) when cows got pregnant were used for embryo survival data. The whole milk produced in the first 60-d corresponding lactation period in which cow got pregnant was used for analysing the production data. Milk production (1826 ± 43, 1820 ± 40, 1886 ± 38, 1908 ± 36, 1821 ± 37, 2052 ± 35, 1999 ± 36, 2034 ± 37, 2030 ± 37 and 2080 ± 36 kg/first 60-d, respectively for years from 1994 to 2003) was significantly different among years ($P<0.0001$). The average postpartum interval for pregnancy to occur (50% of all pregnancies) was 88, 98, 104, 106, 104, 97, 103, 120, 125 and 113 DIM, respectively. Open days and the milk yield were correlated significantly ($P<0.001$). The results showed a significant reduction in reproductive efficiency along with significant increase in milk production. The results also demonstrated that in addition to milk production might be other factors contributed in reduced reproductive efficiency.

Session 22 Poster 19

Effect of media and presence of Corpus Luteum on *In Vitro* maturation of buffalo *(Bubalus bubalis*) oocytes

Marwa S. Faheem, A.H. Barkawi, G. Ashour and Y. Hafez, Cairo University, Faculty of Agriculture, Animal Production Department, Giza, Egypt*

This work is aiming at studying the effect of culture media and presence of Corpus Luteum (CL) on in vitro maturation of buffalo oocytes. Two experiments were conducted to achieve the study goals. Experiment 1 was to test the effect of media (TCM-199 + epidermal growth factor- EGF, TCM-199 + FSH, LH, E_2 hormone and synthetic oviductal fluid- SOF medium) on maturation rate. Each medium was supplemented with 10% FCS and 50µg/ml gentamycin. Experiment 2 was to test the effect of CL presence on maturation rate. Oocytes were collected separately from two types of ovaries (with or without CL) and cultured for 24 h at 38.5°C in 5% CO_2 in air in the same media used in experiment 1. Some oocytes were artificially denuded and their diameters (external, internal diameters and thickness of zona pellucida) were measured in 0 time and 24 h after culturing for maturation. Average (LSM) and percentage were used to analyse data. Results showed that maturation rate of TCM-199 with EGF medium (96.0%) and TCM-199 with hormones (93.9%) were higher ($P<0.05$) than SOF medium (87.6%). Presence of CL had a positive effect ($P<0.05$, 90.8 vs. 85.8% in absence of CL) on maturation rate. External-oocyte diameter and previtelline space were significantly ($P<0.05$) larger after maturation period.

Session 22 Poster 20

Male-female interactions on pubertal events in female goats

C. Papachristoforou[1], A. Koumas[1], C. Photiou[1] and C. Christofides[2], [1]Agricultural Research Institute, P.O. Box 22016, Lefkosia, Cyprus, [2]Department of Agriculture, L. Akrita Ave, 1412, Lefkosia, Cyprus

A number of experiments were conducted with prepubertal goat-kids to investigate male and female effects on the onset of reproductive activity. The results showed that the presence of males advances the initiation of pubertal sexual activity in females by about 4 to 6 weeks regardless of season of birth (autumn, spring) or breed (Damascus, native Machaeras) of females. However, season of birth affected the proportion of Damascus, but not of Machaeras goat-kids responding to the introduction of males, with autumn-born animals being more responsive compared to spring-born ones. Attempts using the male stimulus to further advance the onset of puberty during mid anoestrus in autumn-born Damascus females produced positive results in a lower proportion of females than in late anoestrus. This positive response was observed only during periods of induced (by male introduction) sexual activity in adult goats in the same location leading to the conclusion that besides the effect of males, the effect of oestrous goats on the prepubertal females is required for the onset of reproductive events in the latter.

Session 22 Poster 21

Survey of blood electrolytes changes two weeks before to two weeks after delivery luri Bakhtiari race sheep

M. Faghani and F. Kheiri, Islamic Azad university of sharekord, Iran

In this project, first 30 pregnant ewes of the luri Bakhtiari race were primarily selected. From 2 weeks before up to three weeks after lambing, the ewes blood samples were collected once a week from their jugular veins via blood taking vacuum (venoject) tubes for the separation of blood serum. The values of samples (K and Na) electrolytes were measured using a flame photometer set based on the existing methods of AOAC, and the values of Ca and Mg electrolytes, by spectrophotometry. SAS statistical software was then used to calculate the correlation between the constituents of blood and the process variations were diagramatically detected. Blood Na had the highest value 1 week before lambing, with no significant difference in its value 1 week after lambing. But it was significant base in 2 and 3 week before lambing. Blood K had the highest value 1 week after lambing, with a statistically significant difference in its value in other weeks. Blood Ca had the highest value 2 weeks before lambing, with no significant difference in its value 1 week before, 1 week after and 3 weeks after lambing, but with a significant difference (in its value) 2 weeks after lambing which showed the lowest value. Blood Mg did not have a statistically significant difference in all period of examination.

Session 22 Poster 22

Effects of bombesin neurotransmitter injection on the thyroid hormones concentraction in Sarabi cows

M. Yousef Elahi and E. Baghaei, Islamic Azad University of Azadshahr, Iran

The thyroid hormones are one of the important body hormones which play vital role of body metabolism regulation. There is enough evidence which demonstrates releasing or inhibition of secretion of these hormones are affected by other hormones and neurotransmitters. The aim of this study is determination of bombesin injection effects on the thyroid hormones concentration. Nine Sarabi cows with average body weight of 420 ± 20 Kg were used in a split-plot design study with 3 cows per treatment. Cows divided randomly into 3 groups. Each group received daily injections of either saline or 1, 2, 4 milligrams bombesin per kilogram for 10 days. Blood plasmas were determinated for thyroid hormones by radioimmunoassay. Results represented that there were the most effects of bombesin on the T_3 secretion in the first and third treatments and about the T_4 hormones in the second treatment ($p<0.05$). The obtained different effects from various rates of injection can attribute to especial properties of receptors and with attention to receptors diversity about bombesin, this subject is not far.

Session 23 Theatre 1

Characterization of farms in winter cattle fattening, Argentine

A. Castaldo[1], J. Martos[2], D. Valerio[3], R. Acero[2], A. García[2], and J. Pamio[1], [1]Veterinary School, University of the Pampa (UNLPAM), Argentine, [2]Department of Animal Production, University of Cordoba (UCO), Spain, [3]Agricultural and Forestry Research Dominican Institute (IDIAF), Dominican Republic

The objective of the study is identification and classification of the fattening production bovine systems in pastoral conditions “winter” in the semiarid Argentine Pampa, taking in count their physical, productive, economic characteristics and management. A sampling of 56 farms is made in the Department of Quemu Quemu, province of Pampas. By means of ANOVA techniques and the use of Cluster analysis, three pastoral subsystems are identified. The first, denominated traditional grazing, responds to operations with low levels of productivity and daily average gain. This system designates 56% of the cattle grazing surface to perennial pastures and the wintertime lasts for of 23 months. This group of operations applies to a criterion of minimum cost in the decision making. The third subsystem is denominated technified grazing and show greater levels of productivity and daily average gain; they use greater levels of feed and they dedicate a 77% of the cattle surface to the pastures, and the wintertime is finalized in 17 months. These operations apply to maximum criterion of benefits in the decision making. The second subsystem, denominated of transition marks the evolution between the traditional system and the technified one.

Session 23 Theatre 2

Sustainable technical index for bio-economic farm models in Pampean region Argentine

J. Martos[1], A. García[1], R. Acero[1], D. Valerio[2], V. Rodríguez[1], and J. Perea[1], [1]Department of Animal Production, University of Cordoba (UCO), Spain, [2]Agricultural and Forestry Research Dominican Institute (IDIAF), Dominican Republic

An index of sustainable technical development, based on agrosystem properties (productivity, stability, and sustained growth) and its trade off is proposed. First, system's relative energy is modelled using a Cobb-Douglas function, incorporating physical, technical, and economic variables, for nineteen periods. Level of productivity is analyzed using Greene's absolute frontier, stability and sustained growth using Marshack-Andrews frontier. The results enable the establishment of a scale of sustainable development with two levels. Level A, sustainable farms with 62.22% productivity, 85.90% stability and a sustained growth of 14.40%. Analyzing the trades off, it is observed an inverse relationship between productivity and stability as between stability and sustained growth. Not sustainable farms are grouped in level B, with the following attributes: productivity (48.62%), stability (98.16%) and sustained growth (10.84%). From trade off's analysis we see a direct proportional relationship among productivity, stability and sustainability. We also observe that sustainability index depends, considering time and space, on risk and uncertainty existing.

Session 23 Theatre 3

Animal production systems in Algeria: transformation and tendencies in the area of Sétif

K. Abbas, INRA Algérie, Unité de Sétif, Route des fermes, 19000 Sétif, Algeria*

Algeria has 8 million ha of SAU of which 443 000 ha are irrigated (0.05%). With population of more than 30 million, the country registers 0.25 ha per capita. Agriculture is concentrated in a narrow fringe of the north where dominates a semi-arid climate and a very irregular rainfall. More in the south, a steppe area of more than 30 million ha is devoted mainly to small ruminant production. A large majority of the farms draw a major part of their incomes from cereal - animal association system. The diversity of the climate and the physical environment, on the one hand, and the agrarian policies on the other hand, induced major transformations in these systems. Transformations are also the result of a significant population increase in rural area what generated an increase in the food needs. In this context appear significant stakes on the use of the resources and the system durability. To clarify the actual situation, this study, based on 200 owners investigation, shows that:

- the intensification can constitute a threat on the pastoral resources;
- the brittleness of the farms encourages generalization of practices marking a weak autonomy of the breeding and a bad stock management;
- the animal interspecific integration and partial intensification are positive if they follow models which ensure autonomy, performance and optimal resource managemen.

Session 23 Theatre 4

Evaluation of smallholder pig production systems in North Vietnam considering input, management, output and comparing economic and biological efficiency

U. Lemke[1], L. T. Thuy[2], B. Kaufmann[1], A. Valle Zárate[1], [1]Institute of Animal Production in the Tropics and Subtropics, University of Hohenheim, Stuttgart, Germany, [2]National Institute of Animal Husbandry, Hanoi, Vietnam*

The study aims to assess the suitability of pig breeds for different smallholder production conditions in Vietnam, comparing an indigenous with a Vietnamese improved breed.
Fieldwork was conducted in four villages with different remoteness in North Vietnam from 2001 to 2002, in 64 households keeping the improved Mong Cai or indigenous Ban as sow breeds and crossbreds for fattening. Four visits per farm yielded 234 structured interviews. Individual weights of pigs (n = 755) were obtained. Data were analysed by regression, linear and loglinear models.
Results show daily feed costs to be highest near town. Feed constitutes with 37 to 84% the bulk of variable costs, which were higher near town. The improved local breed yielded higher performances (83 kg piglets weaned/sow/year, ADG of crossbreds 161g/day) than the indigenous breed (31 kg piglets weaned/sow/year, ADG of crossbreds 83g/day). Total liveweight offtake was higher near town and in one village of intermediate location, as was feed efficiency. Cash revenue was higher near town. Gross margins did not differ between villages. Benefit cost ratios were highest for production with the indigenous breed further away from town.
The relative advantages of different pig breeds are discussed along different bio-economic criteria.

Session 23 Theatre 5

Analysis of the beef cattle growth curves in the main livestock systems of the Basque Country

R. Ruiz, A. Igarzabal, N. Mandaluniz, M.E. Amenabar, L.M. Oregui, NEIKER A.B., Apdo. 48, 01080 Vitoria-Gasteiz, Spain*

The increasing importance of beef production in the Basque Country is encouraging farmers' associations to ask for higher demands of technical and advisory services. To do that it was considered that some basic knowledge was required about the farming conditions and their implications on productive yields. The interactions existing between animal growth, nutrition management and reproduction was considered as a key point to begin with. So a project was designed to describe growth curves and bodyweight (BW) changes of pure-bred heifers of Pirenaica, Limousin and Blonde D'Aquitaine during the first 3 years of life in the main livestock systems of the Basque Country. First, a survey was carried out in 37 farms to do the characterisation of the main production systems. Basically, collected information was related to farm size, reproductive management, grazing management, and replacement. Second, serial BW data of heifers ageing less than 3 years-old were measured in the same farms every three months during a whole year. Then a comparison of the following growth curves models was done to find out which one fitted better to the data available: quadratic, logistic, Brody, Von Bertalanfy, Richards and Gompertz. Once a model was chosen, the parameters were calculated for each breed and production system. This paper will focus on the methodology and preliminary results of the project.

Session 23 Theatre 6

Indigenous selection criteria in Ankole cattle and different production systems in Uganda

M. Wurzinger[1], D. Ndumu[1,2], R. Baumung[1], A. Drucker[3], O. Mwai[3], J. Sölkner[1], [1]University of Natural Resources and Applied Life Sciences, Vienna, Austria, [2]National Animal Genetic Resources Centre & Data Bank, Entebbe, Uganda, [3]International Livestock Research Institute, Nairobi, Kenya*

Ankole cattle are kept in South-Western Uganda which is part of the cattle corridor, an area that was traditionally communal grazing land. Currently, the pastoral system undergoes a dramatic change due to land shortage and political reasons. A study was carried out to describe the production system with a focus on indigenous knowledge and the documentation of changes. Four different regions were identified and 30 farmers each were interviewed. In the two areas with more traditional systems main selection criteria in cows and bulls are body characteristics (coat colour, horn size and colour), herds are larger and cattle are the main source of income. Some families still move during the dry season with their cattle in search of water and pasture. In the two other areas farmers are sedentary and both livestock and crop production contribute to the income. Due to increasing population pressure the trend is to keep less but more productive animals. Selection focuses more on production traits like milk yield, growth and fertility. Crossbreeding with exotic cattle breeds is becoming more popular. Farmers mention that Ankole have advantages over exotic breeds in terms of disease resistence, heat tolerance, lower feed requirements and the beauty of the animals.

Session 23 Theatre 7

Concentration of conjugated linoleic acid in grazing sheep and goat milk

E.Tsiplakou, K. Mountzouris and G.Zervas, Department of Animal Nutrition, Agricultural University of Athens, Iera Odos 75, GR-118 55 Athens, Greece*

The conjugated linoleic acid (CLA) content of grazing sheep and goat milk fat, through out their lactation period, was examined. Six sheep and six goat representative farms were selected at random and milk samples were taken at monthly intervals for fatty acids profile determination. Sheep and goat nutrition was based on natural grazing and on supplementary feeding during the winter months. From April onwards, grazing native pastures was the only source of feed for sheep and goats. The University farm was also used as reference, because its sheep are kept indoors all year round. Fifteen individual milk samples were also taken in April from a sheep and goat farm respectively, in order to see the variability of CLA inside the farm. The results showed that: a. the CLA content of grazing sheep and goat milk fat increased significantly in April-May (early growth stage of grass) and then declined while that of indoors kept sheep was more or less constant during the same period, b. the isomers *cis*-9, *trans*-11 and *cis*-12, *trans*-10 of CLA were found in grazing sheep milk fat, while in indoors kept sheep and goats milk fat was found only the *cis*-9, *trans*-11 isomer, c. the CLA content of sheep milk fat was much higher than that of goats and d. there was considerable variation in milk fat CLA content between farms and inside the farm.

Revitalizing livelihoods of tsunami victims in Aceh, Indonesia through flood risk sensitive livestock development

C.B.A. Wollny and Girma Tesfahun, Georg-August University Goettingen, Institute of Animal Breeding and Genetics, Animal Breeding and Husbandry in the Tropics, Kellnerweg 6, 37077 Goettingen, Germany

The magnitude of the human, livestock, and material devastations of the December 2004 Tsunami is massive requiring global efforts to deal with the short term and long term consequences. The densely populated Indonesia's Aceh province was the most affected area with nearly 170,000 people reported dead. Practically everyone of the 4.3 million inhabitants of the province lost relatives and their livelihood base. This paper presents a conceptual framework to invigorate the devastated livelihoods in Aceh with an investment on risk sensitive and quick return livestock production system. In the short-term immediate needs of the survivors such as building confidence and enterprise establishment are on focus whereas in the medium to long-term agricultural systems development would be the main intervention. With a strong argument that livestock are among the few options for fast recovery, the framework outlines the working principles, essential components, and characteristic features of risk sensitive livestock development. The framework presents a comprehensive 5-years plan of action ranging from introducing specific husbandry practices to reestablishing the population through local breeds of the favorite species goats, chicken, ducks and cattle. A prerequisite for success of this master plan would be the institution of a long-term partnership program, which would enable the farmers to draw on the local and global scientific community for assistance.

The development of milk production in dairy farms in Saudi Arabia

M.A. Alshaikh, P.O. box 2460 Department of animal production, college of agriculture king Saud University Riyadh 11451 Saudi Arabia*

This study describes the development in commercial dairy cattle farming in Saudi Arabia during the period 1985 through 1999. Data from Agriculture statistical year books, Saudi Agriculture Bank and date from some commercial dairy herds in Riyadh region were analyzed. The results showed that total milk production was increased five folds during the period of study. Cow number per farm also increased from around 29751 to 77994 during the same period. The results also showed that cow milk yield was improved by 84% during the study period. The number of dairy projects ranged between 32-39 projects, 75% of which are located at Riyadh and Eastern provinces. Riyadh and Eastern provinces were found to supply more than 94% of total milk production in the Kingdom. A characteristics feature of commercial dairy farms in Saudi Arabia in the large size of the dairy herd. Per capita consumption of milk from commercial dairy farms rose from 9.4kg in 1985 to 27.9Kg in 1999. Also the results showed that the interest-free loan provided by the Saudi Agriculture Bank has played a major roll in the development of dairy cattle in Saudi Arabia.

Session 23 Poster 10

An investigation on the angora goat raising in Ankara of Turkey

M.I. Soysal[1] ,E.K. Gürcan[1] ,E. Özkan[1], M. Aytaç[2] and S Özkan[3], [1]Trakya University, Tekirdag Agricultural Faculty , Department of Animal Science, 59030 Turkiye, [2]Lalahan Livestock Central Research Institute, Lalahan, Ankara, Turkiye, [3]Güdül Ankara, Turkiye*

This search had been carried on 40 goat breeders belonged to the 12 different villages of Güdül district of provinces of Ankara. Angora goat breeders had been subjected to the question of inquiry in order to have information about Angora goat raising in region.It is cleared them 65 % of breeders were dealing with animal and plant production both .The rest 35 % breeders were only dealing with animal production. Twenty percent of breeders had no land at all. Thirty four and thirty percent of goat breeders had a land of 0-50 da and 50-100 da respectively. The percentage of goat breeders had land over100 da were % 20. There were 8490 head Angora goat, 1390 head sheep and 250 head cows in this region totally. This research was conducted on the present status of Ankara goat breeders in Ankara of Turkiye. It is aimed to present a glance look on the status and clarify the problems of Ankara goat breeders.

Session 23 Poster 11

HERDYN: a dynamic model to simulate herd dynamics in beef cattle extensive systems

A. Bernués[1], R. Ruiz[2],D. Villalba[3], [1]CITA Gobierno de Aragón, Zaragoza, Spain, [2]NEIKER, Vitoria-Gasteiz, Spain, [3]Universidad de Lérida, Lérida, Spain*

HERDYN is a stochastic dynamic model that simulates beef cattle herd dynamics in the medium/long run, under a wide range of management conditions. It has been developed using STELLA8r software and runs on a daily basis. HERDYN can be linked to a suckler cow-calf model that allows simulating nutrition strategies and their relationship with cow reproductive performance through the Body Condition Score (BCS). Cows are classified into a variable number of management groups according to the length of the postpartum anoestrus (PA), which is related to BCS. After PA, animals begin cycling in 21-days-length periods, and the chances of pregnancy depend on: length of the mating season; number of bulls; number of matings/bull/day; and fertility per mating. Length of pregnancy and probability of abortion are random variables. After calving, cows go through a new PA period, whose length is defined by nutrition management. Non-pregnant cows go to the next mating season or are culled. The model also takes into account random mortality, culling and replacement rates. In this paper, the model structure and application to a hypothetical mountain beef cattle system in the Spanish Pyrenees are described. The integration of herd and cow models will allow simulating the effects of diverse management strategies on the bio-economic efficiency of beef production systems.

Session 23 Poster 12

Comparative study on the behaviour of four goose genotypes during the preconditioning for laying

M. Molnár, I. Nagy, T.Molnár, K. Tisza and F. Bogenfürst, University of Kaposvár, Faculty of Animal Science, Kaposvár, Hungary

In the case of goose liver production adopting the EU regulations, new technologies should be developed instead of force feeding. One of the potential alternatives is the pinguefaction due to periodical feeding, which needs the modification of the feeding behaviour of the goose. In this study the behaviour of four goose genotypes (White and Grey Landes, White meat type, and Hungarian Frizzled goose) was compared during the preconditioning period. The behaviour was recorded 24 hours per week by digital cameras connected directly to a PC. The examined behaviour forms were feeding drinking, resting, social behaviour, preening and playing. The actual behaviour form was monitored in every minute.

Considering the feeding behaviour it can be stated that under intensive conditions the White meat type is preferable because of its higher feed consumption.

The appearing frequency of feeding by the White Landes type was similar to that of the Meat type, but the high ratio of the playing and the preening made its behaviour unfavourable. The increased preening activity of the White Landes genotype originates from a lower stress tolerance and indicate a secondary substitutive activity.

The significant difference from Hungarian Frizzled goose (that is an extensive genotype), and the higher ratio of resting by White meat type suggests the possibility of successful selection on feeding behaviour.

Session 23 Poster 13

Herbage from wet semi-natural meadows: 2. Nutrient content of some dominating species during different parts of the season

Eva Spörndly[1], Zahrah Lifvendahl[2], Åke Berg[2] and Tomas Gustafson[2], Departments of [1]Animal Nutrition and Management and [2]Conservation Biology, Swedish University of Agricultural Sciences*

The wet semi-natural meadows in Sweden are important for bird life. Therefore farmers harvest these areas after July 10th to ensure undisturbed reproduction of birds. A digestibility trial was performed on hay from these unfertilised meadows (poster 1) and it was concluded that the content of metabolisable energy (ME) was exceptionally low. Nutrient contents in some meadow species were studied further. Replicate samples of four species (*Calamagrostis Stricta, Agrostis Canina, Carex Nigra, Carex Aquatilis)* were collected in early season (May 22 - June 6*)*, at hay harvest (July 8) and from late season regrowth (August 28). Dry matter (DM), ME, crude protein and neutral detergent fibre (NDF) of samples were analysed and analysis of variance was performed on the results. The ME content of *C. Stricta, A. Canina, C. Aquatilis* and *C. Nigra* in early season was 11.7, 11.2, 10.9 and 10.3 ME/kg DM, respectively. Corresponding values at harvest time were 10.3, 8.0, 7.7 and 7.9 ME/kg DM and in late season 10.9, 10.2, 9.0, 10.0 ME/kg DM, respectively. Crude protein and NDF content varied between 38 and 160 g/kg DM and 510-666 g/kg DM, respectively. Only C. Stricta had an ME contents that were comparable to table values for well-managed cultivated Swedish pastures (10.5-11.5 ME/kg DM).

Session 23 Poster 14

Herbage from wet semi-natural meadows: 1. Digestibility of hay harvested late to protect bird life

Eva Spörndly, Ingemar Olsson and Kjell Holtenius, Department of Animal Nutrition and Management, Swedish University of Agricultural Sciences, Uppsala, Sweden*

Some of the natural wet meadows in Sweden are particularly interesting for bird life and farmers harvest these areas after July 10th to ensure undisturbed reproduction of birds. Knowledge is scarce about the feeding value of hay produced on these meadows and it has been questioned whether the Swedish routine *in vitro* procedure to determine metabolisable energy (ME) content in herbage can be applied on herbage from natural pastures. Therefore a digestibility trial with heifers using acid insoluble ash as a marker, was performed with late harvested hay from a wet meadow (dominated by *Carex nigra, Agrostis canina, Calamagrostis stricta*) and two control roughages (conventional hay and silage) from cultivated leys dominated by *Phleum pratense* and *Festuca pratensis.* The organic matter digestibility were 47, 64 and 78% and the content of ME determined *in vitro* and *in vivo* was 7.3/6.3, 9.3/9.2 and 11.9/11.2 MJ/kg dry matter for the meadow hay, conventional hay and conventional silage, respectively. For all three feeds the *in vivo* estimates were slightly, but not substantially, lower than ME estimated *in vitro*. The results from the digestibility trial confirm the low ME content in the meadow hay indicated by the *in vitro* estimates. Whether specific species contribute to the low nutrient content of meadow hay was further investigated (poster 2).

Session 23 Poster 15

Reproductive, survival and growth traits of the crossbreeding Belgian Texel x Moroccan local breeds of sheep

M. El Fadili[1], P.L. Leroy[2]. [1]Institut National de la Recherche Agronomique, 10100, Rabat, Morocco, [2]Université de Liège, B-4000, Belgium

An experiment was carried by INRA_Morocco to evaluate performances of Belgian Texel (BT) rams and their progeny when mated to Moroccan local breed ewes. Three BT rams were mated to Timahdite (T=30) and D'man x Timahdite (DT=30) ewes and compared to purebred D'man (D=22) and (T=30) for ewe and lamb pre and post-weaning traits. Results indicated that ewes mated to BT rams showed higher fertility (91%) and productivity per ewe joined at weaning (25.40 kg). Corresponding values for productivity were 20.78 and 17.12 kg for purebred T and D ewes. Lambs born from ewes mated to BT have higher survival rates at birth (93%) and at weaning (86%). Furthermore, lambs sired by the BT rams had superior weaning weight (+3 kg), ADG10-30 (+42g/d) and ADG30-90 (+25 g/d) when compared to purebred lambs. Crossed lambs had higher fattening ADG (225 g/d), less DM intake (1.06 kg) and better conversion feed rate (5.20). Corresponding values were: 207 and 211 g/d and 1.17 and 1.14 kg and 6.42 and 5.31 for purebred lambs D and T. These results indicate that Belgian Texel rams and their progeny have well performed under Moroccan management conditions. Since Belgian Texel is known for its ability to produce higher meat quality, this breed can be considered in crossbreeding to improve sheep meat quality in Morocco.

Comparison of small and large size dairy cows in a pasture-based production system

M. Steiger Burgos[1], R. Petermann[1], P. Hofstetter[4], P. Thomet[1], S. Kohler[1], A. Muenger[2], J.W. Blum[3] and P. Kunz[1], [1]Swiss College of Agriculture, Zollikofen, Switzerland, [2]Swiss Federal Research Station for Animal Production and Dairy Products, Posieux, Switzerland, [3]Inst. of Animal Genetics, Nutrition and Housing, Univ. of Bern, Bern, Switzerland, [4]Schuepfheim Agricultural Education and Extension Centre, Schuepfheim, Switzerland*

To evaluate cow body size as a relevant issue in a pasture-based dairy production system with block-calving in spring, two herds were formed. The first consisted of 13 bigger cows (type B, >700 kg body weight) and the other of 16 smaller cows (type S, <600 kg body weight) of Swiss breeds. Each herd had access to 5.8ha pasture (rotational system), so that the same overall stocking rate was obtained (1700kg/ha). Each herd received 2115 kg of concentrate until the end of the breeding season. Data about body weight, BCS, milk production, milk contents and veterinary costs were collected during 3 years. Herd S produced more milk than herd B. During the same time, the changes in BW and BCS were slightly higher in the B-cows than in the S-cows. Acetone milk content was also higher in the B-cows, especially during year 1 (year 3 not yet available). Veterinary costs per cow were similar in both herds. Differences between herds were small, but some parameters seem to show that B-cows had more difficulties to express their full potential in a system as practised in this trial.

Daytime maintenance and social behaviours during suckling period in jennies reared under semi-extensive conditions

A.G. D'Alessandro[1], D. Casamassima[2], G. Martemucci[1], N. Simone[1], G.E. Colella[2], [1]Dipartimento PRO.GE.S.A., Università degli Studi di Bari, Italy, [2]Dipartimento S.A.V.A., Università degli Studi del Molise, Italy*

Daytime behavioural maintenance and social activities were studied during the suckling period (6 months) in Martina Franca breed jennies. The animals (N. 10) were reared in Southern Italy (40° 37' N) under semi-extensive conditions. The behaviour of jennies was observed in presence of their foals for 120 daylight hours. The following 14 activities were recorded using scan sampling: feeding, drinking, standing, walking, resting (maintenance behaviours), urinating-defecating (eliminative behaviours), grooming, sniffing letter, nasal sniffing, playing between dams and their foals, playing between jennies, bray, snorf, bray-snorf (social behaviours).

The results indicate that the maintenance behaviours are significantly influenced by both the time from foaling (0.05>P<0.01) and individual effects (P<0.01). Significant (0.05>P<0.01) resulted also the effect of a.m. or p.m. hours of day, excepting for standing. Social behavioural activities of jennies showed a significant (0.05>P<0.01) variability during the suckling period.

This study offers some information on behavioural profile in jennies and shows that during the suckling period their behaviour changes in relation to time from foaling. Moreover, daytime maintenance activities are affected also by a.m. or p.m. hours of day.

Session 23 Poster 18

Application of discriminant analysis to the morphostructural differentiation of 7 extensive goat breeds of extensive

M. Luque, E. Rodero, F. Peña, A. García and M. Herrera, Department of Animal Production, University of Córdoba (UCO), Spain

We have studied 7 breeds of goat used extensively for the production of meat in Spain. The application of the discriminate analysis to 18 measures taken on the animals, is an effective method in the morphostructural differentiation, as much between the breeds the eco-types. The Mahalanobis test establishes significant differences between the 7 breeds studied.

Session 23 Poster 19

Size versus beauty: farmers' choices in a ranking experiment with African Ankole Long-Horned Cattle

D. Ndumu[1,2], M. Wurzinger[1], R. Baumung[1], A. Drucker[3], O. Mwai[3], J. Sölkner[1]*, [1]University of Natural Resources and Applied Life Sciences, Vienna, Austria. [2]National Animal Genetic Resources Centre & Data Bank, Entebbe, Uganda. [3]International Livestock Research Institute, Nairobi, Kenya

Ankole cattle are well known for their massive white horns and red coat colour. To identify other phenotypic characteristics important as indigenous selection criteria, an experiment with farmers was carried out at a governmental Ankole nucleus farm in South-Western Uganda. A large number of body measurements were taken from 15 bulls and 35 cows and phenotypic characteristics were described in detail. Twelve groups of 6 to 8 farmers each were invited to rank animals according to liking. Each farmer group was presented with 2 groups of 5 bulls and 4 groups of 4 cows, randomly sampled from the bull and cow cohorts. For analysis, horn orientation, horn tip interval, horn base interval, horn colour, horn length, horn circumference, coat colour and pattern were grouped together to represent "beauty". Body length, body weight, height at withers and heart girth formed the size traits. The coefficient of determination from a linear model was taken as an indicator. In cows, 26% of the variability in ranking was explained by beauty and 13% by size. In bulls the corresponding figures were 53% and 48%, respectively. The repeatability of ranking decisions between farmers was higher for bulls (0.60) than for cows (0.37).

Session 24 Theatre 1

Equine science education in Sweden: ten years of experience of the equine studies program

A-L. Holgersson, L. Roepstorff, J. Philipsson, G. Dalin, S. Lundesjö-Öhrström, K. Morgan, K. Ericson, M. Gottlieb-Vedi, A. Forslid. Department of Equine Studies, Swedish University of Agricultural Sciences (SLU), P.O.Box 7046, S-750 07 Uppsala, Sweden*

The equine studies program at SLU is a two years curriculum program leading to a Diploma of higher Education in Horse Management. The program started in 1994 and includes theoretical studies combined with practical applications and training. After a basic year students specialise to become riding instructors, stable managers or for various positions in the racing sector. Annually 150-200 students apply and after an admission test including riding skills 45-55 students are enrolled.
The equine studies program utilizes scientists of various disciplines at SLU as examiners and partly for the teaching of theoretical subjects, whereas well recognized trainers teach practical subjects. For this reason the program is located at three different equine centres. Two is responsible for the basic year, and all three for the specialized second year. The main challenge in developing this program has been to get the university philosophy of learning into an established system of practical horse training and stable management, i.e. how to apply science in practice. In developing the curriculum various evaluations take place regularly among students, by horse industry representatives and through inquiries to previous students. Job market studies show that practically all students get jobs in the horse sector. Many want continued education for a third year primarily in pedagogics, economics or riding.

Session 24 Theatre 2

Equine science education in Finland

M.T.Saastamoinen, Agri-Food Research Finland (MTT), Equine Research, FI-32100 Ypäjä, Finland

The number of horses and the horse industry in Finland is in a phase of strong growth. This means that there is an increased need for the personnel employed by stables, riding schools, trade and other business of the sector, as well as for other professionals working in the industry, e.g. veterinarians, nutrition consults, advisors, teachers, researches etc. However, it is not only the number of the professionals to be concerned, but also the level of their skills. The highest level of education in equine sciences in Finland is given in two faculties of the University of Helsinki: Faculty of Veterinary Medicine and Faculty of Agriculture and Forestry. The latter is responsible for the sciences of animal (horse) breeding and nutrition. In addition, some other universities have also short courses concerning e.g. husbandry and nutrition of horses. To develop and improve education and research in equine sciences, MTT and the faculty of veterinary medicine has increased their co-work. The facilities and researchers at MTT can offer unique resources in the high-level equine science education in Finland. The second highest level of education is given by Häme Polytechnic which is an multidisciplinary institute of higher education. Vocational education is provided by six schools specialised to equines and horse industry. The vocational education includes study programmes e.g. for riding instructing, training of trotters as well as horse husbandry and horse tourism. The study programmes include a six-month practical work period in various types of equine and equestrian enterprises.

Session 24 Theatre 3

Equine science education in Norway

D. Austbø, Department of Animal and Aquacultural Sciences, Norwegian University of Life Sciences, P.O. Box 5003, N-1432 Aas, Norway

At the Department of Animal and Aquacultural Sciences, Norwegian University of Life Sciences (UMB) there is a Master's degree programme in Animal Science. Within this programme, students can focus on 6 different study branches, including Sport animals and pets which cover mainly horses and dogs. The programme is based on a three year Batchelor degree in animal science with courses in anatomy/physiology (15 credits), etology (10 credits), nutrition (20 credits), breeding (15 credits) and molecular genetics (5 credits) or corresponding courses. In addition, there is a requirement for the following foundation courses or corresponding, with 10 credits of each: Scientific theory, chemistry, math, biology, statistics plus introduction to animal science (20 credits).

For the two-year Master programme, students can take 70-100 credits (out of a total of 120 credits) specializing in horse breeding/genetics, nutrition or ethology. 25 credits must be at 300 level within the study branch field, and the master thesis (30 or 60 credits) must also be within the chosen study branch. A combination of teaching and evaluation methods is being used, including lectures, practicals, independent work, seminars and student prepared lectures. UMB has exchange agreements with a number of foreign universities and students can take a portion of their studies at other universities.

Session 24 Theatre 4

Equine science education at universities and higher technical colleges in Germany

E.W. Bruns, Institut für Tierzucht und Haustiergenetik, Universität Göttingen, Albrecht-Thaer-Weg 3, 37075 Göttingen, Germany

All the eight agricultural faculties at universities in Germany offer some basic lectures on horse production which cover general topics on domestication, evolution, breeds, production, performance, housing and nutrition. However some faculties, Bonn, Gießen, Göttingen, Halle, Hohenheim and Kiel, present full courses on equine sciences in their agricultural bachelor or master programme covering in total between 20 to 110 lecture hours equal to 3 up to 12 ECTS points. Out of the higher technical colleges there are only two colleges (Osnabrück and München/Weihenstephan) providing special training on horse production within their agricultural bachelor programme. Most faculties offer also practicals and excursions. Lecturers either are based at universities or colleges keeping chairs for breeding, nutrition, reproduction etc. or are working in the horse industry (private or public). The four veterinarian faculties offer equine specific courses on surgery, diseases, radiology, reproduction etc. The broadest education based on several years' practical training at studs selected by FN is provided for the professional or amateur rider and breeder, blacksmith and groom.

Session 24 Theatre 5

Equine science education in the UK

A.D. Ellis[1], S.V. Tracey[2], H.C. Owen[1], [1]Nottingham Trent University(NTU), Southwell, NG25 0LZ United Kingdom, [2]Writtle College, Writtle, CM0 3 RR, United Kingdom

Equine Science Education in the UK, apart from traditional veterinary education, includes a range of College and University courses. Foundation degrees (FdSc), Bachelor of Science, Honours (BSc) degrees and more recently a range of Master of Science degrees (MSc) are now on offer under Equine titles such as Science, Sports Science or Behaviour and Welfare. Equine Breeding and Stud Management is also on offer (Writtle College). In 2005 a total of 118 equine university courses are available, of which 45 are science based at 33 different institutions. Innovative recent degrees incorporate human sports science and the horse-human relationship, such as the MSc in Human and Equine Sports Science (Essex University) or a BSc in Equestrian Psychology (NTU). Another novel development due to industry demand is the BSc in Equine Dentistry now offered at Hartpury College. Institutions offering Equine Science Education have close links with veterinary universities and research institutions such as the Animal Health Trust. Some graduates progress into postgraduate education, while many find work within the continuously expanding UK and international equine industry. Foundation degrees incorporate equitation as well as work-based and industrial learning experience and focus on combining science based knowledge with practical application. A range of pre-university equine science education is also available allowing students to specialise from the age of 16 by gaining a National Diploma, equivalent to science A-levels for university entry.

Session 24 Theatre 6

Equine science education in France

C. Drogoul, R. Beaufrère, A. Rousselière, D. Perrin and V. Julliand, ENESAD, BP 87999, 21079 Dijon Cedex, France

Horse Industry in France has developed during this past 20 years. According to a recent study, 58000 equivalent full-time employments exist : 30000 directly connected to horse management or breeding, 7000 "up and down stream" of the horse industry and 21 000 involving administration (education, administration), racing (races courses, betting) or socio-professional organizations. These employments are concentrated mainly in small and medium-sized companies.
In France, diplomas for horse related jobs are delivered by two ministries: The ministry of Agriculture when horse training and stud or stables management is concerned, The ministry of Sports for equestrian education.
At low levels (V, IV), diplomas are specifically designed for the horse industry (stable-lad, jockey, trainers, riding school teachers). At medium levels (III) no specific diploma exists but are included as optional courses in animal production diplomas. Altogether, around 1500 diplomas are delivered by the ministry of agriculture and 500 by the ministry of sports every year.
At a higher level (II and I) apart from riding instructors, no specific diploma exists but students (agricultural engineers or veterinarians) are often offered optional courses concerning horse management or health.
To accompany this evolution, new types of jobs are raising, requiring high qualifications and diplomas that aren't yet delivered. Therefore and in order to anticipate the European harmonization of diplomas, a few specific diplomas have recently been (Bachelor) or could be created (Master).

Session 24 Theatre 7

Equine science education in Italy

N. Miraglia, Molise University, SAVA Dept., Via De Sanctis, 86100 Campobasso, Italy

The paper will contain a synthesis of Horse Breeding and Horse Sport activities in Italy. Then, it will be presented the results of a survey(in progress) carried out in all the Italian research centres and institutions involved in equine research and education. A plan concerning a short description of the programs and of the research project will be presented together with the on line addresses and the name of the program leaders.

Session 24 Theatre 8

Equine science education in Switzerland

D. Burger[1], I. Imboden[2] and P.-A. Poncet[1], [1]Swiss National Stud, Avenches, Switzerland, [2]Vetsuisse Faculty, University of Zürich, Switzerland

Switzerland has a population of approximately 80 000 horses, about 50 % of which are registered and identified. The equine industry employs almost 15 000 people, with many more involved with horses in their free time. The number of horses in the agricultural sector has increased enormously in the last few years (1996 - 2003: +17%). However, this rise has been accompanied by a general drop in horse know-how.

Agricultural colleges are continuously reducing the number of courses aimed specifically at the equine industry. Professional education and training in the Swiss equine industry incorporates apprenticeships in farriery, as a groom, or as a riding instructor.

For riders and drivers, a certificate in basic horsemanship ("Brevet") involves 20 hours of theoretical and practical instruction and an examination. In most sporting disciplines, a license examination is required when competing; this demands a more thorough knowledge of the chosen discipline as well as proof of adequate riding / driving skills.

Until recently, no form of coordinated basic education or training was available to the large number of people involved in horse-breeding or husbandry. To fill this gap, a new educational concept for the equine industry was created by the National Stud in 2004; Equigarde® is a diploma course run over a year, incorporating 21 days of instruction and culminating in a final examination. To further improve communication and the transfer of information between research and horse people, the National Stud is currently introducing a national research Network.

Session 24 Theatre 9

Equine science education in Hungary

S. Mihók[1], I. Bodó[1], J. Posta[1], W. Hecker[2], [1]Debrecen University, H-4032 Debrecen Böszörményi út 138, [2]Kaposvár University, H-7400 Kaposvár, Dénesmajor

Equine science as a subject in form of horse breeding is involved in the curriculum of graduate students at university level all over the country. It is taught at Debrecen, Mosonmagyaróvár, Keszthely, Kaposvár universities. The same topic is facultative at high school level i.e. in Nyíregyháza, Hódmezövásárhely during study years.

Special courses for riding and driving are available at the Debrecen, Kaposvár, Mosonmagyaróvár and Gödöllö universities as well as at Nyíregyháza and Hódmezövásárhely high schools. Students can learn riding in the everyday practice. The trainer, the manage and other facilities are at their disposal here. Riding Academy was established at the Kaposvár University for education of riders and trainers, where special horse trainers and riders are formed. The department of horse breeding at the Research Institute of Herceghalom gave its tasks more and more to the university departments.

Session 24 Theatre 10

HorseConnexion: translating scientific knowledge into easily accessible information for riders, riding teachers and horse owners

M. Zetterqvist Blokhuis[1] and A.D. Ellis[2], [1]HorseConnexion, Lelystad, The Netherlands, [2]Nottingham Trent University, Nottingham, United Kingdom

The horse is an important leisure, sport and companion animal with the equine industry generating an increasing share of income in rural economies all over Europe. More and more people are involved in horse-related activities and their constant education in areas such as behaviour, nutrition and training of the horse is required. It should also be realised that nowadays many horses are used by people with no former equine education at all.

At international meetings, such as the Equine Behaviour Workshop at the 37th Congress of ISAE (2003) or the Equine Group Meeting at recent meetings of the EAAP, it has been highlighted that ordinary horse owner/riders, even those working in the equine industry, can find it very difficult to access and apply knowledge arising from equine research. The gap between scientists and practical horse people makes it difficult for research results to be put into practice.

To solve this problem, the international website HorseConnexion (www.horseconnexion.org) has been set up. HorseConnexion translates scientific knowledge into easily accessible information that can be applied by riders, riding teachers and horse owners to improve the management, training and use of horses in their care.

HorseConnexion is produced in English, Swedish and Dutch, and provides a platform for horse scientists all over the world to disseminate their results to a wider public.

Equine higher education in the United Kingdom

M.J. Kennedy, Anglia Polytechnic University, Department of Life Sciences, Cambridge, United Kingdom

According to the United Kingdom Universities and Colleges Admissions Service (UCAS), there are 35 Higher Education Institutions in the United Kingdom offering 118 courses with an equine theme for a September 2005 start. 69 of these are either Bachelor of Science (BSc) or Bachelor of Arts (BA) degrees, reflecting greater or lesser emphasis on science in these courses respectively. At sub-degree level, there are 24 Higher National Diploma (HND) courses and 24 Foundation Degree (FD) courses available. Degree course titles, indicative of the general emphasis of the courses, include Equine Science, Equine Dental Science, Equine Sports Science, Equine and Human Sports Science, Equine Studies, Equine Studies and Business Management and Equine Breeding and Stud Management. HND and FD courses have similar titles, with the addition of Equine Tourism, Rural Recreation and Tourism Management with Equine Studies, Equine Leisure Management and Equine Sports Performance and Coaching. Only one of these courses, a degree in Equine Science, is listed as being offered by a Russell Group institution (an association of 19 research-intensive United Kingdom Universities); co-incidentally this is the only equine higher education course taught within a United Kingdom veterinary school. In the United Kingdom equine higher education is almost exclusively offered by the 'new' higher education institutions which offer more vocational subjects of study, but which do not as yet have the established reputation for research of the Russell Group institutions.

Session 24 Poster 12

Equine education on bachelor level in Finland

T. Thuneberg-Selonen and K. Paakkolanvaara, Häme Polytechnic, Degree Programme in Agriculture and Rural Industries, Mustialantie 105, FIN-31310 Mustiala, Finland*

Häme Polytechnic (HAMK) is a multidisciplinary university of applied sciences with more than 20 degree programmes and 7 500 students. It offers broad-based, high-quality education, research and development, and strong internationalisation. Natural resources and the environment is one of the education branch of HAMK. Degree programme in Agriculture and Rural Industries (in Mustiala) offers education on polytechnic level in countryside studies (BSc Agric.). There are two possible curriculums of 240 ECTS: Agriculture and Equine option. The number of annual applicants to equine option is about 100 and 12 students are chosen by national entrance examination. There are basic courses common to all Häme Polytechnic, basic courses compulsory for all Mustiala students and special studies which enable a wide choice of courses and projects. Active co-operation with several agricultural colleges and universities in Europe opens for the students opportunities for doing part of their studies abroad. Many of the students do their practical training in a foreign country. During the final, fourth year the students write a thesis on a subject related to their major subject. Equine option offers extensive information of horse management, e.g. nutrition, breeding, veterinary treatment and physiology, production environment, and economy and marketing. Besides equine studies students choose one or two additional subjects from wide variety of subjects. Graduates can work e.g. as a teacher in secondary vocational level, advisors, entrepreneurs and in administrative duties.

Session 24 Poster 13

Equine Science Education in Slovenia

F. Habe, University of Ljubljana, Biotechnical faculty Zootechnical department, Groblje 3, 1230 Domzale, Slovenia

The paper on equine science in Slovenia represents the conditions and organisation of horse breeding and horse sport in Slovenia and their connections with educational and research institutions. Main information on teaching the horse breeding and its programmes at vocational, secondary, university and post-graduate level are given. The addresses of main institutions that carry out education of professionals for these fields at the above levels and an overview of their organisation and facilities are given. The Riding Academy of Slovenia as an educational service of FN with its organisation and programmes is discussed.

Session 24 Poster 14

Equine science education in Croatia

A. Ivankovic[1],Z. Petrovic[2] and M. Baban[3], [1]Facultyt of Agriculture, Svetosimunska 25, 10000 Zagreb, Croatia, [2]Croatian Equestrian Federation, Cimermana 5, Zagreb, Croatia, [3]Faculty of Agriculture, TrgSv. Trojstva 3, Osijek, Croatia

The horse population in Croatia during last two decades has significantly decreased. Warm-blooded breeds have been present with a share of 30% in the total population of about 10 thousand horses. The breed structure of warm-blooded horses is diverse and mostly reflects affinities of breeders and competitors towards certain horse. The development of the equestrian sport is in its initial phase, which can be seen in a small number and insufficient education of personnel involved in the horse industry, small interest of wide audience for equestrian sports, low level of investment into breeding of sport horses and small number of horse events. A complete education is necessary, since most of personnel already involved in the horse industry lack adequate professional education. Previous education of the part of personnel mostly took part in the programmes of university studies at the Faculty of Agriculture, Faculty of Veterinary Medicine and Faculty of Kinesiology, with more emphasis on a theoretical component. The Croatian Equestrian Federation will soon start an adapted professional programme of personnel education which is close to equestrian sports, and in its realisation, all relevant, previously mentioned education potentials will be involved. The equestrian sport in Croatia has been experiencing thorough changes and development and demands the adjustment of the educational base of personnel in the equestrian sport.

Session 24 Poster 15

EuroRide: an international education for riding instructors on level 2

M. Zetterqvist Blokhuis, Ridskolan Strömsholm, 730 40 Kolbäck, Sweden

The riding sport gets more and more international and the number of young people looking for a job abroad is increasing. EuroRide is a new international course for riding instructors on international Level 2 or higher and is a cooperation between Ridskolan Strömsholm in Sweden, Ecole Nationale d´Equitation in Saumur, France and Deutsche Reitschule in Warendorf, Germany. The aim is to provide riding instructors with a broad international insight and understanding of riding technique, teaching methods and horse keeping.
The duration of the course is all together ten months and the students start with three months in Saumur. Then smaller groups of students go to Sweden and Germany. Some periods the students are on the different schools while in other periods they are on placements. The first group of six students finished EuroRide in December 2004 and they organised a closing seminar in Warendorf. At the seminar, the students reported on their experiences and the different aspects of riding, training and teaching in the different countries. The students were in general very pleased with the course and with the different schools and the placements. They have gained lots of experiences and met a lot of interesting people. The course clearly contributed to "open the students' eyes" and hopefully they will be able to find a good career in the future.

Session 25 Theatre 1

Organic Livestock Systems: characteristics and challenges for improvement

John E. Hermansen[1], Troels Kristensen[1], B. Ronchi[2], [1]Department of Agroecology, Danish Institute of Agricultural Sciences, Denmark, [2]Department of Animal Production, University of Tuscia, Italy

There is generally a growing interest for organically produced food in Europe and North America, although in some countries the interest is stagnating. Organic animal productions in some countries make up to 10% of total production. The development of organic farming is driven by a mutual interest among: farmers who are sceptical towards the methods of conventional farming or aim for possible premium prices; consumers perceiving the products to be safer; and authorities seeing organic farming as a tool to reduce environmental load of the livestock farming systems. Existing evidence shows that levels of production experienced in organic production are often moderately reduced compared to conventional production or in some cases considerable lower (piglets (per sow per year), broilers). The organic production methods include elements that may enhance animal welfare, but some animal health problems are highlighted in several species, such as endoparasitic diseases. Only small differences in global warming potential, but a reduced eutrophication potential, per unit of milk produced, may be expected in organic milk compared to conventional milk production. Major challenges for a further development of the organic livestock farming systems include a better understanding of the animal nutrition - animal health - genotype interactions, of the possible positive role of organically produced food in relation to human health, of ways to balance free range rearing with acceptable production control measures, and of the particular benefits organic livestock systems may have in rural development exploiting new types of production and marketing.

Session 25 Theatre 2

Environmental ethics and organic farming

K.K. Jensen, Danish Centre for Bioethics and Risk Assessment, Rolighedsvej 25, DK-1958 Frederiksberg C., Denmark

Organic farming is often believed to be morally superior to conventional farming from the point of view of environmental ethics. Organic farming, it is claimed, attaches intrinsic value to ecosystems, and it will therefore be favoured by the versions of environmental ethics prescribing that nature should be valued for its own sake. However, perhaps surprisingly, closer inspection shows that this conventional wisdom is largely without foundation. It presupposes a rather vague and dubious notion of naturalness, which has been heavily criticised.

The now dominating view among environmental ethicists is that only *wild* nature, i.e. areas not cultivated by humans, should be valued for its own sake. It may well be that organic farming serves this aim better *instrumentally*, because it has fewer effects on the natural areas surrounding it. However, from the point of view of an environmental ethic valuing only wild nature, there is nothing intrinsic valuable *vis-à-vis* nature about farming - be it organic or not.

From this it follows that organic farming, just like conventional farming, will have find its moral foundation within an anthropocentric framework. Along this line, it is suggested that what distinguishes organic farming is adherence to a strong version of the precautionary principle.

Session 25 Theatre 3

Factors causing a higher level of liver abscesses in organic compared with conventional dairy herds

K. F. Jorgensen[1], A.M. Kjeldsen[2], F. Strudsholm[2] & M. Vestergaard[1], [1]Danish Institute of Agricultural Sciences, Foulum, DK-8830 Tjele, Denmark, [2]Danish Cattle Federation, Danish Agricultural Advisory Service, Skejby, DK-8200 Århus N, Denmark*

A data analysis based on the Danish Cattle Data Base shows that organic Holstein-Friesian cows have significantly higher frequencies of liver abscesses (LA) than conventional Holstein-Friesian cows (8% versus 5%). Based on a questionnaire among 91 organic dairy herds, a statistical analysis was made in order to identify which feeding and management factors that were related to the level of LA in the herds. Compared with conventional herds, the feed ration in organic herds had a lower energy and fatty acid level and a higher starch level. During summer season the organic herds had a lower level of digestible cell walls in the ration. Organic herds with a higher grazing level had a higher level of LA (P=0.012). Increasing levels of grain also tended to increase the level of LA. A herd-based analysis (3,102 herds) showed that lower minimum fat percentage and a higher variation in both fat and protein percentage corresponded with higher levels of LA. Organic herds generally had a 0.1 unit lower fat percentage. The results indicate that organic dairy cows, compared with conventional dairy cows, are more exposed to rumen acidosis and liver abscesses due to higher starch levels and unbalanced feeding strategies in particular during grazing.

Session 25 Theatre 4

Comparing sheep for meat production of organic versus conventional farms: structures, functioning, technical and economic results

M. Benoit, and G.Laignel, Laboratoire Economie de l'Elevage INRA Theix 63122 France*

In the centre of France, 20 organic sheep farms (OF) and 36 conventional sheep farms (CF) were compared (year 2003) in lowland (n=27; 7 in OF) and upland areas (n=29; 13 in OF). Under organic management (OM), the farm size was 31% lower in upland and 39% in lowland areas. Stocking rate was 17% lower in upland but comparable in lowland areas. Thanks to the use of rustic breeds, upland organic farmers produced as much in 'counter-season' as upland conventional farmers. The weight of lambs was slightly lower under OM (-2 to -5%); the price of meat was 17% higher under OM in upland but comparable in lowland areas where the price of conventional meat reached a good level (5.4 /kg). Numerical productivity was lower under OM (-15% in upland and -4% in lowland areas) as well as the consumption of concentrates (-14% upland), but the unit price was 65% higher (0.304 vs 0.184 /kg). Finally, the gross margin per ewe was 29% lower in upland and 23% in lowland areas and the income was 27% lower on average. Without the 'agri-environmental' premiums paid in particular for conversion to organic production, the incomes would be 81% lower. The future is uncertain since the premiums for organic production are not perennial in France.

Session 25 Theatre 5

Alternative low input system in sheep milk production: Competitiveness to intensive production system

S. Kukovics[1], P. Kovács[2], S. Nagy[2], G. Csatári[2], J. Jávor[3], [1]Research Institute for Animal Breeding and Nutrition, Gesztenyés u. 1. Herceghalom 2053 Hungary, [2]Bakonszeg Awassi Corporation, Bakonszeg, Hungary, [3]University of Debrecen Centre of Agricultural Sciences, Debrecen, Hungary

Two production systems were compared within the same farm. About 1,800 purebred and crossbred Awassi sheep were kept under intensive production system. Animals were settled in barns and served with feeds (forage and supplements) two times a day by feed mixer machine. Their average milk production level was between 200-600 litres during 170-220 days. The prolificacy of these animals was about 130%. In extensive system (classified as eco/bio/organic system) 2,500 Transylvanian Racka sheep were bred and kept. They had only minimum protection from wind and rain/snow over winter time, and kept on pastures over the whole year. Their nutrition was based on grazing in paddocks on clovers, grass, silage corns over 6 months milking period, and on natural pastures and stables during the other part of the year. Their average milk yield was 60 litres/head and the lambing rate was about 100-110 %.

The milk from the two systems was processed in the own cheese factory of the Corporation. The surplus lambs were sold mainly to export as suckling lambs between 13-24 kg live body weight from both system.

Because of the low input demands of the organic system its profitability was very similar to that of the intensive system.

Session 25 Theatre 6

Low input dairy systems: balancing environment and animal performance

L.B.J. Sebek[1], R.L.M. Schils[1], J. Verloop[2], H.F.M. Aarts[2] and Z. van der Vegte[1]. Wageningen UR, The Netherlands, [1]Animal Sciences Group, P.O. Box 65, 8200 AB Lelystad, [2]Plant Sciences Group, P.O. Box 16, 6700 AA Wageningen*

Dairy production systems have to adjust to increasing public demands and (European) regulations concerning environment, landscape, animal welfare and product quality. These adjustments require an integrated approach to optimise environmental, economical and animal welfare objectives. In The Netherlands, these objectives were subject of research in an integrated whole-farm system approach on experimental dairy farm 'De Marke'. The main objective was to improve nutrient efficiency and to match the production to the potential of the site with minimal reliance on external feed sources and nutrient supplies. Thus, a sustainable low input system for dairy farming was developed and tested under the difficult conditions of a dry sandy soil. Low input refers to low nitrogen (N) and phosphorus (P) inputs from feeds and fertilizers. When restricted to a N-surplus of 125 kg N/ha and a P-surplus of 1 kg P/ha, the milk production potential of dry sandy soils for dairy farming is approximately 12.000 kg milk/ha. 'De Marke' was able to combine low N and P-losses with sufficient economic performance, by optimising crop rotation and manure management. As a result feed management is challenged to adapt to these system restrictions. This paper shows the effect of these restrictions on animal production and puts them in an European setting.

Session 25 Theatre 7

Genotype environment interaction for milk production traits between conventional and organic dairy farming in the Netherlands

W.J. Nauta[1], T. Baars[1], H. Bovenhuis[3], [1]Louis Bolk Institute,Hoofdstraat 24, NL-3972 LA The Netherlands, [2]Animal Breeding and Genetics Group, Wageningen University, P.O. Box 338, NL-6700 AB Wageningen, The Netherlands

Most Dutch organic dairy farmers use breeding bulls from conventional breeding companies because no organic breeding is available. However, in organic farming , for example, the use of concentrates and antibiotics are restricted and roughage contains less energy and protein because chemical fertilizers are prohibited. Due to genotype environment interaction (GxE) we expected that breeding values of conventional bulls are not appropriate for organic farming. Recent findings showed significant changes in milk production traits when farms converted to organic farming. Genetic parameters and genetic correlations between organic and conventional farming were estimated, based on 305-day lactation records of first lactations of Holstein cows of 109 organic (1767 records) and 152 conventional farms (9239 records). Genetic correlations were estimated for kg milk, percentage milk fat and protein using ASREML. Results showed that phenotypic variances for milk production decreased strongly in organic farming and heritability (h^2) of milk production increased from 0.47 for conventional to 0.70 for organic milk production. Genetic correlations between conventional and organic farming environments for milk production and kg milk fat and protein were 0.82 on average. This indicates that the use of conventional bulls in organic farming may result in re-ranking of bulls and less genetic gain for production traits.

Session 25 Theatre 8

<B>Nitrogen self-sufficiency at the suckler cattle farm scale: adaptation of the farming systems, economic consequences<B>

<I>P. Veysset<+><+>, M. Lherm, D. Bébin, INRA Clermont-Theix, Laboratoire d'Economie de l'Elevage, 63122 Saint Genès Champanelle, France*
<I>

Suckler cattle farms of the Charolais region are characterised by a high food self-sufficiency (almost all the fodder is on-farm produced) despite their dependence on nitrogen supply. Concentrates and chemical fertilisers are the main N inputs. The use of a model for optimizing farming systems under constraints will allow us to study the search for nitrogen self-sufficiency at the farm scale under two types of farming systems. In mixed crop-livestock farming, N self-sufficiency can be reached by cropping 10 to 28 ares of protein-rich plants per calving (for the production of lean and fattened animals respectively), introducing more than 20% legumes in grasslands and reducing the part of high N-exporting cash crops for the benefit of fodder areas; considering 50% of legumes in grasslands the optimum is reached with 75% of total area allocated to the herd and 20% to cash crops. In grassland farming, this N self-sufficiency will only be reached if 13.5% of the total area is cultivated with cereal and protein-rich plants (to be used for livestock feeding). In all cases, the economic impact of this search for N self-sufficiency is very low or even slightly negative, which is not inciting if this low input production system does not result in a better sale price of environment-friendly produced agricultural commodities.

Session 25 Theatre 9

Effects of chicory roots on finishing performance and CLA and fatty acid composition in longissimus muscle of Friesian steers

M. Vestergaard[1], H.R. Andersen[1], P. Lund[2], T. Kristensen[1], L.L. Hansen[1] and K. Sejrsen[1]. [1]Danish Institute of Agricultural Sciences, Tjele, Denmark, [2]Danish Technical University, Lyngby, Denmark*

The objective was to study the influence of a bioactive crop, chicory roots, on performance, carcass quality and fatty acid composition of beef. Forty autumn-born Friesian bull-calves castrated at two month were grazing lowland non-fertilized pastures for two summers. The two year-old steers (492 kg) were housed in tie-stalls and finishing-fed for 10 weeks. Steers were allocated to 4 treatments; B-1: 1.7 kg DM of barley, B-2: 3.4 kg DM of barley, C-1: 1.7 kg DM of chicory, and C2: 3.4 kg DM of chicory in a 2x2 factorial design with two types (B vs. C) and two levels (1 vs. 2) of concentrates. Beside the fixed concentrate allowances, steers had free access to clover grass silage. The high concentrate level increased total ME intake ($P<0.001$) and IMF ($P<0.09$). Overall, there was no difference in ADG (887±40 g/d) between treatments, but numerically, C-steers gained 13% ($P<0.12$) less than B-steers. Carcass quality was not affected. Longissimus muscle of C-compared with B-steers had 10% less C18:0 ($P<0.01$) and 7-17% more C16:1, C18:1 and C18:3 ($P<0.03$). Neither treatment affected contents of *cis*-9,*trans*-11-CLA (0.22mg/100 mg FA), vaccenic acid (C18:1,*trans*-11) or linoleic acid (C18:2*n*-6). The results show that steers finishing-fed with 15-30% of ME from chicory roots compared with barley have slightly lower gain, similar carcass quality, and more unsaturated fat but similar CLA content in the muscles.

Session 25 — Theatre 10

Influence of amino acid levels to indoor and outdoor growing/finishing pigs on performance, carcass quality and behaviour

M. Høøk Presto, K. Andersson and J.E. Lindberg, Deptartment of Animal Nutrition and Management, Swedish University of Agricultural Sciences (SLU), P.O. Box 7024, SE-750 07 Uppsala, Sweden*

The aim of this project is to study the influence of different levels of amino acids (recommended, 7% and 14% lower levels) in a phase feeding system, with a low-energy diet provided *ad libitum*, on performance, carcass quality and behaviour of growing/finishing pigs. During two years, a total of 192 outdoor born pigs ((Landrace*Yorkshire) * Hampshire) were raised indoors in conventional pens or outdoors on pastures. Live weight and feed consumption were recorded continuously during the experiment and carcass parameters were assessed conventionally at slaughter. Behaviour studies were done at three occasions, at 60, 110 and 140 days of age. Statistical analyses were performed with the SAS programme. Amino acid level did not influence growth rate, feed conversion ratio, carcass weight or lean meat content but affected the activities performed by the pigs. Rooting activity occurred more often by pigs with lower levels of amino acids ($p<0.001$). Outdoor pigs grew faster than indoor pigs (943 g vs. 832 g; $p<0.001$) however, feed conversion ratio did not differ. Lean meat content was higher for indoor than for outdoor pigs (57.8% vs. 56.6%; $p=0.004$), whereas dressing percentage was lower (73.1 vs. 74.0; $p<0.001$). Social behaviour parameters including sniffing, nibbling, pushing and tail biting occurred more often indoors than outdoors.

Session 25 — Poster 11

Future scenarios for sustainable beef production in Sweden

K.-I. Kumm[1], S. Stern[2], S. Gunnarsson[3], U. Sonesson[6], I. Öborn[4], T. Nybrant[5]. [1]Department of Economics, [2]Department of Animal Breeding and Genetics, [3]Department of Animal Health and Environment, [4]Department of Soil Sciences, [5]Department of Biometry and Engineering, P.O. Box 7044, Swedish University of Agricultural Sciences, SE-750 07 Uppsala, [6]SIK, P.O. Box 5401, SE-42902 Gothenburg, Sweden*

The aim with this study was to formulate and evaluate future scenarios for beef production in Sweden based on FOOD 21 sustainability objectives. A step-wise method was used to create scenarios with focus on different sustainability goals. The resulting scenarios were then evaluated from economic and environmental standpoint.

The scenarios: Nature is based on beef cows with a high degree of grazing, being outdoors all year. Meat quality is based on beef heifers inseminated with female sex-sorted semen. They calve once before slaughter, calving is spread evenly throughout the year to meet the market need. The feeding consists of high-quality roughage and grazing. Environment is based on bulls born by dairy cows inseminated with male sex-sorted beef semen. They are indoors all their life and fed silage complemented with grain and concentrate.

Nature had lowest production costs and contributes most to open landscape and biodiversity, nitrogen surplus being intermediate. Meat quality had the highest nitrogen surplus and production costs per kg meat produced. Environment had lowest level of nitrogen surplus and land requirements. The production costs per kilo meat were intermediate.

Living and slaughtering performance of Piemontese young bulls reared according to organic farming method

C. Lazzaroni, D. Biagini, Department of Animal Science, University of Turin, Via Leonardo da Vinci 44, 10095 Grugliasco, Italy. carla.lazzaroni@unito.it*

For the increasing interest in organic farming, estimated as less intensive, respectful of animal welfare and guaranteed of more healthy production, the possibility to apply such method in fattening Piemontese young bulls was verified. Two groups of 10 calves, from about 8 months of age and 200 kg l.w., were fed in the same environmental conditions according to the organic method compared to the traditional one, with hay and concentrate, up to the slaughtering weight. Monthly individual weights, average daily weight gain, daily feed consumption, and feed conversion rate were recorded, and at the end of the trial the market price was fixed and animals slaughtered. During the first part of the trial, the performances of both groups were comparable, but during the last month the organic-like feeding group showed meagre weight gains (0.6 *vs.* 0.3 kg/d) and reduced hay consumption (-1 kg/d DM), so they didn't reach the proper condition to be priced by butchers. The organic farming method didn't suit to fattening Piemontese young bulls, as it didn't allow to obtain a product appraised on the market: in fact, even if the animals reached normal weight (499 *vs.* 455), they didn't show a proper muscular development and fattening degree. Only after a conversion to a traditional fattening method (about 4 months) they were ready for the market, showing higher live weight (530 *vs.* 580 kg, $P<0.05$) but similar dressing percentage (66.6 *vs.* 68.8).

Session 25 Poster 13

Feeding of layers of different genotypes in an organic feed environment

G. Lagerkvist, K. Elwinger and R. Tauson, Swedish University of Agricultural Sciences (SLU), Uppsala, Sweden

Prohibition on use of synthetic amino acids, GMO or products derived there from, and limited access to approved raw materials restrict possibilities to compose a nutritionally balanced diet in organic egg production. Differences exist between genotypes of birds to resist feather pecking and cannibalism. An experimental genotype (SH), selected for egg production for over 25 generations on low protein diet, was tested against two common commercial hybrids; LSL (Exp. 1) and Hyline (Exp. 2). The birds were fed a control or organic diets, with different sources of protein. LSL and Hyline had a higher production (20-80 weeks) than SH, but production deteriorated more severely in the former genotypes when fed the low protein/methionin diet (2 g methionine/kg). Plumage condition was better in SH, compared to LSL. In Exp. 2, feathering was almost complete throughout the whole cycle. However, all genotypes were inferior when fed the low protein diet. SH used the outdoor runs more than LSL, as well as the groups fed the low protein diets, probably due to an increased feed scavenging. On the contrary, Hyline used the outdoor area twice as much as SH. The experiments showed differences between genotypes but all performed better on a diet with a high protein quality. Consequently, it is difficult to fulfil KRAV´s standards of 100% organic feed stuffs, without supplements with synthetic methionine or fish meal.

Session 25 Poster 14

Housing systems for organic slaughter pigs

A-C. Olsson, J. Svendsen, J. Botermans, & M. Andersson, Swedish University of Agricultural Sciences (SLU), Alnarp, Sweden

An interdisciplinary research program has been initiated to study different methods of organic pig production to increase efficiency, rationalisation and sustainablity. An uninsulated stable for 128 organic slaughter pigs in 8 pens has been built with two different pen designs; four pens with "straw flow" and four with deep straw bedding. All pigs have access to an outside pen area with concrete flooring. Two pens with deep straw bedding and two pens with "straw flow" also have access to a yard with pasture during the summer. One group has access to the entire yard at once, whereas the second has access to a new small pasture area once weekly. Different types of grass mixtures are also being tested. Preliminary results have shown that:

- Trough feeding inside must be used to be able to monitor and confine the animals if necessary, since all the animals will then be inside at feeding.
- During the winter it is necessary to improve the cleaning of the outside solid concrete floors in both pen types.
- There were no clear differences in behaviour between the two pen types.
- Having access to roughage or pasturage did not affect pig behaviour in general, but affected their choice of pen area when active.
- The total activity of the pigs significantly depended on their age and the environmental temperature.

The housing systems not only affected the animals, their production and well-being, but also affected, e.g., manure handling and nitrogen losses.

Session 25 Poster 15

Effect on milk production and vitamin status in cows fed without synthetic vitamins

B. Johansson and E. Nadeau
Swedish University of Agricultural Sciences (SLU), Skara, Sweden

An EU law that forbids the use of synthetic vitamins in organic ruminant production was established in 2000. Further, no use of products made from genetically modified organisms is allowed. This limits the use of naturally occurring vitamins. At present a general exemption applies to the 31st of December 2005. Consequently, it is important to investigate whether dairy cows can maintain their production and health when no synthetic vitamins are added to their diets. Vitamin A and E are of great importance for ruminants. Shortage of the vitamins can depress the immune system and cause reproductive disturbances. In an ongoing experiment 50 cows are divided into two groups of which one group is fed without synthetic vitamins. The other group is the control group which is fed synthetic vitamins according to Swedish recommendations. The study has been going on for one lactation period and will continue for another lactation period. Effects on milk yield, milk composition, milk quality, health and fertility are studied. The vitamin status in cow blood plasma and milk is followed during lactation. Feeds are analysed for nutrient composition and vitamin content. Statistical analyses of variance were made using the GLM procedure of SAS. Preliminary data showed no greater differences between the two experimental groups. The final results will be presented in the early 2006.

Session 25 Poster 16

A historical perspective on low input dairy production

Carin Israelsson, Division of Agricultural History, Department of Economics, Swedish University of Agricultural Sciences, P.O. Box 7013, SE 750 07 Uppsala, Sweden

Two principally different systems for cattle husbandry were practised in Sweden during the second half of the 19th century. Poor country people used a system for subsistence production, characterized by use of a minimum of material resources. Better off farmers practised another system, based on a high flow of resources, aiming at commercial dairy production. Historical sources have exhibited the fact that the subsistence system included most of the herds and a large part of the Swedish population. Despite little or no land for feed production, many households had one or two cows. The animals were fed with scarce quantities of nutrient-poor feeding stuff like by-products from arable farming, and leaves and moss from common wood lands. The scarce feeding resources were unequally distributed over the year, forming a cycle of nutritional undernutrition and compensatory recovering. Estimations based on a reconstruction of contemporary feeding stuff indicate a general severe malnutrition in winter-time. This was extremely obvious with regard to protein, which barely covered the needs for maintenance. Energy input may have balanced the needs for maintenance and production of one or two litres of milk, while higher production caused a negative energy balance.

Session 25 Poster 17

Control of fitness in low input socio-ecological systems

Brigitte A. Kaufmann, Institute for Animal Production in the Tropics and Subtropics, Hohenheim University (480a), Garbenstr. 17, 70593 Stuttgart, Germany

In low input socio-ecological systems, such as pastoral livestock husbandry systems, fitness of the - mostly - local livestock populations is of paramount importance for the sustenance of the system. As fitness parameters have a low heritability, the livestock keepers exhibit a strong influence on the parameter values. However, their ways to control the production process has so far been largely ignored in the analysis. The aim of this paper is to analyse how pastoral livestock keepers control fitness of their livestock, in this case camels and to discuss this in connection with the fitness parameters determined for their camel population.
The analysis of the management followed second order cybernetics. It revealed which traits of the livestock are observed and which distinctions of trait expressions are made by livestock keepers and upon which rules they base their actions. Livestock keepers' target value for age at first calving (AFC) is 5 years and for calving interval (CI) it is 2 years in camels. The target values for AFC and CI were not reached in 33% and 39% of all cases (N= 415 and N=1031 respectively). Overall 41% of the camels under study (N= 467) had at least one calf less than targeted at by the livestock keepers. Based on the analysis of the management, reasons for lack of control are discussed and measures to improve control of livestock keepers in low input systems are proposed.

Session 26 Theatre 1

A systematic approach to the design and enhancement of breeding programmes

D.J. Garrick, Department of Animal Sciences, Colorado State University, Fort Collins, CO 80523-1171, USA*

Animal breeding is the application of knowledge in genetics and other disciplines to improve animal systems. A wealth of courses and texts introduce principles of genetics but few deal comprehensively with the steps involved in enhancing breeding programmes. These steps are consistent across species, and can be applied to systems influenced by multiple traits, polygenes, genetic markers, heterosis and maternal effects. The first step is to define the high-level goal. The next step is to identify the list of traits that influence the goal. The merit of candidates for these traits must be able to be determined from phenotypic or DNA measurements. Decisions need to be made as to which animals should be measured and these may need to be specifically created (eg progeny tests). The selection approach, the dissemination system and the mating plan must be defined. Analysis of the strategic fit of the nucleus through commercial system in relation to the goal is required. The steps interact such that an iterative approach is desirable to compare scenarios. External factors such as marketing and public relations are also critical to success. In practice, the realisation of an improved breeding programme has as much to do with people and personnel management than it has to do with genetics and theoretical aspects of animal selection and mating.

Session 26 Theatre 2

Practical aspects in setting up a national cattle breeding program for Ireland

V.E. Olori[1], A.R. Cromie[1], A. Grogan[1], B. Wickham[1], [1]Irish Cattle Breeding Federation, Highfield house, Bandon, Co Cork, Ireland*

Dairy and beef production in Ireland is primarily based on grass with a focus on profitability through reduced production costs and increased product value post decoupling. Increasing concern for product quality, animal welfare and environmentally sustainable production systems means that future profitability requires robust breeding objectives. Our dairy breeding objective aims to improve production and functional traits. Beef production, carcass quality, calving ease and longevity are the focal points of the beef breeding objective. To achieve optimum progress and produce bulls to meet the domestic needs, a national breeding program with capacity to test bulls locally is essential. The key to success is data availability, accurate genetic evaluation, selection of superior animals, and rapid dissemination of their genes. Major steps in achieving this in Ireland were the establishment of a central database, an on-farm 'animal events' recording system and a genetic evaluation system for all traits across breeds. The new technologies introduced have simplified data recording and reduced costs. The database adds value to data and facilitates accurate and timely genetic evaluation. It plays a pivotal role in identifying superior animals for selection, provision of a mating advisory service and management of inbreeding. Success in the national breeding program is projected to yield gains of about twenty million euro per annum to Irish farmers compared to the present testing system.

Session 26 Theatre 3

Developments in international pig breeding programmes

J.W.M. Merks, E.H.A.T. Hanenberg and E.F. Knol, IPG, Institute for Pig Genetics B.V., P.O. Box 43, 6640 AA Beuningen, The Netherlands*

During the last decades several pig breeding programs have developed from regional and/or national programs into international programmes. This is the consequence of globalisation of the pig industry, increase in scale of production and market driven pork production. This trend requires breeding programmes for a wider range of systems.

In this paper an overview is given of the worldwide developments in scale and regional coverage of pig breeding programmes and the consequences this has for the range of systems that needs to be covered in breeding programmes, especially climate, feed, health level and management. For several of these aspects results will be presented with regard to (potential) genotype x environment interaction: e.g. heritabilities for daily gain ranging from 0.1 to 0.4 and different underlying traits that are triggered in a wide range of systems (like protein deposition and feed intake).

The development of pig breeding programs to cover a wide range of systems will be discussed along the options of (1) combined crossbred and purebred selection (CCPS), (2) the use of information from breeding herds under a wide range of systems and (3) the development of specific lines for specific production systems. The ultimate goal is porcine genetics with a high level of genetic plasticity. In the near future this may be reached by balanced breeding programmes that take a wide range of systems into consideration or are specifically developed for regional markets and systems.

Session 26 Theatre 4

Collaboration of breeding programs with genotype by environment interaction: possibilities and limitations.

H.A. Mulder[1], R.F. Veerkamp[2], P. Bijma[1], [1]Animal Breeding and Genetics Group, Wageningen University, PO Box 338, 6700 AH Wageningen, The Netherlands, [2]Animal Sciences Group, Division Animal Resources Development, PO Box 65, 8200 AB Lelystad, The Netherlands*

Dairy cattle breeding programs are often selecting sires and dams across environments. Genotype by environment interaction (G × E) might limit the possibilities for collaboration of breeding programs in different environments. This study investigated the possibilities for collaboration of breeding programs with G × E. A dairy cattle situation with two breeding programs and two environments was simulated using a deterministic pseudo-BLUP selection index.

Long-term collaboration of breeding programs with G × E was guaranteed, when genetic gain in different breeding programs was equal for performance in one of the environments. A simple selection index overestimated the possibilities of collaboration as a consequence of neglecting reduction of genetic variances and covariances due to selection. Collaboration of breeding programs was possible in the long-term, when the genetic correlation was higher than 0.80-0.90, resulting in up to 15% extra genetic gain. On the contrary, in the first generations breeding programs could mutually select each other sires and dams, when the genetic correlation was higher than 0.40-0.60. With more intense selection, breeding programs were less likely to benefit from collaboration. Small breeding programs had larger benefits from collaboration than large breeding programs and collaboration was possible with lower values of the genetic correlation.

Session 26 Theatre 5

Cattle breeding programs for environments with poor formal information: Integrating traditional breeding knowledge and local breeding strategies - the case of The Gambia

M. Steglich, Kurt.-J. Peters, Animal Breeding in the Tropics and Subtropics, Humboldt University of Berlin, Philippstr.13, 10115 Berlin, Germany*

Breeding programs for disadvantaged tropical production conditions have to cope with weak infrastructures, poor formal information systems and breeding organisations. If the planning for such programs is operated in conventional ways, they are in many cases not sustainable. Difficulties arise particularly due to the inability to sustain formal information systems, the non-involvement of local breeders, and inadequate breeding objectives. Experiences from participatory community-based development approaches in research and development emphasise the importance of local knowledge and the understanding of the existing production system and prerequisites for sustainable development. To facilitate the involvement of local breeders and to incorporate their knowledge about cattle breeding, an adequate methodological approach is required that supports the collaboration between professional breeders and local people. This paper reports on the application of a participatory methodology to investigate the existing breeding strategies and to understand the existing informal/formal institutional and organisational breeding structures together with local breeders. Considerable information about local breeding objectives and breeding practises is gained. Furthermore, the information about traditional breeding institutions, which are in a process of disintegration while new formal institutions do not yet provide the required organisational support, emphasises the importance to allocate sufficient resources into this aspect of a cattle breeding program.

Session 26 Theatre 6

Improving carcass quality of UK hill sheep using Computerised Tomography (CT)

J. Conington, N. Lambe, P. Amer[1], S. Bishop[2], L. Bünger, G. Simm, SAC, W.Mains Rd., Edinburgh, EH9 3JG, Scotland, [1]Abacus Biotech Ltd. Otago House, 475-Moray Pl., PO Box 5585, Dunedin, New-Zealand, [2]Roslin Institute, Roslin, Midlothian, EH25 9P, Scotland, United Kingdom

Hill sheep have a crucial role in the UK sheep industry, to supply breeding females to lowland farms, as well as males for the lamb meat industry. Breeding objectives for these breeds must ensure that genetic improvements made in productivity are not antagonistic to other important traits such as ewe and lamb survival. Research in Scottish Blackface sheep (SBF) is centred on the development of breeding programmes including maternal and lamb performance characteristics, and body composition measured using CT. Since 1999, two breeding indices to improve hill-sheep performance are being tested in two different locations, using three genetic lines of SBF (Selection/S, Control/C, and Industry/I). So far, selection has been successful, i.e. S-animals tend to have higher indices, have heavier weights and their carcasses achieve higher prices compared to C- or I-animals, but with no differences in carcass composition. To address this, current research in hill-sheep is investigating the best use of CT to accelerate changes in carcass quality, and the cost-effectiveness of implementing 2-stage selection. From modelling studies, predicted rates of response to selection differ up to 1.3 s.d. of index value per generation, according to the correlation between the first- and second-stage indices, proportion of animals selected for second-stage measurement and proportion selected for breeding.

Session 26 Theatre 7

Genetic evaluation of multibreed dairy sires in five environmental clusters within New Zealand

J.R. Bryant[*1], N. Lopez-Villalobos[1], J.E. Pryce[2] and C.W. Holmes[1], [1]Institute of Veterinary, Animal and Biomedical Sciences, Massey University, Private Bag 11-222, Palmerston North, New Zealand, [2]Livestock Improvement Corporation, Private Bag 3016, Hamilton, New Zealand*

Knowledge of whether dairy sires of multiple breeds re-rank for milk production traits in different climatic and nutritional environments is important for the correct selection of the best sires for each farm environment. The objective of this study was to determine if re-ranking for multibreed sires occurs within the range of environments in New Zealand (NZ). Lactation records from Livestock Improvement's Sire Proving Scheme (1989-2002) were merged with climatological information. Five environmental clusters were formed based on heat index data, herd mean production level and latitude. In each cluster, sire breeding values for milk, fat and protein were calculated and the rank and genetic correlations between clusters for the most widely used sires were estimated. Rank and genetic correlations were generally high between clusters 1, 2 and 3 which had average milksolids (MS; fat + protein) yields ranging from 240 to 293 kg and similar heat index and latitude values. However, some re-ranking did occur for fat and protein yield when comparing these clusters (1-3), with clusters 4 (337 kg MS/cow, similar latitude and heat index), and 5 (343 kg MS/cow, cooler and more southern). The results also illustrated Jersey were better suited than Holstein Friesian cattle to a low yield, warm environment and vice versa for a high yield, cool environment.

Session 26 Theatre 8

A selection index for Ontario organic dairy farmers

P. Rozzi[1] and F. Miglior[2,3]*
[1]OntarBio, Guelph, ON, Canada, [2]Agriculture and Agri-Food Canada, [3]Canadian Dairy Network, Guelph, ON, Canada

Objectives of the research were to verify if selection priorities were homogeneous within the organic sector, and to compare selection objectives between organic and conventional dairy farmers. Organic standards, verified by certification agencies, require changes in management practices (e.g.: no use of antibiotics, feeding only organic rations), so that health, fertility and overall fitness become more important than in traditional dairy farms. A survey involving 40% of Ontario organic dairy farms was carried out to identify their management systems. Compared to conventional dairy farms, organic farms had lower milk production, lower replacement rate, lower use of concentrates, and grew very little corn and soybean. In 2003, culling rate was 21% in organic farms and the four main causes were: fertility, mastitis, feet and production, while in conventional they were: fertility, production, sickness and mastitis. Based on farmers' subjective scores, the relative weights of production to durability and health traits were 28:47:25, substantially different from those in the Canadian index LPI (57:38:5). The relative emphasis on health traits, five times higher for organic farmers, included udder health, calving ease and persistency. Even though organic farms at different production levels followed different strategies regarding breeds and AI use, their priorities for selection were remarkably similar. Therefore, there was no ground to justify the use of different selection indexes within this sector.

Session 26 Theatre 9

Environmental influences on genetic and phenotypic relationships between production and health and fertility in Dutch dairy cows

J.J. Windig, M.P.L. Calus, B. Beerda, W. Ouweltjes and R.F. Veerkamp, Animal Sciences Group, Wageningen UR, Lelystad, The Netherlands*

High milk production may lead to health and fertility problems. In this study we tested if the effect of high milk production depends on herd environment. Herd environment was defined by 64 variables derived from production records (e.g. average milk production per cow, average survival) and national farm data (e.g. soil type, land area). With Principal Component Analysis this set was reduced to 3 variables: production intensity (production per cow), fertility and scale. In a dataset containing 3904 herds and 456,574 records Somatic Cell Count (SCC) and Days to First Service (DFS) were more favourable in high intensity herds, while Number of Inseminations (NINS) was less favourable. Within herds the phenotypic relationship between production and health/fertility was absent or weak in low intensity herds, and relatively strong in high intensity herds. Estimates from a multitrait reaction norm model demonstrated that genetic correlations between kg milk and SCC increased from 0.12 to 0.45 from low to high intensity herds, decreased from 0.28 to 0.17 from low to high fertility herds and were about constant (0.26) going from small to large herds. Genetic correlations of production with DFS and NINS ranged from 0.13 to 0.51. Correlations of DFS were stronger at low intensity, low fertility and small farms, while trends over environments were opposite for NINS.

Session 26 Theatre 10

Use of structured antedependence models to estimate genotype by environment interaction

M.P.L. Calus[1,], F. Jaffrézic[2], and R.F. Veerkamp[1], [1]Animal Sciences Group (ASG-WUR), Division Animal Resources Development, Lelystad, The Netherlands, [2]INRA Quantitative and Applied Genetics, Jouy-en-Josas Cedex, France*

The objective of this research was to compare structured antedependence models (SAD) to random regression models (or reaction norm models) (RRM) in their ability to estimate genotype by environment interaction. In RRM, random effects for a trait are estimated as a regression on the environments and the (co)variances are modelled by a covariance function. In SAD, random effects for a trait are estimated as a function of the same trait in different environments and the (co)variances are modelled by so-called innovation variances and antedependence parameters. One of the major differences between the models is that the SAD allows the genetic correlations between a trait in different environment to be less than unity in situations where the genetic variance is constant across environments, whilst RRM implicitly model the change in covariances and variances simultaneously. Thus, SAD might be more flexible in estimating genetic correlations between a trait in different environments. Initial results from a simulation study indicate that genetic correlations between a trait expressed in different environments tend to be overestimated when using RRM and underestimated when using SAD.

Session 26 Theatre 11

Possibility and profitability of crossbreeding in Holstein dairy cattle

S. König and H. Simianer, Institute of Animal Breeding and Genetics, University of Göttingen, Germany*

Reports of crossbreeding in dairy cattle have shown evidence of favourable heterosis for several traits and increased profitability. The aim of this study was to evaluate the economic gain of crossbreeding systems in North West Germany compared with purebred Holstein (reference scenario) utilizing actual phenotypic and genetic parameters and current market values for a population wide point of view. An algorithm was developed to define the potential of crossbreeding under a fixed fat quota of 30 Mio. kg for different mating systems involving the Holstein-Friesian, Brown Swiss and Simmental. A second restriction was to ensure the purebred Holstein population and female Holstein x Brown Swiss F_1-crosses considering present replacement rates and taking calve losses into account. In a two-breed discontinuous cross (scenario A) milking Holstein were mated with Simmental to achieve higher prices for fattening calves. The potential of crosses to ensure 91,700 purebred lactating Holstein cows producing the fixed fat quota comprised 20,9% of total matings. The economical advantage was about 1,19 Mio. . In scenario B, a first generation cross (F_1) was generated from matings of Brown Swiss bulls to purebred Holstein cows. Final products (F_2) stemmed from three-way-crosses with Simmental. Higher results in net merit (+3,67 Mio. , +1,62 Mio.) compared with the reference scenario and scenario A, respectively, were essentially because of heterosis effects in the F_1 (= 25% of milking cows).

Session 27 Theatre 1

Alternatives to piglet castration

M. Bonneau and A. Prunier, UMR INRA-Agrocampus Rennes SENAH, 35590 Saint Gilles, France*

Most male piglets are currently castrated, with the main purpose of preventing boar taint, an offensive odour/taste affecting the meat from some entire male pigs, due to the accumulation of skatole and androstenone. Surgery without anaesthesia is by far the most common castration procedure. Because it is clearly painful, it has been recently banned in Norway, and is already submitted to some limitation in EU. There are a number of possible alternatives. Local anaesthesia is effective in reducing pain, however it is costly and time consuming. Intra-testicular injection of chemicals remains to be investigated regarding both efficacy and associated pain. Immunocastration (immunisation against GnRH) is very effective in reducing boar taint and is currently used in commercial practice in Australia. Whether its effectiveness in reducing taint should be assessed on the slaughter-line remains to be established. Raising entire males cannot be envisaged in most countries unless the incidence of boar taint is drastically lowered. Skatole production can be reduced by a combination of management procedures. Recent results suggest that skatole degradation is under genetic control. Androstenone is mostly under genetic control; however, selection is not currently feasible in practice, in the absence of well-defined genetic markers. In the event of entire male pig production, on-line detection of boar taint will be greatly needed; however there is currently no satisfactory method to perform it. On the whole, alternatives to castration still need further investigation.

Session 27 Theatre 2

Experiences with use of local anaesthesia for piglet castration

B. Fredriksen, O. Nafstad, B.M. Lium and C.H. Marka, Norwegian Meat Research Centre, Oslo, Norway*

Piglet castration will be forbidden in Norway from January 2009. From August 2002 and until then, the castration of piglets in Norway is to be performed only by veterinarians, and use of anaesthesia is mandatory. To evaluate the experiences with the current practice, the Norwegian Meat Research Centre performed two queries, one directed to veterinarians and one to swine producers. Both queries covered questions about the routines in connection with castration, complications and an overall evaluation.

The results show that the piglets on average are 10 days old when castrated. The most common anaesthetic was lidocain (2%) with adrenalin, used in undiluted solution by 56% of the veterinarians. A combination of subcutaneous and intratesticular administration was most often used (64%). The effect of the anaesthesia was evaluated to be good by 54% of the veterinarians and 19% of the producers.

Complications after castration were not a common problem, and the frequency seemed to be the same as when the producers performed the castration without use of anaesthesia. Abscesses were the most common complication. The majority of veterinarians and producers had never experienced deaths after castration. Prolapse of the intestines due to undiscovered inguinal hernia was the most common cause of death.

Overall 30% of the producers and 68% of the veterinarians were satisfied or very satisfied with the present scheme.

Session 27 Theatre 3

Estimation of genetic parameters of boar taint; skatol and androstenon and their correlations with sexual maturation

H. Tajet[1,2] and T.H.E. Meuwissen[2], [1]Norsvin, P.O.Box 504, NO-2304 Hamar, Norway, [2]Norwegian University of Life Sciences, Ås, Norway

Boar taint is mainly caused by two components; skatol and androstenon. By castrating the male pigs, boar taint has been avoided. In Norway, castration of pigs will no longer be permitted after 2009. This represents a substantial cost for the Norwegian swine production. Other Norwegian studies have shown that a large proportion of pigs are above the consumer detection limits for these two chemical components. The obvious question for the geneticist arises: Is it possible to select against skatol and androstenon in a breeding programme? Skatol is produced by bacteria in the gut. It is then absorbed in the blood stream. Skatol is either metabolised in the liver or transported and stored in fatty tissue. Androstenon is produced in the testis, and its biochemical pathway is related to the pathway of testosteron. In this study, fatty tissue was collected from the carcasses of Norwegian landrace and duroc boars, and analysed for androstenon and skatol. The length of *glandula bulbo urethralis* was measured on the same animals, as this is regarded as a good indicator of sexual maturation. Heritabilities of androstenon and skatol were substantial. The two components were not genetically correlated. Sexual maturation was also highly heritable. However, correlations to both androstenon and skatol were significantly unfavourable.

Session 27 Theatre 4

Inhibition of CYP2E1 expression by androstenone: relation to boar taint

O. Doran [1], J.D.McGivan[2], F.M.Whittington[1], W.S.Tambyrajah[3], J.D.Wood[1], [1]Department of Clinical Veterinary Science, University of Bristol, BS405DU, UK. [2]Department of Biochemistry, School of Medical Sciences, University of Bristol, BS81TD, UK, [3]Department of Biochemistry, University of Sussex, Brighton, BN19QG, UK.*

Boar taint is an off-odour of some pig meat and is due to accumulation of skatole and androstenone in adipose tissue. Boar taint can be prevented by castration. One of the possible alternatives to castration could be a genetic test for high taint depositors. Development of such a test requires identification of physiological candidate genes. It was previously shown that the level of backfat skatole negatively correlates with the expression of CYP2E1 in liver and with the rate of hepatic skatole metabolism. The present study investigated regulation of CYP2E1 expression by androstenone. The results show that (i)Androstenone inhibits CYP2E1 expression in cultured pig hepatocytes. (ii)The transcription factors HNF-1 and COUP-TF1 activate CYP2E1 promoter activity. Androstenone represses CYP2E1 promoter activity via inhibition of binding of COUP-TF1. (iii)The rate of hepatic androstenone metabolism negatively correlates with the levels of androstenone and skatole in backfat. The enzyme responsible for hepatic androstenone metabolism was shown to be 3-beta-hydroxysteroid dehydrogenase (HSD). The expression of HSD is lower in pigs manifesting high levels of skatole and androstenone. On the basis of these results we suggest that defective expression of HSD leads to accumulation of androstenone in backfat, which represses expression of the CYP2E1 gene and results in high skatole levels.

Session 27 Theatre 5

Rearing of entire male pigs in a "farrow-to-finish-system": effects on boar taint substances and animal welfare

B. Fredriksen[1], O. Nafstad[1], B.M. Lium[1], C.H. Marka[1], Berit Heier[2] and Ellen Dahl[3], [1]Norwegian Meat Research Centre, [2]National Veterinary Institute, [3]Norwegian School of Veterinary Science, [1,2,3]Oslo, Norway*

The aim of the present study was to investigate if keeping the littermates together in stable groups without moving them ("farrow-to-finish-system") would influence the initiation of puberty, and thereby the levels of the boar taint substances androstenone and skatole in the carcasses. For a subgroup of the study population, welfare parameters as skin lesions and aggressive behaviour were also recorded. Pigs raised in a FTF-system were compared to pigs in groups composed of pigs from different litters.

The results showed that the levels of androstenone in Norwegian slaughter pigs are high, especially in crossings including the Duroc breed. Raising entire male pigs in a FTF-system delayed the onset of puberty, and reduced the levels of androstenone in fat compared to entire male pigs in mixed groups. When adjusted for other relevant factors as herd, breed and season, the average levels of androstenone in the two groups were 0.85 and 1.23 µg/g, respectively.

Compared to mixed groups, raising pigs in a FTF-system reduced aggressive behaviour in entire male pigs. The frequency of skin lesions was reduced to the level of non-castrated pigs. This indicates that raising pigs in a FTF-system can be a positive contribution to the animal welfare aspect of raising entire male pigs.

Session 27 Theatre 6

Seasonal production of small entire male pigs raised in one unit system

H.F. Jensen and B.H. Andersen. Danish Institute of Agricultural Sciences, Research Centre Bygholm, DK-8700 Horsens, Denmark*

Animal welfare is given high priority within organic production, and the omission of castration of male pigs may present a welfare benefit. However, there may also be some disadvantages. The maturity of the male pigs may involve risks of undesirable behaviour and meat with off-flavour. An organic production system for pigs was established in one-unit-pens with deep litter bedding in the outdoor area at Research Centre Bygholm. Male pigs from four seasons and 40 different litters grew up together with their littermates from birth to slaughtering without being mixed with unfamiliar pigs. They were slaughtered in the range of 100 to 150 days of age. The levels of skatole, indole and androsterone in the back fat of the pigs were recorded. 10 % of the male pigs had levels of androsterone (>1.,25 ppm) or skatole (>0.15 ppm) that could be detected as off-flavour. The youngest pig with a high concentration of androsterone (1.81 ppm) was 118 days old and the lightest pig with a high level of androsterone (2.19 ppm) had a slaughter weight of 44.0 kg. This pig moreover had the highest level of skatole (0.40 ppm). To ensure that the meat will not have an off-flavour, the male pigs should neither be older than 120 days nor heavier than 55 kg of live weight at the time of slaughtering.

Session 27 Theatre 7

Boar taint in pigs fed raw potato starch

G. Chen[1], G. Zamaratskaia[1], H.K. Andersson[2], K. Andersson[3], K. Lundström[1], Departments of [1] Food Science, [2]Animal Nutrition and Management, [3]Animal Breeding and Genetics Swedish University of Agricultural Sciences, Box 7051, SE-750 07 Uppsala, Sweden

Androstenone, skatole and indole are compounds responsible for boar taint. Androstenone is produced in the testes and indolic compounds are synthesised in the pig's large intestine from tryptophan. Our previous study demonstrated that dietary supplement of raw potato starch (RPS) dramatically decreased skatole levels, whereas androstenone levels were unaffected. However, little is known about the effect of diet on indole levels. In the present study the effect of RPS on the levels of androstenone, skatole and indole in fat and plasma was evaluated. 100 pigs were used in the study. Half of the pigs received 0.6 kg raw potato starch per day, two weeks prior to slaughter. Consumption of the RPS diet resulted in a significant reduction in skatole levels in both fat and plasma ($P<0.001$); however, indole levels were not altered by RPS ($P>0.05$). The effect of RPS on skatole levels was explained by increased production of butyrate, which was suggested to reduce substrate formation for skatole and indole synthesis. The fact that indole levels were not altered by RPS does not support this hypothesis. Surprisingly, plasma androstenone levels were reduced by RPS ($P<0.001$). No effect of RPS on fat androstenone levels was demonstrated ($P>0.05$) suggesting that fat androstenone levels do not ultimately depend on plasma levels.

Session 27 Theatre 8

Influence of chicory roots (*Cichorium intybus L*) on boar taint in entire male and female pigs

L.L. Hansen[1], M.T. Jensen[1], H. Mejer[2], A. Roepstorff[2], S.M. Thamsborg[2], D.V. Byrne[3], A.H. Karlsson[1], J. Hansen-Møller[4] and M. Tuomola[5], [1]Department of Food Science, DIAS, DK-8830 Tjele, Denmark

The aim of this study with entire male (and female) pigs was to find an economically bioactive feed, which could decrease boar taint and maintain eating quality in organic pig production. Three experiments with different feeding before slaughter were carried out. First trial: crude chopped chicory roots were fed for 4 and 9 weeks before slaughter; Second trial; crude and dried chopped roots and inulin were fed for six weeks Third trial; short time feeding with dried roots for 1 and 2 weeks. In all the experiments the control treatments were standard organic and "conventional" fed. In the first trial the skatole concentrations in plasma and backfat were significantly decreased in the pigs fed for 3-9 weeks with feed containing 25% chicory compared to the two control treatments. In the second trial the skatole concentration in plasma and backfat decreased significantly compared to the control treatment. In the third trial feeding male pigs with 25% dried chicory for 7 days and 14 days reduced skatole concentration in blood and backfat close to zero. The androstenone concentration in plasma of the chicory fed entire male pigs showed no significant decrease except when fed 8 weeks.
With respect to eating quality chicory feeding reduced boar taint from a sensory perspective. Moreover, dried chicory had the best potential for development of a commercial product.

Session 27 Theatre 9

Alternatives to antimicrobial growth promoters (AGP)

Caspar Wenk, Animal Science, ETH Zentrum, Zurich, Switzerland

AGP have been successfully used in diets for farm animals over almost 50 years. With the increasing fear of resistant microorganisms against antibiotics in animal and human health care their use is more and more questioned. Organic acids, enzymes, probiotics, prebiotics as well as herbs or botanicals can be seen as potential alternatives to antibiotics. In the young animal feed intake, nutrient content and acid binding capacity are additional factors to maintain gut health.
Organic acids can suppress the growth of undesired microorganisms in the diet and digesta in the upper digestive tract by lowering of the pH or specific antimicrobial effects. Exogenous enzymes are available to enhance digestive capacity especially of young or ill animals and to increase the digestibility of the feed. Probiotics (living microorganisms) on one hand stimulate beneficial microorganisms in the gastrointestinal tract and on the other hand suppress pathogens by competitive exclusion. A new concept is the use of resistant starch or oligosaccharides like fructose- or mannan-oligosaccharides. These additives help to optimize the eubiosis, a balanced healthy micro flora in the digestive tract, improve the digestion capacity and increase the health status of the animal. Herbs and botanicals can beneficially affect feed intake, secretion of digestive juices and the immune system of animals. They may have antibacterial or coccidiostatic activities and particularly antioxidant properties.
The best results in the substitution of AGP can be achieved in the combination of these additives and under good environmental conditions.

Session 27 Theatre 10

Holo-analysis of the effects of genetic, managemental, chronological and dietary variables on the efficacy of a pronutrient mannanoligosaccharide in piglets

G.D. Rosen, Pronutrient Services Ltd., 66 Bathgate Road, Wimbledon, London SW19 5PH, England

All available piglet test data from 21 publications on the efficacy of a yeast mannanoligosaccharide product, Bio-Mos[R] (BM) Alltech Inc., have been analysed by multiple regression of start-to-finish feed intake (FDI), liveweight gain (LWG) and feed conversion ratio (FCR) effects on 12 independent variables, control performance, year of test, duration, dosage, male, slatted floor, pellet, soy, animal protein, added oil/fat, antibacterial and non-USA. Average responses from 57 negatively-controlled tests are +7.6g/day (1.5%) for FDI, +15.9g/day (4.73%) for LWG and -.0588 (3.84%) for FCR using BM at 1-4gBM /kg feed (mean 2.12) in 10-49 day tests (mean 30.6) on 3,178 piglets in 1996-2002. Respective coefficients of variation of these responses are 346, 155 and 213%. Beneficial response frequencies for LWG and FCR are 74 and 70% respectively (58% jointly). Models based on 41 (outlier-free) tests show (i) better FCR effects in less efficient converters; (ii) no significant dose-response relationships; (iii) higher FDI and LWG effects with step-down dosage; (iv) smaller USA LWG and FCR effects; and (v) lower FDI and LWG effects in pellets. Future research when data suffice could include (a) dose-response elucidation; (b) quantification of feed ingredient and nutrient content effects; (c) efficacies of BM with and without other pronutrients; and (d) oligosaccharide comparisons relative to dietary lactose contents.

Session 27 Theatre 11

Compatibility of *B. toyoi* and colistin in post-weaned piglets medicated diets

J. Morales[1], C. Piñeiro[1], G. Jiménez[2] and A. Blanch[3]; [1]PigCHAMP Pro Europa, S.A., Segovia, Spain, [2]ASAHI VET, S.A., Barcelona, Spain, [3]Andersen, S.A., Barcelona, Spain*

Recently the European Public Health Alliance (SANCO, 2004) announced that the use of *B. cereus* var. *toyoi* in medicated diets was not prohibited. The present study was carried out to assess the compatibility of the combination of this probiotic with colistin, a therapeutic antibiotic. A total of 336 piglets were used (28-42 d of age) and distributed in a 2x2 factorial design, based on the administration or not of two additives (*B. toyoi* and colistin). Average daily gain (ADG) and feed intake (FI) were controlled and faeces sampled to determine *Coliforms* and *Lactobacillus* populations. Colistin increased FI ($P<0.01$) and ADG ($P<0.05$). Supplementation with *B. toyoi* also resulted in an increased FI ($P<0.05$). Furthermore, the combination of both products showed a synergetic effect on all productive performance parameters, especially on FI (287 vs 254 g/d; $P<0.05$). Regarding the microbiology of the faeces, colistin decreased *Lactobacillus* concentration and *Lactobacillus:Coliforms* ratio. However, the combination of colistin and *B. toyoi* increased both parameters. In conclusion, the use of *B. toyoi* in combination with colistin in post-weaning medicated diets may have a synergetic effect on growth and FI during this critical period, improving the characteristics of gut bacteria population. However, *B. toyoi* population in faeces was not modified by the inclusion of colistin, suggesting that both products are fully compatible.

Session 27 Theatre 12

Efficacy of benzoic acid in the feeding of weanling pigs

D. Torrallardona[1] and J. Broz[2], [1]IRTA-Centre de Mas Bové, Reus, Spain, [2]DSM Nutritional Products, Basel, Switzerland

Two trials were conducted to evaluate the effects of feeding benzoic acid on the performance of weanling pigs. Ninety-six and forty-eight piglets (4 wk of age) were used in trials 1 and 2 respectively. For each trial half of the piglets were fed a control diet and the other half the same diet with 0.5 % benzoic acid. In both trials benzoic acid improved body weight gain ($P<0.01$) and feed to gain ratio ($P<0.1$) during the pre-starter phase (0-14 days). Similarly, during the starter phase (14-28 days) benzoic acid improved weight gain in trials 1 and 2 ($P<0.05$ and $P<0.1$ respectively) and feed intake in trial 1 ($P<0.1$). Over the whole trial (0-28 days), benzoic acid improved weight gain in trials 1 and 2 ($P<0.05$) and also improved feed intake and feed to gain ratio ($P<0.1$) in trial 1. At the end of trial 1, twenty-four pigs were slaughtered and their urine was sampled. It was observed that benzoic acid significantly lowered ($P<0.1$) the pH in urine and that hippuric acid concentration in urine increased ($P<0.01$) as a result of the metabolism of benzoic acid. It is concluded that the inclusion of 0.5% benzoic acid in pre-starter and starter diets for piglets improves their performance and reduces the pH in urine by increasing its hippuric acid concentration.

Session 27 Poster 13

Season for pigs: new quality of pig meat, experiences with a demonstration sale

B H. Andersen[1], A C. Bech[2] and H F. Jensen[1], [1]Danish Institute of Agricultural Sciences, Research Centre Bygholm, DK-8700 Horsens, Denmark, [2]Jysk Analyseinstitut A/S, Boulevarden 1, Postboks 1533, DK-9100 Aalborg, Denmark*

A new pig production system was developed at the organic research station of Rugballegaard in Denmark. The sows and pigs were kept on deep litter bedding in a one-unit-system with tents and outdoor areas. The pigs grew up with their littermates in the same area from birth to slaughtering. The sows farrowed in the spring and again in the autumn. The small entire male pigs born in the spring were slaughtered from June to July, the result being a pale, soft and lean meat product. The female pigs were slaughtered in the late autumn (November, December), the result being a large carcass and tastier meat. The small entire male pigs were slaughtered at a local slaughterhouse, and the meat was sold by a local supermarket or as meals from a recognized restaurant. The supermarket made a demonstration sale, and the costumers were asked if they liked the meat and if they would consider buying the product. A focus group interview on organic pig meat was made among a selected group of costumers. Their recommendation was that roasted pig meat should be lean and prepared carefully resulting in juicy and tender meat with a slightly pink colour.

Session 27 Poster 14

The effects of different concentrations of hops on the performance, gut morphology, microflora and liver enzyme activity of newly weaned piglets

J. Williams, A. H. Stewart, J. Powles, S. P. Rose and A. M. Mackenzie, Harper Adams University College, Newport, Shropshire, TF10 8NB, United Kingdom

This study investigated the use of hops in diets for newly weaned pigs to enhance growth performance, gut health and liver function. Thirty pigs were allocated to one of three treatments. These were (I) Control diet (with no additives), (II) a diet containing 1 kg/t of hops and (III) a diet containing 10 kg/t of hops. Results showed no significant improvement in growth performance. Although a trend was observed for the improved FCR in the pigs fed 10 kg/t hops (P=0.094) on day 22-26. The level of diarrhoea and number of *E. coli* in the faeces were unaffected by treatments. However, the number of Lactobacillus in the faeces was lower in the pigs on the hop treatments (day 11 and 26). The 10kg/t of hops resulted in a higher level of total volatile fatty acids in the colon (P=0.034), which was mainly due to higher levels of acetic acid. The VFA levels in the caecum were not affected by treatment. The liver weight and function was not affected by treatments. The gut morphology was not affected. These findings suggest that hops fed after weaning may have a potential to improve the FCR and have a modifying effect on hindgut microbial fermentation.

Session 28 Theatre 1

Testing young horses for sport and for genetic evaluations of Swedish riding horses

Å. Viklund[1], J. Philipsson[1], Å. Wikström[1], Th. Arnason[2], E. Thorén[1], A. Näsholm[1], E. Strandberg[1] and I. Fredricson[3], [1]Department of Animal Breeding and Genetics, Swedish University of Agricultural Sciences, S-750 07 Uppsala, Sweden, [2]IHBC, Knubbo, S-744 94 Morgongåva, Sweden, [3]Tärnö säteri, S-610 56 Vrena, Sweden*

In order to get information on young horses to be selected for sport and for genetic evaluations according to more specialized sport breeding objectives, a system for testing Swedish Warmblood horses has been set up. A one day field test for 4-year old riding horses of both sexes was introduced in 1973. The test includes assessment of health, conformation, gaits under rider, jumping ability and temperament. To date nearly 17,000 horses have been tested. The information is used for genetic evaluations based on a BLUP Animal model, which was adopted in 1986. Genetic analyses show moderate heritabilities (0.15-0.35) for conformation and performance traits, and high genetic correlations (0.7-0.9) with competition results at mature age. Based on the experiences of this test the performance tests for stallions were modernized and expanded during the 1980-ies. A simplified one day performance field test for 3-year old horses was developed in the late 1990-ies. Today about half of all young horses are tested at either 3 or 4 years of age. Analyses show that large genetic progress has been achieved. Present activities aim at integrating all sport horse information into the genetic evaluations.

Session 28 Theatre 2

Estimation of genetic parameters for two different performance tests of young stallions in Germany

B., Harder[1], T. Dohms[2], E. Kalm[1], [1]Institute of Animal Breeding and Husbandry, University of Kiel, D-24098 Kiel, Germany, [2]Deutsche Reiterliche Vereinigung e.V. (FN), D-48231 Warendorf, Germany

Since 2001 all young stallions have to do a 30-day performance test at station in order to get their preliminary covering permission. One alternative for the final covering permission is a performance test at station, which lasts 70 days. The data included results from all stallions tested in Germany between 2001-2004. 1048 stallions joined the 30-day-test, and 773 stallions joined the 70-day-test. 212 stallions had results in both tests. In both tests the traits rideability, free jumping, walk, trot, and canter were evaluated, moreover course jumping was considered in 70-day-test. Heritabilities were estimated with a multivariate analysis within test. The results ranged between 0.4-0.71 and 0.45-0.63 for 30-day-test and 70-day-test, respectively. The genetic correlations within 30-day-test were moderate to high (0.33-0.78) except for those between free jumping and all other traits, which were low or negative (-0.27-0.14). The genetic correlations between the traits of the 70-day-test were higher than those from the 30-day-test and ranged between 0.46-0.99, again the jumping traits free jumping and course jumping showed low to negative relationships to the other traits (-0.15-0.16). The genetic correlations between the same traits of the different tests were analysed with a bivariate analysis. The resulting correlations were high in the range of 0,87-0,99.

Session 28 Theatre 3

Genetic parameters for competition traits at different ages of Swedish riding horses

Å. Wikström, Å. Viklund, A. Näsholm and J. Philipsson, Department of Animal Breeding and Genetics, Swedish University of Agricultural Sciences, S-750 07 Uppsala, Sweden*

In Sweden competition results have been registered since the beginning of 1960'ies. Until 2002 results for approximately 38,000 horses were available from dressage, show jumping and eventing competitions. Genetic parameters were estimated for results obtained at different stages in life of the horses. The results were divided in three groups. The first group included results up to six years of age, the second up to nine years and the last included all results the horses had until 2002, i.e. lifetime results. A multi-trait animal model was used for estimation of genetic parameters for traits in dressage and show jumping. Heritabilities were higher for show jumping than for dressage and increased with increasing of age of the horses and amount of information. For dressage heritabilities increased from 0.08-0.10 for the youngest group to 0.15-0.16 for the lifetime results. For show jumping the corresponding values increased from 0.11-0.20 to 0.23-0.38. Genetic correlations between different age groups were high (0.94-1.00). Results from this study show that lifetime results are more accurate to use for genetic evaluations than results achieved at six years of age, but correlations between the results at six years and the accumulated lifetime results are high. It is suggested that lifetime results from competitions should be integrated into the genetic evaluation system.

Session 28 Theatre 4

Application of a 3D morphometric method to the follow-up of conformational changes with growth and to the study of the correlations between morphology and performance

N. Crevier-Denoix[1], P. Pourcelot[1], D. Concordet[2], D. Erlinger[1], A. Ricard[3], L. Tavernier[4], J.-M. Denoix[1], [1]UMR INRA 957, Ecole Vétérinaire d'Alfort, [2]UMR INRA 181, Ecole Vétérinaire de Toulouse, [3]Génétique Quantitative-INRA Toulouse, [4]Centre d'Enseignement Zootechnique de Rambouillet, France

A computerized method of morphometric measurement has been developed. Its main originalities are to use video (each horse is filmed while walking instead of being photographed at standing), to be in 3-D (4 digital cameras are used; they film the horse from its lateral side, from the front and behind), and not to require skin markers (which makes the recording procedure quick). In a preliminary study, 20 international and 20 low-level jumping horses were compared. The height of the main anatomical landmarks of the pelvic limb, the length of the croup and thigh, the length of the fore cannon and shortness of the forearm, are all conformation traits that characterize the international level horses. This method is presently applied in the breeding program of the Selle Français horse through 2 studies: 1. one includes 750 4 and 5 year-old horses genetically affiliated and of known sport levels. The heritability of the main conformation traits has been determined, and the analysis of the correlations with performance is in process; 2. the other consists in the longitudinal follow-up of conformation in 130 horses from 6 months to 3 years.

Session 28 Theatre 5

Group housing exerts a positive effect on the behaviour of young horses

E. Søndergaard and J.W. Christensen, Danish Institute of Agricultural Sciences, P.O. Box 50, 8830 Tjele, Denmark*

To examine the effects of social environment on the behavioural development of young horses 40 Danish Warmblood colts were used in two replicates of 20 horses from weaning until 21/2 years of age. In each replicate, 8 horses were housed singly, and 12 horses were housed in 4 groups of 3 horses during the housing period from September to May. The horses had daily access to paddocks, either singly (singly stabled) or in groups (group stabled). The summer period was spent on pasture in groups. Half of the singly housed horses and half of the group-housed horses were handled for 10 minutes three times per week during each housing period, in total approx. 20 hours of handling. Handling involved leading, tying up, touching etc. Non-handled horses were only handled for monthly weightings, farrier and veterinary treatment. Singly housed horses completed fewer stages in the training program and they also bit and kicked more during training sessions than group housed horses. Previously group stabled stallions had a higher frequency of displacements and submissive behaviours but fewer direct aggressive interactions compared to previously singly housed stallions, implying that group housed stallions had a more well-developed social language. The study shows that group housing is essential for young horses in relation to their social behaviour and also beneficial in relation to their reactions in training situations.

Session 28 Theatre 6

Learning performance in relation to fear in young horses

J.W. Christensen[1,2], K. Olsson[3], M. Rundgren[3], L. Keeling[1], [1]Swedish University of Agricultural Sciences (SLU), Animal Environment and Health, Skara, Sweden, [2]Danish Institute of Agricultural Sciences, Foulum, Denmark, [3]SLU, Animal Nutrition and Management, Uppsala, Sweden*

Responses of horses in frightening situations are important to both horse and human safety. The process, where young horses learn not to react to otherwise frightening stimuli, was studied using 26, 2-year-old, naïve Danish Warmblood stallions. The horses were trained according to three different methods. In the first method, the horses (n=9) were exposed repeatedly to the full stimulus (a moving, white paper bag, 1.2 x 0.75 m) until they met a predefined habituation criterion. In the second method, the stimulus was divided into several less frightening steps, and the horses (n=9) were habituated to each step, before the full stimulus was applied. In the third method, the horses (n=8) were trained to feed from the bag, i.e. to associate the stimulus with a positive reward, before they were exposed to the full stimulus. Five training sessions of 3 minutes were allowed per horse per day. Heart rate and behavioural responses were registered. Surprisingly, there was no significant difference between methods in heart rate during the first training session. Horses that were trained according to method 2 showed fewer flight responses and needed fewer training sessions to learn not to react to the stimulus (ANOVA, mean±se; Met2: 2.6±0.8 vs. Met1: 4.3±2.1 and Met3: 10.3±2.7; $F_{2,23}$=4.05, P=0.03).

Session 28 Theatre 7

Evaluation of breeding strategies against osteochondrosis (OC) in warmblood horses

M. Busche and E. Bruns, Institute of Animal Breeding and Genetics, Georg-August-University Göttingen, Albrecht-Thaer-Weg 3, D 37075 Göttingen*

Osteochondrosis (OC) is a skeletal disease caused by disturbed bone formation and can lead to loose bone fragments in the joints (OCD). Affected horses have a higher risk for locomotion problems and culling (Koenen, 2000). The frequency of OC is up to a quarter in different warmblood breeds. Heritability estimates for OC range from 0,1 to 0,3 and indicate that genetic selection can reduce the frequency of OC (Bruns, 2001). In the present study, breeding strategies against OC are simulated for a breeding population with 5.000 mares and 200 stallions that produces in total 20.000 foals per generation. The selection considers three different traits (Dressage, Jumping, OC). Different genetic models are developed. The first model assumes OC affected by an autosomal gene with two alleles, in the second model OC is assumed as a all-or-none phenotypic trait with an underlying normally distributed genotype. Moreover a two-step selection model is developed. At first horses are selected based on their performance in dressage and jumping. Secondly horses are tested for OC by x-rays. This model is refined to divide the phenotype into the 4 common x-ray classes. Within 5 generations the changes of the genotype frequency of OC due to selection are observed. First results show that the genotype frequency of OC decreases more rapidly by using the first model.

Breed variations in the distribution of osteo-articular lesions in horses at weaning

C. Robert, J.-P. Valette, S. Jacquet and J.-M. Denoix, UMR INRA - ENVA de 'Biomécanique et Pathologie Locomotrice du Cheval', 7 avenue du général de Gaulle, 94704 Maisons-Alfort cedex, France*

Juvenile Developmental Orthopaedic Diseases (DOD) affect most horse breeds and their occurrence is reported to be increasing. In order to establish the prevalence of DOD in French breed horses, 133 Thoroughbred (Tb), 161 French Trotters (FT) and 99 Selle Français (SF) foals kept on 21 stud farms were examined radiographically at weaning. Radiographic files (including views from the front and hind digital and fetlock joints, carpi, hocks and stifle joints) were analysed by 3 veterinarians. For each joint, osteo-articular lesions were graded according to a standardised protocol depending on their severity. The scores of all the images on the 10 radiographs were added to calculate a global radiographic score. Differences in the distribution of the lesions were evaluated using an analysis of variance. SF were globally more severely affected than FT and Tb ($p<0.001$). Considering the regional distribution of the lesions, SF were more severely affected on the front fetlock ($p<0.001$), on the dorsal aspect of the hind fetlock ($p<0.001$) and on the femorotibial joint ($p<0.05$). FT were more severely affected on the plantar aspect of the hind fetlock ($p<0.001$) and the tibiotarsal joint ($p<0.01$). The present study shows that osteo-articular lesions differ between breeds independently of their use and training regimens.

Genetic analyses of radiographic appearance of navicular bones in the Warmblood horse

K.F. Stock, O. Distl, School of Veterinary Medicine Hannover (Foundation), Department of Animal Breeding and Genetics, Bünteweg 17p, 30559 Hannover, Germany*

Radiographic appearance of navicular bones (RNB) was analyzed genetically using the results of a standardized radiological examination of 5,157 Hanoverian Warmblood horses selected for sale at auction. Different categorisation schemes were used to describe RNB. Definition of RNB traits was based on type and/or extent of radiographic findings in the navicular bones of the front limbs. Analyses included pseudo-linear (RNB_{0-3}, RNB_{0-7}) and binary traits ($RNB_{0/1a-e}$). The most specific pseudo-linear trait (RNB_{0-7}) accounting for different types of radiographic findings served as reference trait for RNB. Deviation from normal RNB was classified for defining more general RNB traits. Genetic parameters were estimated multivariately with REML. Breeding values were predicted uni- and multivariately using PEST. Heritability estimates for navicular bone appearance were in the range of $h^2 = 0.10$ to $h^2 = 0.33$. Additive genetic correlations between RNB traits were positive and mostly larger than $r_g = 0.80$. Binary coding based on existence or non-existence of marked radiological alterations in the navicular bones ($RNB_{0/1d}$) resulted in an additive genetic correlation of $r_g = 0.98$ to the reference trait (RNB_{0-7}). Correlation between respective breeding values was $r = 0.91$. Higher heritability of the binary trait ($RNB_{0/1d}$; $h^2 = 0,24$) suggests that classification of RNB and use of binary coding could be beneficial when considering RNB in future breeding plans for the Warmblood horse.

Session 28 Theatre 10

Effect of maximal vs moderate growth on osteoarticular development in the yearling

M. Donabedian[1], Géraldine Fleurance[2], G.Perona[3], Catherine Trillaud-Geyl[2], Celine Robert[4], Sandrine Jacquet[4], J.M. Denoix[4], O. Lepage[5], D. Bergero[3] and W.Martin-Rosset[1], [1]INRA Center of research of Clermont/theix 63122 Saint Genes Champanelle, France, [2]Haras Nationaux, Experimental station 19370 Chamberet, France, Veterinary Schools of: [3]Torino, 1095 Grugliasco, Italy, [4]Paris, 94704Maisons Alfort,France, [5]Lyon, 69280 Marcy l'Etoile, France

Two groups of foals of Anglo arab and Selle Français breeds were subjected during the first postnatal year to maximal (HL, n=20) or moderate (ML, n=19) growth. ML group was fed 100p100 of allowances recommended by INRA 1990. HL group was fed 130 and 150 p100 of total and balanced allowances fed to ML group before and after weaning respectively. At one year of age, body weight (BW) and height at withers (HW) were significantly higher for HL group than for ML group: +88.3kg BW ($p<0.0001$) and +4,5cm HW ($p<0.001$).The discrepancy determined between the two groups at 12 months of age were mainly achieved at weaning for BW(65p100) and HW(100p100). Osteoarticular status was assessed using a thorough radiographic examination of limbs at 6 and 12 months (n=39), and a necropsic examination (n=16) of limb and cervical joints. In both cases, lesions were detected, counted and scored. First analysis of radiographic data suggest that the incidence of osteochondrosis was 2.5 fold higher in HL group. Further data analysis are in progress to refine the influence of high plane of nutrition on osteochondrosis incidence.

Session 28 Theatre 11

Bone spavin in Icelandic horses

Sigridur Björnsdóttir, Icelandic Veterinary Services, Iceland

Bone spavin is an osteoarthrosis (OA) of the distal tarsal joints.

In a field survey, radiographic signs of OA in the distal tarsal joints (RS) were found in 30.3% of horses in the age range of 6-12 years (n=614), significantly correlated to hind limb lameness after flexion test of the tarsus.

The prevalence of RS was strongly correlated to age and tarsal angle. The heritability of age-at-onset of RS, reflecting the predisposition for OA, was by survival analysis estimated to be 0.33.

In order to detect and describe the earliest changes compatible with OA, specimens from the centrodistal tarsal joint of young horses (0-4 year, n=82) were examined by high detail radiography and histology. Histological chondronecrosis was seen in 33% of the joints, located both medially and laterally. Subchondral bone sclerosis was not correlated with the chondronecrosis medially, but laterally the bone sclerosis was considered to be secondary to the cartilage lesions.

The high prevalence of histological findings in the young horses (1-4 year) and radiographical findings in the 6-12 year old horses demonstrated a progressive nature of the disease although the progression may be slow. The initiation of the disease was unrelated to the use of the horses for riding and workload was not found to effect the development of the disease negatively.

It was suggested that poor tarsal conformation or architecture of the distal tarsal joints is the main etiological factor of the disease.

Session 28 Poster 12

Evolution of haematological and biochemical reference values in the growing horse

J.P. Valette, C. Robert, G. Fortier, M.P. Toquet, J.M Denoix, MR INRA-ENVA Biomécanique et Pathologie Locomotrice du Cheval, 7 avenue du Général-de-Gaulle, 94704 Maisons-Alfort, France, laboratoire Frank Duncombe, 14053 Caen, France*

An epidemiologic, longitudinal study was conducted in Normandy between 1997 and 2003. A total of 674 foals then yearlings were followed during 2 years from their birth to the departure to the training centre. The subjects studied all belonged to the three main horse breeds used in horse racing and equine sports: Thoroughbred, French Trotter and Selle Français. They were weighted and measured regularly and at the same time jugular venous blood was taken to determine haematological and biochemical parameters. Horses were divided into 5 age groups: 0 to 30 days, 30 to 90 days, 3 to 6 months, 7 to12 months and 12 to 18 months. Data were analysed by analysis of variance (GLM procedure, SAS institute). There were significant differences between age groups. For example, between birth and 1 year-old, mean value for haematocrit varies from 35.1 to 38.7 %, phosphorus from 73 to 54 mg/l and copper from 0.94 to 1.42 mg/l. There were also differences between breeds in each age group. Thoroughbred and French Trotter horses have higher haematological reference values than Selle Français horses except for leucocytes and fibrinogen. Thoroughbred have higher haematological values than French Trotter only after1year. Selle Français horses usually have higher biochemical values than others horses.

Session 29 Theatre 1

Search for finding ovulation rate responsible mutation in the Booroola fecundity (*FecB*) gene in Iranian Afshari sheep

S. Qanbari[1], M.P. Eskandari Nasab[1], R. Osfoori,[2], A. Javanmard[3], [1]Department of Animal Science, Zanjan University, P.O. Box, 313, Zanjan, Iran, [2]Research Center of Agricultural-Jihad, Zanjan Province, [3]Agricultural Biotechnology Research Center, Tabriz, Iran*

Afshari sheep is one of the most important double-purpose breeds of Iran that have good performance in meat and milk traits. But from prolificacy point of view, this breed has twinning percentage less than 10. As lamb production is an important source of income in all flocks, increasing the fecundity of the Afshari sheep has always been an important breeding goal. Agricultural-Jihad Ministry of Iran has also been planed an introgression program to improve the prolificacy of Afshari sheep breed through inclusion of the B gene of the Booroola sheep. But it is first necessary to knowledge on the absence of the gene in this sheep. For this purpose a direct test to detection $FecB^B$ gene of Booroola Merino sheep was carried out using the PCR based restriction fragment length polymorphism (PCR-RFLP) method. The study was conducted on 31 single birth (at least two records) and 43 twin birth (at least one record) Afshari ewes. There was not any difference between digestion banding patterns between ewes having single and twinning records. This preliminary result indicates that the $FecB^B$ Allele was not present in this breed. This result would assist making decision on starting backcrossing program to introgression of the FecB gene into the Afshari breed.

Session 29 Theatre 2

Selection strategies for body weight and reduced ascites susceptibility in broilers

A. Pakdel[1], P. Bijma[2], B. J. Ducro[2] and H. Bovenhuis[2], [1]Ilam University, P.O.Box 69315-516, Ilam, Iran, [2]Animal Breeding and Genetics Group, Wageningen Institute of Animal Sciences, PO Box 338, 6700 AH Wageningen, The Netherlands

Ascites syndrome is a metabolic disorder in broilers. Mortality due to ascites especially in later ages results in significant economic losses and has a negative impact on animal welfare. The aim of the present study was to evaluate the consequences of alternative selection strategies for BW and resistance to ascites using deterministic simulation. Besides the consequences of current selection, i.e. selection for BW, alternative selection strategies including information on different ascites-related traits measured under normal and/or cold conditions were quantified. Five different schemes were compared based on the selection response for BW, ascites susceptibility and the rate of inbreeding. Traits investigated in the index as indicator traits for ascites were hematocrit value (HCT) and ratio of right ventricle to the total ventricular weight of the heart (RV:TV). The results of base and alternative selection strategies indicated that by ignoring ascites susceptibility in the breeding goal, although achieved gain for BW is relatively high (130 gr), the genetic response for ascites susceptibility is 0.025 unit and rate of inbreeding is 3% per generation. Including information of ascites makes it possible to achieve relatively high gain for BW (111.4 gr) with constant level of ascites susceptibility (zero) at lower rate of inbreeding (2.4% per generation).

Session 29 Theatre 3

Detection of quantitative trait loci for meat quality and carcass composition traits in Blackface sheep

E. Karamichou[1,2], G.R. Nute[3], R.I. Richardson[3], K. McLean[4] and S.C. Bishop[1], [1]Roslin Institute, Edinburgh EH25 9PS, UK, [2]University of Edinburgh, Edinburgh EH9 3JT, UK, [3]University of Bristol, Langford, Bristol BS40 5DU, UK, [4]SAC, Kings Buildings, Edinburgh EH9 3JG, UK*

Quantitative trait loci (QTL) were identified for traits related to carcass and meat quality in Scottish Blackface sheep. The population studied comprised a double backcross between lines of sheep previously divergently selected for carcass lean content (LEAN and FAT lines), comprising nine half-sib families. Carcass composition (600 lambs) was assessed non-destructively using computerised tomography (CT) scanning and comprehensive meat quality measurements (initial and final pH of meat, colour and carcass weight) were taken on 300 male lambs. Lambs and their sires were genotyped across candidate regions on chromosomes 1, 2, 3, 5, 14, 18, 20 and 21. QTL analyses were performed using regression interval mapping techniques. In total, six genome-wide significant and 12 chromosome-wide and suggestive QTL were detected in seven out of eight chromosomal regions. The genome-wide significant QTL (chromosomes 1, 2, 3 and 5) affected meat colour, muscle density, bone density and live weight-related traits. The most significant QTL affected muscle density and was segregating across families on chromosome 2. This study provides information on new QTL affecting meat quality and carcass composition traits in sheep. Verification of these results is now required in independent sheep populations.

Session 29 Theatre 4

Statistical aspects of QTL detection in small samples based on a data set from selective DNA pooling in blue fox

J. Szyda[1], M. Zaton-Dobrowolska[1], H. Wierzbicki[1] and A. Rzasa[2], [1]Department of Animal Genetics, Kozuchowska 7, 51-631 Wroclaw, [2]Department of Veterinary Prevention and Immunology, Grunwaldzki 7, 50-366 Wroclaw; Agricultural University of Wroclaw, Poland*

We analysed data from a selective DNA pooling experiment consisting of 120 unrelated individuals of blue fox (*Alopex lagopus*), which originated from two different types regarding body size. Association between allele frequency and body size was tested using uni- and multivariate logistic regression approach with Odds Ratio and various test statistics from power divergence family. Unfortunately, due to a small sample size and resulting sparseness of the data table, in hypotheses testing we could not rely on the asymptotic distributions of the tests. Instead, we accounted for data sparseness by (i) modifying confidence intervals of Odds Ratio and (ii) by using a normal approximation of the asymptotic distribution of the power divergence tests with different approaches for calculating moments of the statistics. As a result, significant association was observed for markers C03.629 and C05.771 representing dog chromosomes 3 and 5 respectively (map location in the blue fox genome is unknown). Furthermore, we used simulation to assess accuracy of the normal approximation of the asymptotic distribution of the test statistics for small, sparse samples.

Session 29 Theatre 5

The map expansion obtained with recombinant inbred strains and intermated recombinant inbred populations for finite generation designs

F. Teuscher and V. Guiard, Research Unit Genetics and Biometry, Research Institute for the Biology of Farm Animals (FBN) Dummerstorf, Germany

The generation of special crosses between different inbred lines such as recombinant inbred strains (RIS) and intermated recombinant inbred populations (IRIP) is being used to improve the power of QTL detection techniques, in particular fine mapping. These approaches acknowledge the fact that recombination of linked loci increases with every generation, caused by the accumulation of crossovers appearing between the loci at each meiosis. This leads to an expansion of the map distance between the loci. Recently it was postulated in the literature, that the map expansion factors of advanced designs depend on the distance of the considered loci. This was refuted here. It was proofed that map expansion factors, evaluated for small distances, are valid for all distances. While the amount of the map expansion of RIS and IRIP is known for infinite inbred generations, it is not known for finite numbers of generations. This gap was closed here. Since the recursive evaluation of the map expansion factors turned out to be longwinded, a useful approximation was derived.

Session 29 Theatre 6

Optimal haplotype size for combined linkage disequilibrium and co-segregation based fine mapping of quantitative trait loci

M.Z. Firat[1], H. Gilbert[2], R.L. Fernando[3], J.C.M. Dekkers[3], [1]Akdeniz University,Faculty of Agriculture, Antalya, Turkey, [2]Station de Genetique Quantitative et Appliquee, INRA, France, [3]Department of Animal Science, ISU, Ames, IA 50011, USA

Identity by descent (IBD) method to fine map QTL using the linkage disequilibrium (LD) has been extended to combine information from the co-segregation of alleles at linked loci. When data from a single generation were used, it was shown that using a haplotype of four markers resulted in the greatest mapping accuracy. The goal of this study is to determine the optimum haplotype size for fine mapping QTL when mapping is based on the combined method. Different designs with varying number of pedigreed generations and QTL effects are used to compare the ability of this technique. LD is generated by introducing a mutation at the QTL after 100 generations of random mating. Meuwissen and Goddard algorithm is used to compute IBD probabilities (IP) between founders conditional on marker haplotypes of varying sizes. The remaining IBD probabilities are calculated recursively based on conditional segregation probabilities (SP) given marker information. The SP depends on the haplotype size. The haplotype size that is optimum for IP may be different from the size that is optimum for SP. Here, we focus on determining the optimum size for IP. Thus, to simplify the simulation, SP were determined by observing recombination events between markers flanking the putative QTL.

Session 29 Theatre 7

A comparison of regression interval mapping and multiple interval mapping for linked QTL

M. Mayer, Research Instutute for the Biology of Farm Animals (FBN), Dummerstorf, Germany

Regression interval mapping and (maximum likelihood) multiple interval mapping are compared with regard to mapping linked QTL in inbred line cross experiments. A simulation study was performed using genetic models with two linked QTL. Data were simulated for F_2 populations of different size. The comparison includes the aspects of QTL detection and estimation of QTL positions. The criteria for comparison are power of QTL identification and the accuracy of the QTL position and effect estimates. Further the estimates of the relative QTL variance are assessed. There are distinct differences in the QTL position estimates between the two methods. Multiple interval mapping tends to be more powerful as compared to regression interval mapping. Multiple interval mapping further leads to more accurate QTL position and QTL effect estimates. The superiority increased with wider marker intervals and larger population sizes. If QTL are in repulsion the differences between the two methods are very pronounced. For both methods the reduction of the marker interval size from 10 to 5 cM increases power and greatly improves QTL parameter estimates. The use of asymptotic statistical theory for the computation of the standard errors of the QTL position and effect estimates proves to give much too optimistic standard errors for regression interval mapping as well as for multiple interval mapping. Extensions and implications for other non-maximum-likelihood estimation procedures are discussed.

Session 29 Theatre 8

The BovMAS Consortium: Analysis of BTA14 for QTL influencing milk yield, milk composition and health traits in the Italian Holstein-Friesian cattle breed

L. Fontanesi[1], E. Scotti[1], D. Pecorari[1], M. Dolezal[2], P. Zambonelli[1], S. Dall'Olio[1], D. Bigi[1], R. Davoli[1], E. Lipkin[3], M. Soller[3] and V. Russo[1], [1]DIPROVAL, Sezione di Allevamenti Zootecnici, University of Bologna, Reggio Emilia, Italy, [2]Department of Sustainable Agricultural Systems BOKU, University of Natural Resources and Applied Life Sciences, Vienna, Austria, [3]Department of Genetics, The Hebrew University of Jerusalem, Jerusalem, Israel*

Bovine chromosome 14 (BTA14) was scanned for QTL affecting milk yield, protein percentage and yield, fat percentage and yield and somatic cell count. Selective milk DNA pooling strategy was applied in a daughter design for eight Holstein-Friesian sires. For each sire-trait combination, milk pools were constructed comprising ~200 best or worst daughters ranked by their EBV or DYD. The sires were genotyped for 16 microsatellites covering BTA14, for the A232K and VNTR mutations at the *DGAT1* locus and for a SNP of the tyroglobulin gene. The milk pools were genotyped at the microsatellites heterozygous in their sire. Shadow corrected estimates of allele frequencies were calculated and differences in sire allele frequencies between high and low pools were computed. An adjusted false discovery rate was applied to calculate chromosome-wide significant levels and the data were analysed using an approximate interval mapping procedure. The results indicate the presence of one/more QTL at the centromeric end affecting most traits and one/more putative QTL in the middle-distal part influencing mainly protein percentage.

Session 29 Theatre 9

Bayesian analysis of selection with restrictions in beef cattle

L.M. Melucci[1], A.N. Birchmeier[2], E.P. Cappa[2] and, R.J.C. Cantet[2,3], [1]Unidad Integrada Balcarce (FCA, UNdMP - EEA, INTA), [2]Departamento de Producción Animal, FA-UBA, [3]CONICET, Argentina*

Birth (BW) and weaning weights (WW) were collected over a 34-yr period (1960-2003), from 1,921 Hereford calves, from an experimental herd of INTA Balcarce. The selection criterion emphasized on higher weaning weights and lower birth weights. In a first stage (1986-1993), phenotypic selection was performed, whereas from 1994 to 2003 selection was based on BLUP of breeding values (BV). The Bayesian analysis consisted on a Gibbs sampler for a two-trait animal model including maternal effects for BW and WW. Estimated heritabilities for direct BW, maternal BW, direct WW and maternal WW were 0.45, 0.33, 0.07 and 0.24, respectively. The direct-maternal additive correlation was negative (-0.05) for BW but positive (0.35) for WW. The additive correlation between BW and WW was -0.06 for direct effects and 0.50 for maternal effects. Selection response was measured as 1) regression of BLUP of BV on generation coefficients; 2) average BV per generation of selection. The two methods gave similar results: BV for direct effects of WW increased whereas direct BV for BW decreased, though maternal effects for both traits increased through time. The dynamics of (co)variance components with selection showed that direct variances decreased while maternal variances increased. The results show that selection to increase direct WW and to decrease direct BW was effective.

Session 29 Theatre 10

Selection of commercial boars and replacement sows for variable grid pricing systems

V.M.Quinton, J.W. Wilton and J.A.B. Robinson, CGIL, University of Guelph, Guelph, Ontario, Canada*

Market prices for livestock raised for meat such as fish, pigs and beef cattle, are usually determined under contract with a particular buyer, and depend on a price grid set by the buyer as well as the base price at the time the stock is sold. Producers need a flexible method of ranking terminal sires or dams for different price grids. The objective of this paper was to develop a method of calculating economic values for producers selecting commercial parents when carcass price per unit of weight is determined according to a grid and depends on overall carcass weight categories and possibly other traits. The method was illustrated for pig producers when prices are determined by both carcass weight and lean yield categories. Net revenue was defined in a single equation (profit function), and economic values were derived by evaluating the partial derivatives of the profit function at the commercial population mean. Economic values for this situation depend on the phenotypic variance as well as the mean of the carcass weight of the animals sold. Profit functions based on poorly defined grids may have local maxima, and producers should consider this situation, both when choosing a buyer for their stock, and in selecting parents of the commercial animals.

Session 29 Theatre 11

Genetic analysis of survival data from challenge testing of furunculosis in Atlantic salmon: Model comparison using field survival data

J. Ødegård[1], I. Olesen[2], B. Gjerde[2] and G. Klemetsdal[1], [1]Norwegian University of Life Sciences, Department of Animal and Aquacultural Sciences, P.O. Box 5003, N-1432 Ås, Norway, [2]AKVAFORSK (Institute of Aquaculture Research), P.O. Box 5010, N-1432 Ås, Norway

Challenge-test data of Atlantic salmon for survival to *Aeromonas salmonicida* infection were used in genetic analyses using four different sire-dam models, consisting of linear models, threshold models, and proportional hazards models. In these models challenge-test survival was defined as either a cross-sectional binary trait (survival during the testing period (30 days)), repeated measurements of a binary trait (test-day survival), or as time until death. The cross-sectional threshold model and proportional hazards model resulted in the highest heritabilities (0.59 and 0.63, respectively), while the lowest heritability was found for the linear model with repeated records of survival (0.02). The different models were ranked based on their ability to predict full-sib family survival to disease outbreak in field. All models had high correlations between predicted full-sib family effects (sum of additive genetic and common environmental effects) and full-sib family field survival, ranging 0.71 to 0.75. Predictive ability was best for the linear model with repeated records, followed by the cross-sectional threshold model, Weibull frailty model, and the cross-sectional linear model. The Weibull frailty model was unable to estimate a common environmental family effect in addition to additive genetic effects, which may explain the relatively low ranking of this model. However, a high Spearman correlation coefficient (0.98) was found between predicted breeding values from the Weibull model and the linear model with repeated records.

Session 29 Theatre 12

Genetic analysis of racing performance in Irish Greyhounds

H. Täubert, D. Agena, H.Simianer, Institute of Animal Breeding and Genetics, University of Göttingen, Albrecht-Thaer-Weg 3, D-37075 Göttingen, Germany*

Greyhound racing in Ireland became a large growing economy. In 2004, the prize money was 10.3 mio. and the total betting turnover amounted 139.6 mio. Over 1.385 million people attended Greyhound races last year. Animal breeding and selection is yet still based on phenotypic performances, genetic investigations were never done for racing speed. The goal of this investigation was to estimate genetic parameters for Irish Greyhounds and to predict breeding values in order to calculate genetic trends.

The collected data comprised 239'829 performances of 42'880 individual dogs. The races had a length of 480 meters and were collected over 4 years. Racing speed and standardized rank were analysed as traits. Genetic parameters were estimated with an animal model, taking race, age of dog and track into account. Estimated heritabilities were .307 for racing speed and .093 for rank. The genetic correlation was .991. It could be shown, that racing speed can be used as the only selection criterion, because of its high heritability and high genetic correlation to place.

An estimation of BLUP-breeding values showed a constant increase of EBVs for racing speed over the last years. This shows, that selection only on pure racing performance improved the genetics of the animals. It has to be analysed in future, if a BLUP-breeding value estimation can improve genetic progress to breed more successful racing dogs.

Session 29 Poster 13

Cumulative discounted expressions of dairy and beef traits in Ireland

D.P. Berry[1], F.E. Madalena[2], A.R. Cromie[3] and P.R. Amer[4], [1]Teagasc, Moorepark Production Research Center, Fermoy, Co. Cork, Ireland, [2]School of Veterinary Sciences, Federal University of Minas Gerais, Cx.P. 567, 30123-970, Belo Horizonte, MG, Brazil, [3]Irish Cattle Breeding Federation Society Ltd., Shinagh House, Bandon, Co. Cork, Ireland, [4]Abacus Biotech Limited, P.O. Box 5585, Dunedin, New Zealand*

Generic equations were derived using probability and transition matrices to track the flow of genes originating from a purebred or crossbred mating. Cumulative discounted expressions (CDE) were calculated for annual, replacement heifer, cull cow, birth, yearling, and slaughter traits using input parameters observed in the Irish population. Matrices used in the derivation of CDE described the rate and frequency of expression, genetic contribution among generations, a discounting factor, survival rates, the proportion of females becoming self-replacing or terminal, and the probability of crossbreeding occurring. The calculated CDE were sensitive to prevailing parameters especially the probability of a cow surviving from one age to the next. The CDE for annual, replacement heifer, cull cow, birth, yearling and slaughter traits were 0.89, 0.28, 0.19, 1.05, 0.66, 0.59 following an initial purebred mating and 0.24, 0.06, 0.04, 0.66, 0.45, 0.41 following an initial crossbred mating. The relative differences between CDE of trait categories highlights the necessity to investigate further the economic consequences of diverse breeding goals incorporating traits over and above those representing the principle intended use of the sire.

Session 29 Poster 14

Effectiveness of selection for lower somatic cell count (SCC) in herds with different levels of SCC

M.P.L. Calus, L.L.G. Janss, J.J. Windig, B. Beerda, and R.F. Veerkamp, Animal Sciences Group WUR, Division Animal Resources Development, The Netherlands*

Health risks for high producing dairy cows can partly be reduced by selection on health traits, such as SCC. In breeding value estimation, SCC is commonly log-transformed to somatic cell scores (SCS) for statistical reasons. Our objectives were to investigate selection responses for reduced SCC in herds with different levels of SCC and whether breeding values on the SCS scale fully represent those selection responses. Sires breeding values for SCS were estimated using 282,078 lactations of 3379 herds in the Netherlands. Differences in daughter average SCC, between the worst and best bulls for SCS, were respectively 39,600 and 65,800 cells/ml in the 25% of herds with lowest and highest SCC. This scaling effect indicates that selection on SCS is more effective in herds with high SCC, which can be interpreted as a form of genotype by environment interaction. On the SCS scale, ranges of daughter average performance were comparable in herds with different SCC levels, indicating that the log-transformation removed the scaling effect and consequently the information that explains the environment specific selection response. The hypothesis of this work is that reaction norm models, taking this genotype by environment interaction into account directly on the scale of SCC, may provide a better prediction of the true breeding values in different herds.

Session 29 Poster 15

Impact of management and recording quality on the success of community breeding programmes for smallholder dairying in the tropics

B. Zumbach and K.J. Peters, Humboldt University Berlin, Department of Animal Breeding in the Tropics and Subtropics, Philippstr. 13, Haus 9, D-10115 Berlin, Germany*

Improvement of dairy cattle productivity in the tropics could be achieved by the establishment of appropriate breeding programmes. The success of such programmes presupposes basic frame conditions like active involvement of farmers, links between the community of breeders with regional and national institutions and an effective but low cost breeding structure. Therefore, a young sire breeding scheme seems to be most appropriate. On the other hand, factors like management and recording quality may be crucial.

The objective of this paper is to quantify the impact of management and recording quality on genetic gain per year based on results from crossbred cattle in the tropics.

Calculations are based on a population of 1000 cows with natural mating (mating relation 1:40; useful life of bulls: 3 years; pre-selection factor: 2). Management is represented in survival rate of calves and calving interval. Given a survival rate of 80% and a calving interval of 18 months, 83 bull dams are required for the provision of breeding bulls while under bad conditions (survival rate 50%, calving interval 30 months) 222 are needed.

Accuracy implies an adequate recording quality where aspects like animal identification, appropriate recording of performance traits, pedigree and environmental effects and the mating structure play an important role.

Session 29 Poster 16

Genetic evaluation of mothering ability for multiple parities in Iberian pigs

*A. Fernández, J. Rodrigáñez, MC. Rodríguez and L. Silió **
Departamento de Mejora Genética Animal, INIA, Carretera La Coruña km. 7.5, 28040 Madrid, Spain

The usefulness of genetic progress for litter size in pig breeding programmes is limited by mothering ability and piglets survival. It is a critical aspect for breeds used as dam lines in a wide range of management sytems, in which litter weight (LW) at weaning could be an alternative selection goal. Our objective was to investigate the genetic basis of LW at 21days for the six first parities in 1,449 Iberian sows with available records from 2,709 litters without crossfostering. REML genetic parameters were estimated using an animal model with repeatability (RM) or a multi trait animal model (MT). Estimated parameters for LW21d using the RM model were: $h^2 = 0.13$ (SE = 0.02) and $p^2 = 0.12$ (SE = 0.02). Heterogeneity of heritabilities across parities was observed using the MT model, being the h^2 estimates for the first six parities 0.32, 0.08, 0.18, 0.15, 0.01 and 0.35 (SE from 0.01 to 0.04). The genetic correlations between parities ranged from 0.59 to 0.91, most of them were above 0.80. These results suggest that LW21d records from different parities should be considered as different traits. The application of random regression models to set up a system of genetic evaluation for this objective is also discussed.

Session 29 Poster 17

Characterization of four Ethiopian cattle breeds for typical phenotypic features, productivity and trypanotolerance

J. Stein[1], W. Ayalew[3], B. Malmfors[1], E. Rege[2], J. Philipsson[1], [1]Dept. of Animal Breeding and Genetics, Box 7023, S-750 07 Uppsala, Sweden, [2]ILRI, Box 30709, Nairobi, Kenya, [3]ILRI, Box 5689, Addis Ababa, Ethiopia*

Trypanosomosis is the most important livestock disease in the lower altitudes of Ethiopia and causes great losses of livestock and production. A potentially cost effective and sustainable method to reduce the problem of trypanosomosis is to utilise the genetic variation in trypano tolerance that is found among some livestock breeds indigenous to areas where the disease is endemic. In Ethiopia there are a number of cattle breeds that inhabit areas with heavy challenge of tsetse flies and therefore are at high risk of infection. It is suggested that these breeds might have some level of trypanotolerance. The aim was to elicit descriptive statistics about farmers' perception and knowledge regarding diseases, breeding and production. Four Ethiopian cattle breeds (Abigar, Gurage, Horro and Sheko) were included in an on-farm survey in each breeds' natural habitat. Three villages and 20 farmers per village were visited in each habitat. The peak challenge period, in which blood samples are taken, ranges from April to July in the different areas. Farmers' perception of the importance of trypanosomosis varied in the different regions. Most farmers practiced herd mating and paternal pedigrees were often unknown.

Session 29 Poster 18

Systems of breeding in Algeria: transformation and tendencies of evolution. Case of the area of Sétif

Khaled Abbas, INRA Algérie, Unité de Sétif, Route des fermes, 19000 Sétif, Algeria

Algeria has 8 million ha of SAU (Useful Agricultural Surface) of which 443 000 ha are irrigated. With population of more than 30 million, the SAU/population ratio is very weak (0.25 ha per capita). Agriculture is concentrated in a narrow fringe of the north of the country where dominates a semi-arid climate and a very irregular rainfall. More in the south, a steppe area of more than 30 million ha is devoted mainly to small ruminants production. A large majority of the farms associate animal production and cereals and draw a major part of their incomes. The diversity of the climate and the physical environment, on the one hand, and the agrarian policies on the other hand, induced major transformations in the systems of production and the systems of breeding. These transformations are also the result of a significant population increase in rural area what generated an increase in the food needs. In this context appear significant stakes on the use of the resources and the production systems durability. This study, based on investigations of 200 owners of the area of Sétif (east Algeria), shows that: - the intensification can constitute a threat on the pastorales resources; - the brittleness of the farms encourages generalization of practices marking a weak autonomy of the breeding and a bad stock management; - the animal interspecific integration and partial intensification of breeding systems are positive if they follow models which ensure autonomy, performance and optimal management of the resources.

Session 30 Theatre 1

Improving resource use in ruminant systems

J.A. Milne, Macaulay Institute, Craigiebuckler, Aberdeen, United Kingdom

The objectives of ruminant systems are becoming more diverse with dairy systems becoming more efficiency-driven and intensive within the context of a number of external constraints whilst meat-producing systems are becoming more multi-objective and often more extensive. Whilst this divergence has meant that some nutrition and management strategies have also become different, this has not been universal. Dairy systems require increased precision in the nutritive value of feeds and their combinations and the efficiency with which nutrients are utilized. Meat systems from ruminants often seek to use feed by-products more effectively and utilize alternative feeds. In terms of management strategies, diary systems are developing management strategies which minimize the impact of external constraints, such as environmental legislation, whilst meat systems may seek to alter management to obtain the benefits that can arise from environmental management. However, there are common technological advances that all ruminant systems seek to use. Examples of these are the increase in the understanding of the interaction between nutrition, management and disease, and the role of dietary manipulation in providing animal products to meet consumer demands. In dairy systems, laminitis, and in meat systems, intestinal parasites, are examples where the role of nutrition and management strategies are providing options in the control of disease. The manipulation of conjugated linoleic acids in milk and meat through the diet also has potential. Future opportunities for nutrition and management strategies to improve resource efficiency are indicated.

Session 30 Theatre 2

Nutrition and animal management as part of a global strategy for reducing the environmental impact of pig production

M. Bonneau[1], J.Y. Dourmad[1], C. Jondreville[1], P. Robin[2], H. van der Werf[2] and P. Leterme[2], INRA-Agrocampus Rennes [1]SENAH, 35590 Saint Gilles,[2]SAS, 35042 Rennes cedex, France*

In areas with high animal densities, manure production exceeds the fertiliser requirements of the surrounding arable land. Pollution load can be decreased by i) reducing the amount produced by the animals via nutrition or husbandry, ii) processing manure to decrease its pollutant content, iii) optimising the application of manure to the land. Fine tuning of amount fed to animal requirements, as well as the use of fast growing, lean, animals reduces the output of nitrogen and phosphorus. The use of litter instead of slatted floor has a large effect of both the quantity and the form of nitrogen outputs. The output of zinc and copper can be reduced when their use at supra nutritional levels is limited to relevant periods. Microbial phytase improves the availability of phosphorus and zinc and therefore contributes to a reduction of their outputs. At farm level, modelling can contribute to a co-optimisation of the production of nitrogen/phosphorus in animal manure and its use on arable land for crop fertilisation. Modelling at a higher integration level can also be useful for optimising the use of animal manure for crop fertilisation between farms, within a geographical area. Integrated approaches, such as Life Cycle Assessment (LCA), can be used to compare scenarios proposed to reduce pollution load, identify their environmental hot spots and propose options for improvement. LCA can be particularly useful to identify cases of "problem shifting", e.g. reducing a local impact (NO_3 emission) at the cost of an increased global impact (N_2O emission).

Session 30 Theatre 3

Impact of alternative dairy systems on greenhouse gas emission

A.H. Fredeen, M. Main, S. Juurlink, S. Cooper and R. Martin, Department of Plant and Animal Sciences, Nova Scotia Agricultural College, P.O. Box 550, Truro, Nova Scotia, B2N 5E3 Canada*

Environmental impacts of pasture and confinement dairy production systems in Atlantic Canada were evaluated. Effects on greenhouse gas(GHG) emissions will be reported. Respiration chambers and the SF_6 tracer method were employed to isolate effects of feeding a total mixed ration (TMR) based on alfalfa- grass silage or forage from a predominantly grass pasture under Management Intensive Grazing (MIG) on enteric CH_4 emission. Tents were employed over urine patches on pasture to determine N_2O emission. Milk yield was not affected by diet. Emission of CH_4) averaged 16.6 vs. 17.0, and 13.5 vs. 11.5 g CH_4 kg^{-1} milk (MIG vs. TMR) using the chamber and tracer techniques respectively. The MIG system used less grain kg^{-1} milk. N_2O emission was temporarily elevated from urine affected patches created by feeding high N diets, and represented 1 to 3% of total GHG impacts of milk production. Life cycle GHG emissions predicted using the Atlantic Dairy Sustainability Model were 1.02 and 1.17 kg CO_2 equivalents kg^{-1} milk for MIG and TMR-based farms. MIG systems can reduce GHG impact of dairying relative to confinement systems.

Session 30 Theatre 4

Effect of group size on feed intake and growth rate in kids and lambs

D.T.T. Van[*1] and I. Ledin*[2], [1]*Goat and Rabbit Research Centre, Hatay, Vietnam,* [2]*Department of Animal Nutrition and Management, SLU, Uppsala, Sweden*

The effect of group sizes of 1 to 5 animals per pen on performance of kids and lambs was studied in an experiment with 30 kids and 30 lambs in three periods. In a second experiment 36 kids and 36 lambs were allocated to 6 pens with 1 or 5 animals per pen. The animals were fed fresh foliages, sugar cane and a commercial concentrate. The space allowance was 0.73 m^2 per animal.
In Exp.1, kids had a significantly higher daily dry matter intake (DMI) than sheep, 133 g and 126 g/kg$W^{0.75}$, respectively. Group size 1 had significantly lower daily intake compared to group sizes 3 to 5, 96 g and 144 g/kg$W^{0.75}$, respectively. In Exp.2 the group size of 5 animals had significantly higher DMI than group size 1, 102 g and 67 g/kg$W^{0.75}$, respectively. The daily weight gain was higher for lambs than for kids, 89 g and 63 g, respectively, but similar for the two group sizes 1 and 5, 76 g and 77 g, respectively. The feed conversion ratio was significantly lower for lambs compared to kids, 8.8 kg and 12.0 kg DM/kg live weight gain, respectively, and significantly higher for group size 5 than for group size 1, 11.1 kg and 9.80 kg DM/kg live weight gain, respectively.

Session 30 Theatre 5

Effect of feeding different levels of foliage from *Moringa oleifera* to creole dairy cows on intake, digestibility, milk production and composition

**N.R. Sánchez*[1] *and Inger Ledin*[2], [1]*Faculty of Animal Science, Universidad Nacional Agraria, Managua, Nicaragu,* [2]*Department of Animal Nutrition and Management, Swedish University of Agricultural Sciences, Uppsala, Sweden*

An experiment was conducted in Nicaragua to determine the effect of feeding foliage from *Moringa oleifera* Lam to dairy cows on intake, digestibility, milk production and milk composition. The treatments were: *Brachiaria brizantha* hay *ad libitum*, either unsupplemented or supplemented with 2 kg or 3 kg of Moringa on a dry matter (DM) basis. Six cows of the Creole Reyna breed, with a mean body weight of 394 ±24 kg were used in a replicated 3x3 Latin square design. Milk production was recorded and sampled during the last two weeks of each 5 weeks experimental period and digestibility was estimated during the last week of each experimental period. Supplementation with Moringa increased DM intake from 8.5 to 10.2 and 11.0 kg DM day^{-1} and milk production from 3.1 to 4.9 and 5.1 kg day^{-1} for *B. brizantha* hay only and supplementation with 2 kg and 3 kg DM of Moringa, respectively. Milk fat, total solids and crude protein and organoleptic characteristics, smell, taste and colour, was not significantly different between the diets. The apparent digestibility coefficients of DM, OM, CP, NDF and ADF increased ($P<0.05$) in the diets supplemented with Moringa compared with *B. brizantha hay* alone.

Session 30 Theatre 6

Effect of milking frequency and nutritional level on milk production characteristics and reproductive performance of dairy cows

B. O'Brien, D. Gleeson and J.F. Mee, Teagasc, Dairy Production Research Centre, Moorepark, Fermoy, Co. Cork, Ireland*

This study investigated the effect of milking frequency at two nutritional levels (NL) on milk production and reproductive performance.

Sixty spring-calving, pluriparous Holstein-Friesian cows were assigned to treatments after calving (mean=11 March); twice a day milking (TAD) on a high (TH) or low (TL) NL; once a day milking (OAD) on a high (OH) or low (OL) NL. NL was defined by concentrate offered (420 kg or 135 kg) and post-grazing sward height (75 or 55 mm). Milk yield and composition and cow live-weight were recorded. Milk samples (thrice-weekly) were analysed for progesterone to determine the commencement of luteal activity (CLA). Data were analysed using the PROC Mixed procedure in SAS.

OAD milking and a low NL reduced milk yield ($P<0.001$) (15.0 kg/cow per d and 16.1 kg/cow per d) compared to TAD milking and a high NL (20.2 kg/cow per d and 19.1 kg/cow per d). Fat and protein contents of milk were increased ($P<0.001$) with OAD (4.41g/100g and 3.65 g/100g) compared to TAD milking (4.09g/100g and 3.38 g/100g). Live-weight loss was reduced with OAD milking and the high NL ($P<0.01$). OAD cows had better reproductive performance (earlier CLA, higher pregnancy rate) than TAD cows.

This study demonstrated that OAD milking provides a viable management option that may improve labour output within a dairy enterprise.

Session 30 Theatre 7

Effect of milking frequency and nutrition on cow welfare

L. Boyle, B. O'Brien and D. Gleeson, Teagasc, Moorepark Research Centre, Fermoy, Co. Cork, Ireland*

The aim of this study was to assess the effects of milking frequency and nutritional level (NL) on cow welfare. Sixty spring-calving multiparous Holstein-Friesian cows were assigned after calving to four treatments in a 2x2 factorial design: Twice a day (TAD) milking on a high (TH) or low (TL) NL; once a day (OAD) milking on a high (OH) or low (OL) NL. NL was defined by concentrate offered (420 and 135kg) and post-grazing height (75 and 55mm). Lesions to the soles of both hind feet and locomotory ability were scored throughout lactation. Lameness was also recorded. Behaviour was recorded during two 24hr periods in May, July and September. Data were analysed by SAS using mixed and non-parametric procedures as appropriate. The incidence of lameness was 33%, 13%, 13% and 0% in the TH, TL, OH and OL treatments, respectively. OAD cows had higher locomotion scores in early lactation ($P<0.001$) but lower sole lesion scores towards the end of lactation ($P<0.05$) compared to TAD cows. OAD cows spent less time grazing and ruminating but more time idling while lying and engaged in behaviours such as grooming ($P<0.05$). OAD milking had positive implications for hoof health and animal behaviour. However, locomotory disorders in OAD cows in early lactation indicate that they experienced discomfort at this time. The high incidence of lameness in TH cows poses a serious welfare concern.

Session 30 Theatre 8

Dealcoholized beer replacement for water in poultry as a novel alternative approach

S.S. Parlat, I. Yildirim, Selcuk University, Agricultural Faculty, Department of Animal Science, 42031 Konya, Turkey

The study was conducted to evaluate the effects of dealcoholized beer *(Dealcoholized beer is different from alcohol-free beer.)* replacement for water on the growth performance parameters of mixed sexes Japanese quail from 10 to 45 d of age (5 weeks). A total of 40 Japanese quail chicks were divided into 2 experimental groups: Control group (water) or treatment group (dealcoholized beer) each consisting placed individually pens 20 chicks. Basal diet was formulated to meet NRC (1994) nutrient requirements. Beer containing 4% alcohol (v/v) was dealcoholized by HCl acid treatment. The diet, water and dealcoholized beer were available *ad libitum* and lighting was continuous. The data were subjected to analysis of variance, and differences between the groups were analized by t-test. Any deleterious effects of dealcoholized beer on performance parameters were not observed. Dealcoholized beer treatment compared with the control group significantly increased food consumption, body weight gain and resulted in improved food conversion ratio of Japanese quail. Food consumption increased by 15% in quail chicks consuming dealcoholized beer compared with the control groups. Similarly, overall body weight gain and food conversion ratio improved by 22% and 17% in dealcoholized beer treatment compared with the control group, respectively. No mortality was observed in any groups. These results suggest that dealcoholized beer replacement for water in Japanese quail as model animal for poultry effectively improved performance parameters such as food intake, body weight gain and food conversion ratio.

Session 30 Theatre 9

Replacing barley with soy hulls in an automatic milking system

I. Halachmi[1], E. Shoshani[2], R. Salomon[2], E. Maltz [1],J. Miron[1], [1]Agricultural Research Organization (A.R.O.), [2]Extension Service, Bet Dagan, Israel

More milkings mean more intake of concentrate feed while the cow is being milked in the robot. Consumption of a high quantity of starchy grains (barely) within a short period of time could have a negative effect on the cow's appetite, voluntarily milkings and NDF digestibility, leading to further reduction in the voluntary DM and NDF intake.

The potential of soy hulls as a replacement for barley grain in pellets fed as a supplement to lactating cows was measured. Fifty-four cows were divided into two equal groups and fed individually for 3 months one of the two experimental pellet supplements. Both diets were group fed as a basic mixture along the feeding lane, and a pelleted additive containing either barley (B) or soy hulls (S) as a barley replacer was fed individually to each cow via the concentrate feeder in the milking robot or in the self-feeder accessible only after milking.

The B and S diets resulted in a similar number of voluntarily milkings (3.31-3.39 milkings/cow/day), and consequently milk yield and composition were significantly higher (t-Test α=0.05). In the B group the 4% FCM was 31.50 (std=2.03) kg vs. FCM= 33.45 (2.05) kg in the S group.

By-product feeds such as soy hulls are often favourably priced and, as investigated in this study, may reduce the inhibitory effect of the starchy pellets on NDF intake and digestion.

Session 30 Theatre 10

Development of a deterministic model to create a dairy herd

J.I. Nousiainen, L. Jauhiainen, M. Toivakka and Pekka Huhtanen, Agrifood Research Finland, FIN-31600 Jokioinen, Finland*

Dairy herds are complex systems and in the situations, where expensive experimental management studies are difficult to conduct simulations would be a feasible tool. Most common herd simulations have been dynamic stochastic models. Variation in the life cycle of dairy animals have been introduced by altering the factors that affect e.g. culling criteria. A deterministic dynamic herd creation model utilizing the data of culled cows from Finnish dairy herds participating in the national milk recording scheme is described. A iterative macro operating in SAS was developed and data determining the initial herd size was formulated. The macro sorts the initial herd by the culling date and samples a cow from the animal database using the "random sampling with replacement"- method. Then a new cow is created, based on the information of the randomly selected cow. Its first calving date is one day after the culling date of the removed animal. The new cow data is added to the initial data and the culled cow information is deleted. The process is continued until the predetermined moment is reached. The final data contains birth, calving and culling dates. Thus a herd with dairy cows and necessary young stock for a certain time period can be created. The variation in the herd is restricted to the one present in the original recording data.

Session 30 Poster 11

Performance of imported dairy cows under Libyan hot weather conditions

B.M. Belgasem[1] and A.B. El-Magdub[2], [1]University of Omer Mukhtar, Libya, [2]university of El-Fateh, Libya

A dairy cattle farm was used to study the productive and reproductive performance of Eurpean imported dairy cows during first two lactations. Four experiments were conducted.

Exp.I: designed to study the effect of feeding frequency on production and reproduction effeciency. Cows were divided into 2 groups, groupA (250cows) received their daily concenterate allowances during AM and PM milking times .Group B (270cows) received their daily allowances 3 times /day during AM milking , shortly after AM milking and during PM milking.

Exp.II: designed to study the effect of breed origin on the productive and reproductive performance. Group A from Denemark origin (270 cows), group B from Germany origin were used in this expermint.

Exp.III: designed to study the effect of dryng off methods on the subsequent productive and reproductive performance

Exp. IV: Designed to expose the actual millk records to predict validity of using DHIA factors to correct incomplete records under Libyan conditions.

The results indicated that cows fed 3X showed higher milk yield. The breed of Denemark origin showed better performance than the German. No differences were observed in the dry off methods .The use of DHIA factors can be used efficiently after 60 days of lactation.

Session 30 Poster 12

Efficiency of using grape marc in rations of Rahmani lambs

I.M. Awadalla, Animal Production Department, National Research Center, Dokki, Gizza, Egypt

Twenty one Ramani male lambs with an average body weight (BW) of 26.1 ±1.01 kg and 5 months old were divided into three equal groups to be fed on a) clover hay (G1, Control), b) 1:1. Mixture of clover hay and grape marc (G2) and c) grape marc only (G3) in addition to concentrate feed mixture at basis of 2% of their BW. Animals were weighed bi-weekly to determine growth performance and daily feed intake was estimated to calculate the feed conversion. At the end of experiment a digestion trail was conducted on three animals from each group. Data were analyzed using the general liner model procedures (SAS, 1996). The final BW was (47.4, 43.3 and 42.8 kg for G1, G2, and G3, resp.) Feeding on clover hay (G1) showed significantly ($p \leq 0.05$) higher total gain (21.4 kg), relative growth rate (43.5%) and average daily gain (191.3 g) compared to (17.1 kg, 38.1% and 152.4g in G2) and (16.7 kg, 37.2% and 149.2g in G3). This excellence of G1 coincided with significant ($p \leq 0.05$) better feed conversion (7.3 kg DM/kg gain in G1 *vs* 8.5 and 8.7 kg in G2 and G3, resp.). Daily feed intake inG1 was about 100 gm DM more than G2 and G3. This leads to a higher DM and DCP intake. It is concluded that, grape marc can used in rations of Rahmani lambs instead of clover hay by about 50%.

Session 30 Poster 13

The effect of yeast culture (*Saccharomyces cervisiae*) on digestibility of sugar cane bagasse in sheep

F. Kafilzadeh and Ali Paryad, Razi University, Animal Sci., College of Agric. Kermanshah, 67155 Kermanshah, Iran

Sugar cane baggasse (SCB) is produced over 600,000 tons in year in Iran. As a feedstuff this by product is poorly utilized by animals and results in an unsatisfactory performance of animals. The purpose of this study was to determine the effect of yeast culture (Saccharomyces Cervisiae) on DM, OM, CF,CP digestibilty of a pelleted sugar cane bagasse containing 1% urea and 30 percent Molasses. A completely randomized design experiment was conducted, using 5 mature rams receiving SCB either alone or with 3gr yeast per head daily. DM digestibility increased from 36.11 %to 42.29% with addition of YC ($p<0.05$). Mean OM digestibility was higher ($p<0.05$) in YC treatment (47.74%) compared with the control (37.43%). Digestibility of components of OM (CF, EE, and CP) were not improved significantly as the result of YC addition. Feeding YC resulted in a significantly ($p<0.05$) higher digestible energy of SCB (1.59Mcal/kgDM vs. 1.72Mcal/kgDM).

Session 30 Poster 14

Effects of substitution of cottonseed meal by canola meal on milk yield and apparent digestibility of dry matter, organic matter and crud protein in diets of dairy cow

G.R. Ghorbani, S.M. Masumi, Animal Science Department, Isfahan University of Technology, Isfahan, Iran.

A duplicated 3 × 3 Latin square trial (21-d periods) was conducted to determine the effect of the source of protein in diets based on alfalfa hay on milk production and composition. Cottonseed meal (CSM) and canola meal (CM) was substituted for supplemental protein portion. Six Holstein cows averaging 34 kg/d milk production and BW 550 ± 50 kg were used in this trial. Treatments Were: 1) canola meal (100%), 2) cottonseed meal and canola meal (50%, 50%), and 3) cottonseed meal (100%). Treatment 1 in compare to treatments 2 and 3 showed a higher milk protein percentage, (3.08 vs. 2.98 and 3.01), apparent digestibility of organic matter (77 vs. 72.8 and 72.5), dry matter (75.83 vs. 70.42 and 70.67) and crude protein (78.33 vs. 73.17 and 74.5) ($p < 0.05$). However, milk yield, milk fat, fat yield, milk protein yield, milk lactose, DMI, feed efficiency, digestibility of NDF, ADF, concentration of T_3, T_4 and Albumin were not affected by the treatments for cows fed the various protein supplements. It is concluded, that substituting of CSM by CM increase the digestibility of organic matter, dry matter and crude protein and milk protein percentage.

Session 30 Poster 15

Effects of short dry periods on milk yield of Holstein dairy cattle

G.R. Ghorbani, A. Pezeshki, and H.R. Rahmani, College of Agriculture, Isfahan University of Technology, Isfahan, Iran 84156

One-hundred twenty Holstein cows assigned to evaluate the effects of shortened dry periods on milk yield of cows with different BCS and milk yields. Treatments were arranged in a 3×2×2 factorial design that included dry period (35, 42 & 56 d) BCS (BCS<3.2 and BCS≥3.2) and milk yield (yield ≥20 and yield<20). Milk yields and compositions were recorded until 8 wk of postpartum and the data were analyzed with proc mixed procedure of SAS. There was no difference between 35 and 56 days dry (DD) on milk yield (36.51 vs. 38.87; p>0.14). Milk yield for cows with 56 DD was more than 42 DD (38.87 vs. 34.91; p<0.01). No significant differences due to 35 and 56 DD were detected for high BCS (37.44 vs. 41.5; p >0.08) and low BCS cows (35.58 vs. 36.24). There was significant difference between 35 and 56 DD on high producing cows (37.15 vs. 44.43; p<0.003). Milk yield for low producing cows with different DD were not different. There was a significant difference between 35 and 56 DD on high BCS-high producing cows (46.45 vs. 36.8). No significant differences due to DD were detected for milk protein, fat, lactose and SCC yields. Short dry periods have more detrimental effects on high-Producing cows however; net milk income by shortening the dry period may compensate it.

Session 30 Poster 16

Influence of phosphorus level and soaking on phosphorus availability and performance in growing-finishing pigs

K. Lyberg, A. Simonsson and J.E. Lindberg, Swedish University of Agricultural Sciences (SLU), Box 7024, 750 07 Uppsala, Sweden

The effects of one-hour soaking of a pig diet based on wheat and barley with low (4.1 g P/kg) and high (6.8 g P/kg) total phosphorus content on total tract apparent digestibility and performance in growing-finishing pigs were studied. The results showed that, soaking had an increasing affect on the apparent digestibility of P. The digestibility of P did not differ between the high P dry and the low P soaked. The average daily weight gain, final body weight and carcass weight were lower ($p<0.001$), and the energy conversion ratio was higher ($p<0.001$) in the low P dry treatment than in the other treatments. Pigs on the low P diets had lower levels of inorganic P in serum ($p<0.001$) and slightly higher serum Ca values, and there were no effects of soaking. The density of femur was lower ($p<0.01$) in the low P treatments than in the high P treatments, and soaking of the low P diet improved ($p<0.01$) femur density. In conclusion, a one-hour soaking of pig feed in water appears to be sufficient to improve P availability and growth performance.

Session 30 Poster 17

Effect of ruminant livestock systems on grassland condition in Patagonia, Argentina

S. Villagra[1,2], C.B.A. Wollny[1], C.Giraudo[2] and G. Siffredi[2], [1]Georg-August Universitaet Goettingen, Institute of Animal Breeding and Genetics, Animal Breeding and Production in the Tropics, Kellnerweg 6, 37077, Goettingen, Germany, [2]Instituto Nacional de Tecnología Agropecuaria (INTA) C.C. 227. 8400, Bariloche, Argentina

Since the beginning of last century when Europeans settled Patagonia and started the sheep industry, overgrazing and erosion has been seen as an indicator of non-sustainable management. For economic reasons smallholder farmers are forced to exploit the natural grassland to the maximum. Since it is known that mixed grazing increases the stability of the ecosystems, a study was carried out to compare the status of the grassland between farms keeping only sheep (SF) and farms keeping mixed flocks of sheep, goats and cattle (MF). Six indicators were selected for comparison, namely: vegetation cover, forage production, grassland condition, grassland trend, wind erosion and water erosion. Lansat - TM satellite imagines geo-referenced and rectified were used to classify different vegetation types of 16 MF and 6 SF farms. In the field 32 and 97 vegetation census were carried out in SF and MF farms, respectively. No significant differences in all six indicators of grassland status were found between systems. However, the ruminant stocking rate MF was significantly ($p<0.05$) higher than SF. Concluding, MF support higher stocking rate without showing a higher degree of grassland degradation than SF.

Effect of two diets on the growth of the Helix aspersa Müller during the juvenile stage

A. García[1], J. Perea[1], R. Martín[2], R. Acero[1], A. Mayoral[2] and M. Luque[1], Department of Animal Production, University of Córdoba (UCO), Spain, [2]Agricultural Research Institute of Andalucia (IFAPA), Spain

We have been studied the effect of two diets one the growth of the *Helix aspersa* Müller during the juvenile stage under laboratory conditions. Diet I consists of commercial layer's mash. We used ten lots of forty animals of 0.34 ± 0.05 g, assigning five to each diet and keep them during the 6 first weeks of life in the laboratory conditions described by Perea *et al.* (2003) and García *et al.* (2004). The results show that the diet with forage (II) presents low growth rates (81 mg), high variability (24.56%) and asymmetrical distribution of individual (the hypothesis of normality is rejected). On the other hand the diet with commercial feed (I) shows high growth (955 mg), low variability (13.50%) and a normal distribution of the weight.
The diet II with forage does not show a relationship between the weight and the age. In the case of the diet with commercial feed is possible consider a regression model between the weight and the age with a coefficient of determination of 90% and a level of signification of P<0.0001. Also we find significant differences (P<0.001) in the conversion index, being higher in the diet I, with a lower mortality.

Effect of milking frequency and nutritional level on milking characteristics and teat condition of dairy cows

D. Gleeson, B. O'Brien and L. Boyle, Teagasc, Dairy Production Research Centre, Moorepark, Fermoy, Co. Cork, Ireland*

This study investigated the effect of milking frequency (MF) at two nutritional levels (NL) on milking characteristics and teat condition.
Sixty spring-calving, pluriparous Holstein-Friesian cows were assigned to treatments after calving; twice a day milking (TAD) on a high or low NL; once a day milking (OAD) on a high or low NL. NL was defined by concentrate offered (420 kg or 135 kg) and post-grazing sward height (75 or 55 mm). Milk yield, cluster-on time, maximum flow-rate, average flow-rate and time to milk flow were recorded daily. Teats were classified for teat-end hyperkeratosis (HK) monthly on a severity scale from 1 to 5. Milking characteristic and HK data were analysed using the PROC Mixed procedure in SAS and a non-parametric test, respectively.
TAD milking increased daily cluster-on time (P<0.001) (688 sec/cow per d) compared to OAD milking (410 sec/cow per d). MF or NL did not significantly affect cluster-on time, maximum flow-rate, average flow-rate or time to milk flow at the morning milking. However, milk yield for OAD cows was higher (P<0.01) compared to TAD cows at this milking. A significant interaction with stage of lactation was observed for all milking characteristics (P<0.001). In conclusion, MF and NL had no effect on milking rate or on teat condition as measured by HK.

Comparison of pelleted vs mash-feed form administration in nursery-finisher pigs

J. Morales, L.M. Ramírez and C. Piñeiro, PigCHAMP Pro Europa, S.A., Segovia, Spain*

Pelleting of feed presented many advantages in pig production, including easy management, better digestibility, increased intake and improved efficiency. However, pelleting increases the cost of feed manufacturing and the process may have negative effects on certain micro-ingredients stability. Moreover, farmer's perception is that feed presented as mash diminishes the incidence of enteric disorders. Three studies in different locations were performed to compare the feed administration in pellets or mash form. Two studies were conducted during the nursery period (28 to 63 d of age) and one during the growing-finishing phase (80 to 148 d). Feed intake (FI), average daily gain (ADG) and feed:gain ratio (FGR) were controlled every two weeks. In the nursery period, feed as mash had worse FGR than feed as pellets (1.47 vs 1.35 g/g; $P<0.05$), but no differences were observed for FI, suggesting that differences were due to an increase in feed wastage. However, at the immediate post-weaning phase, mash promoted higher ADG and FI than pellets, although only significantly in one of the experimental farms. In the growing period, pellets promoted higher ADG due to better FGR and higher FI. A reduction in FGR represented a reduction in production cost of about 6% throughout the productive life of the pig. As conclusion, pelleting of feed improve the economics of pig feeding throughout the productive cycle, except during the week immediately after post-weaning.

Performance and carcass quality of broilers fed diets based on three fibre types

E.A. Iyayi[1,2] O. Ogunsola[2] and R.A. Ijaiya[2], [1]Institut für Enährungswissenschaften, Martin-Luther-Universität, 06108 Halle, Germany, [2]Department of Animal Science, University of Ibadan, Ibadan, Nigeria*

One hundred and twenty 1-d-old chicks were allocated to 4 diets - 1 Basal and 3 diets each containing 40% of corn bran (CB), brewer's dried grain (BDG) and palm kernel meal (PKM) with 5 replicates of 6 birds each. They were fed the test diets from the 5th to the 12th week. Birds were withdrawn for slaughter at the end of the 4th, 8th and 12th week for carcass measures and quality. Period but not dietary treatment significantly ($p<0.05$) affected the feed intake of the birds. Weight gain, FCR, nutrient digestibility, cooking loss and feed cost were significantly ($p<0.05$) reduced by the CB, BDG, PKM-based diets and by period. Dietary treatment, period and their interaction significantly ($p<0.05$) affected the carcass measures and organ relative weight of the birds, produced significantly ($p<0.05$) less tender meat and meat with higher value of shearforce. Results suggest that inclusion of CB, BDG and PKM up to 40% in diets of broiler finishers reduces the rate of weight gain when birds are fed for long periods but with a reduction in feed cost. The meat from such birds have higher value of shearforce, are less tender but with reduced cooking loss.

Session 30 Poster 22

Dynamics of pastures and fodder crops for Mirandesa cattle breed

L. Galvão[1], O.C. Moreira[2], R. Valentim[1], J. Ramalho Ribeiro[2], V. Alves[3], [1]ESAB-Escola Superior Agrária de Bragança, 5301-855 Bragança, Portugal, [2]INIAP- Estação Zootécnica Nacional, 2005-048, Vale de Santarém, Portugal, [3]UTAD-Universidade de Trás-os-Montes e Alto Douro, 5000-911, Vila Real, Portugal

Mirandesa cattle are a local breed from the Northeast region of Portugal, playing an important role on the maintenance of the rural spaces, contributing to the fixation of the populations and to the environmental preservation. The aim of this study is to characterise the feed resources available along the year in this farming system.

The animals graze natural pastures in Spring and beginning of Summer, being after fed with hays (of natural pasture or oat) and straws (oat, barley or wheat) and complemented with local feeds like squash or potatoes.

Samples of feeds were taken from three different farms in two consecutive years and analysed for crude protein (CP), cell wall components, minerals and *in vitro* organic matter digestibility (OMD). Data were evaluated using the ANOVA statistical approach. Seasonal variations were observed in natural pastures with decreases of CP from Spring to Summer (16.0 to 9.4 % DM) and increases of NDF content from 32 to 41% DM. Consequently a reduction of OMD from 69 to 58 % was observed. Regarding hays composition, differences were observed for CP content which was lower for oat hay, compared with that from natural pasture (2.4 vs. 4.6 % DM). Straw quality varied between farms and type of straw.

Session 30 Poster 23

Reduction of nitrogen environmental impact by diet manipulation in swine

O. Moreira[1], O. Oliveira[1], J. Martins Santos[1], M.A. Castelo Branco[1], F. Calouro[1], S. Sousa[2], A.S. Monteiro[3], J.R. Ribeiro[1], [1]INIAP- Estação Zootécnica Nacional, 2005-048 Vale de Santarém, Portugal, [2]ISQ, Av. Prof. Dr. Cavaco Silva, 33, 2780-920, Porto Salvo, Portugal, [3]FPAS, Av. António Augusto de Aguiar, 179, r/c Esq, 1050-014, Lisboa, Portugal

Diet manipulation became an alternative to reduce nitrogen excretion and the negative impact of wastes to the environment. The purpose of this experiment was to quantify the effects of reducing the protein content of pig diets on the excretion of nitrogenous compounds.

Two experimental diets with 18 (diet T) and 15 (diet N) % crude protein were studied in metabolic trials with growing/finishing pigs (35-100 kg). Digestibility and balance of dietary N, as well as the fractionation of N excretion were evaluated.

N digestibility was near 88% in the two diets. For diet N, nitrogen intake was 19 g/day less, but N retentions were 7% higher ($P<0.05$). Faecal N excretion was 12% of the intake with both diets. Differences were observed in urinary excretion ($P<0.05$). Regarding NH_3 excretion (g/animal/day) the observed values were 8.36 (3.05 kg/year) for diet T and 2.65 (0.97 kg/year) for diet N, meaning a reduction of 2.08 kg/animal/year with diet N. Transposing these value to the Portuguese indigenous swine population, a total reduction of NH_3 emission of 2 715 ton/year was estimated. It was concluded that a reduction of dietary N would be a valuable proposal to minimize the environmental impacts resulting from traditional production systems.

Session 31 — Theatre 1

Swedish horse breeding and sport

C. Olsson[*1], D.-A. Danielsson[2] and J. Philipsson[3]. [1]Swedish Trotting Association, S-161 89 Stockholm, Sweden, [2]Swedish Horse Board, Herrskogsvägen 2, S-730 40 Kolbäck, Sweden, [3]Department of Animal Breeding and Genetics, Swedish University of Agricultural Sciences, S-750 07 Uppsala, Sweden*

The Swedish horse population has increased in the last three decades from a low 70 thousand horses to a present number of approximately 275 thousand horses. The breed structure has changed completely from primarily coldblood draft horses to sport horses. That reflects the strongly increased interest among people for various horse sports. The most common breeds today are Standardbred trotters and Swedish warmblood horses. They constitute about 30% each of the total number of mares covered. Shetland ponies and Icelandic horses have gained most in numbers in the last decade among all the 33 breeds registered in Sweden, and constitute about 8% each of the breeding stock. Swedish horse breeding has since long been characterized by a thorough evaluation and strong selection of stallions to be licensed for breeding. Licensing is still regulated by law, whereas the breed societies conduct the tests and evaluations. BLUP Animal models are applied for genetic evaluations of the trotter, riding horse and Icelandic horse populations. Coldblood trotters are jointly evaluated with the Norwegian population and for Icelandic horses there is a joint evaluation for several countries including Iceland. Other breeds practice phenotypic evaluations of conformation, soundness and performance. For the major breeds considerable genetic progress has been achieved in the various sport disciplines.

Session 31 — Theatre 2

The economic importance of the Swedish horse industry

D. Johansson[*1], H. Andersson[2] and A. Hedberg[3]. [1]ATG Hästklinikerna, Gråbrödragatan 6, S-532 31 Skara, Sweden, [2]SLU, Departement of economics, Box 7013, S-750 07 Uppsala, Sweden, [3]Department of Agriculture, Fredsgatan 8, S-103 33 Stockholm*

This study quantifies the economic importance of the Swedish horse sector on a national level. The Swedish horse sector consists of enterprises selling goods and services aimed towards horse owners and horse-related activities. The parameters which are used to capture the economic importance of the sector include turnover, employment (measured by full time equivalents, FTE), tax revenues and contribution to Gross Domestic Production (GDP). Input/Output-analysis has been used to facilitate the calculations of the direct economic impact of the sectorFor the first time an overview of the economic significance of the Swedish horse sector on an aggregate level is estimated. The use of Input/Output-analysis as an instrument to examine the sector as a whole has not been done before in Sweden or abroad. The outcome of the study reveals that the annual turnover of the Swedish horse sector is 20 billion SEK, of which about 10 billion SEK can be attributed to gambling. The contribution to GDP from the sector makes up 0.34 % of total GDP. Tax revenues from the sector are slightly above 4 billion SEK, of which 1.1 billion SEK is derived from taxes on gambling. The sector gives rise to employment equivalent to 9 500 FTE.

Session 32 — Theatre 1

How important are regulatory mutations for genetic variation in multifactorial traits?

L. Andersson, Swedish University of Agricultural Sciences (SLU) and Uppsala University, Uppsala, Sweden

The molecular basis for genetic variation in multifactorial traits is still poorly understood. It is an open question to which extent Quantitative Trait Loci (QTLs) are caused by regulatory or structural mutations. Structural mutations clearly dominate among the limited number of QTLs that has been revealed today. However, it is very likely that this sample is strongly biased because it is much easier to identify and verify a structural mutation than a regulatory mutation. Moreover the evolutionary significance of regulatory sequences is obvious from the fact that more than 50% of the sequences that is well conserved between distantly related species constitute non-coding sequences. In this paper I will review examples of regulatory mutations affecting phenotypic traits in farm animals and discuss how such mutations can be revealed in the future using genetic or expression analysis or a combination of the two approaches.

Session 32 — Theatre 2

Bioinformatic tools in analysing molecular genetic data for breeding and genetics

L.L.G. Janss, Animal Sciences Group, Wageningen UR, P.O. Box 65, 8200AB Lelystad, The Netherlands

Animal breeding is now strongly focused on identification of genes in order to increase efficiency of selection, especially for quality and robustness traits. QTL mapping alone is not sufficient to unambiguously select causative mutations and to give insight in the biological background of gene effects - items which are valuable for efficiency and confidence of DNA-based selection. Therefore, other approaches, notably *functional genomics* approaches such as gene expression studies, are to be added. This requires analysis tools to integrate more data and more types of data (markers, gene expressions, phenotypes, sequence data, literature), which may be referred to as "bioinformatics". A pertinent statistical issue in this field is handling of errors in pipelines where sequential corrections are made, and in combining different sources of data. Statistical tools to select differentially expressed genes have introduced the False Discovery Rate (FDR) to better handle tests errors. FDR will also be valuable for QTL mapping where now approaches are used with high false negative error rates, but further work is needed to combine multiple (erroneous) test results. Insight in biological background can be improved by adding pathway information (e.g., from literature) to the analysis of gene expression through Bayesian model priors - but here too external data may be erroneous. Further research on approaches for meta-analysis combining data from multiple sources with various levels of error will therefore be warranted.

Session 32 Theatre 3

Reassessing quantitative genetic theory in the light of modern molecular genetic ideas

G.E. Pollott, Department of Agricultural Sciences, Imperial College London, Wye campus, Ashford, Kent, TN25 5AH, United Kingdom

The infinitesimal model and the body of theory developed around it has been the cornerstone of most genetic improvement in quantitative traits for half a century. The limitations of this model have been clearly recognised for a number of years and the recent use of a quantitative trait locus approach to marker assisted selection and direct selection on specific molecular genotypes have been documented and implemented. However, these newer approaches, whilst valuable, have only used a limited number of ideas arising from genome analysis. More recently the linking of the genome, through proteomics and metabolomics to the physiological manifestation of quantitative traits suggests that a complete overhaul of the infinitesimal model is needed in order to integrate new and old ways of looking at quantitative traits. In particular these approaches have called into question the exact nature of the environmental effects on a quantitative trait and may suggest a different underlying model for their understanding and analysis. This paper attempts to integrate both approaches to quantitative traits and investigates the implications of the new approach on the full range of quantitative genetic tools; breeding value estimation, genetic parameters, genotype-by-environment interactions and selection decisions.

Session 32 Theatre 4

Estimable genetic variance components under mixed additive mendelian and imprinted inheritance

N. Reinsch and V. Guiard, Forschungsinstitut für die Biologie landwirtschaftlicher Nutztiere FBN, 18196 Dummerstorf, Wilhelm-Stahl-Alle 2, Germany*

In the presence of genomic imprinting a quantitative trait may be determined by genes of three expression classes: biparentally expressed (equal expression levels of paternal and maternal alleles), paternally expressed (maternally imprinted) and maternally expressed (paternally imprinted). Three random genetic effects, each corresponding to the combined effect of genes from these expression classes, can be estimated with an appropriate mixed linear model and proper variance components. This model was termed the explicit model. By incorporating awareness on partial imprinting and developmentally different imprinting into a simplified model genome it could be shown that these three random genetic effects are correlated, which requires six ("causal") genetic covariance components to be estimated. A so-called compact model was then introduced, comprising only three genetic covariance components (functions of the "causal" ones), but still allowing to assess the existence of genomic imprinting. It turned out that the six covariance components of the explicit model are not estimable, while the three components of the compact model are. The compact model is well suited for population analyses of genetic variance components, genetic evaluation under all additive patterns of imprinted inheritance and can even be adopted to the detection of imprinted quantitative trait loci in extended pedigrees.

Session 32 Theatre 5

Strategies for selective DNA pooling for multiple traits

H. Schwarzenbacher[1], P. Visscher[2], M. Dolezal[1], M. Soller[3] and J. Sölkner[1], [1]Department of Sustainable Agricultural Systems, BOKU - University of Natural Resources and Applied Life Sciences, Vienna, Austria, [2]Institute of Cell, Animal and Population Biology, School of Biological Sciences, University of Edinburgh, West Mains, Road, United Kingdom, [3]Department of Genetics, Hebrew University of Jerusalem, Israel

Selective DNA pooling is a very powerful method for QTL mapping markedly reducing genotyping costs while maintaining high statistical power. Nevertheless, when mapping is focused on multiple traits the relative feasibility of selectively pooling DNA samples of extreme animals is reduced. One approach of dealing with this limitation is to construct an index by weighting traits of interest. We performed a simulation study evaluating the power of pooling for single traits in dairy cattle such as milk yield (mkg), protein percent (p%), maternal fertility (fert) and somatic cell count (scc) versus the incorporation of a selection index by accounting for single traits with different economic weights. The results show that this strategy is able to reduce genotyping costs seriously, while maintaining a satisfactory statistical power. Furthermore, the power to identify QTL for single traits when pools are constructed on an index is affected rather by the correlation structure among the traits in the index than the correlation of each single trait with the selection index. This may be advantageous when the index includes traits which are negatively correlated to each other, such as mkg and p%.

Session 32 Theatre 6

Testing candidate genes using a Bayes Factor

L. Varona, Area de Producció Animal. Centre UdL-IRTA. 25198. Lleida, Spain

The general approach for QTL mapping and characterization is the first step to locate the genes that affect the traits of interest. A natural follow up is to identify the causal mutation explaining these phenotypic differences. Usually, candidate genes are selected according to the physiological mechanism of the trait, and they are located in the same region of the QTL. After sequencing those genes to find polymorphism, the relevant question is to determine if the detected mutation is the one that causes the phenotypic difference. However, this is not an easy task, because linkage disequilibrium between genes located in the same region. Zhao *et al.*, (2003) and Varona *et al.* (2005) have proposed alternative methods that consider the marker information to validate candidate genes. Here, the Bayes Factor is suggested as an alternative to these procedures. Under the classical statistics paradigm, models should be nested to be compared, and, in contrast, the Bayes Factor allows comparing between non-nested models. An algorithm to calculate the Bayes Factor between a model with the candidate gene and a model with the QTL effects is presented. The performance of the method is evaluated under a simulation scheme with causal mutations of several sizes and candidate genes within a range of linkage disequilibria with the causal mutations.

Benefits from marker assisted selection under an infinitesimal model

B. Villanueva[1], R. Pong-Wong[2], J. Fernández*[3] and M.A. Toro[3], *[1]SAC, West Mains Road, Edinburgh EH9 3JG, UK, [2]Roslin Institute (Edinburgh), Roslin, Midlothian EH25 9PS, UK, [3]INIA, Carretera La Coruña km. 7, 28040 Madrid, Spain*

Computer simulations were used to investigate the effect of using marker information in genetic evaluations under the assumption of an infinitesimal model. Marker information was used in conjunction with pedigree information to compute the numerator relationship matrix (NRM) used in BLUP evaluations, thus improving the accuracy of overall effects of genes with small effect (polygenes). The trait under selection was controlled by 2000 loci of additive small effect and evenly distributed in c chromosomes of one Morgan each ($c = 5, 10, 20, 30$). Markers ($n = 1, 2, 5, 10, 20, 40$) were also evenly distributed. The NRM was computed by averaging the IBD matrices conditional on marker information and calculated at different positions across the genome. The NRM was accurately estimated by averaging IBD matrices computed every 10cM. For all the genome sizes considered, the response from MAS was higher than that from schemes using standard BLUP where the NRM is only pedigree based. At generation ten the response from MAS was 11, 9, 7 and 5% higher than that from non-MAS for genomes with 5, 10, 20 and 30 chromosomes, respectively. One marker every 10cM was enough to achieve most of the benefit. In conclusion, for genome sizes typical of livestock populations, extra gains from MAS were still detectable.

Gene expression profiling for meat quality in swine

C. Gorni[1], C. Garino[1,2], S. Iacuaniello[1,2], B. Castiglioni[3], G.L. Restelli[2], A. Stella[1], G. Pagnacco[2], P. Mariani[1]*

[1]PTP, Livestock Genomics 2, Lodi, Italy, [2]VSA, Veterinary School, University of Milan, Milan, Italy, [3]IBBA-CNR, Milan, Italy

The Suppressive Subctractive Hybridisation (SSH) technique was used to identify candidate genes for meat quality in pig. SSH libraries (forward and reverse) were created from skeletal muscle tissue obtained by pools of Landrace and Large White pigs. Pools were obtained by selecting individuals with extreme adjusted phenotypes. At the same time a control experiment was carried out to check that each step of the library construction was successfully taken. A total of 310 positive clones were identified. So far two-hundred seven clones with inserts ranging between 150-1200 bp, have been analysed. Sequences were alligned and analysed with the MAP program. The estimated libraries redundancy is 16% (forward) and 24% (reverse) respectively. Sequences present in both libraries and thus considered as background represent only 3% of the overall positive clones. Around 25% of the analysed sequences shows significant homology to previously described swine ESTs or known genes related to muscle metabolic pathways. Comparative mapping *in silico* analyses were carried out by comparing the sequences with positive hits to the results of previous QTL studies and the ESTs produced in the present study confirmed as candidate genes.

Haplotype structure of casein genes in Norwegian goats and effects on production traits

N. Hagesæther, B. Hayes, T. Ådnøy, G. Pellerud and S. Lien, Department of Animal and Aquacultural Sciences, Norwegian University of Life Sciences, N-1432 Ås, Norway*

In goat milk the most abundant proteins are the caseins, α_{S1}, β, α_{S2} and κ. Mutations have been identified within these genes affecting gene expression and milk production traits. There is some evidence to suggest that if the effect of a mutation is to increase the level of expression of a particular casein, the other caseins may be downregulated. This is an argument for pursuing a haplotype approach, where combinations of alleles across the caseins are selected for simultaneously, in order to increase total protein production. The aim of this study was to detect polymorphisms (SNPs) in the caseins of Norwegian milking goats, resolve the haplotype structure among these SNPs and assess effects of the haplotypes on milk production traits. In all 462 Norwegian bucks were genotyped for 36 polymorphic sites. Nineteen haplotypes were identified in the population, suggesting considerable linkage disequilibrium. Haplotypes had significant effects on milk, protein and fat yield. An interesting feature of the Norwegian population is the high frequency of a polymorphism leading to low expression of α_{S1}. This locus had significant effects on protein and fat contents. However, for marker-assisted selection the haplotypes across the casein loci will be more useful, as favourable effects on protein yield can be selected for across the four loci simultaneously.

Session 32 Theatre 10

Effect of *IGF2* on growth characteristics of F2 Meishan X White crossbreds

H.C.M. Heuven, H. Bovenhuis, Animal Breeding and Genetics group, Wageningen-UR, P.O. Box 338, 6700AH Wageningen, The Netherlands*

Recently, a mutation in a regulatory element of the paternally expressed *IGF2* gene was identified as the causative mutation affecting muscle growth and backfat. In this study the effect of this QTN on growth characteristics is studied for a F2 Meishan X White crossbred population. Pigs that inherited the wild type allele from their sire showed a lower weight at birth, at weaning, at start of the test, end of test and in the slaughter house. The effect of the paternally inherited IGF2-allele was highly significant for all traits except for weaning weight, probably due to management interference during the suckling period. The least square estimate of the difference between the wild type and mutant allele was 65, 125, 760, 2040 and 2840 grams for birth-, weaning-, start-, end- and slaughter weight respectively. The effect on early growth (0-25 kg), test growth (25-100 kg) and life growth was 9.4, 24.0 and 12.1 gr/day respectively. The estimated difference of the paternally inherited allele on ultrasonic backfat, HGP-backfat and HGP-muscle depth was 1.69, 1.89 and -2.24 mm respectively. Due to these large effects most commercial sire lines where this gene is segregating will become homozygous. The selective advantages of the mutant allele in specialized dam lines will depend on the effect of *IGF2* on dam related traits such as litter size, mothering ability, interval weaning-oestrus.

Session 32 Theatre 11

Fishy taint in chicken eggs is associated with a substitution within a conserved motif of the FMO3 gene

M. Honkatukia[1], K. Reese[2], R. Preisinger[3], M. Tuiskula-Haavisto[1], S. Weigend[2], J. Roito[1], A. Mäki-Tanila[1], and J. Vilkki[1], [1]MTT Agrifood Research Finland, 31600 Jokioinen, Finland, [2]Federal Agricultural Research Centre, 31535 Neustadt, Germany, [3]Lohmann Tierzucht GmbH, Cuxhaven, Germany*

Fishy odour of urine and other secretions characterises trimethylaminuria in humans, resulting from loss-of-function mutations in the flavin-containing mono-oxygenase *FMO3*. The odour is caused by an elevated level of excreted trimethylamine (TMA), due to deficient oxidation by *FMO3* of TMA derived in the gut from the diet. A similar phenotype exists in cattle, where a nonsense mutation in the bovine orthologue causes fishy off-flavour in cow's milk. We report the mapping of a similar disorder (fishy taint of eggs) and the chicken *FMO3* gene to chicken chromosome 8. One substitution of an evolutionary highly conserved amino acid in the chicken *FMO3* gene was found to be homozygous in all affected hens. No differences in the expression of *FMO3* were found among individuals with different taint-associated genotypes, indicating that the trait is not caused by altered expression of the gene. The results support the importance of the evolutionary conserved motif that has been speculated to be a substrate recognition pocket of the enzyme. The mutation is associated with elevated levels of TMA and fishy taint in the egg yolk in several chicken lines, thus providing a marker for selection against the trait in breeding programs.

Session 32 Poster 12

Mapping of quantitative trait loci for leg conformation traits in Danish Holstein

A.J. Buitenhuis, M.S. Lund, B. Thomsen and B. Guldbrandtsen, Danish Institute of Agricultural Sciences, Department of Genetics and Biotechnology, P.O.Box 50, 8830, Tjele, Denmark*

Strength of legs and feet has a positive effect on herd life of dairy cows. Many countries include measures of leg and foot conformation traits in their breeding programs, often as early predictors of longevity. Five leg conformation traits were measured on grand-daughters of 19 Danish Holstein sire families with 33 to 105 sons. The traits measured were: rear legs side view, rear legs rear view, hock quality, bone quality and foot angle. A genome scan for QTL was performed. Out of the 19 sire families, 12 families were genotyped for 29 autosomes with micro-satellite markers, while 3 families were typed for 28 autosomes, 2 families were typed for 27 autosomes and one family for 23 autosomes. Data were analysed across and within families for QTL effecting leg conformation traits. The variance component method was used to estimate QTL positions and variances. Seven chromosome wide significant QTL were detected across families for rear legs side view, 5 for rear legs rear view, 5 for hock quality, 4 for bone quality and 1 for foot angle. In total, eleven sire families were segregating for QTL effecting rear legs side view, 7 families for rear leg rear view, 5 families for hock quality, 8 families for bone quality and 2 families for foot angle.

Session 32 Poster 13

QTL analysis for eight milk production traits in the German Angeln dairy cattle population

K. Sanders[1], J. Bennewitz[1], N. Reinsch[2], E.-M. Prinzenberg[3], and E. Kalm[1], [1]University of Kiel, Institute of Animal Breeding and Husbandry, D-24098 Kiel, Germany, [2]Research Institute for Biology of Farm Animals, D-18196 Dummerstorf, [3]University of Giessen, Institute of Animal Breeding and Genetics, D-35390 Giessen, Germany*

Mapping QTL for milk production traits was a main objective of marker assisted investigations in dairy cattle populations. Some of these traits (e.g. milk yield) are rather complex and the additional information of QTL for related traits show a more clear physiological background and might help to explain the milk production QTL better. In this study we mapped QTL for the milk production traits milk, fat, protein, and energy yield, fat, protein, and energy content, somatic cell count, and the physiological traits lactose yield and lactose content on five different chromosomes (BTA06, 14, 16, 18, 27) in the German Angeln dairy cattle population. The red Angeln breed is a small dairy cattle breed located in the North of Germany. The daughter design included five families with a total of 805 daughters. Forty-three microsatellites were genotyped with an average distance of 20 cM. Chromosomewise significant QTL were found for lactose content and protein content on BTA06. Furthermore, on BTA18 significant QTL segregated for energy yield, lactose yield, protein yield, and lactose content. These results can be used for further investigations of the physiological background of QTL for milk production traits.

Session 32 Poster 14

Selection of the habitat in the rest phase of the *Helix aspersa* under laboratorial conditions

J. Perea[1], M. Herrera[1], A. García[1], A. Mayoral[2], M. Luque[1], E. Felix[1], and C. Pérez[1], Department of Animal Production, University of Córdoba (UCO), Spain. [2]Agricultural Research Institute of Andalucia (IFAPA), Spain

The objective of this study is to determine the incidence of light and the shape of the raising box on the preferred habitat during the rest phase of rest of the *Helix aspersa*. To do this experiment have taken a sample of 500 animals with an average weight of 1.80±0.05 g and distributed them in 20 raising boxes. The study took place between December and January with decreasing diurnal phase and average temperature of 21.9±0.09 °C. The rearing boxes were made of translucent plastic, 14.5 x 14.5 x 7.5 cm, and were cleaned daily to avoid the negative effects of excreta, mucus and density. The study lasted of 30 days. Each day resting place, and the orientation with respect to the light, of each snail was recorder. The results indicate that the snails, during the phase of rest, display positive phototropism. They prefer the high places of the box ($p<0.0001$), and they prefer the illuminated places rather than the dark ($p<0.0001$). With respect to the form of the raising box they show preference by the corners of the front walls or ceiling of the box ($p<0.0001$).

Session 32 Poster 15

The association between *CSN3* genotypes and milk production parameters in Czech Pied cattle

J. Kucerová[1], E. Nemcová[1], M. Stípková[1], O. Jandurová[1], A. Matejícek[2] and J. Bouska[1] [1]Research Institute of Animal Production, Prague - Uhrineves, Czech Republic, [2]University of South Bohemia, Ceske Budejovice, Czech Republic*

The aim of this study was to detect an association between marker genotypes and milk production parameters. Data on *CSN3* genotypes in Czech Pied sires and on milk production parameters (milk yield, protein yield, fat yield, protein content and fat content) of their daughters were available. Three genotypes (AA, AB and BB) of the kappa-casein marker (*CSN3*) were observed in the investigated population. Genotype BB was associated with higher protein and fat content, but with lower milk production. In constract to this, genotype AA was associated with higher milk production, but lower protein and fat content. Further analyses using a granddaughter design population structure will follow this study.

This study was supported by project NAZV 1G46086.

Session 32 Poster 16

Genetic variability of *MYF3* and *MYF4* genes in Large White and Landrace breeds of pigs in the Czech Republic

J. Verner, T. Urban, Department of Animal Morphology, Physiology and Genetics, Mendel University of Agriculture and Forestry Brno, Czech Republic

The aim of the present study was to determine the genetic variation in two loci of *MYOD* family and assess their associations with performance traits in pigs. The *MYF3* and *MYF4* genes belong to the class of muscle specific regulatory factors, which control development of muscle cell line. In intron 1 of the *MYF3* gene and at 3´ side of the *MYF4* gene, the *Dde*I and *Msp*I polymorphic sites occur, respectively. The genotypes were detected by PCR-RFLP method in two breeds of pigs, Large White and Landrace. The associations of the two polymorphic sites with production traits (daily gain, back-fat thickness, weights of neck, loin, shoulder and ham) were assessed. For the analyses, the mixed models procedure of SAS 8.2 was used. The significant associations between *MYF3* gene and studied traits were found out, however to find and evaluate associations between performance traits and *MYF4* gene polymorphism it would be necessary to analyse larger population of pigs.

This work was supported by MSMT Czech Republic No. MSM 432100001, Czech Science Foundation No. 523/03/H076 and FRVS MSMT Czech Republic No. 239/2005.

Session 32 Poster 17

SNPs analysis in selected candidate genes in pigs using resequencing

A. Knoll, J. Verner and Z. Vykoukalová, Department of Animal Morphology, Physiology and Genetics, Mendel University of Agriculture and Forestry Brno, Zemedelská 1, 613 00 Brno, Czech Republic

Single nucleotide polymorphism (SNP) detection is needed for identification of economically important loci (ETLs) by means of positional candidate gene approach and for gene mapping. A key step in all strategies for causative gene identification is the resequencing of candidate genes or other genomic regions of interest in phenotype divergent animals to identify those single nucleotide polymorphisms associated with a certain phenotype. Based on sequences (porcine or orthologous) deposited in genomic databases PCR primers were designed and gene fragments of samples of different animals synthesized. The direct automatic fluorescent sequencing of PCR products was performed using 3100 Avant Genetic Analyzer (Applied Biosystems). The raw sequencing data were processed using DNA Sequencing analysis Software Ver. 5.1 and SeqScape Software Ver. 2.1.
This resequencing approach was applied to detection of polymorphisms in *MYF6* (myogenic factor 6; herculin) and *IGF2* (insulin-like growth factor 2) genes. DNA sequences of Large White, Landrace, Piétrain, Meishan and Wild pig were aligned and nucleotide differences were detected, but none of founded polymorphisms were connected to amino acid change. Study of association between these markers and meat production and quality traits are in progress. This research was supported by the Grant Agency of the Czech Republic no. 523/02/D026 and presentation by the project 523/03/H076.

Session 32 Poster 18

Relationship between the myogenin gene (*myf4*) and litter size of large white sows

P. Humpolícek, T. Urban and J. Verner, Department of Animal Morphology, Physiology and Genetics, Mendel University of Agriculture and Forestry Brno, Czech Republic

The objective of this study was to estimate the effects of myogenin gene (*MYF4*) on the litter size of Large White sows in the investigated nucleus herd. In total, 86 sows were genotyped (PCR-RFLP) and analyzed to evaluate whether the gene polymorphisms influence the litter size. The association analysis was carried out by mixed linear model in SAS for Windows 8.2. The association was determined separately on the first, the second, the first to the fourth, and the second to the fourth litter. The associations of the myogenin gene with a total number of born piglets (NB), number of piglets born alive (NBA) and number of weaned piglets (NW) were determined. The significant differences in the efficiency of sows with variable genotypes of *MYF4* gene were found. This suggests that this gene may influence the development of reproduction traits and therefore, the bigger attention to this gene should be given.

This work was supported by MSMT Czech Republic No. MSM 432100001, Czech Science Foundation No. 523/03/H076.

Session 32 Poster 19

Isolation of differentially expressed genes related with ham salting loss in Italian heavy pig

V. Russo, D. Bigi, P. Zambonelli and M. Colombo, Sezione Allevamenti Zootecnici, DIPROVAL, University of Bologna, Via Fratelli Rosselli 107, I-42100, Reggio Emilia, Italy*

We used Fluoro Differential Display (FDD) technique to identify differentially expressed genes candidate for ham salting loss (SL) in Italian heavy pig. A group of 277 Large White, 183 females and 94 castrated males were slaughtered. Four unrelated animals (2 females and 2 castrated males) with the lowest SL genetic index and 4 with the highest SL index were selected. Total RNA was extracted from individual samples of semitendinosus muscle. An equal amount of each sample was used to obtain pools for high and low SL. The FDD gave fingerprints of about 50 well distinct bands in the size range 200-1500 bp. On the whole, about 5000 cDNAs fragments were analysed. We found 12 strongly differentially-expressed bands, always over expressed on the pool with highest SL. The DNA extracted from 11 bands was sequenced and similarity was found to 6 known genes, to 4 genes with uncharacterized function, and to one L1 interspersed element. Four out of six known genes, Sarcolipin (*SLN*), Triadin (*TRDN*), Titin (*TTN*) and Down syndrome critical region 1 (*DSCR1*), are mostly expressed in skeletal and cardiac muscles.

Session 32 Poster 20

Allelic frequencies of MC1r and ASIP genes in Iberian horses

L.J. Royo[1], I. Álvarez[1], I. Fernández[1], M. Valera[2], J. Jordana[3], A. Beja-Pereira[4], E. Gómez[1], and F. Goyache[1], 1SERIDA, Gijón, Spain, 2Universidad de Sevilla, Spain, 3Universitat Autònoma de Barcelona, Spain, 4 CIBIO-UP, Vairão, Portugal*

Coat colour affects differentiation of populations in breeds. The relative amounts of eumelanin (black/brown) and phaeomelanin (yellow/red) are controlled by the Extension (E) and Agouti (A) loci. In horses, two different alleles have been described in the MC1r gene, the wild type allele (E+) and the recessive allele (C901T; e), which determines the chestnut coat colour when homozygous. In the agouti gene also two alleles have been described, the wild type (A+) and the black recessive allele (ADEx2; Aa), determining the black coat color when homozygous. The MC1r gene is epistatic to agouti gene.

Iberian horses are classified in two different groups 'Celtic' and 'Iberian' that can be well differentiated by means of morphological traits. Here, we show the allelic frequencies of MC1r and ASIP genes in a representative sample of breeds belonging to 'Celtic': Asturcón (45), Caballo de Corro (8), Mérens (19), Losino (12) and Garrano (10); and 'Iberian': Carthusian (10), Andalusian (7) and Marismeño (10).

The black recessive allele is predominant in Celtic horses, being all Asturcón, Merens and Losino samples homozygous. The chestnut allele is mostly found in Iberian horses and Garrano (>30%), and at a very low frequency (<10%) in the Asturcón and Mérens breeds.

This work is founded by INIA-RZ03-011.

Session 32 Poster 21

In silico inference of multi-locus genotypes from SSCP markers

F. Panzitta[1,2], P. Mariani[1], G.C. Gandini[2], P.J. Boettcher[3], and A. Stella[1] PTP-CERSA, Lodi, Italy, [2]VSA, University of Milan, Italy, [3]IBBA-CNR, Milan, Italy*

Single-Stranded Conformation Polymorphisms (SSCPs) are a cost-efficient type of genetic markers. However, depending on the eventual information desired, sequencing of the resulting fragments may be necessary, and this can be a time-consuming procedure. Traditional approaches for estimation of genetic distances are often based on genotypes at specific sites of polymorphism. Within the SSCP fragments obtained by PCR amplification, several Single Nucleotide Polymorphisms (SNPs) can be present. The objective of the study was to develop an *in silico* method for inference of multi-locus genotypes from SSCP markers. Two critical steps were involved. First, the lengths of the fragments expected to be visualised after electrophoresis, were obtained based on the embedded restriction sites. Second, the three-dimensional structure of each potential fragment was predicted by applying energy minimization algorithm. Multiple suboptimal candidate structures were also predicted. The fragment lengths and 3D structures were used to determine the relative distance each fragment was expected to migrate on a gel. The best set of candidate structures was then determined by comparison with the observed band segregation profile. The approach was tested on publicly available SSCP data and on SSCP data from the GH gene in different Italian goat breeds. The correct genotypes were determined in at least 80% of the cases tested.

Session 32 Poster 22

The use of phenotypic information for refinement of haplotype reconstruction using the Expectation-Maximization algorithm

P.J. Boettcher[1,2], and A. Stella[3]. [1]IBBA-CNR, Milan, Italy, [2]Animal Production and Health Section, IAEA, Vienna, Austria, [3]PTP-CERSA, Lodi, Italy*

When genetic studies involve the effects of multiple linked polymorphisms, analysis of haplotypes may be preferred over the analysis of the individual loci. The number of statistical tests will be reduced and statistical power may be increased. Unfortunately, standard genotyping methods do not yield all the information necessary to identify haplotypes. Therefore, a number of in silico approaches have been developed for haplotype reconstruction. A popular approach is based on maximum likelihood and the Expectation-Maximum (EM) algorithm. Like all computer based approaches, the EM method cannot reconstruct haplotypes with 100% certainty. In cases where haplotypes have a real effect, the phenotypic information could be used to increase the accuracy of reconstruction, and estimation of haplotype effects could be done simultaneously. The objective of this study was to develop an algorithm, based on EM, for this specific purpose. Stochastic simulation was used to test the algorithm on a for a variety of haplotype configurations and a range of haplotype effects.

Consideration of phenotypes increased the accuracy of haplotype reconstruction, but the benefit depended on the size of the haplotype effect. The haplotype effect had to account for at least 10% of the phenotypic variance for discernable effects on accuracy.

Session 32 Poster 23

A pipeline for automatic detection of SNPs in goat polymorphic sequences

B. Lazzari[1], J. Nardelli-Costa[2], F. Panzitta[1,3], A. Stella[1], A.R. Caetano[4] and P. Mariani[1]*
[1]PTP-CERSA, Lodi, Italy, [2]Catholic University of Brasília, Brazil, [3]University of Milan, Milan, Italy, [4]Embrapa Cenargen, Brasilia, Brazil

A number of amplified fragments of the Growth Hormone gene (GH) was obtained using three different sets of primers (GA, GB and 2-3) on goat genomic DNA extracted from a total of 231 animals from ten different Italian breeds. The same DNAs were also amplified with another set of primers specific for the Luteinizing Hormone Receptor (LHR). All of the obtained amplicons were purified and sequenced both in the forward and reverse directions.
About 1400 electropherograms were produced and processed with the program Phred, to generate sequence and quality files from all the trace files. Multiple sequences from each amplicon were assembled with Phrap to generate contigs of aligned reads. To allow automatic detection of SNPs in these sequences the PolyBayes software was run on the Phrap contigs. Based on a Bayesian-statistical formulation, PolyBayes produces a list of candidate polymorphic sites, each with an associated SNP probability score. Multiple alignments marked up with SNP information can be viewed directly with the Consed sequence viewer.
The analysis revealed a high level of polymorphism within both the GH and the LHR genes, and a number of high score putative SNP sites was identified.

Session 32 Poster 24

QTL mapping for teat number in an Iberian by Meishan pig intercross

M.C. Rodríguez[1], A. Tomás[2], E. Alves[1], O. Ramírez[2], M. Arque[3], C. Barragán[1], L. Varona[3], L. Silió[1], G. Muñoz[1], M. Amills[2], J.L. Noguera[3]*
[1]Departamento de Mejora Genética Animal, INIA, 28040 Madrid, [2]Departament de Ciència Animal i dels Aliments. Facultat de Veterinària, UAB. 0819 Bellaterra9, [3]Àrea de Producció Animal, Centre UdL-IRTA, 25198 Lleida, Spain

The aim of this study was to investigate chromosomal regions affecting the number of teats in pigs. An experimental F_2 cross between Iberian and Chinese Meishan lines was used for this purpose. These two breeds proceed from independent domestication processes and present great differences in teat number. A genomic scan was conducted with 115 markers covering the 18 porcine autosomes. Linkage analyses were performed by interval mapping, using the animal model to estimate QTL and additive polygenic effects. Complementary analyses with models fitting two QTL were also carried out. The results showed three genomewide significant QTL, mapping on chromosomes 5 (29 cM), 10 (72 cM) and 12 (60 cM), whose joint action control up to 30% of the phenotypic variance of the trait. All the three Meishan alleles had a positive additive effect on teat number. Two positional candidate genes have been identified for QTL on chromosomes 5 and 10, and their molecular analysis could improve the knowledge of the genetic architecture of teat number.

Session 32 Poster 25

Analysis of effects of genes differentially expressed during myogenesis on pork quality

E. Murani[1,8], M.F.W. te Pas[2], K.C. Chang[3], R. Davoli[4], J.W.M. Merks[5], H. Henne[6], R. Wörner[6], H. Eping[7], S. Ponsuksili[1,8], K. Schellander[1], N. da Costa[3], D. Prins[5], B. Harlizius[5], Egbert Knol[5], M. Cagnazzo[4], S. Braglia[4] and K. Wimmers[1,8,], [1]University of Bonn, 53115 Bonn, Germany, [2]Wageningen University and Research Centre, Animal Sciences Group, 8200 AB Lelystad, The Netherlands, [3]University of Glasgow, Glasgow G611QH, UK, [4]DIPROVAL University of Bologna, 42100 Reggio Emilia, Italy, [5]IPG, 6641 SZ Beuningen , The Netherlands, [6]BHZP Lueneburg, 21335 Lueneburg, Germany, [7]LRS, 53115 Bonn, Germany, [8]Research Institute for the Biology of Farm Animals, 18196 Dummerstorf, Germany*

Genes regulated during myogenesis may be involved in the development and control of muscle-(structure) and consequently may have an effect on meat quality. Transcription profiles of embryonic (presumptive) and foetal M. longissimus dorsi were compared between Pietrain and Duroc breeds at 7 key stages of myogenesis employing microarrays, SSH and DD-RT-PCR. Fifty three differentially expressed genes were selected for further study. For 35 genes DNA polymorphisms were detected. The association between DNA variation of 23 candidates and meat quality and content was analysed in four Duroc and Pietrain based commercial lines and one Duroc × Pietrain experimental cross. The most interesting effects were found for genes on chromosomes 2, 4, 5 and 14 in regions harbouring QTLs for muscle structure and meat quality traits. This work is part of an EU-funded project (PorDictor - QLK5-2000-01363).

Session 32 Poster 26

Leptin polymorphism and its association with milk production and plasma glucose in early lactating cows

A. Heravi Moussavi[2], M. Ahouei[1], M. R. Nassiry[2], E. Jorjany[1], M. Salary[1] and A. Javadmanesh[2], [1]Dept. of Animal science, Agriculture Faculty, University of Zabol, [2]Dept. of Animal Science, Agriculture Faculty, Ferdowsi University of Mashhad, Iran*

Since the goal of animal breeding is to improve the economic traits in farm animal, studying the genes which have main effects on economic traits is very crucial. Previous studies have shown that polymorphisms on leptin gene are associated with energy balance, milk production, fertility, immune system function and feed consumption. The study was designed to study leptin polymorphism in Iranian Holstein dairy cows and its association with milk production and plasma glucose in early lactating cows. In total, blood samples were collected from one hundred and twenty Holstein cows via venipuncture from coccygeal vessels. DNA extraction was done on the blood samples using guanidium thiocyanate-silica gel. PCR-RFLP method was used to detect the polymorphism of a 423 bp fragment from intron 2 of leptin gene. Two genotypes, AA and AB have been distinguished which have the frequency of 0.86 and 0.14, respectively. From d 1-80 postpartum, fifty cows were selected and fed a same diet. Milk production and composition were not affected by the genotypes. Weekly plasma glucose concentrations were similar between the two genotypes. Results from this experiment demonstrated that leptin polymorphisms had not any apparent effect on milk production and composition and also plasma glucose.

Session 32 Poster 27

Effects of the bovine *DGAT1* (K232A) polymorphism in Swedish dairy cattle

J. Näslund, F. Fikse, G. Pielberg, A. Lundén, Department of Animal Breeding and Genetics. Swedish University of Agricultural Sciences, Uppsala, Sweden*

A quantitative trait nucleotide (QTN) for milk production traits has been found in the *DGAT1* gene located on the centromeric region of the bovine chromosome 14. The QTN is a dinucleotide substitution replacing lysine with alanine (K232A) in the amino acid sequence of the enzyme acyl-CoA:diacylglycerol acyltransferase1 (DGAT1). The mutation results in an increase in protein and milk yield and a decrease in fat yield, fat and protein percentage. The high yielding dairy herd at the Swedish University of Agricultural Sciences includes cows of the Swedish Red and White (SRB) and the Swedish Holstein breeds. Since 1985 the SRB cows have been selected for either high (HF) or low (LF) fat content with the two lines having an equal total milk energy production. In total, 279 cows were genotyped for the *DGAT1* polymorphism and the alanine variant was found to be the most frequent (0.88) among the cows. Weekly registrations on milk yield and milk composition from the genotyped cows were used to estimate the effect of genotype on yield and composition traits. For the statistical analysis we used the PEST package and a model that adjusts for systematic environmental effects and the relationship between the animals. *DGAT1* genotype accounted for a considerable part of the phenotypic variation in milk fat content and showed an additive gene effect.

Session 32 Poster 28

Identification of mutations and mapping of candidate genes for lysosomal proteinase and esopeptidase activities of dry cured hams in pigs

L. Fontanesi, R. Davoli, S. Galli and V. Russo, DIPROVAL, Sezione di Allevamenti Zootecnici, University of Bologna, Via F.lli Rosselli 107, 42100 Reggio Emilia, Italy*

Lysosomal proteinases and esopeptidases of the skeletal muscle play an important role during the curing process and influence the final quality of dry cured hams. High activities of lysosomal proteinases are related to defects of the hams and aminopeptidases are involved in the generation of free amino acids that contribute to the development of flavours in the cured products. Here we investigated several candidate genes (cathepsin D, *CTSD*; cathepsin L, *CTSL*; cystatin C, *CST3*; alanyl-aminopeptidase, *NPPEPS*) with the final aim to identify DNA markers that could be used in association studies with these quality traits in pigs. The 3'-UTR of two genes (*CTSL* and *CST3*) was analysed by PCR-SSCP and sequencing. Three alleles, due to two SNPs, were identified for *CTSL* and allele frequencies at this locus were studied in Large White, Landrace and Duroc animals. For *CST3*, four alleles, that differed by 6 SNP and by a 6 or 8 bp indel, were identified. This polymorphic region was used to confirm the linkage map of *CST3* to SSC17. *CTSD* and *NPPEPS* were mapped using a somatic cell hybrid panel to SSC9 and SSC12, respectively. Other studies are under way to identify more polymorphisms and to investigate possible associations with meat quality traits in pigs.

Session 32 Poster 29

Creating high density Radiation Hybrid maps to fine map QTLs in cattle

John L. Williams[1] Stephanie D. McKay[6] Nicola Hastings[1] Oliver Jann[1] Michal Janitz[2], Steffen Hennig[2], Sandrine Floriot[3], Paolo Ajmone-Marsan[4], Elisabetta Milanesi[4] Alessio Valentini[5], Cinzia Marchitelli[5], Marilina Savarese[5] Stephen S. Moore[6] and Andre Eggen[3]. [1]Roslin Institute, Midlothian, Scotland. [2]Max-Planck-Institute for Molecular Genetics, Berlin, Germany. [3]INRA, Jouy-en-Josas, France. [4]Università Cattolica del S. Cuore Piacenza, Italy. [5]Universita della Tuscia, Viterbo, Italy. [6]University of Alberta, Edmonton, Alberta, Canada

In order fine map QTL it is necessary to increase the density of markers available on the chromosome to which the QTL maps. With the exponential growth of information available for the bovine genome there are many sources of information to generate new genetic markers. We used BAC-end sequence data from the International BAC Mapping consortium, ESTs sequences from the EC BovGen project and public data-bases, and data on bovine sequence containing SNPs from the International Bovine Sequencing Project, and aligned this data with the human genome sequence. Using comparative mapping information, bovine sequences mapping to human chromosomes with predicted conservation of synteny with the target regions in cattle were identified and used to design primers to align the sequences on the bovine genome using an RH mapping approach. High density maps have been built for several chromosomes revealing errors in the current BAC contig assemblies, and differences in the order of genes between human and bovine genomes. A re-sequencing strategy has been adopted to identify or confirm the presence of SNPs in sequences mapping at target QTL regions.

Session 32 Poster 30

Genetic and genomic approaches to improve production in water buffalo

M. Strazzullo[1,2], C. Campanile [2], M. D'Esposito[1] and L. Ferrara[2], [1]Institute of Genetics and Biophysics 'Adriano Buzzati Traverso'- CNR, Naples Italy [2]Institute for animal production system in mediterranean environment - CNR, Naples Italy*

Water buffalo represents an important specie for Italian food industry, especially in Southern Italy. A way to obtain an optimization in meat and milk production is to approach a systematic study of genetic and genomic features of the involved tissues. In this light we are performing the construction and analysis of tissue specific cDNA libraries. We isolated and characterized B.bubalus myostatin, MyoD1 and myogenin genes involved in mammalian muscle development. These genes have been identified by comparative genomics from a skeletal muscle cDNA library. Northern analysis has been performed to reveal possible alternative mRNAs. A search for DNA polymorphisms is currently undergoing by using different breeds to obtain a wide comprehension of water buffalo genetic variability.

Here we present the results of polymorphisms and expression profile analysis.

Session 32 Poster 31

Analysis of allelic variants in exon 5 of the growth hormone gene in Sistani and Dashtiari breeds of cattle

Hossein Emrani[1], Adam Torkamanzehi[2], [1]Animal Science Research Institute, Karaj, Iran. [2]Sistan and Baluchestan University, Iran

Genetic variability is the basis for phenotypic differences in many population, reproduction and disease resistance traits, as well as, fitness and adaptation to harsh environment between zebu and European cattle. Studying genetics differences and their relationship with various characteristics, at the DNA level, in these breeds can reveal the basis for these discrepancies. GH gene, because of the physiologic effects of its product on several production traits such as milk production, reproduction, growth and immune response, has attracted considerable attention as a candidate gene for these characteristics. In this project, polymorphism was studied at the exon 5 of the GH gene in two Iranian indigenous breeds of zebu cattle. DNA samples were extracted from blood of 85 Sistani and 79 Dashtiari individuals and a 404 bp fragment from exon 5 was amplified using the PCR technique. Using PB-RFLP , polymorphism was studied at two points in this fragment. Digestion of the 404 bp fragment by AluI restriction enzyme showed that all studied samples had the genotype AluI (+/+). This allele is reported to code for the amino acid Leucine at the position 127. Digestion by DdeI enzyme revealed two alleles. The frequency of DdeI(-), was 0.86 and 0.84 in Sistani and Dashtiari samples, respectively. These genetic differences may be related to production traits in these breeds.

Session 32 Poster 32

Type I DNA markers (*ATP1A2*, *CA3* and *DECR1*) on porcine chromosome 4 for carcass and meat quality traits in pigs

R. Davoli[1,], S. Braglia[1], I. Nisi[1], L. Fontanesi[1], L. Buttazzoni[2] and V. Russo[1], [1]DIPROVAL, University of Bologna, 42100 Reggio Emilia, Italy, [2]Associazione Nazionale Allevatori Suini, 00161 Roma, Italy*

We have investigated the genes *ATP1A2*, *CA3* and *DECR1*, mapped on porcine chromosome 4 in a region where QTLs for fat deposition, carcass traits and average daily gain were localised. We analysed by PCR-RFLP technique two biallelic polymorphisms for *ATP1A2* and *DECR1* already described in literature. By SSCP we found a new polymorphism at *CA3* locus and we identified the mutation (C>T) by sequencing analysis. Allele frequencies of the three genes were studied in 11 pig breeds. For *CA3*, we performed linkage and radiation hybrid mapping confirming the localisation on porcine chromosome 4 as previously obtained by somatic cell hybrid panel. In order to investigate the relationship between the studied genes and carcass and meat quality traits, we genotyped Italian Large White and Duroc pigs with extreme breeding values for important productive traits. Allele frequencies comparison between the groups of divergent pigs showed, for *CA3*, statistically significant differences for lean cut weight in Italian Large White pigs and for intermuscular visible fat in Duroc animals. Considering also its metabolic role and map position, *CA3* could be considered as a candidate gene useful for further investigations on association studies with meat quality and carcass traits.

Session 32 Poster 33

Molecular charactrization of CD_{18} gene and identification carriers of BLAD disorder in Iranian Holstein bulls

N. Asadzadeh[1], M. Esmaeilzadeh[2], F. Sarhaddi,M[1].A. Javanmard[3], Mollasalehi[4], [1]Animal Science Research Institute,Iran, [2]Razi vaccine & serum ResearchInstitute, Iran, [3]Biotechnology Department East Azarbijan,Iran, [4]Animal Breeding Center of Iran

Bovine leukocyte adhesion deficiency is a autosomal recessive hereditary disease characterized by greatly reduced expression of the heterodimeric beta-2 integrin adhesion molecules on leukocytes or white blood cells, resulting in multiple defects in leukocyte function. Most affected calves die within a few months after birth. While it is possible for some of the animals to live past two years of age, they are typically severely stunted in growth and suffer from recurrent various infectious conditions of the skin, gastrointestinal and respiratory tracts. Single nucleotide changes (point mutations) responsible for the genetic disorders were detected by polymerase chain reaction coupled with restriction fragment length polymorphism assays (PCR-RFLP). 277 blood and semen samples were collected from semen bank of animal breeding center related to Iranian Holstein bulls and analyzed. The allele frequency and the carrier percentage were estimated 0.00544 annd 1.087%,respectively.

Session 33 Theatre 1

Scrapie: an overview; policy issues and potential eradication measures

Danny Matthews, TSE Programme Manager, Veterinary Laboratories Agency, New Haw, Surrey, KT15 3NB, United Kingdom

Scrapie has been recognised as a disease of sheep for 250 years. It has been endemic in several European countries, at unknown prevalence, and introduced into other countries through trade in live animals. Whether incursions were into flocks or countries, the long incubation, low prevalence of clinical disease and absence of a live diagnostic test made detection and eradication of infected animals difficult. At times the disease was thought to be entirely genetic in origin.

The advent of BSE in bovines brought with it concern that it had spread to small ruminants. This highlighted our lack of understanding of the biology of scrapie, and left policy makers with a dilemma. While embarking on a search for BSE in small ruminants, precautionary measures were introduced to reduce the risk to consumers. Research into live animal diagnostic testing was progressed, but meanwhile evidence that sheep were not uniformly susceptible to scrapie was used as the basis of a policy aimed at eradicating scrapie by breeding for resistance. This policy was based solely on the relationship between genotype and the occurrence of clinical disease. Unfortunately, no sooner were breeding programmes started than active surveillance detected evidence that sheep with genotypes associated with resistance to scrapie were potentially also susceptible to prion infection.

Session 33 Theatre 2

Rare sheep breeds and breeding for scrapie resistance in the Netherlands

L.M.T.E. Kaal, J.J. Windig, Animal Sciences Group, Wageningen UR, Lelystad, The Netherlands*

Concern exists whether scrapie eradication programmes may lead to unacceptably high inbreeding rates and losses of diversity. The Dutch scrapie eradication programme is one of the strictest in Europe. Since 2004 only ARR/ARR rams can be used for breeding. Small holders with less than 10 animals and rare breeds with low frequencies of ARR are exempt from this obligation. However, herd books and individual farmers with rare breeds have to apply for this status and use a special breeding programme that balances inbreeding rates and elimination of non-ARR alleles. Despite this possibility most herd books have decided not to opt for exemption. Consequently other measures are needed to restrict inbreeding rates. These measures are described here and their effectiveness is analysed with the help of computer simulations. Lessons from the Dutch eradication programme may help to design eradication programmes in other European countries.

Session 33 Theatre 3

Simulation of different strategies for breeding towards scrapie resistance

F. de Vries[1,2], H. Hamann[2], C. Drögemüller[2], and O. Distl[2], [1]Clinic for Cattle, University of Veterinary Medicine Hannover, Foundation, Germany, [2]Institute for Animal Breeding and Genetics, University of Veterinary Medicine Hannover, Foundation, Germany*

The objective of the study was to assess the effects of different strategies to breed towards the scrapie resistant ARR/ARR homozygous genotype of the ovine prion protein gene in sheep on inbreeding, drift variance, possible negative side effects, bottleneck effects and breeding costs. A simulation programme was developed, in which different population structures could be used.
The input for each simulation were the allele frequencies of male and female founder animals, the population size, the age structure, the mating ratio, the effect of a QTL, a polygenic component associated or not with the QTL, genetic distance between QTL and prion protein gene locus. Breeding strategies were optimized for a large range of population structures and ARR allele frequencies, based on the mean inbreeding coefficients, the genetic distribution of founder rams to later generations, and the distributions of phenotypic and breeding values.
Based on the results, the strategy for breeds with a small population size, a low initial ARR allele frequency, or with a negatively correlated side effect should be to breed initially towards ARR heterozygous sheep until a threshold value of 30 % for the ARR allele is reached to avoid a genetic bottleneck. After this the strategy should change and only ARR homozygous sheep should be selected.

Session 33 Theatre 4

PrP allele frequencies in non-infected Valle del Belice and infected cross-bred flocks

J.B.C.H.M. van Kaam[1], R. Finocchiaro[2], M. Vitale[1], B. Portolano[2], F. Vitale[1], S. Caracappa[1], [1]Istituto Zooprofilattico Sperimentale della Sicilia "A. Mirri", Via Gino Marinuzzi 3, 90129 Palermo, Italy; [2]Dipartimento S.En.Fi.Mi.Zo.-Sezione Produzione Animale, Università degli Studi di Palermo, Viale delle Scienze, 90128 Palermo, Italy*

The EU is implementing policies to avoid human health risks for Creutzfeldt-Jakob disease by selecting against alleles of the PrP gene, which make sheep susceptible to scrapie. An initial step is the investigation of the allele frequencies in various populations. In order of increasing susceptibility the following alleles are distinguished: ARR, AHQ, ARH, ARQ and VRQ. Initial results for PrP allele frequencies in Sicilian Valle del Belice dairy sheep are presented here and compared with infected cross-bred flocks. Allele frequencies of 578 animals in four non-infected flocks were: ARR (36.9%), AHQ (4.0%), ARH (1.1%), ARQ (56.8%) and VRQ (1.2%). Infected flocks are always genotyped, results from four flocks with scrapie outbreaks were ARR (41.1%), AHQ (2.8%), ARH (2.5%), ARQ (51.0%) and VRQ (2.6%). These flocks consisted of 1766 Sicilian cross-bred sheep. Allele frequencies are very similar between these populations. However the VRQ allele had a more than twice as high frequency in the outbreak farms. This confirms that the VRQ allele shows the largest risk. The second most susceptible allele, ARQ, was the most frequent allele in both populations. Several generations will be needed to eradicate this allele.

Session 33 Theatre 5

Blue tongue in sheep: brief overview of the disease, impact on production and current epidemiological situation in Europe, control and prevention

J. Casal, Centre de Recerca en Sanitat Animal (CReSA) / Departament de Sanitat i Anatomia Animals. Edifici V. Universitat Autònoma de Barcelona. 08193-Bellaterra (Barcelona), Spain

Blue tongue is a disease of ruminants produced by an RNA virus of the Togaviridae family. The disease is transmitted by vectors of the genus *Culicoides*. The disease is particularly severe in sheep being the main clinical signs mucosal inflammation, haemorrhages, oedema and ulcerations of digestive and respiratory mucosae. In cattle, the infection is usually inaparent, but bovines can be viremic for a long time and act as reservoirs of the virus.

Blue tongue is an emerging disease in the Mediterranean basin. Since 1998, outbreaks of the disease caused by several serotypes (2, 4, 9 and 16) have appeared in a number of countries. Usually the introduction of the virus into previously free areas is attributable to animal movements, semen imports or air streams containing infected *Culicoides*.

The main impact on production is through the indirect costs caused by trade restrictions. The direct impact on production is variable, and can rank from a severe disease with a high mortality to a very slight disease without a significant impact on productivity.

In affected areas the control of the disease is difficult and is based on vaccination, restriction of animal movements and control of vectors. In disease-free areas the main preventive measures are quarantine, control of animal movements, control of vectors and epidemiological surveillance.

Session 33 Theatre 6

The problem of anthelmintic resistance in nematode parasites of sheep and goats, and the prospects for non-chemotherapeutic methods of control

P.J. Waller, Department of Parasitology (SWEPAR), National Veterinary Institute and Swedish University of Agricultural Sciences, SE-751 89 Uppsala, Sweden

Collectively, nematode parasites of domestic ruminants continue to be the greatest disease problem in grazing livestock systems worldwide, despite the fact that the chemotherapeutic arsenal available for their control is powerful and extensive. Widespread development of anthelmintic resistance in nematode parasites of small ruminants, particularly in the hot and humid regions of the world, is escalating at such a rate that all countries with significant sheep and goat numbers must consider that tackling this problem is an urgent priority. Additionally, the increasing trend towards organic farming, where there is a prohibition in the prophylactic use of all chemical compounds, means that non-chemotherapeutic alternatives to parasite control are urgently needed. Researchers have responded to this challenge and a variety of quite different approaches have been the subject of intense investigation in many countries, now for several decades. These include grazing management strategies, breeding for worm resistance, targeted nutritional strategies, worm vaccines and biological control. These vary in relation to their stage of development for "on-farm" use. However it is important to recognize that no single method on its own can be expected to provide satisfactory, sustainable parasite control more-or-less indefinitely. The challenge therefore is to utilize a combination of these strategies, but importantly to recognize that there should be pragmatic flexibility in future parasite control programmes for sheep and goats that allow for the possible need of occasional, selective use of effective anthelmintics.

Session 33 Theatre 7

Sustainable internal parasite control in Australian Merino sheep

L.J.E. Karlsson and J.C. Greeff, Great Southern Agricultural Research Institute, 10 Dore Street, Katanning, WA 6317, Australia

For the last 50 years internal parasite control has come to rely heavily on chemical control options. In some environments this has been target at periods when most of the parasite population will be exposed to the treatment, this was initially considered to be very effective. However, it is now recognised that this results in a very high selection pressure on the parasite population in favour of resistant genotypes. Anthelmintic resistance is now at a very high level and poses a major threat to sustainable parasite control. It is now recognised that sustainable parasite control should be based on Integrated Parasite Management (IPM). One component of IPM is to increase the host immune competence. The Rylington Merino resource flock was established in 1987. The main selection criterion for the first ten years was low worm egg count (WEC). The other economically important traits were also recorded. In the last five years important production traits have been incorporated into an overall selection index. The annual genetic response to a reduced WEC is over 2% without any adverse penalties in the production traits. However, a recent analysis of the data found a significant increase in hypersensitivity diarrhoea. The current breeding objective now aims at a reduction in both WEC as well as hypersensitivity diarrhoea as two independent traits.

Session 33 Theatre 8

Analyses of udder health in Valle del Belice dairy sheep using SCC

B. Portolano[1], V. Riggio[1], M.T. Sardina[1], R. Finocchiaro[1], J.B.C.H.M. van Kaam[2], [1]Dipartimento S.En.Fi.Mi.Zo.- Sez. Produzioni Animali, Viale delle Scienze, 90128 Palermo, Università degli Studi di Palermo, Italy, [2]Istituto Zooprofilattico Sperimentale della Sicilia "A. Mirri", Via Gino Marinuzzi 3, 90129 Palermo, Italy*

Intramammary infections (IMI) are a complex of inflammatory diseases of the mammary gland. Mastitis is one of the most frequent IMI affecting small dairy ruminants. Direct selection against clinical mastitis is difficult because mastitis is not widely recorded; therefore, somatic cell count (SCC) is promoted as selection criterion for mastitis resistance. A dataset containing 2.475 first-lactation Valle del Belice ewes from 14 flocks recorded from 1998 to 2003 was analysed. In total 116 sires with at least four daughters with a record were included in the pedigree file. Test-day IMI events were coded as a binary trait using a SCC cut-off value of 750.000 cells/ml. Analyses of risk factors for the binary IMI trait were based on a logistic regression model. The herd effect was the major risk factor for IMI. The risk of culling due to IMI increased for late season of lambing. A single trait threshold sire model was applied. The heritability calculated using a logit function was 0,16±0,072 very close to the expected value of 0,14 calculated with Dempster and Lerner's formula treating IMI as a continuous variable.

Session 34 Theatre 1

The link between animal health planning and record keeping for UK farm assurance schemes

Andy Butterworth, Clinical Veterinary Science, Langford, N Somerset, BS40 5DU, UK

Animal health planning is beginning to be perceived as a valuable aid to both animal management for the farmer, and as a tool in the inspection of farms by certification bodies and government agencies in the UK. Health planning takes its root from the records that some farmers keep to structure their calendars for events such as vaccination, worming, weaning, and management procedures such as clipping, dehorning etc. From these roots have grown the more complex health plans which have become mandatory for farms which sell their products into some retail chains, or who are certified to UK certification body standards. Health plans now record strategies for common farm health and production issues such as mastitis, lameness, pneumonia - and create a visible record of the success (and failure) of the planned activities with the aim of tackling a range of production challenges - and so enabling both the farmer and also the 'inspector' to gauge how effective previous health planning was, and so to move the health plan on and make it a 'living' and useful device. This presentation outlines the value and costs of health planning, and describes some of the problems and limitations of formal health planning experienced to date in the UK.

Session 34 Theatre 2

New behaviour based developments to collect health information in dairy cattle farming

C. Winckler, University of Natural Resources and Applied Life Sciences (BOKU), Department of Sustainable Agricultural Systems, Vienna, Austria

Dairy cows experience a high incidence of several diseases which reduce productivity, cause high costs and impair the animal's welfare state. Traditional health related informations such as body condition scoring and the analysis of milk performance testing records can be regarded as mid- to long-term indicators and mainly refer to the nutritional status. However, there is an increasing need for the early detection of signs of disease including lameness, metritis, mastitis or metabolic disorders which enables the farmer to minimize or to control clinical outbreaks. Changes in behaviour may indicate animals at risk but are often difficult to recognize due to reduced contact times per animal and/or uncertainty of subjective diagnoses.

This paper will review current promising approaches for automatically monitoring health in commercial dairy cattle farming based on the electronically supported measurement of behavioural indicators. For example, automatic lameness detection systems include force plate or load cell measurements obtained from single cows in a walk-through area. Accelerometer technology allows activity monitoring with regard to lying, standing and stepping and may provide useful information on the general health status. Furthermore, transponder based feed alley monitoring systems allow for passive recording of feeding behaviour of individual cows and have been shown to identify transition cows at risk for metritis. Computerized sound monitors are currenly developed to even monitor rumination time and patterns in order to early detect metabolic diseases.

Session 34 Theatre 3

Health and welfare management of pigs based on slaughterline records

A. Velarde[1], E. Fàbrega[1], X. Manteca[2], [1]Institut de Recerca i Tecnologia Agroalimentaries (IRTA) Monells, Spain, [2]Universitat Autònoma de Barcelona, (UAB) Bellaterra, Spain*

Health and welfare management of pigs are relevant issues to optimise productivity. Diseases and injuries are important elements when monitoring health and welfare. On farm, disease is assessed by observation of symptoms and behavioural signs. Evaluation of injuries includes inspection of skin lesions or tail and ear wounds due to aggression or biting, respectively. However, these measures are taken in large groups, in dirty animals or when insufficient light is available. Theses constraints may compromise sometimes reliability and feasibility.

At slaughter, carcass and viscera examination, permit the evaluation of skin lesions, and tail and ear wounds, and the identification of diseases. In infection, affected lymphatic nodes become swollen and abnormal in colour. Conditions such as pneumonia or porcine atrophic rhinitis have characteristic lesions. Validity and reliability are high. However, to be a feasible and valid method, carcass identification should be kept throughout the process.

Determination of acute plasma proteins (APPs) in blood after sticking gives valuable information on clinical and even subclinical disease on farms. Furthermore, several reports have suggested that APPs could be good indicators for the assessment of animal welfare.

At the slaughterline, animals from several farms can be sampled on the same day, reducing the risk of disease transmission. However, to use the slaughterline records to improve health and welfare, a feedback system to the farm should be installed.

Session 34 Theatre 4

Pathogen records as a tool to manage udder health

M. Koivula, E.A. Mäntysaari, MTT, Agrifood Finland, Animal Production Research, Animal Breeding, FIN-31600 Jokioinen, Finland

Mastitis remains one of the major diseases in dairy herds causing profound economic losses to whole milk production chain. Strategies to reduce mastitis are important in decreasing costs and improving the quality of production, as well as for animal welfare reasons. Still Nordic countries are the only countries with national health recording system and, for example, the only countries including clinical mastitis directly into selection objectives of dairy cattle. In addition to health recordings where mastitis cases are reported, a database containing pathogen specific information of mastitis has been established in Finland. There bacterial information about type of mastitis is recorded on individual cow basis. Recording practice starts when veterinarians send milk samples from infected cows to laboratories. After the analyses, bacteriological results are automatically sent to the database to be joined with other recording data. After the first year of full operation information from about 47000 milk samples was entered to the data. This included of 27000 cows from 4000 herds. According to the collected pathogen data, *Staphylococcus aureus* (15.6%) and CNS bacteria (21.3%) cause most of the mastitis infections in Finland, but also different streptococci and coliform species are important udder pathogens. In the future combining of health, production and somatic cell count records to the pathogen information will give a more effective tool for the udder health program and also for selection of more resistant animals.

Session 34 Theatre 5

Can changes in milk composition at udder quarter level be used to detect udder disturbances?

K. Svennersten-Sjaunja[1], I.Berglund[1], G. Pettersson[1], K. Östensson[2]*

[1]Kungsängens research Centre, Swedish University of Agricultural sciences, Uppsala, [2]Box 70XX, Uppsala, Sweden

The aim of the present study was to investigate whether milk composition is changed in relation to a moderate increase in milk SCC in separate udder quarters. During 13 weeks, 4158 quarter milk samples from 68 cows were collected and analysed for milk SCC and milk composition. For calculations three groups of cows were formed according to their SCC value, group 1 was considered healthy, group 2 had one udder quarter with an increased SCC lasting for one sampling occasion while for group 3 the increase remained for more than one consecutive sampling occasion. Differences within pairs of udder quarters in the parameters fat, protein and lactose were calculated, and tested if the difference was separated from 0. All differences between udder quarters were calculated within rear and front quarters respectively. For both group 2 and 3 cows the lactose content decreased significant simultaneously with the increase in SCC. For group 3 cows, the lactose levels remained depressed after the initial increase in milk SCC. In conclusion, differences in lactose content within front and rear quarters, respectively, may be a useful tool for detection of increased SCC values in separate udder quarters and can therefore be recommended in on-line systems, since lactose is the milk component with the lowest relative day-to-day variation.

Session 34 Theatre 6

The use of a database for genetic evaluation and to manage health in dairy cows

S. Karsten[1], E. Stamer[2], W. Junge[1], W. Lüpping[3], E. Kalm[1], [1]Institute of Animal Breeding and Husbandry, Christian-Albrechts-University, 24098 Kiel, Germany, [2]TiDa Tier und Daten GmbH, Brux, Germany, [3] Landwirtschaftskammer Schleswig-Holstein, Kiel, Germany*

The quantity of data recorded for each animal will continue to increase in the future. Records computerised in management programs can be used for operational analyses but the information systems are of limited use for research and the evaluation of experiments. Problems will arise especially if analyses are performed across farms. Various software systems and data sources and unstandardised data codes make evaluations difficult.
A database has been developed to organise data of numerous farms. Records of parentage, feeding, fertility, health and individual performance are electronically transferred between management programs and the database. Irregularly occuring data can be entered on-line via an electronic form. An accurate and consistent data retention is guaranteed by validity checks. The data can be compiled, exported and analysed easily because of a uniform coding system. A user interface will be provided for on-line access. Data processing by a statistical program package is also be possible. Health records can be analysed easily for example by control charts. In the moment the database is applied on two research farms. On one farm data of trials like feeding experiments are collected, on the other farm functional traits are measured within the scope of a performance test and evaluated genetically.

Session 34 Theatre 7

Mastitis detection in dairy cows by application of Fuzzy Logic

D. Cavero[1], K.-H. Tölle[1], C. Buxadé[2], and J. Krieter[1], [1]Institute of Animal Breeding and Husbandry, Christian-Albrechts-University of Kiel, D-24098 Kiel, Germany, [2]Department of Animal Production, ETSIA, Polytechnic University of Madrid (UPM), 28040 Madrid, Spain*

The aim of the present research was to develop a fuzzy logic model for classification and control of mastitis for cows milked in an automatic milking system. A data set of 420,083 milkings from 478 cows were accumulated. The data were divided in two data sets, training and test data.
Cases of mastitis were determined according to three different definitions, by taking into account udder treatments and somatic cell counts.
The alerts were generated by a fuzzy logic model connecting electrical conductivity, milk yield, milk flow rate and time between milkings. Sensitivity for a mastitis case was set to a minimum of 80%. The parameters specificity and error rate were applied for evaluation of the reliability of the detection model. Specificities ranged between 91.3% and 71.2% and the error rate varied between 97.3% and 41.6% depending on mastitis definition. The results for the test data not only verified those for the training data, but were also slightly better.
The obtained results were satisfying for sensitivity and specificity. However the error rate still remained very high. Fuzzy logic has been used to develop a detection model for mastitis, which could be used in the future as an easy system for the farmer, without need of great expertise.

The relative day-to-day variation in milk yield and composition for cows milked two or three times daily

K. Svennersten-Sjaunja[1], U. Larsson[1],J.Bertilsson[1] and L-O Sjaunja[2], [1]Swedish University of Agricultural Sciences, Kungsängens Research Centre, 753 23 Uppsala, [2]L-O Sjaunja AB, Kungsgatan 119, 753 18 Uppsala, Sweden

Management decision on farm is based on data from the monthly official milk recordings scheme. Do these recordings provide enough information for management purpose? To answer this the relative day-to-day variation of milk yield and components has to be known, for both twice and thrice daily milkings. The relative day-to-day variation was studied during five weeks. Cows from the university herd (Swedish Red and White breed) participated in the study, 19 cows were milked 2x and 24 3x daily. The relative day-to-day variation was estimated within cow and week and was for cows milked twice daily 5.3%, 5.6%, 1.6%, 1.4% and 3.1% for yield, fat, protein, lactose and citric acid respectively, while for cows milked three times daily it was 4.9%, 5.3%, 2.4%, 1.3% and 4.2%. To give an example, the obtained results for twice daily milking implies that one monthly recording for a cow with actual production of 30 kg milk with 4% fat and 3% protein can have records for yield in the range of 26.8-33.2 kg, fat content 3.6-4.5% and protein 2.9-3.1%. Deviations from the range can be used as diagnostic tools. For management decisions frequent recordings can be recommended since the relative day-to-day variation is quite large.

Milk flow and udder health in dairy cows

K.-H. Tölle[1], U. Tölle[2] and J. Krieter[1], [1]Institute of Animal Breeding and Husbandry, Christian-Albrechts-University of Kiel, Germany, [2]Landeskontrollverband Schleswig-Holstein e.V. Kiel, Germany*

Milk flow parameters (LactoCorder) and somatic cell counts were recorded between January 2002 and Mai 2004 on 393 farms in Schleswig-Holstein from 352,521 milkings of 24,900 German Holstein dairy cattle. Data set was randomly divided into 10 independent subsets for statistical analysis. A mixed model for somatic cell score (SCS) was performed separatly for all subsets with the random effect of the cow and the fixed effects milk yield, stage of lactation, number of lactation, testday within herd, average milk flow (alternative the duration of milking was tested as well) and bi-modality (yes/no) of milk flow curve.

All fixed effects in the model were high significant ($p < 0.001$). Only for bi-modality of milk flow curve no significant influence was found.

The most conspicuously fixed effect on SCS was milk yield. From lowest milk yield class (< 7 kg) to highest class (> 20 kg) SCS was halved. Comparably obvious was the effect of lactation number. SCS clearly increased from 2.2 in first lactation to 3.8 in lactations ≥6.

The average milk flow was significant but the differences of SCS between milk flow classes were moderate. Only cows with lowest average milk flow (< 2 kg/min) had evident higher SCS. Results for the effect duration of milking were comparable to average milk flow.

Session 34 Poster 10

Oestrus detection in dairy cattle using ALT-Pedometer and Fuzzy-Logic

St. Pache[1], C. Ammon[2], U. Brehme[3], H.J. Rudovsky[1], R. Brunsch[3], J. Spilke[2], U. Bergfeld[1]
[1]Saxon State Institute for Agriculture, Department of Animal Production, Am Park 3, 04886 Köllitsch, [2]Martin-Luther-University Halle, Agricultural Faculty, Ludwig-Wucherer-Straße 82-85, 06108 Halle/Saale, [3]Institute of Agricultural Engineering Bornim, Max-Eyth-Allee 100, 14469 Potsdam, Germany

Oestrus intensity, oestrus duration and cycle length are negative correlated to the milking traits. In order to support the oestrus detection a new developed four canal pedometer system was tested in practise which measures the number of steps, the lying position and the temperature.
Ten cows were sampled after calving and equipped with the so called ALT-Pedometer. These cows were observed for a period of 6 month. The parameters of the system were stored each 15 minutes. All together 18.432 records per cow were collected.
For oestrus detection five different fuzzy-logic-models with up to four combinations of input variables (steps, lying periods, days after heat and percentage of activity) were developed. The validation of the models was performed on the basis of detailed observations of the oestrus and of the date of conception. The quality of the models was evaluated with the rate of oestrus detection and the error rate. With two models a oestrus detection rate of 84% with a maximum error rate of 17% were arrived.
Further applications of the system to predict the health situation or an expected calving are possible.

Session 35 Theatre 1

Modelling repeated measures of ejaculate volume of Holstein bulls using a Bayesian random regression approach

M. Serrano, M.J. Carabaño, C. Díaz, Dpto. Mejora genética Animal, INIA Ctra de La Coruña Km 7,5, 28040 Madrid, Spain*

Data of 8524 ejaculates from 213 Holstein bulls and collected between 400 and 1000 days of age from year 1988 to 2002 were used in this study. Repeated measures per bull ranged between 7 and 83 observations. The pedigree file included 1140 animals. All models used to describe the ejaculate volume at different ages included a year-month of collection effect and fixed and random Legendre polynomial regressions on days of age at collection. Other systematic effects such as year and age at first collection and different degree polynomials were considered in the alternative models analysed. Inferences on the unknown parameters of interest were carried out in a Bayesian framework. Bayes factors (BF) and cross validation predictive densities (PD) were used as criteria to compare models. Results showed small differences among models in terms of genetic parameter estimates or in the statistics measuring goodness of fit and predictive ability. Estimates of heritability of ejaculate volume along time were around 0.20 for most of the studied period and showed values up to 0.32 in the final stages. The model including year-month of collection, fixed regression fourth order polynomial fitted within year of entry and fourth order polynomials fitted to genetic and permanent environmental components was best according to the BF and PD criteria.

Session 35 Theatre 2

Genetic and environmental effects on semen quality of Austrian Simmental bulls

B. Gredler[1,2], C. Fuerst[2], B. Fuerst-Waltl[1], H. Schwarzenbacher[1] and J. Sölkner[1], BOKU - University of Natural Resources and Applied Life Sciences Vienna, Department of Sustainable Agricultural Systems, Gregor Mendel Str. 33, 1180 Vienna, Austria, [2] ZuchtData EDV-Dienstleistungen GmbH, Dresdner Str. 89/19, 1200 Vienna, Austria*

Semen production data from an Austrian AI centre collected between 2000 and 2004 were analysed. In total, 13,377 ejaculates from 301 AI bulls were examined considering different effects on ejaculate volume, sperm concentration, percentage of viable spermatozoa in the ejaculate, total spermatozoa per ejaculate and motility. The model included the fixed effects of age of bull, collection interval, number of collections on collection day, bull handler, semen collector, day of week, temperature on day of semen collection, average temperature during epididymal maturation, average temperature during spermatogenesis and length of day, and a random additive genetic component. Age of bull, collection interval and number of collection on collection day significantly affected most semen quality traits. The collection team (bull handler and semen collector) had relevant effects on semen traits. All semen production traits were moderately heritable and correlated. Heritabilities for volume, concentration, percentage of viable spermatozoa, total spermatozoa and motility were 0.30, 0.38, 0.24, 0.23 and 0.10, respectively. Correlations between estimated breeding values of sperm quality traits and routinely estimated breeding values for male fertility were low and ranged from 0.10 to 0.20.

Session 35 Theatre 3

Effect of body condition on dairy and reproductive performance in Holstein-Friesian cows

E. Szücs[1], J. Püski[2], Tran Anh Tuan[1], A. Gáspárdy[1], and J. Völgyi-Csík[3], [1]Szent István University, Gödöllö, [2]Formilk Ltd., Telekgerendás, [3]Research Institute for Animal Breeding and Nutrition, Herceghalom, Hungary*

Body condition scores (BCS) were registered in Holstein-Friesian dairy cows at monthly intervals within the framework of National Milk Recording Scheme using a 1-5 scale established by *Wildman et al.* (1982). The division size between scores was reduced to 0.25 to improve accuracy. The change of BCS was characterized by SD of recordings throughout the lactation lasting for 345 days in milk-Distribution of cows among 1st to 4th parities were 186, 136, 93, and 36; the overall means as well as SE_ms (N=452) for milk yield adjusted for 305 day standard lactation, butterfat and milk protein percentage, average and peak daily yield, and persistency were 8788±82.8; 3.49±0.024; 3.26±0.178; 29.4±0.53; 39.1±1.06, and 68.5±0.55, respectively. Mean values for conception rate (CR) and days open (DO) were 1.89±0.043; 99.2±1.74. In this study, curvilinear relationships between BCS and production and reproduction traits have been established at high level of probability with significant linear and quadratic effects ($P<0.001$, $P<0.01$, in a few cases $P<0.05$). Regression analysis reveal that for milk yield the optimum range of BCS in the post partum and midlactation period varies between 3.0-4.0, and 3.5-4.5 just prior to dry period. Favourable effect in cows with lower BCS in all phases of reproduction cycle has been established, as well.

Session 35 Theatre 4

The relationship between lactation persistency and reproductive performance in New Zealand dairy cattle

N. Lopez-Villalobos[1], L.R. McNaughton[2], R.J. Spelman[2], [1]Massey University, Palmerston North, New Zealand, [2]Livestock Improvement Corporation, Hamilton, New Zealand*

The objective of this study was to investigate the relationship between lactation persistency and fertility traits in first lactation grazing cows in New Zealand. Milk herd-test records from 810 first lactation cows were used to obtain individual lactation curves using a random regression model fitting five knots splines. All cows were second cross Friesian x Jersey from a crossbreeding experiment established for quantitative trait loci identification. Lactation persistency was defined as the proportion of milk yield from day 121 to day 180 with respect to milk yield from day 1 to day 60. Cows were classified into one of four quartiles evenly distributed according to lactation persistency. Measurements of reproductive performance, typical of seasonal grazing systems were calculated for each quartile. Compared to low persistency cows (quartile 4), high persistency cows (quartile 1) had higher total milk yield (3174 vs. 2900 kg; $P<0.001$), lower peak yield (15.1 kg vs. 16.3; $P<0.001$), higher proportion of cows cycling at 42 days after calving (71 vs. 36%; $P<0.001$) and at the start of mating season (PSM) (98 vs. 84%; $P<0.001$) but they had similar proportion of pregnant cows at 24 days after PSM (70 vs. 66%) and at the end of the mating season (93 vs. 92%). These results suggest that there is a relationship between reproductive performance and lactation persistency in New Zealand dairy cattle.

Session 35 Theatre 5

Normal and atypical progesterone profiles in Swedish dairy cows

K.-J. Petersson[1], H. Gustafsson[2], E. Strandberg[1] and B. Berglund[1], [1]Dept. of Animal Breeding and Genetics, Swedish Univ. Agric. Sci., PO Box 7023, SE-750 07 Uppsala, Sweden, [2]Swedish Dairy Assoc., PO Box 7039, SE-750 07 Uppsala, Sweden*

Decreasing fertility and increasing proportion of atypical progesterone profiles in dairy cows have been reported from many countries. The aim of this study was to investigate if milk progesterone analysis in the regular milk recording could be used for reproductive decisions about individual cows and to improve the genetic evaluation for fertility. Data were collected from the university experimental herd during 1987-2002 and included 1049 *post partum* periods from 509 cows of two different breeds (Swedish Holstein and Swedish Red and White). Progesterone samples were taken twice a week and individual progesterone profiles were classified into four different categories, with distribution in brackets: normal cyclicity (70%); delayed cyclicity (16%); cessation of cyclicity (7%); and prolonged luteal phase (7%). Increased probability of atypical profiles were observed for Holsteins compared to Red and White cows, first parity cows compared to older cows, cows that calved during the winter compared to summer season, cows in tie stalls compared to those in loose housing and cows treated for lameness compared to healthy cows. The proportion of samples with luteal activity within the first 60 days after calving was related to the four different profiles and is a promising measure for further investigation.

Session 35 Theatre 6

Oestrus detection in dairy cows using control charts and neural networks

J. Krieter, E. Stamer, W. Junge, Institute of Animal Breeding and Husbandry, Christian-Albrechts-Universität, D-24098 Kiel, Germany*

Exponentially weighted moving average control charts and neural networks were used for oestrus detection in dairy cows. The analysis involved 372 cows, each with one verified oestrus event. Model inputs were the traits activity, measured by pedometer, and the period (days) since last oestrus. In total 10,386 records were available, which were partitioned into training and validation subsets to train and test the neural network (multifold cross-validation). When the trained network was applied to the validation sets (acticity and period since last oestrus), the averaged sensitivity (SE), specificity (SP) and error rate (ER) were 77.5, 99.6 and 9.2 %, respectively. If the input of the model was restricted to the trait activity, SE, SP and ER were 75.3, 99.3 and 17.8 %. Performance for the same data with the univariate control chart was less successful (66.9%, 99.3%, 18.8%). Neural networks are useful tools to improve computerised oestrus detection in dairy cows.

Session 35 Theatre 7

Bioenergetic factors affecting conception rate in Holstein cows

J. Patton[1, 2], D. Kenny[2], F. O'Mara[2], J.F Mee[1] and J.J. Murphy[1], [1]Teagasc, Dairy Production Research Centre, Moorepark, Fermoy, Co. Cork, Ireland, [2]Faculty of Agriculture, University College Dublin, Ireland*

The objective of this study was to examine associations between early lactation energy balance (EB), milk production, and net energy intake (NEI), and the likelihood of conception to first service (CON1) in Holstein cows. Data from 96 cows, collated over two studies, were analysed using logistic regression. Milk production, milk composition and NEI were recorded daily for the first 28 days of lactation. First service occurred after a voluntary waiting period of 65 days. Continuous variables were quartiled and reference categories defined. Cows with the most negative EB (<-5.3 UFL/d) had a lower likelihood of CON1 (Odds Ratio (OR)=0.19, P=0.034) compared to those with the most positive EB (>-1.94 UFL/d). The group with highest NEI had a higher likelihood of CON1 than each of the other groups (OR>10, P<0.05). The likelihood of CON1 was lower for cows in the lowest milk protein concentration group (<32.0g/kg) (OR=0.24, P=0.043) compared to those in the reference group (>35.5g/kg). Neither milk yield nor milk energy output were associated with the likelihood of CON1. In conclusion, EB status during early lactation is positively associated with conception rate at first service. This may be mediated to a greater degree by NEI rather than milk yield. Milk protein percentage may be useful for identifying cows at risk of poor fertility.

Effect of honeybee royal jelly on the nuclear maturation of bovine oocytes in vitro

M. Kuran[1], E. Sirin[1], E. Soydan[1], A.G. Onal[2], Department of Animal Sciences, Universities of [1]Ondokuz Mayis, Samsun, Turkey and [2]Mustafa Kemal, Hatay, Turkey*

Serum supplementation to culture media has been shown to be detrimental to embryonic development and considered as one of the factors causing Large Offspring Syndrome following the transfer of in vitro derived bovine embryos. Therefore, this study aimed to investigate whether honeybee royal jelly, which stimulates embryonic development and metamorphosis in honeybees, can be used as a protein source instead of serum in in vitro maturation of cattle oocytes. Bovine oocytes were aspirated from 2 to 8 mm diameter ovarian follicles and oocytes were matured either in 10% (v/v) fetal calf serum (n=229 oocytes) or 1% (w/v) honeybee royal jelly (n=239 oocytes) supplemented TCM-199 under a humidified atmosphere of 5% CO_2 at 38.6 °C for 22 h. The ratio of oocytes reaching Metaphase-II stage of nuclear maturation did not differ between treatment groups ($P>0.05$) and it was 79.6% in serum-supplemented group and 81.4% in royal jelly supplemented group. Following the parthenogenetic activation of a limited number of metaphase-II stage oocytes, 36% (22/62) of oocytes matured in royal jelly supplemented media and 42% (25/59) of oocytes matured in serum supplemented media cleaved to 2-cell and beyond at 48 h post activation on a granulosa cell monolayer under mineral oil. The cleavage rate did not differ between treatment groups ($P>0.05$). These results show that 1% royal jelly can be used successfully as a protein source in in vitro maturation of bovine oocytes.

Authors index

	Page
A	
Aarnink, A.J.A.	236
Aarts, H.F.M.	312
Abbas, K.	293, 345
Abd El-Razek, S.	280
Abdella, A.	169
Abdoli, H.	62, 166
Abdou, T.A.	70
Abdullah, A.Y.	277, 283
Abdullah, M.	148
Aberle, K.	85
Abou-El-naga, T.R.	70
Ábrahám, Cs.	229, 230, 251, 258
Abrescia, P.	142, 287
Acero, R.	292, 293, 354
Ács, T.	152
Ådnøy, T.	270, 276, 362
Agata, M. D'	256, 256
Agena, D.	342
Ahadi, A.H.	177
Ahangari, Y.J.	157, 284
Ahola, V.	10
Ahouei, M.	370
Aihara, M.	209
Ait-Yahia, R.	205
Aizinbud, E.	183
Ajmone-Marsan, P.	372
Akbar, J.	148
Albarrán-Portillo, B.	260
Alghamdi, A.S.	126
Alikhani, M.	171, 181
Alipour, D.	149
Allam, A.M.	168
Allen, P.	218, 230
Al-Momani, A.Q.	277, 283
Alshaikh, M.A.	296
Altmann, M.	137
Alvarez, F.D.	110
Alvarez, J.C.	184
Álvarez, I.	94, 95, 98, 367
Alves, E.	87, 369
Alves, T.C.	172
Alves, V.	356
Amanlo, H.	134
Amarger, V.	87
Amenabar, M.E.	294
Amer, P.	320, 342
Amills, M.	369
Aminafshar, M.	97
Amini, J.	168
Ammon, C.	383
Anacker, G.	44
Andersen, A.H.	193
Andersen, B.H.	326, 329
Andersen, H.R.	313
Andersson, H.	72, 357
Andersson, H.K.	326
Andersson, I.	178
Andersson, K.	76, 314, 326
Andersson, L.	11, 358
Andersson, M.	10, 237, 316
Andersson, M.A.	234
Andrásofszky, E.	135
Andrieu, S.	165, 278
Antonescu, C.	206
Antunovic, Z.	133, 279
Aouissat, M.	102
Aparicio, M.	238
Aragaw, K.	69
Arakane, T.	208
Aral, F.	281
Araújo, J.P.	251, 262
Arendonk, J.A.M. van	11, 194
Arnason, Th.	330
Arque, M.	369
Arranz, J.	272
Arshami, J.	145
Asadi-khoshoei, E.	209, 210
Asadzadeh, N.	89, 374
Ashour, G.	290
Ask, B.	197
Asmini, E.	114
Austbø, D.	303
Avendaño, L.	110, 196
Awadalla, I.M.	351
Ayalew, W.	69, 344
Aygun, A.	244
Ayllón, S.	238
Aytac, M.	202, 297
Azaga, I.A.	281

	Page
Azamel, A.A.	107
Azarfar, A.	62
Azari, M.	134
Azimifar, B.	89
Azor, P.J.	88

B

	Page
Baars, T.	312
Baas, T.J.	74, 75, 120
Babaei, M.	217
Baban, M.	28, 308
Babot, D.	7
Bacila, V.	180, 225
Baculo, R.	142
Báder, E.	250, 261, 262
Báder, P.	262
Bækbo, P.	128
Baghaei, E.	160, 292
Bagheri, M.	170
Bagnato, A.	253
Bahamonde, A.	94
Bailoni, L.	147
Balogh, K.	230
Bán, B.	25
Banabazi, M.H.	101
Barac, Z.	97
Baranyai, G.	271
Bareille, N.	66
Barkawi, A.H.	290
Barone, A.	106, 112, 113, 266, 267
Barra, R.B.	196
Barragán, C.	14, 87, 369
Bartelt, J.	243
Barton, L.	248, 255
Bartos, Á.	150
Bauchart, D.	257
Baulain, U.	227, 275
Baumung, R.	83, 85, 265, 295, 301
Båvius, A.K.	178
Beard, A.P.	136
Beaudeau, F.	66, 67
Beaufrère, R.	304
Bébin, D.	313
Bech, A.C.	329
Beerda, B.	174, 245, 322, 343
Beghelli, D.	106, 113
Beglinger, C.	2
Beja-Pereira, A.	94, 367
Belgasem, B.M.	350
Belliard, M.	67
Beltrán de Heredia, I.	272
Bene, Sz.	250
Benkel, B.F.	186
Bennewitz, J.	46, 85, 212, 364
Benoit, M.	311
Berg, Å.	298
Berg, P.	264
Bergero, D.	335
Bergfeld, U.	49, 274, 383
Berglund, B.	46, 72, 385
Berglund, I.	380
Bergman, M.	185
Bergsten, C.	175
Bernabucci, U.	176
Bernard, L.	161
Bernués, A.	297
Berry, D.P.	53, 342
Bertilsson, J.	59, 382
Bertoni, G.	106, 176
Bethge, M.	52
Biagini, D.	315
Bidanel, J.P.	12, 122, 213
Biffani, S.	197
Bigaran, F.	270
Bigeriego, M.	41, 41
Bigi, D.	340, 367
Bijma, P.	119, 121, 319, 337
Bijttebier, J.	224
Billon, Y.	12
Binder, S.	78
Bini, P.P.	71, 274
Bionaz, M.	106
Birchmeier, A.N.	340
Birgele, E.	189, 189
Bishop, S.C.	320, 337
Bitti, M.P.L.	8
Bizhannia, A.R.	195
Björnsdóttir, S.	335
Blanch, A.	328
Blanco Roa, E.N.	117
Blasco, A.	92
Blobel, K.	31
Blokhuis, H.J.	66
Blouin, C.	27
Blum, J.W.	300
Blüthgen, A.	2

	Page
Bobillier, E.	*241*
Bodin, L.	*80, 282*
Bodó, I.	*17, 25, 204, 306*
Boehme, H.	*60, 150, 242*
Boettcher, P.J.	*368, 368*
Bogdanovic, V.	*51, 199*
Bogenfürst, F.	*298*
Bolet, G.	*80*
Bonanno, A.	*266*
Bonekamp, P.R.T.	*80*
Bonin, M.N.	*172*
Bonneau, M.	*323, 346*
Bonnet, M.	*248, 257*
Boonyanuwat, K.	*13*
Borda, E.	*154*
Borell, E. von	*137, 222*
Boros, N.	*262*
Bossche, K. Van den	*72*
Bosselman, F.	*247*
Bossen, D.	*144, 247*
Botermans, J.A.M.	*234, 316*
Bottura, C.	*251*
Boulesteix, P.	*257*
Boushaba, N.	*205*
Bouska, J.	*365*
Bovenhuis, H.	*11, 194, 312, 337, 362*
Boyle, L.	*348, 354*
Bozkurt, Y.	*58*
Brabander, D.L. de	*133, 134, 167, 233, 234, 239, 249*
Brade, W.	*52, 275*
Bradic, M.	*97*
Braglia, S.	*370, 373*
Brandt, H.	*269*
Brandt, Y.	*108*
Brehme, U.	*25, 383*
Brem, G.	*100*
Brøkner, C.	*59*
Broz, J.	*329*
Bruckmaier, R.M.	*261*
Bruessow, K.P.	*104*
Bruneteau, E.	*280*
Bruns, E.W.	*303, 333*
Brunsch, R.	*383*
Bryant, J.R.	*321*
Bryedl, E.	*261*
Brzostowski, H.	*276*
Buchor, Y.	*25*

	Page
Buckley, F.	*119*
Budelli, E.	*99*
Buduram, P.	*96*
Buitenhuis, A.J.	*363*
Bujdosó, M.	*254*
Bünger, L.	*263, 320*
Buraczewska, L.	*155*
Bures, D.	*248, 255*
Burger, D.	*305*
Busche, M.	*333*
Buttazzoni, L.	*373*
Butterworth, A.	*378*
Buxadé, C.	*381*
Buys, N.	*219, 224*
Buzás, Gy.	*37*
Byrne, D.V.	*327*

C

Cabaraux, J.F.	*138*
Caelenbergh, W. van	*249*
Caetano, A.R.	*369*
Cagnazzo, M.	*370*
Caja, G.	*6, 7, 40*
Calamari, L.	*176*
Calin, I.	*180, 225*
Calouro, F.	*356*
Calus, M.P.L.	*174, 245, 322, 343*
Camacho-Morfin, D.	*158*
Campanile, C.	*372*
Campeneere, S. de	*133, 134, 249*
Campo, J.L.	*202*
Canario, L.	*122*
Canavesi, F.	*197*
Canepa, S.	*282*
Cantalapiedra, J.	*251, 262*
Cantet, R.J.C.	*340*
Cappa, E.P.	*340*
Cappai, M.G.	*9*
Caput, P.	*33*
Caputi Jambrenghi, A.	*162*
Carabaño, M.J.	*213, 383*
Caracappa, S.	*376*
Carballo, J.A.	*255*
Carcangiu, V.	*71, 116, 274*
Caritez, J.C.	*12*
Carlén, E.	*44, 47*
Carnevali, A.	*109*
Carvalheira, J.	*211*

	Page
Casabianca, F.	114
Casal, J.	376
Casamassima, D.	300
Casimiro, T.R.	172, 172
Cassar-Malek, I.	257
Castaldo, A.	292
Castelo Branco, M.A.	356
Castiglioni, B.	204, 361
Catalano, A.L.	32
Catillo, G.	94, 164
Cavallina, R.	112
Cavero, D.	381
Cecchi, F.	87
Cefis, A.	184
Cengarle, L.	116
Cenkvári, E.	135
Cerolini, S.	12
Cervantes, I.	26
Cervantes, M.	152
Chang, K.C.	370
Chang, Y.M.	45
Chardon, P.	9
Chaudhry, A.S.	141, 149
Chemineau, P.	282
Chen, G.	326
Cheon, S.S.	187
Chessa, S.	204
Chilliard, Y.	136, 161, 257, 268, 280
Chiofalo, B.	28, 227
Chiofalo, V.	28, 227
Cho, I.C.	93, 93
Choi, H.C.	187
Choi, Y.L.	93
Christensen, J.W.	332, 333
Christensen, L.G.	192
Christofides, C.	291
Ciampolini, R.	87
Ciani, E.	87
Ciprovica, I.	180
Cirera, M.	236
Claeys, E.	138, 219
Clausen, S.	34
Cobuci, J.A.	50, 196
Cocca, C.	279
Colaço, J.	251
Colella, G.E.	300
Colombo, M.	367
Comellini, F.	8

	Page
Concordet, D.	332
Congiu, G.	94
Conill, C.	6
Conington, J.	320
Conte, G.	204
Cooper, S.	346
Copado, F.	152
Cornelissen, S.	11
Correa, A.	110, 196
Costa, C.N.	50, 196
Costantini, M.	24
Coulon, C. de	20
Crepaldi, P.	95, 109
Crevier-Denoix, N.	332
Cringanu, I.	212
Cromie, A.R.	55, 318, 342
Crosby, E.J.	278
Crosby, F.	165
Csapó, J.	132, 257
Csatári, G.	311
Csató, L.	220
Cuello, C.	129
Curik, I.	23, 28, 83, 85, 124
Cushion, M.	35
Custura, I.	156
Czarnecki, R.	225, 238
Czyrska, K.	22

D

da Costa, N.	370
Dahl, E.	325
Dákay, I.	55
D'Alessandro, A.G.	300
Dalin, A.-M.	76, 103
Dalin, G.	302
Dall'Olio, S.	340
Dalton, G.E.	65
Danesh Mesgaran, M.	58, 146, 289, 290
Danfær, A.	56
Danieli, P.	176
Danielsson, D.-A.	357
Das, G.	272
Daszkiewicz, T.	185
Dávila, S.G.	202
Davoli, R.	340, 370, 371, 373
Davtalabzarghi, A.	268
Dawson, L.E.R.	139
De Cadolle, D.	16, 18

	Page
De Lorenzo, A.	142
de Nicolo, G.	265
Dedieu, B.	38
Dekkers, J.C.M.	339
Delikator, B.	225, 238
Delmonte, E.	95
Delogu, G.	8, 9
Demeure, O.	12
Dempfle, L.	27, 198
Denoix, J.M.	332, 334, 335, 336
Depuydt, J.	98
Derecka, K.	13
Derrar, A.	205
D'Esposito, M.	372
Desvousges, A.	126
Dettori, M.L.	116, 274
Dhali, A.	286, 287, 288
Dhimi, L.	102
Díaz, C.	213, 383
Díaz, R.	110
Dikic, M.	228
Dimov, D.	115
Dinita, G.	156, 206, 210
Dinparast, N.	96
Dinu, C.	111
Diskin, M.G.	252
Distl, O.	85, 199, 334, 375
Diverio, S.	106, 112, 113, 266, 267
Djedovic, R.	51, 199
Dobson, H.	103
Dodenhoff, J.	50
Dohms, T.	15, 331
Dolezal, J.	147
Dolezal, M.	118, 340, 360
Dolezal, P.	147
Domacinovic, M.	279
Domokos, Z.	251, 254, 257
Donabedian, M.	335
Donn, S.	13
Doran, O.	325
Dourmad, J.-Y.	131, 346
Dransfield, E.	131
Drennan, M.J.	61
Drögemüller, C.	375
Drogoul, C.	304
Drucker, A.	295, 301
Druet, T.	213
Druml, T.	23, 85

	Page
Dublecz, K.	150
Dubois, C.	22
Dubravska, J.	115
Ducro, B.J.	21, 337
Ducrocq, V.	121
Dufrasne, I.	138
Dugo, P.	28
Dunne, W.	35, 38
Durand, D.	161
Duru, M.	39

E

Early, B.	177, 179
Edel, C.	27
Edriss, M.A.	203
Edvardsson, A.	178
Edwards, S.A.	136
Eggen, A.	372
Egger-Danner, C.	42, 54
Eik, L.O.	276
Eikje, L.S.	270
Eila, N.	171
Einarsson, S.	108
Ekkel, E.D.	130
Ekman, S.	175
El Fadili, M.	299
El-Afifi, T.M.	244
El-Arian, M.N.	217
El-Barody, M.	280
El-Battawy, K.A.	285
Elder, R.	240
Ellen, E.D.	119
Ellis, A.D.	57, 304, 306
El-Magdub, A.B.	350
Elo, K.	100
Elsaid, R.	48
El-Sherbiny, A.E.	244, 245
El-Waziry, A.M.	151
Elwinger, K.	315
Elyasi-Zarringabayi, G.	90, 91
Emanuelson, M.	64
Emanuelson, U.	44, 72
Emmans, G.C.	264
Emmerling, R.	50
Emrani, H.	373
Ender, K.	108
Engblom, L.	76
Eping, H.	370

	Page
Erhardt, G.	188, 269
Ericson, K.	302
Eriksen, L.	59
Eriksson, T.	64
Erlinger, D.	332
Ernst, K.	108
Eskandari Nasab, M.P.	88, 336
Eslinger, K.M.	222
Esmaeelkhanian, S.	96, 101
Esmaeilzadeh, M.	91, 374
Estany, J.	231
Ettle, T.	243

F

Fàbrega, E.	131, 379
Faghani, M.	291
Faheem, M.S.	290
Fallon, R.J.	139, 142
Falocci, N.	112, 113, 266, 267
Farhangfar, H.	200
Farhoodi, M.	70
Farid, A.	186
Farkas, J.	220
Farrag, F.H.	217
Faulconnier, Y.	161, 257
Fébel, H.	135, 229, 230
Feizi, R.	58
Fekete, S.	135
Feldmann, A.	270
Felix, E.	364
Ferlay, A.	136, 161, 268, 280
Fernández, A.	14, 344
Fernández, I.	94, 95, 98, 367
Fernández, J.	361
Fernàndez, X.	131
Fernández-Combarro, E.	94
Fernando, R.L.	339
Ferrandi, B.	109
Ferrara, L.	142, 144, 287, 372
Ferreira, A.V.	169
Ferreiro, J.M.	262
Ferri, N.	8
Fésüs, L.	273
Fève, K.	12
Fiaz, M.	105
Fiedler, I.	108
Fiedlerova, M.	260
Fiems, L.O.	249

	Page
Figueroa, J.L.	152
Fikse, F.	371
Filippini, S.	32
Finley, G.G.	186
Finnerty, N.	17
Finocchiaro, R.	99, 266, 376, 378
Fiore, G.	7
Fiorelli, C.	38
Firat, M.Z.	339
Fischer, E.	145, 246
Fischer, K.	130
Fischer, R.	49, 274
Fiske, W.F.	42
Flachowsky, G.	60, 242
Fleurance, G.	335
Flint, A.P.F.	13
Flori, L.	9
Floriot, S.	372
Flury, C.	84, 120
Fontanesi, L.	340, 371, 373
Fooley, M.	165, 278
Fornarelli, F.	95
Foroughi, A.R.	146
Forslid, A.	302
Fortier, G.	336
Fourichon, C.	66
Foury, A.	131
Fredeen, A.H.	346
Fredricson, I.	330
Fredriksen, B.	324, 325
Freitas, A.F.	50, 196
Freyer, G.	145, 246
Fries, R.	78
Fröberg, S.	254
Fuerst, C.	54, 384
Fuerst-Waltl, B.	265, 384
Füssel, A.-E.	18

G

Gaborit, P.	280
Gäde, S.	190
Gajewska, A.	253
Galletti, S.	156, 165, 287
Galli, S.	371
Galvao, L.	356
Gandini, G.C.	253, 368
Gandotra, V.	103
Gaouar, S.	102

Page

Garádi, Z. 150
García, A. 292, 293, 301, 354, 364
Garino, C. 361
Garreau, H. 80
Garrick, D. 318
Gart, V.V. 233
Gáspárdi, A. 384
Gauly, M. 188, 269, 272
Gautier, S. 19, 19
Geers, R. 6, 66
Geerts, N.E. 134
Georges, M. 224
Georgescu, Gh. 30
Gera, I. 204
Gergácz, Z. 261, 262
Gerken, M. 163
Gessa, J.A. 26
Ghahramani, A. 177
Ghanbari, F. 284
Gharadaghi, A.A. 90
Ghassemi, Sh.J. 139
Ghirardi, J.J. 6, 7, 40
Ghodratnama, A. 58
Ghoibaighi, A. 48
Gholami, M.R. 195
Ghoorchi, T. 148, 162, 163, 284
Ghorasi, A. 91
Ghorbani, A. 216
Ghorbani, G.R. 71, 148, 167, 168, 170, 171, 181, 352
Ghuman, S.P. 103
Giannico, F. 162
Gianola, D. 45
Gilbert, H. 339
Giraudo, C. 353
Gjerde, B. 341
Gjerlaug-Enger, E. 77
Gladyr, E. 100
Gleeson, D. 348, 354
Glodek, P. 242
Gogué, J. 12, 101, 122, 213
Gomez, A.G. 239
Gómez, E. 95, 98, 367
Gómez, M.D. 26
Goodwin, R.N. 74, 75
Gorgulu, M. 283
Gorni, C. 361
Gottlieb-Vedi, M. 302

Page

Götz, K.U. 78
Goyache, F. 26, 29, 94, 95, 98, 367
Grabarevic, Z. 133
Grabner, A. 31
Gredler, B. 384
Greeff, J.C. 377
Gresch, P. 2
Grigoli, A. Di 266
Grindflek, E. 14
Groenen, M.A.M. 11
Groeneveld, E. 274
Grogan, A. 61, 318
Gudmundsson, G. 56
Guiard, V. 81, 338, 359
Guibert, X. 16, 18, 19
Guillouet, P. 268, 280
Guldbrandtsen, B. 363
Gundel, J. 152
Gunn, G.J. 4
Gunnarsson, S. 314
Gürcan, E.K. 202, 297
Gustafson, T. 298
Gustafsson, A.H. 56, 127, 385
Gutiérrez, J.P. 92, 95
Guzey, Y.Z. 252
Gyarmathy, E. 115
Györkös, I. 250, 261, 262

H

Haas, Y. de 45
Habe, F. 308
Habier, D. 198
Hafez, Y. 290
Hagberg, C. 182
Hagesæther, N. 362
Häggblom, P.E. 4
Hagiya, K. 209
Hagnestam, C. 72
Halachmi, I. 184, 349
Halekoh, U. 222
Hamann, H. 199
Hamann, J. 375
Hammuda, Y.A.F. 245
Hamza, A.S. 244
Han, S.H. 93, 93
Han, Y.K. 155
Hanenberg, E.H.A.T. 319
Hansen, C.F. 243

	Page
Hansen, L.L.	313, 327
Hansen-Møller, J.	327
Harder, B.	46, 331
Harighi, M.F.	216
Harlizius, B.	370
Harmegnies, N.	224
Harris, B.L.	53
Hartwell, B.W.	267
Hasani, S.	284
Hassan, H.M.A.	245
Hassanabadi, A.	188
Hastings, N.	13, 372
Haugen, J.E.	276
Hausberger, M.	30
Hayder, M.	284
Hayes, B.	362
Hazeleger, W.	104
Hecker, W.	306
Hedberg, A.	357
Hedberg, Y.	103
Heeschen, W.	2
Hegedüs, E.	73
Heier, B.	325
Heissenhuber, A.	35
Hellinga, I.	23
Hemati, B.	171
Hemery, D.	30
Henne, H.	370
Hennig, S.	372
Henno, M.	182, 259
Henry, S.	30
Heravi Moussavi, A.	289, 290, 370
Heringstad, B.	45
Herlin, A.H.	29, 181, 182
Hermán, A.	152
Hermansen, J.E.	309
Hernández, P.	286
Hernández-Jover, M.	6, 7
Herrera, M.	29, 88, 301, 364
Herrloff, A.	254
Herskin, M.S.	261
Heuven, H.C.M.	125, 362
Hielscher, A.	269
Hiemstra, S.J.	282
Hinrichs, D.	46, 83
Hiramoto, K.	203
Hocquette, J.F.	248, 257
Hoereth, R.	227

	Page
Hofer, A.	224
Hoffmann, H.	35
Hofstetter, P.	300
Hogan, S.	235
Holecova, M.	26
Holgersson, A.-L.	302
Holló, G.	132, 254, 257, 258
Holló, I.	132, 254, 258
Holm, B.	77
Holmberg, M.	11
Holmes, C.W.	321
Holmes, D.	15
Holt, M.	83
Holtenius, K.	73, 299
Holtsmark, M.	24, 194
Homann, T.	188
Hommel, B.	150
Homolka, P.	143
Honkatukia, M.S.	363
Høøk Presto, M.	314
Hoque, A.	200, 203
Horcada, A.	26
Hornick, J.-L.	138
Horvai Szabó, M.	73
Houska, M.	141
Huba, J.	117
Huerou-Luron, I. Le	241
Huether, L.	150
Huguet, A.	241
Huhtanen, P.	56, 258, 350
Hulsegge, B.	118
Hultgren, J.	175
Humpolicek, P.	366
Hurtaud, J.	80
Hussein, A.M.	217
Husvéth, F.	25, 150
Húth, B.	258
Hyánková, L.	215
Hymøller, L.	61

I

Iacuaniello, S.	361
Iacurto, M.	249
Iannuccelli, N.	12
Ibáñez, N.	92
Igarzabal, A.	294
Iglesias, A.	251, 262
Ilgaza, A.	189

	Page
Illik, J.	140
Imboden, I.	305
Ingram, C.D.	137
Iniguez, L.	267
Iqbal, A.	148
Ishii, K.	209
Israelsson, C.	317
Istasse, L.	138
Ivankovic, A.	33, 308
Ivankovic, S.	102
Iyaya, R.A.	355
Iyayi, E.A.	355

J

Jacquet, S.	334, 335
Jafari, A.	167
Jaffrézic, F.	49, 322
Jagusiak, W.	51, 201
Jahn, O.	372
Jakobsen, J.H.	42
Jalilian, R.	171
Jamchi, M.	289, 290
Jandurova, O.	365
Janiszewski, P.	185
Janitz, M.	372
Janss, L.L.G.	11, 118, 125, 343, 358
Janssens, S.	98
Jansson, K.	47
Jauhiainen, L.	350
Javadmanash, A.	90, 370
Javanmard, A.	89, 90, 91, 336, 374
Javanrouh Aliabad, A.	96
Jávor, J.	311
Jegorova, J.	189
Jemeljanovs, A.	179, 180
Jensen, H.F	326, 329
Jensen, J.	82
Jensen, K.K.	310
Jensen, M.B.	222
Jensen, M.T.	327
Jeon, J.T.	93
Jezierski, T.	6
Ji, F.	240
Jílek, F.	208, 260
Jiménez, G.	328
Jiskrová, I.	26
Johansson, B.	316
Johansson, D.	357
Johansson, K.	46
Joly, A.	67
Joly, T.	80
Jondreville, C.	346
Jones, R.B.	66
Jongbloed, R.	236
Jordana, J.	367
Jørgensen, B.	132
Jørgensen, E.	222
Jørgensen, K.F.	144, 310
Jorjani, H.	21, 42, 47
Jorjany, E.	370
Jovanovic, J.	86
Józsa, Cs.	25
Julliand, V.	304
Junge, W.	52, 190, 381, 386
Junqueira, F.S.	259
Jurgens, A.H.	169
Juric, I.	228
Jurie, C.	248, 257
Juska, R.	186, 229
Juskiene, V.	186, 229
Juurlink, S.	346

K

Kaal, L.M.T.E.	375
Kaam, J.B.C.H.M. van	99, 266, 376, 378
Kaart, T.	182
Kadarmideen, H.N.	45, 49, 78
Kadlecik, O.	54
Kadowaki, H.	126, 207
Kafilzadeh, F.	351
Kalm, E.	46, 52, 190, 212, 220, 331, 364, 381
Kamada, H.	128
Kamaldinov, E.V.	89
Kamyczek, M.	225, 238
Kandil, S.A.	107
Kang, H.S.	93, 187
Kanis, E.	130
Kanitz, E.	104
Kaps, M.	28, 124
Karacaören, B.	49
Karamichou, E.	337
Karimi, Z.	163
Karlsson, A.H.	327
Karlsson, L.J.E.	377
Karnóth, J.	150
Karolyi, D.	228

	Page
Karsten, S.	381
Kasarda, R.	54
Kashki, V.	268
Katsuki, P.A.	172
Kaufmann, B.A.	294, 317
Kawecka, M.	238
Keane, M.G.	252
Keeling, L.	333
Kehr, C.	49
Keidane, D.	189
Kennedy, M.J.	307
Kennel-Hess, R.	2
Kenny, D.	386
Kenyon, P.R.	265
Kermanshahi, H.	188
Kerry, J.P.	218, 230, 235
Keski-Nisula, S.	100
Khalifah, M.	169
Khanian, S.E.	88
Khattab, A.S.	217
Kheirabadi, M.	164
Kheiri, F.	158, 291
Kianzad, M.R.	268
Kibenge, F.S.B.	186
Kiiman, H.	182
Kijora, C.	166
Killen, L.	119
Kim, S.W.	240
Kim, Y.K.	187
Kindahl, H.	103, 127
Kjeldsen, A.M.	310
Klaver, J.	16
Klein, P.	140
Klemetsdal, G.	24, 45, 194, 270, 341
Klimas, R.	223, 226
Klimiene, A.	223, 226
Kloareg, M.	218
Klunker, M.	49
Knap, P.W.	220
Knauer, M.	74
Knaus, W.	267
Knebel, C.	154
Knol, E.F.	11, 80, 130, 319, 370
Knoll, A.	366
Knox, R.V.	114
Ko, M.S.	93, 93
Kobeisy, M.	284
Koenen, E.P.C.	21, 23

	Page
Kohler, S.	300
Koivula, M.	380
Kompan, D.	115, 270
König, S.	247, 323
Konik, A.	237
Konjacic, M.	33
Konosonoka, I.H.	179, 180
Korn, C.	7
Korn, S. von	275
Kornfelt, L.F.	243
Kornmatitsuk, B.	127
Korotkevich, O.S.	89, 232, 233
Korsgaard., I.R.	193
Kotze, A.	96
Kouar, B.	102
Koukolová, V.	143
Koumas, A.	291
Kovács, A.	250, 261, 262
Kovács, G.	150
Kovács, K.	250
Kovács, P.	311
Kratz, R.	242
Krejcová, M.	208, 248, 255
Kridli, R.T.	277, 283
Krieter, J.	31, 223, 381, 382, 386
Kristensen, T.N.	82, 309, 313
Krupa, E.	117
Kübarsepp, I.	259
Kucerová, J.	365
Küchenmeister, U.	108
Kuchtik, J.	273
Kudrna, V.	140, 248
Kuipers, A.	5
Kukovics, S.	115, 271, 277, 311
Kume, K.	115
Kumm, K.-I.	314
Kunieda, T.	200
Kunz, P.	300
Kuran, M.	107, 283, 387
Kvame, T.	263
Kwon, D.J.	187

L

Láczó, E.	271, 275
Lagerkvist, G.	315
Laignel, G.	311
Lallès, J.P.	153
Lambe, N.R.	263, 320

	Page
Lang, A.	*108*
Lang, P.	*140*
Langlois, B.	*27*
Larsen, M.	*56*
Larsson, U.	*382*
Larzul, C.	*101*
Lassen, J.	*192*
Lastovkova, J.	*140, 215*
Lawlor, P.G.	*218, 230, 235*
Lawlor, T.J.	*191*
Lazarevic, M.	*153*
Lazzari, B.	*12, 369*
Lazzaroni, C.	*315*
Le, P.D.	*236*
Le Bellego, L.	*155, 236, 243*
Le Roy, P.	*101*
Lebret, B.	*131*
Lebreton, Y.	*241*
Lebzien, P.	*60*
Ledin, I.	*347*
Lee, J.H.	*155*
Lelkova, H.	*140*
Leloutre, L.	*268, 280*
Lemke, U.	*294*
Lende, T. van der	*130*
Lengyel, Z.	*55, 204*
Lens, L.	*72*
Lent, M. van	*16*
Lepage, O.	*335*
Leroux, C.	*161, 257*
Leroy, P.L.	*299*
Leveziel, H.	*87*
Lewis, R.M.	*264*
Leyva, C.	*110*
Lherm, M.	*313*
Liamadis, D.	*151, 160*
Lidfors, L.	*254*
Lien, S.	*14, 362*
Lifvendahl, Z.	*298*
Liker, B.	*133, 228, 279*
Lin, C.Y.	*122, 123*
Lindberg, J.E.	*242, 314, 353*
Liotta, L.	*227*
Lipkin, E.	*340*
Lister, C.J.	*141*
Liu, Y.	*125*
Lium, B.M.	*324, 325*
Livshin, N.	*183*

	Page
Loeschcke, V.	*82*
Lombardelli, R.	*106, 176*
Looft, H.	*220*
López Rodriguez, D.G.	*158*
Lopez-Villalobos, N.	*321, 385*
Lorentzon, S.	*181*
Lovendahl, P.	*190*
Lund, M.S.	*48, 363*
Lund, P.	*313*
Lundberg, M.	*29*
Lundeheim, N.	*76, 191, 221*
Lundén, A.	*371*
Lundesjö-Öhrström, S.	*302*
Lundström, K.	*326*
Lungu, S.	*7*
Lüpping, W.	*381*
Luque, M.	*29, 88, 112, 301, 354, 364*
Luron, I.	*153*
Luther, H.	*224*
Luyten, T.	*224*
Lyberg, K.	*353*
Lynch, P.B.	*218, 230, 235*

M

Maagdenberg, K. van den	*219, 224*
Mabry, J.W.	*74, 75*
Macfarlane, J.M.	*264*
Machado, H.	*251*
Mackenzie, A.M.	*330*
Madalena, F.E.	*259, 342*
Madsen, P.	*48, 82, 192, 193*
Maglaras, G.	*114*
Maglione, G.	*144*
Magnusson, M.	*175, 181, 182*
Mahmoud, M.A.	*70*
Main, M.	*346*
Maj, A.	*253*
Mäki-Tanila, A.	*219, 363*
Malek Mohammadi, H.A.	*71*
Malmfors, B.	*69, 344*
Malpaux, B.	*282*
Maltecca, C.	*253*
Maltz, E.	*183, 349*
Mandaluniz, N.	*294*
Manni, K.	*258*
Manrique, E.	*36*
Manteca, X.	*379*
Manteuffel, G.	*108*

	Page
Mantovani, R.	*147*
Mäntsysaari, E.A.	*43, 54, 211, 380*
Marchi, E.	*8*
Marchitelli, C.	*372*
Mariani, P.	*12, 361, 368, 369*
Marilli, M.	*95, 109*
Mark, T.	*42*
Marka, C.H.	*324, 325*
Markowski, W.	*22, 31, 32*
Marmandiu, A.	*30, 111, 206, 210, 269*
Marounek, M.	*141, 159*
Márquez, A.P.	*196*
Martemucci, G.	*300*
Martin, R.	*346, 354*
Martínez, A.	*239*
Martínez, E.A.	*129*
Martinez-Puig, D.	*154*
Martin-Rosset, W.	*335*
Martins Santos, J.	*356*
Márton, D.	*55*
Martos, J.	*292, 293*
Martuzzi, F.	*32*
Martyniuk, E.	*115*
Marusi, M.	*197*
Marzouk, K.M.	*281*
Massimiliano, A.	*251*
Masumi, S.M.	*352*
Mata, C.	*239*
Matejícek, A.	*365*
Mateo, R.D.	*240*
Mathur, P.K.	*125*
Matlova, V.	*115*
Matthews, D.	*374*
Mattsson, B.	*221*
Matysiak, B.	*237*
Maxa, J.	*264*
Mayer, M.	*339*
Mayoral, A.	*354, 364*
Mazzanti, E.	*87*
McCoy, M.A.	*139*
McGee, M.	*61*
McGivan, J.D.	*325*
McGlone, J.	*3*
McKay, S.D.	*372*
McLean, K.A.	*263, 337*
McMillan, I.	*86*
McNaughton, L.R.	*385*
Medina, C.	*26*

	Page
Mee, J.F.	*386*
Meggiolaro, D.	*109*
Mehla, R.K.	*288*
Mehlqvist, M.	*56*
Meijer, L.	*234*
Meineri, G.	*24*
Mejer, J.	*327*
Mekkawy, W.	*124*
Melchior, D.	*155, 236*
Mele, M.	*204*
Melin, M.	*173, 178*
Melodia, L.	*162*
Melucci, L.M.	*340*
Mendel, C.	*270*
Merks, J.W.M.	*319, 370*
Mertens, J.	*72*
Meunier-Salaün, M.C.	*131*
Meuwissen, T.H.E.	*14, 83, 85, 117, 324*
Meyer, U.	*60*
Mézes, M.	*229, 230*
Mezoszentgyorgyi, D.	*39*
Miele, M.	*66*
Miesner, K.	*15*
Migaud, M.	*282*
Miglior, F.	*321*
Mihok, S.	*17, 25, 306*
Mijic, P.	*33*
Mikko, S.	*20*
Milan, D.	*12*
Milán, M.J.	*40, 40*
Milanesi, E.	*372*
Milgen, J. van	*218*
Milis, Ch.	*151, 160*
Milne, C.	*4*
Milne, C.E.	*65*
Milne, J.A.	*345*
Mioc, B.	*97*
Miraei Ashtiani, S.R.	*91, 101, 209*
Miraglia, N.	*24, 305*
Miraii Ashtiani, S.	*216*
Miron, J.	*349*
Misar, D.	*26*
Mishra, D.P.	*288*
Misztal, I.	*191, 193*
Mitloehner, F.M.	*222*
Mizubuti, I.Y.	*172, 172*
Mlázovská, P.	*140*
Moe, M.	*14*

	Page
Moeini, M.	134, 164, 177, 289
Mohamadi, A.	90
Mohamed, M.A.	244, 245
Mohammadalipour, M.	137, 181
Moharrery, A.	60
Moioli, B.	94, 164
Molina, A.	26, 29, 88
Molina, L.	110, 112
Mollasalehi,	374
Mollasallalehi, M.R.	91
Molnár, A.	271, 277
Molnár, M.	298
Molnár, T.	298
Momani-Shaker, M.	277, 283
Mondal, M.	286, 287
Mondello, L.	28
Moniello, G.	8, 9
Monserrat, L.	255
Montalvo, G.	41
Monteiro, A.S.	356
Montgomerie, W.	53
Moore, S.S.	372
Moors, E.	272
Moradi Shahrebabak, M.	216
Moradi Sharbabak, M.	91
Morales, J.	238, 328, 355
Morales, M.	110
Moreira, F.B.	172
Moreira, O.C.	356
Morel, P.C.H.	265
Moreno, A.	213
Moreno, T.	255
Morfín-Loyden, L.	158
Morgan, K.	302
Mori, H.	208
Morris, R.	103
Morris, S.T.	265
Moscati, L.	113
Mountzouris, K.	295
Mourot, J.	218
Moussavi, M.E.	289, 290
Mroczkowski, S.	214, 215
Muenger, A.	300
Mugurevics, A.	189, 189
Muir, W.M.	119, 121
Mulder, H.A.	319
Mullane, J.	218, 230
Müller, M.	100

	Page
Müller, U.	274
Mulligan, F.J.	57
Munksgaard, L.	190, 261
Muñoz, G	14, 369
Mura, M.C.	71, 274
Murai, M.	128
Murani, E.	109, 370
Murphy, J.J.	57, 386
Musavaya, K.	198
Mwai, O.	295, 301

N

Naber, C.H.	114
Nadeau, E.	316
Naeemipour, H.	200
Nafstad, O.	324, 325
Nagamine, Y.	209
Nagata, A.	126
Nagy, B.	204, 273
Nagy, I.	220, 298
Nagy, L.	258
Nagy, S.	311
Napel, J. ten	74
Napolitano, F.	94
Nardelli-Costa, J.	369
Nardone, A.	176
Naserian, A.A.	145, 146, 278
Näsholm, A.	46, 69, 116, 330, 331
Nasirimoghadam, H.	145
Näslund, J.	371
Nassiry, M.R.	90, 91, 370
Nauta, W.J.	312
Navajas, E.A.	263
Navidshad, B.	157
Ndumu, D.	295, 301
Nedelcu, C.I.	30
Nedelcu, M.M.	30, 210
Negussie, E.	43
Neil, M.	79, 242
Nemcová, E.	195, 208, 260, 365
Németh, T.	271, 277
Newman, P.	15
Nicolae, C.	212, 285
Niederhäusern, R. v.	25
Nielsen, H.M.	192
Nielsen, T.S.	135
Nielsen, U.S.	82
Nieto, B.	92

	Page
Nieuwland, M.	11
Niewczas, J.	31
Nikkhah, A.	63, 64, 139, 167
Nilsson, C.	175
Nishida, A.	123, 207
Nisi, I.	373
Noblet, J.	218
Noguera, J.L.	369
Noorbakhsh, R.	289, 290
Norberg, E.	264
Nørgaard, P.	59, 144, 243
Nosrati, M.	90
Nousiainen, J.I.	54, 350
Nowrozi, M.	268
Nute, G.R.	337
Nybrant, T.	314

O

Öborn, I.	314
O'brien, B.	348, 354
Ochoa, S.C.	196
Ødegård, J.	83, 193, 341
Odensten, M.O.	73
O'Donovan, M.A.	57
Oeckel, M.J. van	233, 234, 239
Ogink, N.W.M.	236
Ogunsola, O.	355
Oh, M.Y.	93
Ohtomo, Y.	123
Oikawa, R.	200
Oikawa, T.	126, 203, 208
Ojala, M.	100
Olaizola, A.	36
Olesen, I.	341
Oliveira, O.	356
Ollier, A.	136
Olori, V.E.	55, 318
Olsen, D.	77
Olsen, H.F.	24
Olsson, A.-C.	316
Olsson, C.	357
Olsson, I.	254, 299
Olsson, K.	333
Olsson, V.	57
O'Mara, F.	57, 386
Omer, E.A.	217
Onal, A.G.	283, 387
Oravcová, M.	117
Oregui, L.M.	294
O'Riordan, E.G.	177, 179
Orrù, L.	94, 164
Orzechowska, B.	231, 232
Osfoori, R.	88, 336
Ossensi, C.	147
Östensson, K.	380
Otten, W.	104
Otto, G.	220
Ouweltjes, W.	174, 245, 322
Óvilo, C.	14
Owen, H.C.	304
Owsianny, J.	237
Özkan, E.	202, 297
Özkan, S.	202, 297
Ozkaya, S.	58
Özpinar, A.	222

P

Paakkolanvaara, K.	307
Pace, V.	164
Puche, St.	383
Páchová, E.	246
Padeanu, I.	115
Paepe, M. De	233, 234, 239
Pagnacco, G.	204, 228, 361
Pailleux, J.-Y.	38
Pajor, F.	271, 275
Pakdel, A.	337
Pál, L.	150
Pamio, J.	292
Panicke, L.	145, 246
Panzitta, F.	368, 369
Papachristoforou, C.	291
Park, K.M.	155
Parlat, S.S.	349
Parmeggiani, A.	71
Parmentier, H.K.	11
Parvu, M.	111
Paryad, A.	351
Pas, M.F.W. Te	370
Pashmi, M.	91
Pastoret, P.P.	3
Pataki, B.	17
Patrashkov, S.P.A.	232, 233
Patton, J.	386
Pavic, V.	97
Pavlík, J.	226

	Page
Pavlovic, J.	*102*
Pecorari, D.	*340*
Peddie, S.	*4*
Pedersen, J.	*82, 264*
Pedersen, K.S.	*82*
Pedersen, L.J.	*222*
Pedron, O.	*184*
Peet-Schwering, C.M.C. van der	*236*
Pellerud, G.	*362*
Pelliccia, C.	*106, 113*
Pena, I.	*202*
Peña, F.	*29, 301*
Perea, J.	*112, 293, 354, 364*
Pereira, E.S.	*172, 172*
Perez, F.	*154*
Pérez, A.	*110*
Pérez, C.	*364*
Pérez, C.C.	*112*
Pérez, N.	*110, 255*
Pérez Dosta, J.I.	*158*
Perisic, P.	*51, 199*
Perona, G.	*335*
Perrin, D.	*304*
Perrot, C.	*34*
Persson, Y.	*175*
Persson Waller, K.	*73*
Peskovicová, D.	*117*
Petermann, R.	*300*
Peters, K.J.	*166, 320, 343*
Petersen, S.	*31*
Peterson, H.	*33*
Peterson, K.	*282*
Petersson, K.-J.	*385*
Pethick, D.	*248, 257*
Petim-Batista, F.	*251*
Pétro, T.	*250*
Petron, M.J.	*138*
Petrovic, Z.	*308*
Pettersson, G.	*173, 380*
Petukhov, V.L.	*89, 232, 233*
Peura, J.	*211*
Pezeshki, A.	*170, 181, 352*
Pflimlin, A.	*34*
Philipsson, J.	*21, 42, 46, 69, 72, 302, 330, 331, 344, 357*
Photiou, C.	*291*
Piasentier, E.	*270*
Piccolo, D.	*227*
Pickard, R.M.	*136*
Piedrafita, J.	*121, 131*
Pielberg, G.	*371*
Pieta, M.	*22, 32*
Pieterse, M.	*282*
Pietruszka, A.	*225, 238*
Pijl, R.	*52*
Pilmane, M.	*179*
Piñeiro, C.	*41, 238, 328, 355*
Pinna, W.	*8, 9*
Pinto, A.P.	*172*
Piqueras, P.	*92*
Pires, J.	*251, 262*
Pîrvu, M.	*210*
Pizzi, F.	*253*
Plastow, G.S.	*220*
Poel, J. van der	*11*
Pohn, G.	*257*
Poikalainen, V.	*6*
Polák, P.	*117*
Polidori, M.	*24*
Polimeno, F.	*142*
Pollot, G.E.	*36, 81, 260, 359*
Ponce, J.F.	*196*
Poncet, P.-A.	*305*
Pong-Wong, R.	*361*
Ponsuksili, S.	*109, 370*
Pool, M.H.	*55, 118*
Popa, R.A.	*225*
Popescu-Miclosanu, E.	*156*
Porcelli, F.	*109*
Portolano, B.	*99, 266, 376, 378*
Pösö, A.R.	*288*
Posta, J.	*17, 306*
Póti, P.	*271, 275*
Poulsen, H.D.	*132*
Pourcelot, P.	*332*
Powles, J.	*330*
Prabhaker, S.	*103*
Prakash, B.	*286, 287*
Prakash, B.S.	*286, 287*
Praks, J.	*6*
Preisinger, R.	*363*
Prendiville, D.J.	*177, 179*
Presciuttini, S.	*87*
Preziuso, G.	*256*
Pribyl, J.	*205*
Pribylova, J.	*205*

	Page
Prins, D.	370
Prinzenberg, E.-M.	364
Prunier, A.	323
Pryce, J.E.	321
Pulkrabek, J.	226
Puppe, B.	108
Püski, J.	384
Pyrochta, V.	147

Q

Qanbari, S.	88, 91, 336
Qotbi, A.	195
Quevedo, J.R.	94
Quinn, N.	119
Quinton, V.M.	86, 341

R

Radnóczi, L.	220
Raducuta, I.	180, 210, 269
Raes, K.	138
Ragni, M.	162, 279
Rahimi, G.	157
Rahimi, S.	148
Rahmani, H.R.	71, 137, 158, 168, 170, 181, 203, 352
Rahmany, H.	171
Raisianzadeh, M.	58, 268
Rajkhowa, C.	286, 287
Ramalho Ribeiro, J.	356
Rambeck, W.A.	154
Ramelli, P.	12
Ramírez, L.M.	238, 355
Ramírez, O.	369
Ramos, B.M.O.	172, 172
Rantzer, D.	234, 237
Rave, G.	223
Raynal-Ljutovac, K.	268
Reese, K.	363
Rege, E.	344
Rege, J.E.O.	69
Regius, A.	152
Rehák, D.	260
Reiners, K.	187
Reinsch, N.	81, 359, 364
Reis, G.L.	259
Reixach, J.	231
Relandeau, C.	243
Renard, C.	9

	Page
Restelli, G.L.	228, 361
Rezaeian, M.	148
Ribeiro, E.L.A.	172, 172
Ricard, A.	22, 332
Richard, M.-A.	30
Richardson, R.I.	337
Riek, A.	163
Riera, M.	184
Riggio, V.	378
Rikabi, F.	274
Ringdorfer, F.	115, 270
Rinne, M.	258
Riquet, J.	12
Rivera, F.	110
Rizzi, R.	253
Robert, C.	334, 335, 336
Robin, P.	346
Robinson, J.A.B.	341
Roca, J.	129
Rocha, M.A.	172, 172
Rodero, A.	88
Rodero, E.	29, 88, 301
Rodrigáñez, J.	344
Rodríguez, C.	14
Rodríguez, M.C.	344, 369
Rodríguez, V.	293
Rodríguez-Estévez, V.	239
Rodriguez-Martinez, H.	108
Roehe, R.	220
Roepstorff, A.	327
Roepstorff, L.	302
Rogel Gaillard, C.	9
Rogers, G.W.	43
Roito, J.	363
Ronchi, B.	176, 309
Roozbahan, Y.	157
Ros, G. de	270
Rose, S.	25
Rose, S.P.	330
Rosen, G.D.	328
Rosendo, A.	213
Rosner, F.	44, 52
Roth, A.	174
Roth, F.X.	243
Rouel, J.	161, 268, 280
Roughsedge, T.	84
Roura, E.	147, 235, 241
Rousselière, A.	304

	Page
Routel, A.	102
Roux, M.	87
Rouzbehan, Y.	139, 149
Roy, A.	136
Royo, L.J.	94, 95, 98, 367
Rózycki, M.	225, 238
Rozzi, P.	321
Ruane, J.	24
Rudloff, E.	150
Rudovsky, H.J.	383
Ruiz, R.	272, 294, 297
Rullo, R.	144
Rundgren, M.	29, 333
Rundo Sotera, A.	99
Russo, C.	256, 256
Russo, V.	340, 367, 371, 373
Ruusunen, M.	288
Rydhmer, L.	191, 221
Rzasa, A.	338

S

Saa, C.	40, 40
Saastamoinen, M.T.	302
Sabatini, A.	176
Sabre, D.	259
Sabry, A.	48
Sadeghi, A.A.	63, 97
Sáfár, L.	273
Safarzadeh Torghabeh, H.	162
Safus, P.	205
Saidi-Mehtar, N.	102, 205
Salajpal, K.	228
Salary, M.	370
Salehi, A.R.	91
Salehi-Taba, R.	91
Salem, I.A.	284
Salem, S.M.	159
Salgado, C.	92
Salimei, E.	28
Salmazo, R.	172
Salomon, R.	349
Samie, A.	203
Sanchez, M.P.	12
Sánchez, J.M.	112
Sánchez, L.	251, 255
Sánchez, L.	262
Sánchez, N.R.	347
SanCristobal, M.	101

	Page
Sandberg, K.	20
Sanders, K.	212, 364
Sanjabi, M.R.	48, 177, 289
Sano, T.	208
Santamarina, C.	7
Santos, R.	110
Sardina, M.T.	99, 378
Sarhaddi, F.	374
Sarubbi, F.	142, 287
Sarwar, M.	148
Sasaki, O.	209
Satoh, M.	206
Sauer, W.	152
Sauerwein, H.	137
Savarese, M.	372
Savary, G.	241
Saveli, O.	53, 182, 259
Savic, M.	86
Schaeffer, L.R.	270
Schafberg, R.	44
Schellander, K.	10, 109, 370
Schils, R.L.M.	312
Schlote, W.	99
Schmidt, T.A.	99
Schneider, F.	104
Schneider, M. del P.	174
Schnitz, A.L.	222
Schoen, A.	275
Schoonheere, N.	138
Schrooten, C.	105
Schulz, E.	242
Schwab, C.R.	120
Schwarzenbacher, H.	42, 118, 360, 384
Scossa, A.	164
Scotti, E.	340
Seabrook, M.F.	65
Seal, C.J.	136
Sebek, L.B.J.	312
Secchiari, P.	204
Sedda, P.	8, 9
Seegers, H.	66, 67
Seenger, J.	229, 230
Seidavi, A.R.	195
Sejrsen, K.	135, 313
Sekerden, Ö.	252
Sencic, D.	133, 279
Senturk, B.	68, 68
Sepponen, K.	288

	Page
Seregi, J.	132, 258
Serenius, T.	74, 75
Seric, V.	133
Serlet, S.	98
Serra, A.	204
Serrano, M.	213, 383
Servidio, M.	142
Settineri, D.	249
Sève, B.	153
Sevón-Aimonen, M-L.	219
Shaat, I.	124
Shahzeidi, M.	171
Shaker, Y.M.	107
Shawrang, P.	63, 64, 97
Shibata, T.	126, 207
Shivazad, M.	157
Shoshani, E.	349
Sieme, H.	199
Siffredi, G.	353
Silió, L.	14, 344, 369
Simianer, H.	84, 120, 247, 323, 342
Simm, G.	263, 264, 320
Simone, N.	300
Simonsson, A.	353
Sinjeri, Z.	228
Sirin, E.	107, 283, 387
Sirohi, S.K.	288
Sironen, A.	10
Sironi, L.	12
Sitkowska, B.	214, 215
Siukscius, A.	111
Siwek, M.	11
Sjaunja, L-O.	382
Skeríková, A.	143
Skrivanova, V.	141, 159
Slorach, S.A.	1
Sluyter, F.	20
Smet, S. de	138, 219, 224
Smith, R.F.	103
Sobotkova, E.	26
Soch, M.	140
Soede, N.M.	104
Søland, T.M.	59
Solanes, F.X.	231
Solá-Oriol, D.	235, 241
Solgi, H.	289
Solinas, I.L.	8, 9
Sölkner, J.	23, 42, 83, 85, 118, 124, 246, 265, 295, 301, 360, 384
Soller, M.	340, 360
Sonck, B.	72
Søndergaard, E.	332
Sonesson, A.K.	194
Sonesson, U.	314
Song, J.I.	187
Sorensen, D.	79, 82, 92
Sørensen, A.C.	82
Sørensen, J.G.	82
Sørensen, M.K.	192
Sørensen, M.T.	135
Sørheim, O.	276
Sosnowska, A.	237
Sossidou, E.	5, 6
Souri, M.	134, 164
Sousa, S.	356
Soydan, E.	107, 387
Soysal, M.I.	202, 297
Spagnuolo, M.S.	287
Spallholz, J.E.	240
Spelman, R.J.	385
Speranda, M.	133, 279
Speranda, T.	133
Spilke, J.	383
Spirito, F.	164
Spörndly, E.	185, 298, 299
Sprengel, D.	50
Sprenger, K.U.	18
Srinivasa Rao, D.	129
Stachurska, A.	22, 31, 32
Stadník, L.	205
Stalder, K.J.	74, 75, 120
Stålhandske, L.	237
Stamatis, D.	5
Stamer, E.	52, 190, 381, 386
Staufenbiel, R.	145, 246
Steen, K.	64
Steglich, M.	320
Steiger Burgos, M.	300
Stein, J.	344
Steinbock, L.	46
Steiner, Z.	133, 279
Stella, A.	228, 361, 368, 369
Stella, S.	156
Stern, S.	314
Stewart, A.H.	330
Stiller, Sz.	150

	Page
Stinckens, A.	219, 224
Stipková, M.	67, 183, 205, 208, 260, 365
Stock, K.F.	334
Stojanovic, S.	115
Stollberg, U.	25
Stone, D.G.	36
Stott, A.W.	4, 65
Straarup, E.M.	135
Strandberg, E.	44, 47, 77, 174, 330, 385
Strandén, I.	43, 211
Strazzullo, M.	372
Strudsholm, F.	310
Studnitz, M.	222
Suklim, A.	13
Sulkers, H.	282
Susic, Z.	221
Suwanlee, S.	83
Suwanmajo, S.	13
Suzuki, K.	123, 126, 207
Svendsen, J.	237, 316
Svennersten-Sjaunja, K.	173, 178, 254, 261, 380, 382
Svobodova, S.	27
Swalve, H.H.	44, 52, 207
Swiech, E.	155
Szabó, F.	37
Szabó, F.	204, 250
Szebestová, Z.	215
Szelényi, M.	152
Szentléleki, A.	251, 254
Szreder, T.	214
Szücs, E.	5, 6, 229, 230, 250, 384
Szwaczkowski, T.	99
Szyda, J.	338

T

Tabet-Aoul, N.	205
Taghizadeh, A.	62, 166
Tahmasbi, A.	62, 166
Tähtinen, J.	100
Tajet, H.	14, 324
Takeya, M.	206
Tambyrajah, W.S.	325
Tami, G.	106, 113, 266, 267
Tamminga, S.	62
Tanski, Z.	276
Tapaloaga, P.	225
Tapki, I.	283

	Page
Tarres, J.	121
Tartwijk, J.M.F.M. van	21
Täubert, H.	120, 342
Tauson, R.	315
Tava, A.	144
Tavakoli, H.	268
Tavakolian, J.	89
Tavernier, L.	332
Taylor, W.	141
Tazari, M.	86
Tedesco, D.	156, 165, 287
Tegegne, F.	166
Teixeira, N.M.	50, 196
Telezhenko, E.	175
Tesfahun, G.	296
Tesfaye, D.	10
Teslík, V.	248, 255
Teuffert, J.	223
Teuscher, F.	338
Teyssier, J.	282
Thaller, G.	78
Thamsborg, S.M.	327
Theau, J.P.	39
Theau-Clement, M.	80
Thénard, V.	39
Thodberg, K.	221
Tholen, E.	227
Thomet, P.	300
Thomsen, B.	10, 363
Thorén, E.	21, 330
Thrasivoulou, A.	151
Thuneberg-Selonen, T.	307
Thuy, L.T.	294
Thymann, T.	243
Tibau, J.	131, 231
Tibbo, M.	69
Tietze, M.	120
Tijani, A.	201
Tillocca, G.	116
Tinsky, M.	183
Tisza, K.	220, 298
Togashi, K.	122
Toivakka, M.	54, 350
Toldi, Gy.	277
Tolle, K.H.	31, 381, 382
Tölle, U.	382
Tomás, A.	369
Tomiyama, M.	126, 208

	Page
Toquet, M.P.	336
Tor, M.	231
Torkamanzehi, A.	91, 373
Toro, M.A.	87, 361
Török, M.	37, 55, 204, 250
Torrallardona, D.	235, 236, 241, 329
Torshizi, R.V.	48
Tőszér, J.	251, 254, 257, 258, 275
Toteda, F.	279
Tóth, Zs.	17
Towhidi, A.	289
Tracey, S.V.	304
Trailovic, R.	86
Tran Anh Tuan,	384
Trevisi, E.	106, 176
Trevisi, P.	153
Tribout, T.	101, 122
Trillaud-Geyl, C.	335
Trinácy, J.	143
Tripaldi, C.	164
Troedsson, M.H.T.	126
Tsiplakou, E.	295
Tsuruta, S.	191
Tuchscherer, M.	104
Tudorache, M.	156, 206
Tuiskula-Haavisto, M.	363
Tuomola, M.	327
Turi, L.	162, 279
Turini, J.	165
Tuyttens, F.	72
Twigge, J.	139, 142
Tyra, M.	231, 232
Tyrolova, Y.	141

U

Uchida, H.	207
Udroiu, A.	180
Ueda, Y.	128
Uitdehaag, K.A.	130
Ulutas, Z.	107
Urban, T.	365, 366
Urdes, L.D.	212, 285
Usmani, R.H.	105
Ustyanich, A.E.	69
Ustyanich, E.P.	69
Ustyanich, M.A.	69

V

	Page
Vacca, G.M.	71, 116, 274
Vaccari Simonini, F.	32
Vaez Torshizi, R.	96
Vahmani, P.	145, 278
Valentim, R.	356
Valentini, A.	372
Valera, M.	26, 88, 367
Valerio, D.	292, 293
Valette, J.P.	334, 336
Vali, N.	203
Valipoor, K.	70
Valis, L.	226
Valizadeh, R.	146
Valle Zárate, A.	198, 294
Valloto, A.A.	196
Van, D.T.T.	347
Vanacker, J.M.	133, 134, 233, 234
Vandepitte, W.	98
Vangen, O.	83, 263
Vanthemsche, P.	1
Varga-Visi, E.	132, 257
Varona, L.	124, 286, 360, 369
Vasconcelos, J.	211
Vatzias, G.	114
Vázquez, J.M.	129
Veerkamp, R.F.	55, 174, 245, 319, 322, 343
Vegte, Z. van der	312
Veissier, I.	66
Velarde, A.	379
Veldman, W.A.J.	105
Venerus, S.	270
Ventorp, M.	175, 181
Vereijken, A.L.J.	194
Veresegyházi, T.	135
Verhees, F.J.H.M.	5
Verità, P.	256
Verloop, J.	312
Verner, J.	365, 366
Verstegen, M.W.A.	236
Vervaele, K.	138
Vesela, Z.	205
Vestergaard, M.	310, 313
Veysset, P.	313
Vicenti, A.	162, 279
Vidu, L.	180, 225, 269
Viinalass, H.	259
Viklund, Å.	330, 331

	Page
Vilkki, J.	10, 363
Villagra, S.	353
Villalba, D.	297
Villanueva, B.	84, 361
Vincze, Zs.	37, 55
Visscher, P.M.	118, 360
Vitale, F.	376
Vitale, M.	376
Vitali, A.	176
Vítek, M.	226
Vladu, M.	225
Voegeli, P.	78
Volden, H.	56, 56
Volek, J.	260
Volek, Z.	159
Völgyi Csík, J.	250, 384
Vonghia, G.	162, 279
Voore, M.	53
Vries, A. de	16
Vries, F. de	375
Vykoukalová, Z.	366
W	
Waagepetersen, R.	92
Waaij, E.H. van der	194
Wagenhoffer, Zs.	37, 39
Wágner, L.	150
Wähner, M.	127, 130
Walker, S.	103
Wallenbeck, A.	221
Waller, P.J.	377
Walther, R.	274
Warnants, N.	233, 234, 239
Weber, M.	230
Wehr, U.	154
Weigend, S.	363
Weisbjerg, M.R.	61, 144, 247
Wenk, C.	327
Wensch, J.	207
Wensch-Dorendorf, M.	52, 207
Werner, C.	187
Wesolowska-Janczarek, M.	32
West, D.M.	265
Whittington, F.M.	325
Wicke, M.	187
Wickham, B.	318
Wicks, H.C.F.	139, 142
Wierzbicki, H.	201, 338

	Page
Wiese, M.	227
Wigren, I.J.	242
Wiking, L.	178
Wikström, Å.	330, 331
Wiktorsson, H.	173, 178
Willam, A.	42
Williams, B.A.	62
Williams, F.	4
Williams, J.	330
Williams, J.L.	372
Wilton, J.W.	86, 341
Wimmers, K.	109, 370
Winckler, C.	379
Windig, J.J.	174, 245, 322, 343, 375
Windig, J.J.	
Winkelman, A.M.	53
Witte, I.	223
Wittenburg, D.	81
Woelders, H.	282
Wofova, M.	205
Wolc, A.	99
Wolcott, M.L.	254
Wolf, J.	67, 183, 195
Wolfová, M.	67, 183
Wollny, C.B.A.	296, 353
Wood, J.D.	325
Woolliams, J.A.	13, 83, 84, 194
Wörner, R.	370
Wouters, V.	167
Wredle, E.	261
Wu, G.	240
Wurzinger, M.	267, 295, 301
Wyk, J.B. van	96
Y	
Yalcin, C.	68, 68
Yánez, J.	152
Yang, C.B.	187
Yates, D.	15
Yavuzer, Ü.	281
Yeon, K.Y.	187
Yildirim, I.	244, 349
Yoo, Y.H.	187
Yoon, I.K.	240
Young, M.J.	264
Yousef Elahi, M.	160, 292

	Page

Z

Zahedifar, M.	*58*
Zahrádková, R.	*248, 255*
Zajicova, P.	*273*
Zamaratskaia, G.	*326*
Zambonelli, P.	*340, 367*
Zándoki, R.	*251, 254, 257*
Zareh Shahne, A.	*157*
Zarnecki, A.	*51*
Zaton-Dobrowolska, M.	*338*
Zavadilová, L.	*195, 246*
Zeinali, A,	*163*
Zelenika, A.	*102*
Zeman, L.	*147*
Zenali, S.	*89*
Zenhom, M.	*284*
Zerehdaran, S.	*194*
Zervas, G.	*295*
Zetteler, P.E.	*80*
Zetterqvist, M.	*306, 309*
Zhao, Y.	*222*
Zheltikov, A.I.	*89*
Zimmerman, D.R.	*114*
Zinovieva, N.	*100*
Zisis, T.	*151*
Zitare, I.	*179*
Zuidberg, C.A.	*282*
Zumbach, B.	*343*
Zumbo, A.	*99, 227*
Zwierzchowski, L.	*214, 253*

Printed in the United States
by Baker & Taylor Publisher Services